T0231332

High Performance Embedded Computing Handbook

A Systems Perspective

High Performance Embedded Computing Handbook

A Systems Perspective

Edited by

David R. Martinez
Robert A. Bond
M. Michael Vai

Massachusetts Institute of Technology
Lincoln Laboratory
Lexington, Massachusetts, U.S.A.

CRC Press
Taylor & Francis Group
Boca Raton   London   New York

CRC Press is an imprint of the
Taylor & Francis Group, an **informa** business

CRC Press
Taylor & Francis Group
6000 Broken Sound Parkway NW, Suite 300
Boca Raton, FL 33487-2742

© 2008 by Taylor & Francis Group, LLC
CRC Press is an imprint of Taylor & Francis Group, an Informa business

No claim to original U.S. Government works

Library of Congress Cataloging-in-Publication Data

High performance embedded computing handbook : a systems perspective / editors, David R. Martinez, Robert A. Bond, M. Michael Vai.
    p. cm.
   Includes bibliographical references and index.
   ISBN 978-0-8493-7197-4 (hardback : alk. paper)
   1. Embedded computer systems--Handbooks, manuals, etc. 2. High performance computing--Handbooks, manuals, etc. I. Martinez, David R. II. Bond, Robert A. III. Vai, M. Michael. IV. Title.

TK7895.E42H54 2008
004.16--dc22
                                                 2008010485

Visit the Taylor & Francis Web site at
http://www.taylorandfrancis.com

and the CRC Press Web site at
http://www.crcpress.com

Dedication

This handbook is dedicated to MIT Lincoln Laboratory for providing the opportunities to work on exciting and challenging hardware and software projects leading to the demonstration of high performance embedded computing systems.

Contents

SECTION I   Introduction

SECTION II   Computational Nature of High Performance Embedded Systems

SECTION III   Front-End Real-Time Processor Technologies

SECTION IV   *Programmable High Performance Embedded Computing Systems*

SECTION V   *High Performance Embedded Computing Application Examples*

Chapter 21 A Sonar Application.. 411
W. Robert Bernecky, Naval Undersea Warfare Center

Chapter 22 Communications Applications... 425
Joel I. Goodman and Thomas G. Macdonald, MIT Lincoln Laboratory

SECTION VI   Future Trends

Preface

Over the past several decades, advances in digital signal processing have permeated many applications, providing unprecedented growth in capabilities. Complex military systems, for example, evolved from primarily analog processing during the 1960s and 1970s to primarily digital processing in the last decade. MIT Lincoln Laboratory pioneered some of the early applications of digital signal processing by developing dedicated processing performed in hardware to implement application-specific functions. Through the advent of programmable computing, many of these digital processing algorithms were implemented in more general-purpose computing while still preserving compute-intensive functions in dedicated hardware. As a result of the wide range of computing environments and the growth in the requisite parallel processing, MIT Lincoln Laboratory recognized the need to assemble the embedded community in a yearly national event. In 2006, this event, the High Performance Embedded Computing (HPEC) Workshop, marked its tenth anniversary of providing a forum for current advances in HPEC. This handbook, an outgrowth of the many advances made in the last decade, also, in several instances, builds on knowledge originally discussed and presented by the handbook authors at HPEC Workshops. The editors and contributing authors believe it is important to bring together in the form of a handbook the lessons learned from a decade of advances in high performance embedded computing.

This HPEC handbook is best suited to systems engineers and computational scientists working in the embedded computing field. The emphasis is on a systems perspective, but complemented with specific implementations starting with analog-to-digital converters, continuing with front-end signal processing addressing compute-intensive operations, and progressing through back-end processing requiring intensive parallel and programmable processing. Hardware and software engineers will also benefit from this handbook since the chapters present their subject areas by starting with fundamental principles and exemplifying those via actual developed systems. The editors together with the contributing authors bring a wealth of practical experience acquired through working in this field for a span of several decades. Therefore, the approach taken in each of the chapters is to cover the respective system components found in today's HPEC systems by addressing design trade-offs, implementation options, and techniques of the trade and then solidifying the concepts through specific HPEC system examples. This approach provides a more valuable learning tool since the reader will learn about the different subject areas by way of factual implementation cases developed in the course of the editors' and contributing authors' work in this exciting field.

Since a complex HPEC system consists of many subsystems and components, this handbook covers every segment based on a canonical framework. The canonical framework is shown in the following figure. This framework is used across the handbook as a road map to help the reader navigate logically through the handbook.

The introductory chapters present examples of complex HPEC systems representative of actual prototype developments. The reader will get an appreciation of the key subsystems and components by first covering these chapters. The handbook then addresses each of the system components shown in the aforementioned figure. After the introductory chapters, the handbook covers computational characteristics of high performance embedded algorithms and applications to help the reader understand the key challenges and recommended approaches. The handbook then proceeds with a thorough description of analog-to-digital converters typically found in today's HPEC systems. The discussion continues into front-end implementation approaches followed by back-end parallel processing techniques. Since the front-end processing is typically very compute-intensive, this part of the system is best suited for VLSI hardware and/or field programmable gate arrays. Therefore, these subject areas are addressed in great detail.

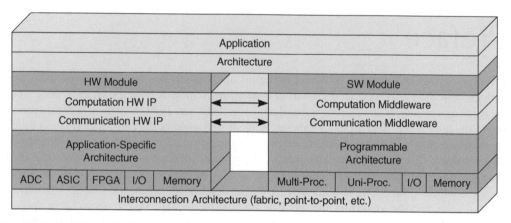

Canonical framework illustrating key subsystems and components of a high performance embedded computing (HPEC) system.

The handbook continues with several chapters discussing candidate back-end implementation techniques. The back-end of an HPEC system is often implemented using a parallel set of high performing programmable chips. Thus, parallel processing technologies are discussed in significant depth. Computing devices, interconnection fabrics, software architectures and metrics, plus middleware and portable software, are covered at a level that practicing engineers and HPEC computational practitioners can learn and adapt to suit their own implementation requirements. More and more of the systems implemented today require an open system architecture, which depends on adopted standards targeted at parallel processing. These standards are also covered in significant detail, illustrating the benefits of this open architecture trend.

The handbook concludes with several chapters presenting application examples ranging from electro-optics, sonar surveillance, communications systems, to advanced radar systems. This last section of the handbook also addresses future trends in high performance embedded computing and presents advances in microprocessor architectures since these processors are at the heart of any future HPEC system.

The HPEC handbook, by leveraging the contributors' many years of experience in embedded computing, provides readers with the requisite background to effectively work in this field. It may also serve as a reference for an advanced undergraduate course or a specialized graduate course in high performance embedded computing.

David R. Martinez
Robert A. Bond
M. Michael Vai

Acknowledgments

This handbook is the product of many hours of dedicated efforts by the editors, authors, and production personnel. It has been a very rewarding experience. This book would not have been possible without the technical contributions from all the authors. Being leading experts in the field of high performance embedded computing, they bring a wealth of experience not found in any other book dedicated to this subject area.

We would also like to thank the editors' employer, MIT Lincoln Laboratory; many of the subjects and fundamental principles discussed in the handbook stemmed from research and development projects performed at the Laboratory in the past several years. The Lincoln Laboratory management wholeheartedly supported the production of this handbook from its start. We are especially grateful for the valuable support we received during the preparation of the manuscript. In particular, we would like to thank Mr. David Granchelli and Ms. Dorothy Ryan. Dorothy Ryan patiently edited every single chapter of this book. David Granchelli coordinated the assembling of the book. Also, many thanks are due to the graphics artists—Mr. Chet Beals, Mr. Henry Palumbo, Mr. Art Saarinen, and Mr. Newton Taylor. The graphics work flow was supervised by Mr. John Austin. Many of the chapters were proofread by Mrs. Barbra Gottschalk. Finally, we would like to thank the publisher, Taylor & Francis/CRC Press, for working with us in completing this handbook. The MIT Lincoln Laboratory Communications Office, editorial personnel, graphics artists, and the publisher are the people who transformed a folder of manuscript files into a complete book.

About the Editors

Mr. David R. Martinez is Head of the Intelligence, Surveillance, and Reconnaissance (ISR) Systems and Technology Division at MIT Lincoln Laboratory. He oversees more than 300 people and has direct line management responsibility for the division's programs in the development of advanced techniques and prototypes for surface surveillance, laser systems, active and passive adaptive array processing, integrated sensing and decision support, undersea warfare, and embedded hardware and software computing.

Mr. Martinez joined MIT Lincoln Laboratory in 1988 and was responsible for the development of a large prototype space-time adaptive signal processor. Prior to joining the Laboratory, he was Principal Research Engineer at ARCO Oil and Gas Company, responsible for a multidisciplinary company project to demonstrate the viability of real-time adaptive signal processing techniques. He received the ARCO special achievement award for the planning and execution of the 1986 Cuyama Project, which provided a superior and cost-effective approach to three-dimensional seismic surveys. He holds three U.S. patents.

Mr. Martinez is the founder, and served from 1997 to 1999 as chairman, of a national workshop on high performance embedded computing. He has also served as keynote speaker at multiple national-level workshops and symposia including the Tenth Annual High Performance Embedded Computing Workshop, the Real-Time Systems Symposium, and the Second International Workshop on Compiler and Architecture Support for Embedded Systems. He was appointed to the Army Science Board from 1999 to 2004. From 1994 to 1998, he was Associate Editor of the *IEEE Signal Processing* magazine. He was elected an IEEE Fellow in 2003, and in 2007 he served on the Defense Science Board ISR Task Force.

Mr. Martinez earned a bachelor's degree from New Mexico State University in 1976, an M.S. degree from the Massachusetts Institute of Technology (MIT), and an E.E. degree jointly from MIT and the Woods Hole Oceanographic Institution in 1979. He completed an M.B.A. at the Southern Methodist University in 1986. He has attended the Program for Senior Executives in National and International Security at the John F. Kennedy School of Government, Harvard University.

Mr. Robert A. Bond is Leader of the Embedded Digital Systems Group at MIT Lincoln Laboratory. In his career, he has focused on the research and development of high performance embedded processors, advanced signal processing technology, and embedded middleware architectures. Prior to coming to the Laboratory, Mr. Bond worked at CAE Ltd. on radar, navigation, and Kalman filter applications for flight simulators, and then at Sperry, where he developed simulation systems for a Naval command and control application.

Mr. Bond joined MIT Lincoln Laboratory in 1987. In his first assignment, he was responsible for the development of the Mountaintop RSTER radar software architecture and was coordinator for the radar system integration. In the early 1990s, he was involved in seminal studies to evaluate the use of massively parallel processors (MPP) for real-time signal and image processing. Later, he managed the development of a 200 billion operations-per-second airborne processor, consisting of a 1000-processor MPP for performing radar space-time adaptive processing and a custom processor for performing high-throughput radar signal processing. In 2001, he led a team in the development of the Parallel Vector Library, a novel middleware technology for the portable and scalable development of high performance parallel signal processors.

In 2003, Mr. Bond was one of two researchers to receive the Lincoln Laboratory Technical Excellence Award for his "technical vision and leadership in the application of high-performance embedded processing architectures to real-time digital signal processing systems." He earned a B.S. degree (honors) in physics from Queen's University, Ontario, Canada, in 1978.

Dr. M. Michael Vai is Assistant Leader of the Embedded Digital Systems Group at MIT Lincoln Laboratory. He has been involved in the area of high performance embedded computing for over 20 years. He has worked and published extensively in very-large-scale integration (VLSI), application-specific integrated circuits (ASICs), field programmable gate arrays (FPGAs), design methodology, and embedded digital systems. He has published more than 60 technical papers and a textbook (*VLSI Design*, CRC Press, 2001). His current research interests include advanced signal processing algorithms and architectures, rapid prototyping methodologies, and anti-tampering techniques.

Until July 1999, Dr. Vai was on the faculty of the Electrical and Computer Engineering Department, Northeastern University, Boston, Massachusetts. At Northeastern University, he developed and taught the VLSI Design and VLSI Architecture courses. He also established and supervised a VLSI CAD laboratory. In May 1999, the Electrical and Computer Engineering students presented him with the Outstanding Professor Award. During his tenure at Northeastern University, he performed research programs funded by the National Science Foundation (NSF), Defense Advanced Research Projects Agency (DARPA), and industry.

After joining MIT Lincoln Laboratory in 1999, Dr. Vai led the development of several notable real-time signal processing systems incorporating high-density VLSI chips and FPGAs. He coordinated and taught a VLSI Design course at Lincoln Laboratory in 2002, and in April 2003, he delivered a lecture entitled "ASIC and FPGA DSP Implementations" in the IEEE lecture series, "Current Topics in Digital Signal Processing." Dr. Vai earned a B.S. degree from National Taiwan University, Taipei, Taiwan, in 1979, and M.S. and Ph.D. degrees from Michigan State University, East Lansing, Michigan, in 1985 and 1987, respectively, all in electrical engineering. He is a senior member of IEEE.

Contributors

Bilge E. S. Akgul
Georgia Institute of Technology
Atlanta, Georgia

James C. Anderson
MIT Lincoln Laboratory
Lexington, Massachusetts

Masahiro Arakawa
MIT Lincoln Laboratory
Lexington, Massachusetts

W. Robert Bernecky
Naval Undersea Warfare Center
Newport, Rhode Island

Nadya T. Bliss
MIT Lincoln Laboratory
Lexington, Massachusetts

Lakshmi N. Chakrapani
Georgia Institute of Technology
Atlanta, Georgia

Robert A. Coury
MIT Lincoln Laboratory
Lexington, Massachusetts

Stephen Crago
University of Southern California
Information Sciences Institute
Los Angeles, California

Joel I. Goodman
MIT Lincoln Laboratory
Lexington, Massachusetts

Preston A. Jackson
MIT Lincoln Laboratory
Lexington, Massachusetts

Jeremy Kepner
MIT Lincoln Laboratory
Lexington, Massachusetts

Hahn G. Kim
MIT Lincoln Laboratory
Lexington, Massachusetts

Helen H. Kim
MIT Lincoln Laboratory
Lexington, Massachusetts

Pinar Korkmaz
Georgia Institute of Technology
Atlanta, Georgia

James M. Lebak
The MathWorks
Natick, Massachusetts

Miriam Leeser
Northeastern University
Boston, Massachusetts

Thomas G. Macdonald
MIT Lincoln Laboratory
Lexington, Massachusetts

Janice McMahon
University of Southern California
Information Sciences Institute
Los Angeles, California

Theresa Meuse
MIT Lincoln Laboratory
Lexington, Massachusetts

Huy T. Nguyen
MIT Lincoln Laboratory
Lexington, Massachusetts

Krishna V. Palem
Georgia Institute of Technology
Atlanta, Georgia

Albert I. Reuther
MIT Lincoln Laboratory
Lexington, Massachusetts

Glenn E. Schrader
MIT Lincoln Laboratory
Lexington, Massachusetts

Brian M. Tyrrell
MIT Lincoln Laboratory
Lexington, Massachusetts

William S. Song
MIT Lincoln Laboratory
Lexington, Massachusetts

Wayne Wolf
Georgia Institute of Technology
Atlanta, Georgia

Kenneth Teitelbaum
MIT Lincoln Laboratory
Lexington, Massachusetts

Donald Yeung
University of Maryland
College Park, Maryland

Section I

Introduction

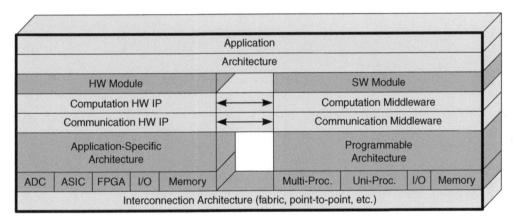

Chapter 1   A Retrospective on High Performance Embedded Computing
David R. Martinez, MIT Lincoln Laboratory

This chapter presents a historical perspective on high performance embedded computing systems and representative technologies used in their implementations. Several hardware and software technologies spanning a wide spectrum of computing platforms are described.

Chapter 2   Representative Example of a High Performance Embedded Computing System
David R. Martinez, MIT Lincoln Laboratory

Space-time adaptive processors are representative of complex high performance embedded computing systems. This chapter elaborates on the architecture, design, and implementation approaches of a representative space-time adaptive processor.

Chapter 3   System Architecture of a Multiprocessor System
David R. Martinez, MIT Lincoln Laboratory

This chapter discusses a generic multiprocessor and provides a representative example to illustrate key subsystems found in modern HPEC systems. The chapter covers from the analog-to-digital converter through both the front-end VLSI technology and the back-end programmable subsystem. The system discussed is a hybrid architecture necessary to meet highly constrained size, weight, and power.

Chapter 4   High Performance Embedded Computers: Development Process and Management Perspective
Robert A. Bond, MIT Lincoln Laboratory

This chapter briefly reviews the HPEC development process and presents a detailed case study that illustrates the development and management techniques typically applied to HPEC developments. The chapter closes with a discussion of recent development/management trends and emerging challenges.

1 A Retrospective on High Performance Embedded Computing

David R. Martinez, MIT Lincoln Laboratory

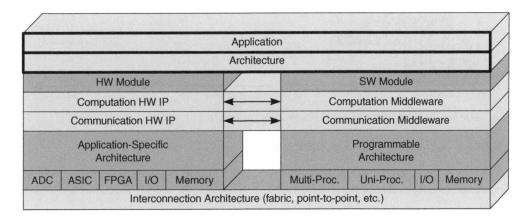

This chapter presents a historical perspective on high performance embedded computing systems and representative technologies used in their implementations. Several hardware and software technologies spanning a wide spectrum of computing platforms are described.

1.1 INTRODUCTION

The last 50 years have witnessed an unprecedented growth in computing technologies, significantly impacting the capabilities of systems that have achieved their unmatched dominance enabled by the ability of computing to reach full or partial real-time performance. Figure 1-1 illustrates a 50-year historical perspective of the progress of high performance embedded computing (HPEC).

In the early 1950s, the discovery of the integrated circuit helped transform computations from antiquated tube-based computing to computations performed using transistorized operations (Bellis 2007). MIT Lincoln Laboratory developed the TX-0 computer, and later the TX-2, to test the use of transistorized computing and the application of core memory (Freeman 1995; Buxton 2005). These systems were preceded by MIT's Whirlwind computer, the first to operate in real time and use video displays for output; it was one of the first instantiations of a digital computer. This innovative Whirlwind technology was employed in the Air Force's Semi-Automatic Ground Environment (SAGE) project, a detection and tracking system designed to defend the continental United States against bombers crossing the Atlantic Ocean. Though revolutionary, the Whirlwind had only the computational throughput of 20 thousand operations per second (KOPS). The TX-2 increased the

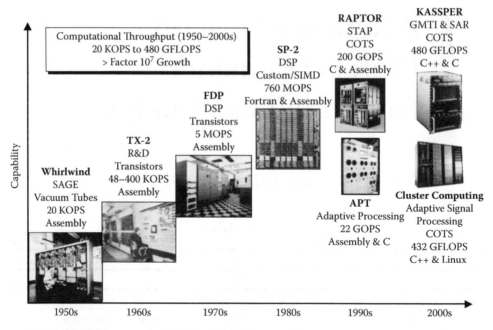

FDP: Fast Digital Processor   SP-2: Synchronous Processor 2

FIGURE 1-1   Historical perspective on HPEC systems.

computational throughput to 400 KOPS. Both were programmed in assembly language. Most of the computations performed required tracking of airplane detections and involved simple correlations (Freeman 1995).

The 1960s brought us the discovery of the fast Fourier transform (FFT) with a broad range of applications (Cooley and Tukey 1965). It was at this time that digital signal processing became recognized as a more effective and less costly way to extract information. Several academic and laboratory pioneers began to demonstrate the impact that digital signal processing could have on a broad range of disciplines, such as speech, radar, sonar, imaging, and seismic processing (Gold and Rader 1969; Oppenheim and Schafer 1989; Rabiner and Gold 1975).

Many of these applications originally required dedicated hardware to implement functions such as the FFT, digital filters, and correlations. One early demonstration was the high-speed FFT processor (Gold et al. 1971), shown in Figure 1-1 and referred to as the Fast Digital Processor (FDP), with the ability to execute 5 million operations per second (MOPS). Later in the 1970s, manufacturers like Texas Instruments, Motorola, Analog Devices, and AT&T demonstrated that digital signal processors could perform the critical digital signal processing (DSP) kernels, such as FFTs, digital filters, convolutions, and other important DSP functions, by structuring the DSP devices with more hardware tuned to these functions. An example of such a device was the TMS320C30 programmed in assembly and providing a throughput of 33 MFLOPS (millions of floating-point operations per second) under power levels of less than 2 W per chip (Texas Instruments 2007).

These devices had a profound impact on high performance embedded computing. Several computing boards were built to effectively leverage the capabilities of these devices. These evolved to where simulators and emulators were available to debug the algorithms and evaluate real-time performance before the code was downloaded to the final target hardware. Figure 1-2 depicts an example of a software development environment for the Texas Instrument TMS320C30 DSP microprocessor. The emulator board (TI XDS-1000) was used to test the algorithm performance on a single DSP processor. This board was controlled from a single-board computer interfaced in a VME chassis, also shown in Figure 1-2.

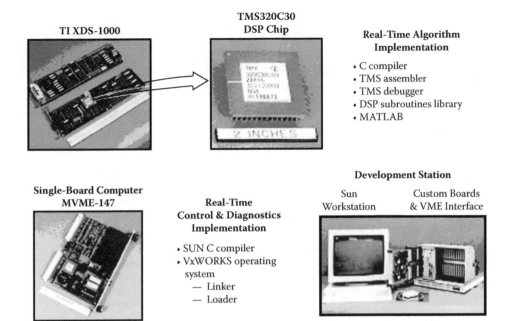

FIGURE 1-2 TI TMS320C30 DSP microprocessor development environment.

One nice feature of this hardware was the ability to program it in C-code and complement compute-intensive functions with assembly subroutines. For those cases in which the DSP-based systems were not able to meet performance, a dedicated hardware tuned to the digital processing functions was necessary. In the next chapter, an example of an HPEC system illustrates a design that leveraged both dedicated hardware and programmable DSP devices needed to meet the real-time performance.

Today a mix of dedicated hardware solutions and programmable devices is found in applications for which no other approach can meet the real-time performance. Even though microprocessors such as the PowerPC can operate at several GHz in speed (IBM 2007), providing a maximum throughput in the gigaflops class, several contemporary applications such as space systems, airborne systems, and missile seekers, to name a few, must rely on a combination of dedicated hardware for the early signal processing and programmable systems for the later processing. Many of these systems are characterized by high throughput requirements in the front-end with very regular processing and lower throughputs in the back-end, but with a high degree of data dependency and, therefore, requiring more general-purpose programming. In Figure 1-3 a spectrum of classes of computing systems is shown, including the range in billions of operations per second per unit volume (GOPS/liter) and billions of operations per second per watt (GOPS/W).

The illustration in Figure 1-3 is representative of applications and computing capabilities existing circa 2006. These applications and computing capabilities change, but the trends remain approximately the same. In other words, the improvements in computing capabilities (as predicted by Moore's Law) benefit programmable systems, reconfigurable systems, and custom hardware in the same manner. This handbook addresses all of these computing options, their associated capabilities and limitations, and hardware plus software development approaches.

Many applications can be met with programmable signal processors. In these instances, the platform housing the signal processor is typically large in size with plenty of power, or, conversely, the algorithm complexity is low, permitting its implementation in a single or a few microprocessors. Programmable signal processors, as the name implies, provide a high degree of flexibility since the algorithm techniques are implemented using high-order languages such as C. However, as discussed in later chapters, the implementation must be rigorous with a high

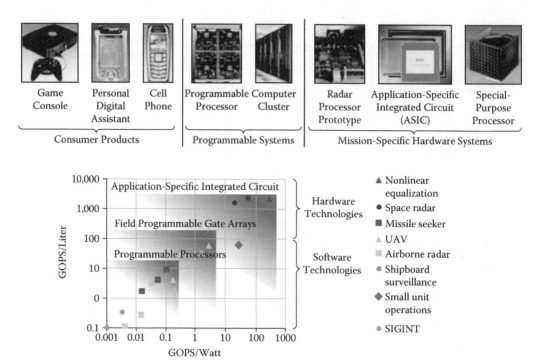

FIGURE 1-3　(**Color figure follows page 278.**) Embedded processing spectrum.

degree of care to ascertain real-time performance and reliability. Reconfigurable computing, for example, utilizing field programmable gate arrays (FPGAs) achieves higher computing performance in a fixed volume and power when compared to programmable computing systems. This performance improvement comes at the expense of only having flexibility in the implementation if the algorithm techniques can be easily mapped to a fixed set of gates, table look-ups, and Boolean operations, all driven by a set of programmed bit streams (Martinez, Moeller, and Teitelbaum 2001). The most demanding applications require most of the computing be implemented in custom hardware to meet capabilities for cases in which trillions of operations per second per unit volume (TOPS/ft^3) and 100s GOPS/W are needed. Today such computing performance demands custom designs and dedicated hardware implemented using application-specific integrated circuits (ASICs) based on standard cells or full-custom designs. These options are further described in more detail in subsequent chapters. Most recently, an emerging design option combines the best of custom design with the capability to introduce the user's own intellectual property (IP), leveraging reconfigurable hardware (Flynn and Hung 2005; Schreck 2006). This option is often referred to as structured ASICs and permits a wide range of IP designs to be implemented from customized hard IP, synthesized firm IP, or synthesizable soft IP (Martinez, Vai, and Bond 2004). FPGAs can be used initially to prototype the design. Once the design is accepted, then structured ASICs can be employed with a faster turnaround time than regular ASICs while still achieving high performance and low power.

　　The next section presents examples of computing systems spanning almost a decade of computing. These technologies are briefly reviewed to put in perspective the rapid advancement that HPEC has experienced. This retrospective in HPEC developments, including both hardware systems and software technologies, helps illustrate the progression in computing to meet very demanding defense applications. Subsequent chapters in this handbook elaborate on several of these enabling technologies and predict the capabilities likely to emerge to meet the demands of future HPEC systems.

1.2 HPEC HARDWARE SYSTEMS AND SOFTWARE TECHNOLOGIES

Less than a decade ago, defense system applications demanded computing throughputs in the range of a few GOPS consuming only a few 1000s of watts in power (approximating 1 MOPS/W). However, there was still a lot of interest in leveraging commercial off-the-shelf (COTS) systems. Therefore, in the middle 1990s, the Department of Defense (DoD) initiated an effort to miniaturize the Intel Paragon into a system called the Touchstone. The idea was to deliver 10 GOPS/ft^3. As shown in Figure 1-4, the Intel Paragon was based on the Intel i860 programmable microprocessor running at 50 MHz and performing at about 0.07 MFLOPS/W. The performance was very limited but it offered programming flexibility. In demonstration, the Touchstone successfully met its set of goals, but it was overtaken by systems based on more capable DSP microprocessors. At the same time, the DoD also started investing in the Vector Signal and Image Processing Library (VSIPL) to allow for more standardized approaches in the development of software. The initial instantiation of VSIPL was only focused on a single processor. As discussed in later chapters, VSIPL has been successfully extended to many parallel processors operating together. The standardization in software library functions enhanced the ability to port the same software to other computing platforms and also to reuse the same software for other similar algorithm applications.

Soon after the implementation of the Touchstone, Analog Devices came out with the ADSP 21060. This microprocessor was perceived as better matched to signal processor applications. MIT Lincoln Laboratory developed a real-time signal processor system (discussed in more detail in Chapter 3). This system consisted of approximately 1000 ADSP 21060 chips running at 40 MHz, all operating in parallel. The total peak performance was 12 MFLOPS/W. The system offered a considerable number of operations consuming very limited power. The total consumed power was about 8 kW requiring about 100 GOPS of peak performance. Even though the system provided flexibility in the programming of the space-time adaptive processing (STAP) algorithms, the ADSP 21060 was difficult to program. The STAP algorithms operated on different dimensions of the incoming data channels. Several corner turns were necessary to process signals first on a channel-

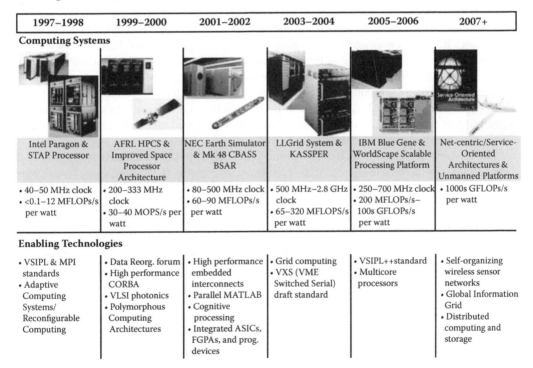

1997–1998	1999–2000	2001–2002	2003–2004	2005–2006	2007+
Computing Systems					
Intel Paragon & STAP Processor	AFRL HPCS & Improved Space Processor Architecture	NEC Earth Simulator & Mk 48 CBASS BSAR	LLGrid System & KASSPER	IBM Blue Gene & WorldScape Scalable Processing Platform	Net-centric/Service-Oriented Architectures & Unmanned Platforms
• 40–50 MHz clock • <0.1–12 MFLOPs/s per watt	• 200–333 MHz clock • 30–40 MOPS/s per watt	• 80–500 MHz clock • 60–90 MFLOPs/s per watt	• 500 MHz–2.8 GHz clock • 65–320 MFLOPS/s per watt	• 250–700 MHz clock • 200 MFLOPs/s–100s GFLOPs/s per watt	• 1000s GFLOPs/s per watt
Enabling Technologies					
• VSIPL & MPI standards • Adaptive Computing Systems/ Reconfigurable Computing	• Data Reorg. forum • High performance CORBA • VLSI photonics • Polymorphous Computing Architectures	• High performance embedded interconnects • Parallel MATLAB • Cognitive processing • Integrated ASICs, FGPAs, and prog. devices	• Grid computing • VXS (VME Switched Serial) draft standard	• VSIPL++standard • Multicore processors	• Self-organizing wireless sensor networks • Global Information Grid • Distributed computing and storage

FIGURE 1-4 Approximately a decade of high performance embedded computing.

by-channel basis. The output results were corner-turned again so that the signal processor could operate on radar pulses and, finally, another corner turn was necessary for operation across multiple received digital processed beams.

These data reorganization requirements resulted in significant latency, leading the Defense Advanced Research Projects Agency (DARPA) to begin investing in a project referred to as Data Reorganization. A forum was created to focus on techniques to achieve real-time performance for applications demanding data reorganization (Cain, Lebak, and Skjellum 1999). About the same time, the HPEC community began testing the use of message-passing interfaces (MPI), but again for real-time performance (Skjellum and Hebert 1999).

For many years, DARPA has been a significant source of research funding focused on embedded computing. In addition to its interest in the abovementioned software projects, DARPA recognized the advancements emerging as large numbers of transistors were available on a single die. The Adaptive Computing Systems and Polymorphous Computing programs were two examples focused on leveraging reconfigurable computing offering some flexibility in algorithm implementations but with higher performance than afforded by general-purpose microprocessors. Several chips were demonstrated with higher performance than reduced instruction set computer (RISC) microprocessors. The RAW chip was targeted at 250 MHz with an expected performance of 4 GOPS for the 2003 year time frame (Graybill 2003). The MONARCH chip in comparison was predicted to deliver 85 GOPS operating at 333 MHz in a 2005 prototype chip (Granacki 2004).

The late 1990s (as shown in Figure 1-4) were also characterized by the implementation of the then newly available PowerPC chip family. This RISC processor was fully programmable in C and delivered respectable performance. The Air Force Research Laboratory designed a system based on the Motorola PowerPC 603e, delivering 39 MFLOPS/W and also targeted at implementations such as the STAP algorithms (Nielsen 2002). Notice the factor of over 3× improvement from the STAP processor developed using the Analog ADSP 21060. The performance improvement was a result of increased throughputs at lower power levels. The PowerPC was also significantly easier to program than was the ADSP21060 device and, therefore, was often used in many subsequent real-time systems as both Motorola and IBM continued to advance the PowerPC family.

From the early to mid-1990s, the HPEC community benefited from the availability of both high performance RISC processors and reconfigurable systems (e.g., based on FPGAs). However, most real-time performance was limited by the availability of commensurate high performance interconnects (Carson and Bohman 2003). Several system manufacturers joined forces to standardize several interconnect options. Examples of high performance embedded interconnects were the Serial Rapid IO and the InfiniBand (Andrews 2006). These interconnects were, and still are, crucial to maintaining an architecture well balanced between the high-speed microprocessors and the intrachassis and interchassis communications.

The experiences gained from the last several years helped put in perspective the advances the HPEC community has seen in microprocessor hardware, interconnects, memory, and software. Many of these advances are a direct result of not only exploiting Moore's Law, which manufacturers have consistently kept pace with, but also evolving the real-time software and interconnects to preserve a balanced architecture. As we look into the future, the HPEC requirements will continue to advance, demanding faster and better performing systems. System requirements will progress toward 10s of GFLOPs/W, in some cases approaching TeraOps/W. The distinctions between floating-point versus fixed-point operations are ignored for purposes of depicting future requirements since the operation type will depend on the chosen implementation. However, the throughput requirements will be significantly higher than those experienced in recent years. This increase in system requirements is a direct consequence of wanting to make our defense systems more and more capable within a single platform. Because the platform costs typically dominate the onboard HPEC system, it is highly desirable to make this system highly capable when integrated on a single platform. All predictions indicate that for the next several years these requirements will be met with a combined capability of ASICs, FPGAs, and programmable devices efficiently integrated into computing systems. These systems will also demand real-time performance out of the interconnects, memory hierarchy, and operating systems. This handbook addresses the details and techniques employed to meet these very high performance requirements, and it also covers the full spectrum of design approaches to meet the desired HPEC system capabilities.

Before embarking into the architecture and design techniques found in the development of HPEC systems, it is useful to briefly review the generic structure of a multiprocessor system found in many HPEC applications. Reviewing the canonical architecture components will help in understanding the key system capabilities necessary to develop an HPEC system. The next section presents an example of a multiprocessor system architecture.

1.3 HPEC MULTIPROCESSOR SYSTEM

To understand an HPEC system, it is worthwhile to first understand the typical classes of processing performed at the system level. Then, from the classes of operations performed, it is best to look at the computing components used (in a generic sense) to meet the processing functions. The subsequent chapters in this handbook present the state of the art in meeting the processing functions as well as the implementation approaches commonly used in embedded systems.

In several defense applications today, the systems are dominated by significant analog computing prior to the analog-to-digital converter (ADC). Therefore, the computing performed is achieved with very unsophisticated processors since the processing post the ADC is limited. However, as we look at evolutions in system hardware, more and more of the computing will be done in the digital format, thus making the HPEC hardware complex. The system architectures are relying on moving the ADC closer and closer to the front-end sensor (in a radar system this is the antenna). Figure 1-5 illustrates a typical processing flow for a phased-array active electronically scanned antenna (AESA) for an advanced radar system envisioned in the future. Later chapters illustrate other applications demanding complex HPEC systems such as sonar and electro-optics. The processing functions for these other applications are different from the processing flow illustrated in Figure 1-5. However, the radar sensor example is used to show the typical processing flow since it is also very demanding and characterizes a very complex data and interconnection set of constraints, thereby serving to illustrate the complexity of demanding HPEC systems.

The advances in antenna technologies are evolving at a pace to enable multiple channels (also commonly referred to as subarrays, depending on the antenna topology). These channels feed a front-end set of receivers to condition the incoming data to be properly sampled by high-speed ADCs. Typical ADC sampling varies from large numbers of bits and lower sampling rates

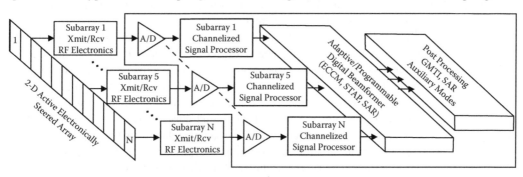

- Billion to trillion operations per second
- 10s of Gbytes per second after the analog-to-digital converters
- Real-time performance with 10s of milliseconds latency
- Mix of custom ASICs, FPGA, and programmable DSPs
- Distributed real-time software
- Portable software

FIGURE 1-5 Next-generation digital radar system for a surface surveillance application.

(e.g., 14 bits and 40–100 MHz sampling) to fewer bits but higher sampling rates (e.g., 8 bits and 1 GHz sampling). The output of the ADCs is then fed into a front-end processing system. In Figure 1-5, this is represented by a subarray channelized signal processor. Typical functions performed within the front-end processing system are digital in-phase and quadrature sampling (Martinez, Moeller, Teitelbaum 2001), channel equalization to compensate for channel-to-channel distortions prior to beamforming, and pulse compression needed to convolve the incoming data with the transmitted waveform. These are representative processing functions for the front-end system. However, they all have a common topology: all processing is done on a channel-by-channel basis leading to the ability to parallelize the processing flow. The actual signal processing functions utilized to perform these classes of front-end processing steps depend on the details of the application. However, FFTs, convolvers, and FIR filtering, to name a few, are very representative of signal processing functions found in these processing stages. Since these front-end processing functions operate on very fast incoming datasets (typically several billions of bytes per second), the processing is regular but very demanding, reaching trillions of operations per second (TOPS).

In these complex systems, the objective is to operate on the ADC data to a point where the signals of interest have been extracted and all the interfering noise has been mitigated. So, one way to think of an HPEC system is as the engine necessary to transform large amounts of data into useful information (signals of interest). Therefore, if the output of the ADCs is low, it might be more cost-effective to send the processing down to a ground processing site via wireless communication links. However, the communication links available today and expected in the foreseeable future are not able to transmit all the data flowing from the ADCs for many systems of interest. Furthermore, several systems require the processed data on board to effect an action (such as placing a weapon on a target) in real time. Furthermore, in many cases the user cannot tolerate long latencies.

Following the front-end processing the data are required to be reorganized to continue additional processing. For the radar example illustrated in Figure 1-5, some of these functions include converting the data from channel space (channel-by-channel inputs) to radar beams. In this process, the typical representative functions include intentional and/or unintentional jamming suppression and clutter mitigation (typically found in surface surveillance and air surveillance systems). From the perspective of an HPEC system, these processing stages require the manipulation of the data such that the proper independent variables (also commonly referred to as degrees of freedom) are operated on. For example, to deal with interfering jammers, the desired inputs are independent channels and the processing involves computation of adaptive weights (in real time) that are subsequently applied to the data to direct all the energy in the direction of interest while at the same time placing array nulls in the direction of the interferers.

The computation of the adaptive weights can be a very computationally intensive function that grows as the cube of the number of input channels or degrees of freedom. In some applications, the adaptive weight computation also requires larger arithmetic precision than, for example, the application of the weights to the incoming data. Typical arithmetic precision ranges from 32 to 64 bits, and it is primarily a result of having to invert an estimate of the cross-correlation matrix containing information on the interfering jammers and the background noise. This cross-correlation matrix reflects a very wide dynamic range representative of the sampled data from the ADCs.

The process of jamming cancellation can result, as a byproduct, into a set of output beams. The two-step process described here is representative of the demanding processing flow. There are other algorithms that combine the process of jammer nulling with clutter nulling or perform these operations all in the frequency domain (after Doppler processing). These different techniques are all options for the real-time processing of incoming signals, and the preferred option depends on the specifics of the application (Ward 1994). However, the sequential processing of jammer nulling followed by clutter nulling is very representative of challenges present in radar systems (for both surface surveillance and air surveillance).

Similar to jamming nulling, clutter cancellation presents significant processing complexity challenges. The clutter nulling, referred to as ground moving-target indication (GMTI) in Figure 1-5, involves a corner turn (Teitelbaum 1998). After converting the data from channel data (element space) to beams (beam space), the data must be corner-turned to pulse-by-pulse data. This is particularly the case if the clutter nulling is done in the Doppler domain. Prior to the clutter nulling, data are converted from the time domain (pulse-by-pulse data) to the frequency or Doppler domain. This operation involves either a discrete Fourier transform or, more commonly, an FFT. The FFT, for example, must meet real-time performance. However, because the data in this example are formed by a number of beams, the processing is very well matched to parallel processing. Furthermore, the signal processor system can be operating on one data cube while another data cube is stored in memory. Another technique is to "round-robin" multiple data cubes across multiple processors. Round-robin means one set of processors operate on an earlier data cube (consisting of beams, Doppler frequencies, and range gates) while a different set of parallel processors operate on a more recent data cube. The number of processors is chosen, and the process is synchronized, such that once the earlier processors finish the processing of the earlier data cube, a new data cube is ready to be processed.

In a similar way to jammer nulling, the clutter nulling also involves the computation of a set of weights, and these adaptive weights must be applied to the data to cancel clutter interference competing with the targets of interest. This weight computation also grows as the cubic of the available degrees of freedom. The application of the weights is also very demanding in computation throughput but very regular in the form of vector-vector multiplies.

For very typical numbers of beams formed and gigabytes of processed data, the total throughput required will range from 100s of GigaOps to TeraOps. This computational complexity must be met in very constrained environments commensurate with missiles and airborne or satellite systems. The next chapter provides examples of HPEC prototype systems built to perform these types of processing functions.

The other representative processing functions worth addressing as an example of HPEC processing functions are target detection and clustering. These are of particular interest because they belong to a different class of functions but are illustrative of the classes of processing functions found on contemporary HPEC real-time systems. Target detection and clustering functions (which sometimes are also combined or followed by target tracking) require a very different processing flow than does front-end filtering or interference nulling described earlier. Since after front-end filtering and interference nulling the data have been expected to only contain signals of interest in the presence of incoherent noise, the processing is much more a function of the expected number of targets (or signals) present. The processing can also be parallelized as a function of, for example, beams, but computation throughput will depend on the number of targets processed. The computation throughput is often much less than in earlier processing but not as regular, requiring processing functions like sorting and centroiding.

Figure 1-6 shows an example of computation throughputs, memory, and communication goals for the processing flow described earlier. Figure 1-7 illustrates different examples of hardware computing platforms. For the same set of algorithms, the choice of computing platform or technology, ranging among full-custom designs, FPGAs, and full programmable hardware, will highly depend on the available size, weight, and power. As shown in Figure 1-7, there can be a factor of 3× between FPGAs and fully programmable hardware in computational density. The differences can be more pronounced between a full-custom, very-large-scale integration (VLSI) solution and a programmable DSP-based processor system. The full-custom VLSI system can be two orders of magnitude more capable in computational density than the programmable DSP-based system.

The very demanding processing goals of HPEC systems must be met in very constrained environments. These goals are unique to these classes of applications and are not achievable with general high performance commercial systems often found in large complex building-sized systems.

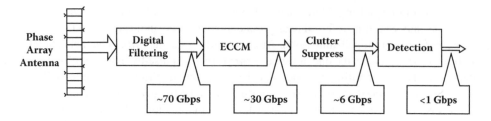

Processor Throughput

Throughput	16 Channels	20 Channels	24 Channels
Fixed Point (GOPS)	1,471	1,622	1,773
Floating Point (GFLOPS)	47	47	48
Total (GOPS)	1,518	1,669	1,821

Processor Memory

Memory	65 PRIs	195 PRIs
Total (MBytes)	2,979	5,966

Communication Bandwidths

	Digital Filtering	ECCM	Clutter Suppression	Detection
Compressed Data Rate (Gbit/s)	69.1	27.6	6.0	<1

FIGURE 1-6   Computation, memory, and communication goals of a challenging HPEC system.

	Full-Custom VLSI	Field Programmable Gate Array	Programmable DSP Processor
Throughput per Chassis* (Total Power)	TeraOps (20 W)	1 TeraOps (700 W)	1 TeraOps (2 kW)
Computational Density**	50 GOPS/W	1.5 GOPS/W	0.5 GOPS/W (Peak)
Processor Type	Custom Xilinx (1 GHz, 130 nm)	Virtex 8000 (400 MHz, 130 nm)	PowerPC 7447 (1 GHz, 130 nm)
Power per Processor	2 watts (100 GOPS/W)	20 watts (3 GOPS/W)	8 watts (1 GOPS/W peak)
Input Control	Coefficients	Reconfigurable	Fully Programmable
*Power assumes 50% dedicated to peripherals, memory, and I/O **Weights = Full Custom ~4 kg; FPGAs ~25 kg; Programmable System ~150 kg			

FIGURE 1-7   Examples of hardware computing platforms.

Later chapters will, therefore, address in detail the implementation approaches necessary to meet the processing goals of complex HPEC systems.

1.4 SUMMARY

This chapter has presented a retrospective, particularly a systems perspective, of the development of high performance embedded computing. The evolution in HPEC systems for challenging defense applications has seen dramatic exponential growth, for the most part concurrent with and leveraging advances in microprocessors and memory technologies that have been experienced by the semiconductor industry and predicted by Moore's Law. HPEC systems have exploited these enabling technologies, applying them to real-time embedded systems for a number of different applications. Furthermore, in the last 15 years, we have seen the evolution of complex real-time embedded systems from ones requiring billions of operations per second to today's systems demanding trillions of operations per second on the same equivalent form factor. This three-orders-of-magnitude evolution, at the system level, has tracked Moore's Law very closely. Software, on the other hand, continues to lag behind in limiting the ability to rapidly develop complex systems.

Subsequent chapters will introduce readers to various applications profiting from advances in HPEC and to several examples of prototype systems illustrating the level of hardware and software complexity required of current and future systems. It is hoped that this handbook will provide the background for a better understanding of the HPEC evolution and serve as the basis for assessing future challenges and potential opportunities.

REFERENCES

Andrews, W. 2006. Switched fabrics challenge the military and vice versa. *COTS Journal*. Available online at http://www.cotsjournalonline.com/home/article.php?id=100448.

Bellis, M. 2007. *Inventors of the Modern Computer. The History of the Integrated Circuit (IC)—Jack Kilby and Robert Noyce*. About.com website. New York: The New York Times Company. Available online at http://inventors.about.com/library/weekly/aa080498.htm.

Buxton, W., R. Baecker, W. Clark, F. Richardson, I. Sutherland, W.R. Sutherland, and A. Henderson. 2005. Interaction at Lincoln Laboratory in the 1960's: looking forward—looking back. *CHI '05 Extended Abstracts on Human Factors in Computing Systems*. Conference on Human Factors in Computing Systems, Portland, Ore.

Cain, K., J. Lebak, and A. Skjellum. 1999. Data reorganization and future embedded HPC middleware. High Performance Embedded Computing Workshop, MIT Lincoln Laboratory, Lexington, Mass.

Carson, W. and T. Bohman. 2003. Switched fabric interconnects. *Proceedings of the 7th Annual High Performance Embedded Computing Workshop*. MIT Lincoln Laboratory, Lexington, Mass. Available online at http://www.ll.mit.edu/HPEC/agenda03.htm.

Cooley, J.W. and J.W. Tukey. 1965. An algorithm for the machine computation of complex Fourier series. *Mathematics of Computation* 19: 297–301.

Flynn, M. and P. Hung. 2005. Microprocessor design issues: thoughts on the road ahead. *IEEE Micro* 25(3): 16–31.

Freeman, E., ed. 1995. Computers and signal processing. *Technology in the National Interest*. Lexington, Mass.: MIT Lincoln Laboratory.

Gold, B. and C.M. Rader. 1969. *Digital Processing of Signals*, New York: McGraw-Hill.

Gold, B., I.L. Lebow, P.G. McHugh, and C.M. Rader. 1971. The FDP—a fast programmable signal processor. *IEEE Transactions on Computers* C-20: 33–38.

Granacki, J. 2004. MONARCH: next generation supercomputer on a chip. *Proceedings of the Eighth Annual High Performance Embedded Computing Workshop*, MIT Lincoln Laboratory, Lexington, Mass. Available at http://www.ll.mit.edu/HPEC/agenda04.htm.

Graybill, R. 2003. Future HPEC technology directions. *Proceedings of the Seventh Annual High Performance Embedded Computing Workshop*. MIT Lincoln Laboratory, Lexington, Mass. Available online at http://www.ll.mit.edu/HPEC/agenda03.htm.

International Business Machines. IBM PowerPC 7XX and 6XX Microprocessors. Was available online at http://www-306.ibm.com/chips/techlib/techlib.nsf/products/PowerPC_7XX_and_6XX_Microprocessors.

Martinez, D.R., T.J. Moeller, and K. Teitelbaum. 2001. Application of reconfigurable computing to a high performance front-end radar signal processor. *Journal of VLSI Signal Processing* 28: 63–83.

Martinez, D.R., M. Vai, and R. Bond. 2004. Tutorial on real-time embedded computing for signal and image processing. IEEE International Radar Conference, Philadelphia. Abstract available online at http://www.radar04.org/program/index.cgi?paper=51.

Nielsen, P. 2002. *Perspective on Embedded Computing.* Keynote address. Sixth Annual High Performance Computing Workshop, MIT Lincoln Laboratory, Lexington, Mass.

Oppenheim, A.V. and R.W. Schafer. 1989. *Discrete-Time Signal Processing.* Upper Saddle River, N.J.: Prentice Hall.

Rabiner, L.R. and B. Gold. 1975. *Theory and Application of Digital Signal Processing.* Upper Saddle River, N.J.: Prentice Hall.

Schreck, R. 2006. Structured ASICs, FPGAs work in tandem. *EE Times* 14 August 2006 edition: 52.

Skjellum, A. and S. Hebert. 1999. MPI/RT-Pro: commercial-grade implementations of MPI/RT for COTS platforms: RACE and CSPI. High Performance Embedded Computing Workshop, MIT Lincoln Laboratory, Lexington, Mass.

Teitelbaum, K. 1998. Crossbar tree networks for embedded signal processing applications. *Proceedings of the 5th International Conference on Massively Parallel Processing Using Optical Interconnections:* 200.

Texas Instruments: Technology for Innovators. Available online at http://focus.ti.com/docs/prod/folders/print/tms320c30.html. Accessed 11 July 2007.

Ward, J. 1994. *Space-Time Adaptive Processing for Airborne Radar.* Technical Report #1015. Lexington, Mass.: MIT Lincoln Laboratory.

2 Representative Example of a High Performance Embedded Computing System

David R. Martinez, MIT Lincoln Laboratory

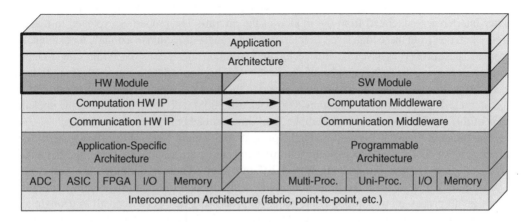

Space-time adaptive processors are representative of complex high performance embedded computing systems. This chapter elaborates on the architecture, design, and implementation approaches of a representative space-time adaptive processor.

2.1 INTRODUCTION

This chapter presents a detailed description of a prototype real-time test bed developed to perform surveillance from an airborne early warning system. This application is particularly interesting in the context of a high performance embedded computing (HPEC) system perspective because it exemplifies a hybrid implementation between dedicated hardware for very regular processing functions in the front-end of the prototype system and programmable hardware for back-end processing demanding less computation but significantly more complexity in software.

The processing throughput is a few billions of operations per second (GOPS), representative of a computational loading that was challenging in the early 1990s when the system was developed. Even though today's computational throughputs are much more capable [in the trillions of operations per second (TOPS) for similar applications and form factor], the real-time test bed presented here serves as a model for the development of real-time systems in terms of approach, design trade-offs, development rigor, and system integration and testing. This development approach is equally applicable today as it was during the development of this test bed; the principal difference is the available computation in a single microprocessor back then.

Section 2.2's description of the application driver is included to allow readers to appreciate the overall system complexity and the generic parameters that drive this complexity. Subsequent sections provide an architecture overview and address design trade-offs among a class of available implementation approaches. The discussion of specific hardware enablers illustrates how and where they fit in the overall system architecture. The chapter concludes with the system development techniques that were very successfully used during the fabrication and integration of the system. The following material is presented as lessons learned to guide future development of complex real-time systems.

2.2   SYSTEM COMPLEXITY

The system application illustrated here is an airborne early warning prototype for detecting targets buried in jamming and clutter interference. Figure 2-1 illustrates the target and interference scenario (Ward 1994). Multiple channels are input into the real-time signal processor to focus the energy in the direction of the target while simultaneously canceling the intentional jamming interference and the clutter interference generated from the motion of the aircraft. The properties exploited to make the target extraction from these high levels of interference are angle (direction in azimuth, for example) and Doppler. These properties have led to this technique's common reference as *space-time* (angle-Doppler) *adaptive processing* (STAP). A number of excellent books, papers, and technical reports in the adaptive sensor array processing for radars literature describe the STAP theoretical and mathematical framework in great detail (Guerci 2003; Klemm 2006; Richards 2005). This section addresses the implementation of this capability via an actual, complex, real-time test bed built as a proof of concept.

The target in Figure 2-1 is assumed to be present on the main beam of the array at zero-degree azimuth angle. Therefore, the competing jamming and clutter are assumed to be competing with the target at a given Doppler but arriving from a different angle. These assumptions are well matched to a computing system since the problem can be nicely divided into multiple beams and multiple Dopplers, all processed in parallel. If the requisite processing had to be done all in a single arithmetic processing unit, the throughput would be very difficult to meet, dominated as it is by both

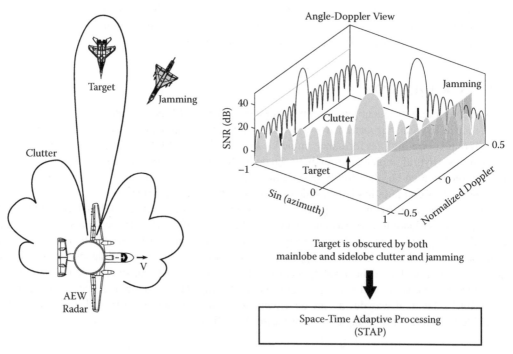

FIGURE 2-1   Airborne early warning radar (target and interference scenario).

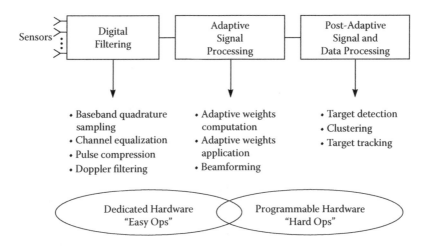

FIGURE 2-2 Signal processor functions.

the operations and the access in and out of memory in real time. The ability to break up the problem into many small subproblems makes the implementation manageable but nevertheless difficult. However, before this stage in the processing flow, it is necessary to perform a significant amount of data preconditioning upstream of the requisite interference suppression.

A way to describe the important signal processing functions necessary to ultimately achieve the goal of detecting a target is to separate these functions into three successive categories: digital filtering (used to precondition the data), adaptive signal processing, and post-adaptive signal and data processing. These signal processor functions are illustrated in Figure 2-2.

Figure 2-2 depicts the processing flow after the data are digitized through a bank of analog-to-digital converters. The data are first processed through the stages of baseband quadrature sampling to focus the data to the bandwidth of interest. Channel equalization is another example of front-end digital filtering necessary to equalize all the incoming channels to have the same filter response and, therefore, compensate for any channel-to-channel imperfections prior to the stage of digital beamforming. Pulse compression and Doppler filtering are performed to match the data to the waveform used in the radar and to map the data to the Doppler domain [as described by Jim Ward (1994) in the taxonomy of STAP algorithms]. These operations are also qualified as "easy ops" because the arithmetic required involves the product of filter taps against the incoming data. Therefore, a fast multiply-accumulator is sufficient to perform these operations. However, the total computational throughput is typically large and involves many of these regular multiply-accumulators. Furthermore, since the operations are performed on a channel-by-channel basis, the processing is particularly conducive to a parallel architecture. If the total operations are in the TeraOps, the best design choice is to use very-large-scale integration (VLSI) solutions for a very constrained size, weight, and power (SWAP) airborne platform. If, on the other hand, the computation is manageable using either field programmable gate arrays (FPGAs) or fast digital signal processors (DSPs), then these latter design solutions permit more flexibility in filter taps coefficient changes (for FPGAs) and programmability (for DSPs or high-speed microprocessors). However, the latter approaches come at the expense of higher power consumption as discussed in Chapter 1. If this digital filtering is implemented using fast Fourier transforms (FFTs), the same trade applies since there are a number of FFTs developed specifically to achieve high computational throughput with minimum power. The main issue in terms of design choice hinges on how well the filtering requirements are known at design time and the available SWAP for implementation.

There are many variants of STAP algorithms, for example, pre-Doppler versus post-Doppler techniques. In this section, the post-Doppler technique is chosen to illustrate representative process-

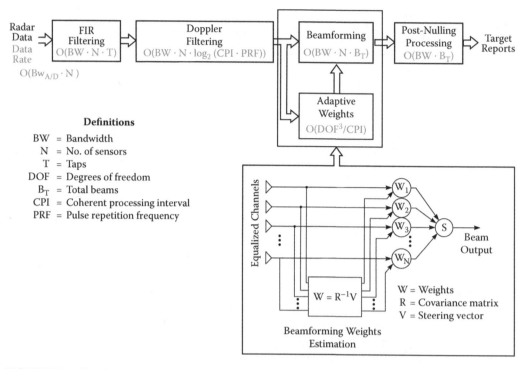

FIGURE 2-3   Signal processor system complexity.

ing stages and their respective implementation challenges. Since many of the signal processing steps in the processing flow are all linear operations, the order of processing is driven by the architecture (and design choices) best matched to maintaining hardware simplicity and real-time performance. The complexity is driven by the processing throughput and the amount of data arriving in real time. For example, if the data arriving in real time from multiple incoming channels is low and the processing throughput is high, then memory requirements would be low but the computation performed per byte would be high. Most of the applications, representative of systems described in this chapter, demand about one-to-one complexity in incoming data per operation. Typical computation per data throughput is only a few operations per byte. Figure 2-3 depicts, in a parametric form, the typical computation complexity for several of the end-to-end stages. The order of complexity for the front-end filtering is usually a direct function of signal bandwidth (BW), number of sensor channels (N), and taps (T) used in the finite impulse response (FIR) (when the filtering functions are implemented using an FIR approach instead of an FFT). If the implementation is done using FFTs, then the computation throughput grows linearly relative to BW and N, but logarithmically in terms of the number of taps. Similarly, as illustrated in Figure 2-3, for Doppler filtering implemented using FFTs (for a post-Doppler STAP algorithm) the computation growth is also linear as a function of signal BW and number of sensors N, but logarithmic in terms of number of pulses over the coherent processing interval (CPI). The number of pulses is simply the product of CPI duration and the radar pulse repetition frequency (PRF).

In contrast to digital filtering, in which the operations are performed against each of the radar pulses proceeding in real time, the Doppler filtering is done across all pulses. It is still done on a channel-by-channel basis, but a corner turn is necessary to gather the pulse data per channel. This step demands an architecture well balanced between memory access and processing. As the processing increases, the memory access time continues to lag behind. The memory speed to access bits of information is continuing to be much slower than the microprocessor speeds, leading to what is often referred to as the memory access wall. This problem continues to be exacerbated as pointed out by Patterson et al. (1997), but is ameliorated by placing a significant amount of memory on a single die. One implementation of this approach is the IRAM (Intelligent RAM) (Patterson et al. 1997), developed as a proof of capability. An IRAM prototype designed in 1996 was targeted at 1 gigabit DRAM of memory integrated on the same die with the arithmetic processing unit.

The case example illustrated in Figure 2-3 is of particular interest because the architecturing challenge was not just computational throughput. Instead, the challenge also involved memory size and memory banks to receive incoming data in real time while at the same time processing earlier data. One approach implemented was a "ping-pong" memory approach: one side of the memory bank was used to access and issue data to the processing units while another side collected the data. At the end of the processing interval, the roles of respective processing banks were switched. In addition to memory and computation balance, the architecture also involved a careful treatment of communication among processing units, among processing boards, and across processing chassis.

After the Doppler filtering shown in Figure 2-3, the data had to undergo yet another corner turn to gather all the channel data prior to the adaptive beamforming stage. The corner turn significantly stresses the capabilities of any parallel processor, so a critical measure of signal processor capability is what is referred to as system bisection bandwidth (Teitelbaum 1998). In simple terms, bisection bandwidth is a measure of how much data flows from one half of the processor to the other if the system is figuratively "bisected." For many of the classes of complex processing described in this chapter, the desired bisection bandwidth (as a rule of thumb) is about 1/10th in bytes per second of the total system computation in operations per second. For example, if a system requires a total system computational throughput of 1 TeraOps, then the approximate minimum bisection bandwidth is approximately 100 gigabytes/s. This is only an empirical rule of thumb that will vary from application to application, but it serves as a general metric of expected capability from the HPEC system, useful for efficiently balancing computation with real-time communication.

Figure 2-3 illustrates in more detail the throughput required as a function of critical system parameters for the adaptive signal processing depicted in Figure 2-2. Since the signal processor must calculate antenna weights in real time to adapt to the changing intentional and/or nonintentional jamming and clutter interference, a very popular approach is to find the best least-squares solution that minimizes the noise and maximizes the signal of interest. This solution simply involves the inversion of an estimate of the interference and noise covariance (Ward 1994). The formulation, as shown in Figure 2-3, can be deceivably straightforward, but the implementation can be quite complex. It is beyond the scope of this chapter to address all the different arithmetic options to calculate the inverse of a covariance matrix (Golub and Van Loan 1996). From the perspective of a complex HPEC system, the computational complexity grows as a function of the DOF[3] [degrees of freedom (DOF) in this case are the incoming channels] and is inversely proportional to the coherent processing interval (CPI). The longer the CPI, the longer is the time to calculate the weights.

This processing step must be balanced against the application of the weights in the process of adaptive beamforming. A very common way to achieve this balance is to compute the weights on new data while prior weights calculated from older data are applied to their respective data. Another approach is to apply the adaptive weights using a systolic architecture (Teitelbaum 1991; Gentleman and Kung 1981). This latter approach involves applying the adaptive weights computed concurrently to the same data used for computation of the weights. The process of adaptive beamforming grows linearly as a function of signal BW, N, and the number of beams formed. However, the architecture is very conducive to a parallel configuration by operating individually on a Doppler-by-Doppler basis. Some algorithms demand sharing of Doppler data, and this requirement imposes sharing of data across processing units. Here, again, there is a critical need to balance memory, communication, and processing across processing elements and often across processing boards and chassis.

Once the adaptive signal processing stage is completed for a given CPI, the processed data continue to the post-adaptive signal and data processing shown in Figures 2-2 and 2-3. Between the adaptive interference suppression and the post-nulling processing, another corner turn is often necessary to organize all the data on a beam-by-beam basis, thereby permitting parallel processing of each respective beam independently. Unfortunately, this stage of computation is somewhat nonde-terministic and highly dependent on the number of targets hypothesized in the incoming data. This stage of processing typically involves target detection, target clustering, and, at times, target tracking. Therefore, the processing throughput depends on the total number of targets prosecuted at any given time. Because of the irregularity in processing, these types of operations are referred to as "hard ops" in contrast to the earlier processing referred to as "easy ops." The "hard ops" functions in the signal processing flow are often best implemented utilizing programmable hardware that permits the ability to change the algorithms and/or modifies the algorithm parameters as the specific application might demand. The only deterministic part of the processing is the number of input adapted beams and the total signal bandwidth. The latter establishes the data volume flowing through the system in real time. Finally, the output of the signal processor is typically a set of target reports. The number of target reports again is a function of the total targets detected (including false alarms). Any given target report may consist of several bytes identifying critical target information such as beam direction, range, Doppler, signal-to-noise ratio estimates, radar-specific parameters for the respective processed CPI, etc. The amount of data reported is typically considerably small compared to the real-time data input into the signal processor at the output of the analog-to-digital converters. Therefore, the signal processor system can be thought of, in simple terms, as a funnel of data in which the processing chain starts with a wide bandwidth of incoming data and results in a much smaller data amount at the end of the processing chain. All the requisite processing reduces the total volume of data while at the same time extracting the target (signal of interest) from the competing intentional and nonintentional jamming, ground clutter, and background uncorrelated noise.

Section 2.3 presents a case example of the implementation approaches chosen to demonstrate the viability of a real-time signal processor constrained to its contemporary processing engines, memory, communication, and SWAP limitations. The example serves as a successful demonstra-tion of the approaches undertaken, including the architecture, design, and a rigorous integration and testing methodology. Interestingly, this discussion is as valid today as it was when this space-time adaptive processing system was developed; the main difference is today's availability of more capable building blocks—processing, memory, and communication hardware. However, since the requirements have also increased in all these dimensions, the design challenges remain similar.

2.3 IMPLEMENTATION TECHNIQUES

Section 2.2 described a generic signal processing flow for a complex real-time STAP HPEC system. This generic description was formulated based on a set of key system parameters illustrating the data rates and processing throughputs and how these parameters require an architecture balanced among processing elements, memory hierarchy, and communication technologies while at the same time meeting SWAP constraints. Even though not explicitly addressed in Section 2.2, the architec-ture trade-offs must also take into account the complexity of the real-time software. The real-time software is addressed in this section as part of a specific HPEC system instantiation of the generic system described in Section 2.2.

Figure 2-4 illustrates the top-level architecture and how different processing functions, described earlier, map to specific processing technologies. The front-end of the HPEC signal processor system was designed to achieve 19 GOPS (channel equalization, pulse compression, and Doppler filtering) in a form factor limited to four VME chassis. Each chassis contained 21 boards and each VME board was 9U × 220 mm. The total computation throughput was 21.5 GOPS, as shown in Fig-ure 2-4. Again, the computational throughput is not very large by today's standards, but it *was* very large when the system was designed (circa 1990s). However, what is important for this discussion is to understand the architecture and design choices since the system requirements today would be scaled to more complex throughputs, therefore still imposing similar design challenges.

Real-time data from the analog-to-digital converters (ADCs) can be thought as an incoming cube consisting of three dimensions—channels, range gates, and radar pulses—per CPI. The data

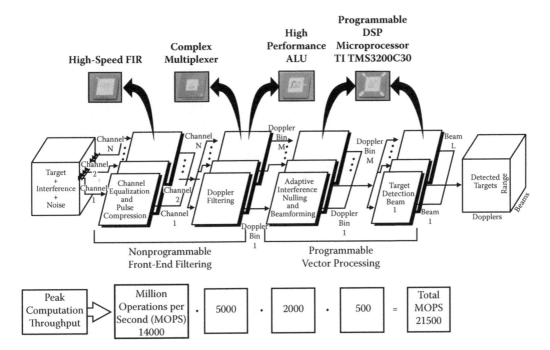

FIGURE 2-4   Real-time signal processor architecture.

cube is distributed across multiple processing boards, each board processing an input channel and all boards operating simultaneously in real time across all channels. This architecture is modular and scalable; adding more channels only requires adding additional processing boards. Each board contains commercial off-the-shelf (COTS) processing chips (INMOS A100) performing high-speed FIR filtering to implement channel equalization and pulse compression.

The choice of FIR filtering chip technology was based on available proven parts that would meet the real-time requirements. Since the processing was regular, as described in Section 2.2, each processing board per channel required a set of input buffers and output buffers to synchronize the overall system clock across all boards without calling for extreme care in clock matching; enough synchronization to preserve the continuity of input and output data flow is all that is needed. Furthermore, once a single board was designed and verified to meet the design requirements, detailed preliminary and critical design reviews were undertaken to make sure pieces would fit in the context of overall system requirements, including system control and radar input and output synchronization. A full slice of data containing all range gates and pulses was processed by each respective board; thus, the whole incoming data cube was distributed across multiple boards within a chassis and across chassis, but with very limited communication to maintain system simplicity. The output of this first stage of processing was then required to undergo a corner turn, and data had to move across the system backplane to arrange the incoming cube such that a Doppler filtering board had all the channels and pulses. A processing board was designed to calculate a Doppler bin independent of other Doppler bins. This approach was more efficiently done using a discrete Fourier transform (DFT) than an FFT since an FFT would calculate all Dopplers together. Instead, the chosen architecture made the system much simpler by independently calculating each Doppler and delivering the respective Doppler data to the subsequent adaptive interference nulling and beamforming stage. This was an important design choice in which higher computation throughput in the Doppler filtering stage (by implementing a DFT instead of an FFT) led to a very regular processing flow of multiple Dopplers processed independently of each other.

A set of weights was computed on a board for a given Doppler to cancel any intentional and non-intentional jamming followed by clutter nulling independent of all other Dopplers. One important

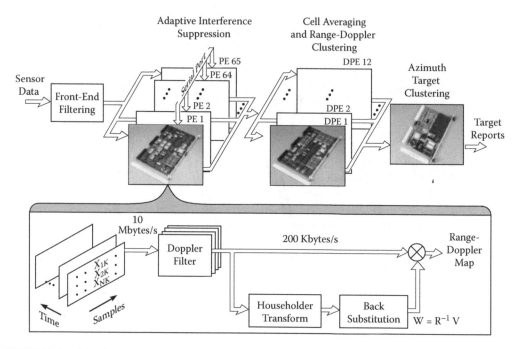

FIGURE 2-5 Adaptive processing and target detection architecture for STAP processor.

design benefit was in the development of the hardware board. Once a processing element board was designed and fabricated, the process of integration proceeded with full testing of this single board. Once the testing was completed, multiple boards were replicated, resulting in an extremely efficient development and system integration approach.* Any board failure was easy to identify by simply determining which Doppler in the output data was showing erroneous results. This rapid identification of failed hardware saved an enormous amount of debugging, integration, and testing time. This method exemplifies the advantage of selecting system simplicity to facilitate ease in debugging and integration. HPEC systems are often required to balance the full design spectrum from architecture through integration and testing phases.

The processing element (the processing board responsible for the Doppler filtering and adaptive signal processing) employed two classes of processing engines. The DFT was implemented using a COTS chip built by Plessey (PDSP 16112) and IDT (IDT 7381) together with a programmable Texas Instrument's DSP TMS320C30 microprocessor. The architecture was very scalable, permitting growth in Dopplers through the addition of more processing elements (PEs). The combination of dedicated COTS DFT chips for the complex multipliers and logic arithmetic unit plus the programmable DSP TMS320C30 provided flexibility in algorithm implementation. The DSP chips were fully programmable. However, the demands of real-time performance required that the software be developed in C and the critical subroutines for the computationally intensive functions (e.g., matrix inversions) be implemented in assembly.

The output of the adaptive signal processing required another corner turn to distribute all Dopplers to a single detection processing element (DPE) per beam. Thus, each DPE operated in parallel on independent beams, thereby lending itself to a very structured architecture. Figure 2-5 depicts the architecture decomposition and the respective boards used in the implementation of the adaptive jamming and clutter suppression (adaptive interference suppression) performed by the PEs, and shows the target detection (consisting of cell averaging, range-Doppler clustering, and azimuth target clustering) performed by the DPEs. The DPE processing involved target detection on each

---

* The design of this system is credited to Bob Pugh, Joe MacPhee, and David Wojick of MIT Lincoln Laboratory.

FIGURE 2-6   STAP signal processor system.

TABLE 2-1
Overall Hardware Complexity

Component Type	Total Size
IC (LSI and VLSI)	≈25,000
Unique Custom Board Designs	13 Board Designs I/O 288 pins ≈200 ICs 125 in^2
Total Boards	134 Boards 113 printed circuit 21 wire wrap
Total Chassis	14 Chassis 11 real-time signal processing 3 control 21 slot 9U VME

beam, also using a DSP TMS320C30. Many of the same design tools and techniques used in the PE boards were leveraged for the DPEs, including design and debugging tools, hardware chips, and software libraries. Finally, the outputs of respective DPEs were sent to a set of single-board computers (SBCs) to complete the creation of final target reports.

Figure 2-6 illustrates a subset of the full hardware system, depicting 7 of the 11 VME processing chassis. The system employed a total of 25,000 integrated chips (ICs), 13 unique board designs, and 134 total VME boards, all contained in 14 VME chassis (including 11 real-time signal processing chassis plus three chassis for control); the overall hardware complexity is summarized in Table 2-1.

2.4   SOFTWARE COMPLEXITY AND SYSTEM INTEGRATION

The development of a complex HPEC system demands special attention to the real-time software and the end-to-end hardware and software integration. The example in this section illustrates software solutions, addressing the complexity of real-time control software and the required real-time

TABLE 2-2

Software Development Effort for STAP Signal Processor

Software	Executable LOSC	Man-Years	Duration (years)	Average Production/ Man-Month (LOCS)
Operational Software				
Real-Time Control	39K	2.0	0.6	1600
DSP Algorithms	5K	2.5	0.8	170
System Testing Software				
Diagnostics	60K	5.5	0.9	900
Processing Verification	23K	3.0	0.7	640
Acceptance Test Software	6K	1.0	0.2	500
Total	133K	14.0	3.2	3810

algorithms. A discussion of the system integration is included in this section because a significant portion of the total developed software involved system testing software.

Table 2-2 highlights the software development effort divided into two classes of software: operational software and system testing software. The real-time control is the software required to operate the signal processor system. The signal processor must (1) be synchronized with the overall radar system in timing, input data, coefficients downloaded, and initialization of memory banks; (2) provide the output data to a display environment; and (3) continuously check the health of the system.

The real-time control software for typical HPEC systems dominates the total operational software. The case example described in Table 2-2 required a total of 39K lines of software code (LOSC) for the real-time control software. It is also interesting to look at the effort involved in terms of man-years, duration, and average productivity per man-month. In this case example, it took two man-years to develop the 39K LOSC over the course of 0.6 years utilizing the equivalent of 24 man-months, or an average of approximately 1600 LOSC per man-month. Though at first glance this seems a large number of LOSC, the hardware was purposely designed to have a high degree of modularity between subsystems so that the software could be easily reused with a minimum of changes, while at the same time adhering to similar interfaces, control parameters from the radar (PRI, PRF, CPI, etc.), and key real-time diagnostics (such as in the monitoring of hardware registers).

In complete contrast to the real-time control, the DSP algorithm software was only 5K LOSC but about an order of magnitude slower to develop in average productivity per man-month. The respective software productivity parameters are illustrated in Table 2-2. The primary reason for the reduction in productivity was the need to ascertain that the real-time software could meet the stringent real-time constraint. The front-end of the system did not incorporate any DSP algorithms. The DSP algorithms were incorporated into the adaptive signal processing stage and the detection subsystems, and these algorithms were implemented in the DSP TMS320C30. The software was primarily developed in C, but utilizing many of the already developed assembly subroutines (provided by the chip manufacturer, Texas Instruments). Only a few subroutines were developed specifically for this application.

The DSP algorithms in the adaptive signal processing stage of the HPEC system involved the matrix inversion, adaptive weight computation, and weight application. The detection subsystem, which involved the CFAR and centroiding of targets, was more difficult to develop in regard to software because of the high degree of logic needed to associate different targets with a centroided set. The computation of target parameters also required the identification of key parameters such as beam direction, Doppler, and range. In addition to these key parameters, the system also had

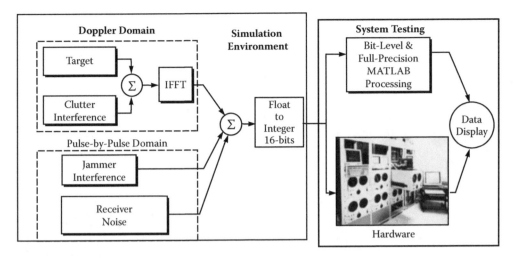

FIGURE 2-7   Simulation and testing environment.

the ability to receive commands to output a single range-Doppler dataset (map) for a given beam direction. This provided a very important diagnostic tool for monitoring the health of the system in real time.

The system testing software was a crucial aspect of the successful development of the system. As shown in Table 2-2, this software dominated the total LOSC for the system. However, the careful, methodical approach used in the development of the software tools contributed significantly to the system's working successfully the first time the signal processor was interfaced with the radar transducer. The diagnostic software was designed to probe critical hardware points in the systems: memories, registers holding coefficients, memory storing the algorithm code, input and output buffers, and hardware health status included in the monitoring of temperature, voltage, etc.

Processing verification was also an important tool to ensure first-time success. The productivity was much higher than for the DSP algorithm software developed for the TMS320C30 because it was written in MATLAB. A simulation environment was built all in MATLAB. As shown in Figure 2-7, the simulation environment included a Doppler domain simulation of targets and clutter interference plus a pulse-by-pulse simulation of jammer interference and receiver noise.

These simulation domains were then summed together and converted to 16-bit integer arithmetic to simulate the real-time input to the signal processor system. The same data were input to a bit-level replica of the arithmetic performed by the signal processor and also to a full-precision replica to the algorithms running in the system. The same data were also input to the actual hardware, and the outputs were compared to the bit-level simulator outputs for validity that the hardware was generating the proper results. These results were then compared to the output of the full-precision MATLAB simulation to ascertain that the differences between the bit-level arithmetic and the desired full-precision results were within acceptable error margins. If at any point the hardware was not generating the proper outputs, it was straightforward to quickly narrow down what portions of the system were in fault because of the high degree of instrumented diagnostic and testing tools.

The graphs in Figure 2-8 illustrate a comparison of the results generated by the hardware to those of the full-precision simulation. The examples illustrated in Figure 2-8 represent the output of the front-end hardware (1) after pulse compression but prior to adaptive nulling and (2) after adaptive nulling to suppress both jammer and clutter interference. The system verification was done at the system level where the performance was quantitatively characterized at the expected signal-to-noise level. Note that the target illustrated in the figure is at the expected signal-to-noise ratio (SNR) of 40 dB for both the full-precision simulation (in black) and the actual output of the hardware (in gray). The hardware output was obviously generated in real time.

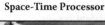

Simulation Environment

	Power Ratio (dB)
Target-to-Noise	−9
Clutter-to-Noise	51
Jammer-to-Noise	42
Processing Gain	49
Expected SNR	40

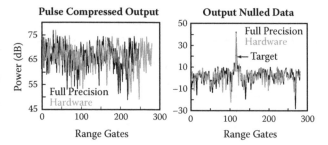

FIGURE 2-8   Signal processor system performance from simulated data.

It is not uncommon to find HPEC systems failing to meet schedules and cost because significant time is spent in debugging the system after it is completed. Little effort is dedicated in the development of diagnostic, verification, and acceptance software tools (system testing software). Some developers consider these software efforts additional overhead (in cost and personnel). However, it is proven time and again that in the long run this disciplined and rigorous methodology leads to successful development and system deployment. The system testing software is not throw-away software; rather it becomes the "gold standard" with the gold test vectors developed through the Doppler and pulse-by-pulse simulators. Any time there is a problem with the system during a mission, this gold standard software and the gold test vectors are input into the system and compared with previously calculated and known results, leading to a fast and efficient way to narrow down problems and rapidly fix the system to minimize mission downtime.

2.5   SUMMARY

The development of a complex HPEC system involves careful attention to the overall system design. The architecture must be very regular and modular in structure to leverage both hardware and software modules and functions developed through the system. This chapter has addressed these techniques by way of an example of a complex HPEC system. The hardware system spanned different classes of technologies, including a wide range of real-time fixed hardware processing units and a programmable COTS DSP microprocessor designed into dedicated boards. This particular system employed the TMS320C30 DSP chips as an example of a high performance real-time programmable chip. The hardware complexity was significant but, because of the modular architecture, it involved a finite set of unique board designs replicated many times. The benefit of this approach is that once a board was deemed fully working, it could be replicated enough times to populate the subsystem for either the front-end digital filtering, the adaptive signal processing, or the detection subsystems.

Two classes of software were critical to the successful development of this complex system. One class had to operate in real time and formed the core of the system operational software. The second class was the system testing software, which did not need to operate in real time, but which was vital to ascertaining the system's proper operation when it interfaced to the front-end radar transducer.

The development effort described in this chapter demonstrates a rigorous development approach that assured success with the real-time system. The lessons learned and techniques described can be applied to any contemporary HPEC real-time system. The full development evolved from chip-level verification using debuggers and a simulation environment (at the board level) developed by the chip manufacturer, to board-level tools, and then to full-system end-to-end testing and verification environments. The latter were done at the system level, incorporating the expected data at the input and comparing the expected outputs from the hardware at the bit level and full-precision level against the actual hardware. These testing tools were carried forward as part of the full-system deliverable and used to diagnose the system when any part of it was identified as failing.

In the future, more platforms will require autonomous operation with very little to no intervention by a human operator during a mission. For example, as unmanned aerial vehicles and satellites proliferate, the HPEC system will need to be made very robust to failures and be equipped for remote diagnostics. The techniques described in this chapter will not only facilitate successful development, integration, and testing, but also provide tools that can be used remotely from the control centers via wireless connections to platforms.

REFERENCES

Gentleman, W.M. and H.T. Kung. 1981. Matrix triangularization by systolic arrays. *SPIE* 298, Real-Time Signal Processing IV: 19–26.

Golub, G.H. and C.F. Van Loan. 1996. *Matrix Computations*. Baltimore: Johns Hopkins University Press.

Guerci, J.R. 2003. *Space-Time Adaptive Processing for Radar*. Norwood, Mass.: Artech House.

Klemm, R. 2006. *Principles of Space-Time Adaptive Processing*. London: Institution of Electrical Engineers.

Patterson, D., T. Anderson, N. Cardwell, R. Fromm, K. Keeton, C. Kozyrakis, R. Thomas, and K. Yelick. 1997. A case for intelligent RAM. *IEEE Micro* 17(2): 34–44.

Richards, M.A. 2005. *Fundamentals of Radar Signal Processing*. New York: McGraw-Hill.

Teitelbaum, K. 1991. A flexible processor for a digital adaptive array radar. *Proceedings of the IEEE National Radar Conference*: 103–107.

Teitelbaum, K. 1998. Crossbar tree networks for embedded signal processing applications. *Proceedings of the 5th International Conference on Massively Parallel Processing Using Optical Interconnections*: 200.

Ward, J. 1994. *Space-Time Adaptive Processing for Airborne Radar*. Technical Report #1015. Lexington, Mass.: MIT Lincoln Laboratory.

3 System Architecture of a Multiprocessor System

David R. Martinez, MIT Lincoln Laboratory

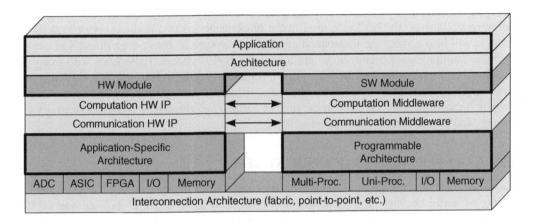

This chapter discusses a generic multiprocessor and provides a representative example to illustrate key subsystems found in modern HPEC systems. The chapter covers from the analog-to-digital converter through both the front-end VLSI technology and the back-end programmable subsystem. The system discussed is a hybrid architecture necessary to meet highly constrained size, weight, and power.

3.1 INTRODUCTION

Many contemporary high performance embedded computing (HPEC) systems are based on an architecture comprising a large number of multiple parallel processors. A parallel processor architecture is very well matched to typical real-time sensor processing because in many instances the sensors create data from parallel channels. These parallel channels of data are then distributed across a parallel architecture in order to meet stringent requirements in computational throughputs, latencies, communication, and memory. One can think of a parallel system architecture as a system designed to break up the dataset into smaller datasets all operating simultaneously in real time.

System requirements continue to grow more stringent as the analog-to-digital converters (ADCs) are brought closer to the sensor transducer, thus minimizing the front-end analog hardware. The trend is for more computation to be done digitally. Digital hardware offers more robust system stability, avoiding unnecessary calibrations characteristic of analog hardware resulting from component drifts with time and temperature (Martinez, Moeller, and Teitelbaum 2001). However, the digital system, in addition to meeting stringent requirements in computational and data throughputs, must also be of small size, low weight, and low power. Furthermore, the hardware must also comply with demanding shock, vibration, humidity, and temperature specifications. Examples of platforms requiring these capabilities are today's airborne early warning radars (an example was described in Chapter 2), unmanned aerial vehicles (UAVs), fighters, and spaceborne surveillance and targeting radars.

This chapter presents a generic architecture representative of HPEC capabilities for the above classes of systems. The generic architecture is then validated against an actual multiprocessor HPEC system to specifically highlight the important architecture building blocks making up the parallel system architecture. The system chosen is ideal for characterizing the key multiprocessor elements because it consists of both a custom parallel processor design and commercial off-the-shelf (COTS) programmable hardware.

Section 3.2's description of a generic multiprocessor system establishes the elements driving the architecture requirements. The major elements of a complex system demand an architecture balanced among computing elements, memory, and interconnections to meet the real-time performance. Following the discussion of the generic multiprocessor system, a specific example is presented to illustrate the hardware used in the implementation of a multiprocessor system. This example will help cement the understanding of what a system architect goes through in mapping a generic multiprocessor architecture into a complex HPEC system. Experience with many generations of multiprocessor systems shows that a common generic architecture is a valid metric for explaining important design elements. The main difference among these multiple generations of systems is the technology available at the time of design (e.g., computation engines, memory components, and interconnection technologies).

3.2   A GENERIC MULTIPROCESSOR SYSTEM

Sensor parameters at the system level drive the complexity of the system. For a radar system, the number of sensor channels, the transducer bandwidth, and the coherent processing interval (CPI) (for a coherent radar) are some of the system parameters determining the data input into a multiprocessor system. Very careful design choices must be made prior to the ADC in order to ensure that the incoming data are preconditioned prior to the ADC. However, this handbook addresses sensor data at the interface with the ADC and the following digital processing. In this chapter, the generic multiprocessor, specifically the example described in Section 3.3, assumes data are input from a set of parallel ADCs. Later chapters describe classes and technologies used for the ADC.

A generic multiprocessor system adapted to HPEC systems is illustrated in Figure 3-1 [system is adapted from references (Hennessy and Patterson 1990; Patterson 2001)]. Data are received from a set of ADCs and input into a high-speed interconnection network. The interconnection network is a very important system component of the overall HPEC system because it is responsible for distributing the incoming data across all the parallel processor engines, as well as for synchronizing the processing element and memory banks, inputting system parameters, and monitoring system health status. All of these data must flow in real time.

The figure also illustrates the desired system performance capabilities found in many real-time HPEC systems. In many real-time sensor applications, the transducer is a very costly part of the overall system. The transducer may be used for multiple generations of Moore's Law cycles; however, the desired computing processor should be scalable to more demanding computational throughputs. Today it is not uncommon to find transducers with a few subarray channels output from the ADCs but capable of later being advanced to generate many more channels. A limited channel number is often driven by the need to maintain low cost, but the actual hardware transducer is designed to be scalable to larger numbers of channels. Therefore, the HPEC system must be scalable in computational throughput. Higher computational throughputs have a domino effect on the increased performance desired from the interconnection network and the increased total system memory.

In many sensor systems applications, a balanced HPEC architecture demands a bisection bandwidth in bytes/s approximately equal to 1/10 the computational throughput in operations/s. This is a general rule of thumb. The bisection bandwidth can be interpreted as a measure of data flow moving at the boundary from one half of the processor to the other half. Imagine the

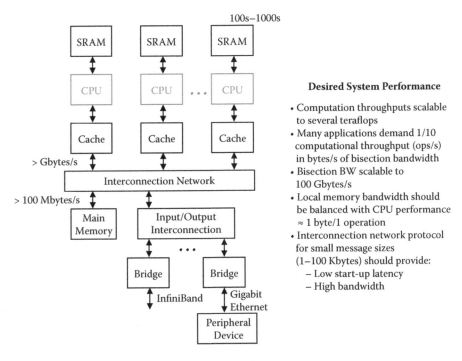

FIGURE 3-1 Generic multiprocessor system.

generic processor shown in Figure 3-1 is split in half and a data flow meter is placed at the interface between the two halves of the processor. The amount of data flowing in real time across the boundary represents the bisection bandwidth. The required bisection bandwidth will obviously depend on the specifics of the system application. Teitelbaum addresses interconnection networks in a later chapter in this handbook. For example, for a TeraOps system, a typical bisection bandwidth is on the order of 100s of Gbytes/s. As the total system computation increases, the bisection bandwidth must also scale accordingly. It is important to realize that the bisection bandwidth is a general performance metric. To meet this bisection bandwidth, the architecture must be very well balanced across the communication between system chassis, across boards inside a computing chassis, among processing on a single board, and among the memory, cache, and registers adjacent to the central processing units (CPUs).

As shown in Figure 3-1, the CPUs are supported by high-speed memory implemented using static random access memory (SRAM) banks or dynamic random access memory (DRAM). DRAM is typically slower than SRAM by a factor of four; however, SRAM is usually more expensive. Closer to the CPUs are several levels of cache to supply data rapidly to the processor to maintain real-time performance. The hardware often has a main memory bank attached to the high-speed network to permit diagnostics, monitor the system health, and store system parameters (e.g., filter coefficients). This part of the system is often implemented using DRAM. Many real-time systems also need to record data in real time to allow for post-mission processing. For this purpose, the system architecture might also have an input/output interconnection to disks or other peripheral devices such as high-speed recorders.

Another important system performance desired of real-time HPEC systems is the ability to fetch and issue data words at a rate commensurate with the computational capabilities of the CPUs. It is very typical in these sensor applications to demand few operations per byte of data. In some applications, this system parameter is about one byte of data per one real operation. Again, this is also a very general rule derived from experience with HPEC systems. The important point is that the CPU and adjacent memory must be able to support vast amounts of sensor data from memory as the CPU is processing in real time. This system performance capability is very different from that of applications found, for example, in modeling and simulation where, after the simulation parameters are defined, the processing proceeds with a large number of operations before the CPU needs to access or download a byte to cache or adjacent memory (large number of operations on a few bytes of data).

The interconnection network found in real-time HPEC systems must be controlled via a real-time operating system. This real-time interconnection network protocol must be able to support low start-up latencies while at the same time supporting high communication bandwidth. In many of the sensor applications of interest, the message sizes range from 1 to 100,000 bytes. So the operating system must minimize the time devoted to formatting the communicated messages. This is one of the important reasons why HPEC systems require real-time operating systems (e.g., Real OS from Wind River systems or real-time Linux). For cases in which the sensor data flow is so high in bytes/s that there would be very little intervention from a real-time operating system, the communication protocol is done in hardware by controlling memory banks, cache, and registers via hardware circuits.

Subsequent sections present a specific example of a complex HPEC system incorporating many of the important architecture components shown in Figure 3-1. Section 3.3 illustrates a custom system with large computational requirements and interfaced to a peripheral real-time recorder. Section 3.4 discusses a system containing 1000s of CPUs all operating simultaneously in parallel and based on a COTS programmable processor system. Sections 3.3 and 3.4 are taken from Martinez, D.R., T.J. Moeller, and K. Teitelbaum, "Application of Reconfigurable Computing to a High Performance Front-End Radar Signal Processor," *Journal of VLSI Signal Processing Systems* 28(1–2): 64–69 © 2001 Kluwer Academic Publishers and are used with the kind permission of Springer Science and Business Media.

3.3   A HIGH PERFORMANCE HARDWARE SYSTEM

To give the reader an understanding of the challenges involved in the implementation of HPEC systems, this section presents a real-time system as a hardware example of the generic architecture shown in the Section 3.2. The section addresses a custom hardware implemented using a custom design based on a very-large-scale integrated (VLSI) chip. A more detailed description of this system and a discussion of a candidate implementation using field programmable gate arrays can be found in Martinez, Moeller, and Teitelbaum (2001).

A typical radar signal processing flow starts at the output of the ADC with real-time data processed through a set of digital filters. The digital filters are needed to first convert the data from real ADC samples to complex digital in-phase and quadrature (DIQ) samples. This operation is commonly known in the radar community as DIQ sampling (Martinez, Moeller, and Teitelbaum 2001). The DIQ sampling is necessary to preserve the target Doppler information. From the Doppler information, one can discern the target velocity.

The radar data after the ADC will exhibit a set of spectrum replicates with periodicity equal to the Nyquist sampling (1/2 the sampling rate). Because the data are real samples, at this stage the spectrum maintains equal images with respect to 0 Hz. The DIQ filtering will extract one of the sidebands, map the sideband spectrum to baseband, and filter all remaining images, including any DC offsets introduced by the ADC. The filter coefficients will vary depending on the characteristics of the bandwidth present in the transmitted radar pulse. The resulting output contains a single sideband spectrum replicated every increment of the sampling frequency. The single sideband represents the complex signal (in-phase and quadrature components) characteristic of a Hilbert transform performed in a demodulation operation (Oppenheim and Schafer 1989).

The analog receiver filtering, prior to the ADC converter, is selected such that the information bandwidth is preserved without aliasing. The signal instantaneous bandwidth is typically several factors below the ADC sampling frequency. This oversampling leads to a very simple DIQ archi-

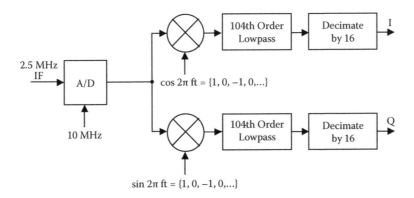

FIGURE 3-2 Digital in-phase and quadrature sampling architecture. (From Martinez, D.R., T.J. Moeller, and K. Teitelbaum. Application of reconfigurable computing to a high performance front-end radar signal processor. *Journal of VLSI Signal Processing Systems* 28(1–2): 65, Figure 2 © 2001 Kluwer Academic Publishing, with kind permission of Springer Science and Business Media.)

tecture. For example, if the ADC sampling is four times the instantaneous frequency of the radar (Teitelbaum 1991), the mapping of the spectrum to baseband reduces to multiplication by +1, 0, and –1 for the in-phase component, and 0, +1, and –1 for the quadrature component. Once the data are mapped to baseband, the complex signal can be filtered and decimated to match the signal bandwidth. This simplification leads to the typical DIQ sampling architecture shown in Figure 3-2. Two filter banks of equal length, each operating on the in-phase and quadrature samples, respectively, are used; the odd samples are used to form the in-phase component and the even samples are used to form the quadrature components.

In the HPEC system discussed in this section, the ADC used was a DATEL ADS945 sampling at 10 MHz, with a precision of 14 bits. After the digital low-pass filtering, shown in Figure 3-2, the radar data were decimated by a factor of 16 or, equivalently, a factor of 8 complex samples. Thus, the output sampling rate out of the DIQ filter was down to 0.625 MHz complex data. Although this output data rate was twice what was needed to maintain the signal information bandwidth, it simplified the following filtering stage past the DIQ sampling, resulting in a simpler filter design with a smaller number of taps.

On a per-channel basis, the computational throughput is proportional to the data bandwidth and the number of filter taps. For 208 complex filter taps and an input data rate of 10 MHz, the DIQ filter must perform at least 2.08 giga (billion) operations per second (GOPS), based on the architecture shown in Figure 3-2. The reduction in sampling rate by a factor of 8 could have allowed the processing of only the samples needed at the output. However, the hardware designer concluded that the custom VLSI implementation was more regular and easier to lay out by processing all the incoming samples and decimating at the end (Greco 1996). For a 48-channel system, the total computational throughput was 100 GOPS.

3.4 CUSTOM VLSI IMPLEMENTATION

During the preliminary design phase, a detailed search was done to determine if either general-purpose hardware or commercially available dedicated chips could be used to solve the DIQ sampling requirements. The conclusion was that no digital signal processor (DSP) or reduced instruction set computer (RISC) chips would be suitable to meet the processing requirements, particularly when one accounts for the loss in processing efficiencies caused by programmable hardware. Several commercial specialized chips designed specifically to perform finite impulse response (FIR) filtering (e.g., Gray chip GC2011, Harris HSP43168, GEC Plessey PDSP16256) were available; none of them met the requirements in all the critical dimensions, such as input data precision bits, coeffi-

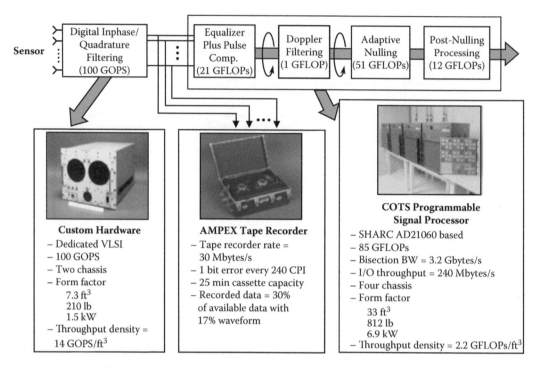

FIGURE 3-3   Airborne signal processor subsystems. (From Martinez, D.R., T.J. Moeller, and K. Teitelbaum. Application of reconfigurable computing to a high performance front-end radar signal processor. *Journal of VLSI Signal Processing Systems* 28(1–2): 67, Figure 3 © 2001 Kluwer Academic Publishing, with kind permission of Springer Science and Business Media.)

cient precision bits, internal arithmetic accuracy, and computational throughput at low power. The decision was to proceed with the design of a custom VLSI chip based on a mix of standard cells and datapath multiply-accumulate (MAC) cells.

The final front-end signal processor was integrated into two 9U-VME chassis together with additional control electronics, a single-board computer (SBC), and interface boards to the follow-on processing and an instrumentation recorder. The airborne signal processor subsystems are shown in Figure 3-3. The generic multiprocessor system shown in Figure 3-1 is encapsulated by the front-end custom hardware and, as a separate subsystem, by the COTS programmable signal processor. The latter is further discussed in the Section 3.5. As shown in Figure 3-1, there was also a peripheral device that was implemented using an AMPEX tape recorder responsible for recording 30% of the data at a rate of 30 Mbytes/s. The interconnection network was based on the VERSAmodule Eurocard bus (VMEbus) interface with parts of the signals dedicated to radar-specific controls. The VME interconnection was interfaced via the backplane to a set of Mercury Computer boards serving as the bridge between the incoming data output from the custom hardware and the AMPEX tape recorder.

The DIQ hardware subsystem consisted of three unique board designs. There were two 9U-VME chassis; each chassis performed 50 GOPS. A 9U-VME chassis housed 10 boards. The internal architecture of this subsystem is shown in Figure 3-4. One board slot was dedicated to a SBC control computer used to download radar coefficients to the processing boards and to initialize all the board registers. Another slot was dedicated to a distribution board used for subsystem control. This board interfaced to the radar control processor and provided control signals to each of the subsequent boards in the chassis. There were six DIQ processing boards. Each DIQ board handled four input channels. The outputs from two DIQ boards were then sent to a data output board (DOB). Each DOB board packed the data and provided the same interface outputs to the back-end COTS

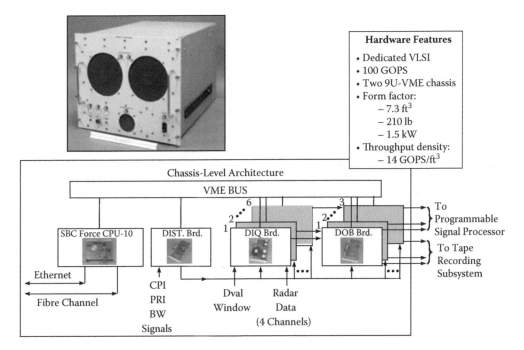

FIGURE 3-4   DIQ subsystem architecture. (From Martinez, D.R., T.J. Moeller, and K. Teitelbaum. Application of reconfigurable computing to a high performance front-end radar signal processor. *Journal of VLSI Signal Processing Systems* 28(1–2): 68, Figure 4 © 2001 Kluwer Academic Publishing, with kind permission of Springer Science and Business Media.)

programmable processor and to the AMPEX tape recorder. These custom boards were based on a 14-layer printed circuit board (PCB) design of size 9U × 220 mm.

The VME bus and direct board interconnections (using the P3 custom hardware connector) represented the implementation equivalent of the interconnection network shown in the generic multiprocessor architecture depicted in Figure 3-1. The direct hardware interconnection between the DIQ boards and the DOB boards was necessary to permit fast communication between the processing boards and the memory contained within the DOB boards. There were 24 channels input to each DIQ subsystem chassis. The full-custom hardware subsystem consisted of two VME chassis. The total data rate to one chassis was 420 Mbytes/s or, equivalently, 840 Mbytes/s for two VME chassis. These data per chassis were then distributed to six DIQ processing boards over the custom backplane. Each DIQ board received 70 Mbytes/s of data. There were four custom VLSI chips per DIQ board; each chip processed one channel of data. The input data was mapped to 18-bit data on the DIQ board prior to going into the VLSI chip. Therefore, each chip processed (70/4 × 18/16) Mbytes/s with a processing throughput of about 2 GOPS. The output of the chip was a 24-bit data word. A 16-bit word was then selected and sent over the custom backplane to the DOB board. Thus, the output of the DIQ board was reduced down to 10 Mbytes/s (four channels, 16-bit complex words, at 0.625 MHz of in-phase and quadrature samples). The aggregate data rate at the output was 120 Mbytes/s (48 channels, 16-bit complex words, at 0.625 MHz of in-phase and quadrature samples). The equivalent bisection bandwidth can be interpreted as the total data following into both of the chassis, which amounted to 840 Mbytes/s (48 channels, 14-bit ADC real samples, at 10 MHz) and a total of 100 GOPS in computational throughput. The ratio between bisection bandwidth and processing was close to 100 (instead of the factor of 10 rule of thumb described earlier). Similarly, the ratio between operations to bytes was also 100 ops/byte. The increase in computation relative to a byte of data was a good compromise between large computational throughput and a simpler hardware chip design.

The standard VME interconnections (P1 and P2 connectors) were not fast enough to support the 840 Mbytes/s data rates; so the custom connector available in the P3 VME standard was used instead. The overall characteristics of this subsystem for two 9U-VME chassis were

- 100 GOPS
- Size = 7.3 ft^3
- Weight = 210 lb
- Power = 1.5 kW
- Chassis throughput/power = 67 MOPS/W
- Throughput density = 14 GOPS/ ft^3
- Power density = 205 W/ ft^3

3.4.1 CUSTOM VLSI HARDWARE

The VLSI custom front-end chip, designed using 1996 process technology, is shown in Figure 3-5. The chip was fabricated at the National Semiconductor facility in Arlington, Texas, under the C050 process. The key features of this chip were

- 2.08 billion operations per second
- 18-bit input data and coefficients; 24-bit output data
- 585 mil × 585 mil die size
- 1.5 million transistors
- 0.65 μm feature size
- CMOS (complementary metal oxide semiconductor) using three-layer metal
- Designed for 4 watts power dissipation; measured 3.2 watts in operation
- Throughput/power = 0.65 GOPS/W

The design was based on a combination of standard cells and datapath blocks (Greco 1996). The standard cells were used in the chip control interface, barrel bit selector, and downsampler. The datapath blocks formed the multiply and add (MAC) cells. A total of 64 MACs available on the chip could be used either as 64 complex taps or as a maximum of 512 real taps. Since the objective was to calculate the in-phase (I) and quadrature (Q) data from real ADC data, each MAC processed up to 128 real taps for the I samples and, in parallel, another up to 128 taps for the Q samples, at 10 MHz.

For this specific technology demonstration, ADC samples arrived at 10 MHz. Every other sample went to the I computation and the Q computation, respectively (see Figure 3-2). Thus, two sets of outputs were computed at 10 MHz (5 MHz for I and Q), each applying 104 real taps. Since there were two real operations per tap (a multiply and an add), the total computational throughput for these parameters was 2.08 GOPS. These operations were performed on 18-bit sign-extended data (from 14-bit ADC samples) using 18-bit coefficients. The resulting significant 16-bit outputs (I and Q samples) were selected on the DIQ board from a 24-bit data output from the VLSI chip.

Figure 3-5 illustrates a die photograph. The packaged chip used was a pin grid array (PGA) with 238 pins. Figure 3-5 also shows the results of processing simulated radar data. The performance results contain three plots. Two of the curves illustrate the output from the hardware superimposed over the output of MATLAB, respectively. MATLAB simulates the filtering operation using full-precision arithmetic. The third plot shows the difference between the hardware output and the MATLAB results. The difference between the hardware output and the MATLAB full-precision simulation is about 100 dB relative to the square waveform peak. These results are very good and commensurate with what one would expect using 18-bit finite precision arithmetic.

If one compares the characteristics of the DIQ subsystem, described in the Section 3.3, against the characteristics of the custom VLSI chip, several observations can be made. The DIQ subsystem

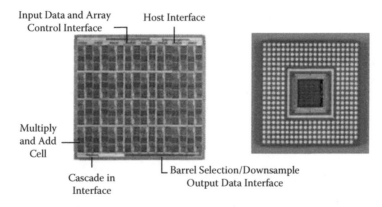

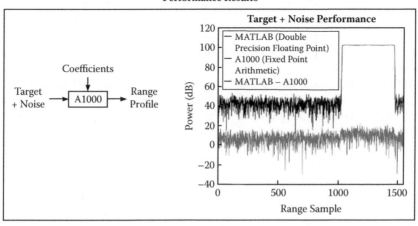

FIGURE 3-5 Custom VLSI front-end chip. (From Martinez, D.R., T.J. Moeller, and K. Teitelbaum. Application of reconfigurable computing to a high performance front-end radar signal processor. *Journal of VLSI Signal Processing Systems* 28(1–2): 69, Figure 5 © 2001 Kluwer Academic Publishing, with kind permission of Springer Science and Business Media.)

has a computational throughput per unit power of only 67 MOPS/W. The custom VLSI chip has a throughput per unit power of 0.65 GOPS/W. This decrease of one order of magnitude in throughput/power performance is typical of hardware integrated as part of an overall subsystem, in contrast to a single-chip performance. Additional power is needed for the control and interface boards, backplane drivers, and cooling. Another important observation is the lag between the time when the VLSI chip was designed (circa 1996) to the time when the system was deployed (circa 1998). This two-year lag was used to integrate the overall system. Therefore, at deployment time, the system was over one generation old in feature size process technology. Furthermore, for fabrication orders consisting of few wafer lots (circa 1996), the only available process technology was 0.65 μm. This feature size was sufficient for this one-of-a-kind prototype technology demonstration. Fabrication of larger wafer lots, often found in commercial chip designs, has access to more advanced fabrication process lines or, equivalently, feature size.

3.5 A HIGH PERFORMANCE COTS PROGRAMMABLE SIGNAL PROCESSOR

Figure 3-3 depicts the back-end subsystem following the DIQ filtering described in the previous section. The implementation for this part of the overall HPEC airborne signal processor consisted of COTS programmable hardware from Mercury Computer Systems. The system contained a total of 1000 SHARC AD21060 DSP chips from Analog Devices. The key characteristics of the system were

- Approximately 85 GFLOPs (floating-point arithmetic)
- Bisection BW = 3.2 Gbytes/s
- I/O throughput = 240 Mbytes/s
- Four VME chassis
- 33 ft^3
- 812 lb
- 6.9 kW
- Throughput density = 2.2 GFLOPs/ft^3

In this particular implementation, the ratio between total operations to bisection bandwidth was approximately 27, not as stringent as the more conservative general goal of 10 stated earlier but sufficient for this application. The interconnection network was based on the RACE++™ with the ability to communicate data across the backplane at a rate sufficient to keep the 1000 processors operating continuously in real time. The interconnection topology was a crossbar switch (Duato, Yalamanchili, and Ni 1997).

The AD21060 CPUs had a level 1 and 2 cache with a size of 512 KBytes and a total DRAM memory attached to each processing unit of 64 Mbytes (Mercury SHARC Node). The system was controlled utilizing the real-time proprietary software from Mercury Computer Systems, referred to as MCOS. All of the software was implemented utilizing C code supplemented with Mercury-specific signal processing library functions.

Several important observations can be made after comparing the performance of the front-end processing subsystem and back-end COTS programmable signal processor subsystem. The first is the dramatic difference in throughput density (a factor of approximately 7). This is an important trade-off between achieving higher throughputs in a smaller form factor at the expense of no programming flexibility. The other important difference is in power density. The front-end custom hardware delivered 67 MOPS/W. This is in contrast to a power density of 12 MFLOPS/W. Even though the VLSI chip alone delivered 650 MOPS/W and the AD21060 was rated at 38 MFLOPS W, the much lower performance for either one was due to all the additional memory and control hardware needed to build up a full-up system. These system implementation realities are important to take into account during the architecture and design phases of any complex HPEC system. The HPEC systems of interest and discussed in this handbook often are very constrained in size, weight, and power. Therefore, the capabilities of the system must be very carefully addressed at the system level relative to the available technologies at the time of design. In some cases, the algorithm requirements must be dramatically simplified to meet the overall system constraints. If these factors are not taken into account at the architecture and design phases, the results are a system requiring major changes and program overruns in cost and schedule.

Since the development of this HPEC system (circa 1996), several improvements have been made in technologies and standards. Many of these are described later in this handbook. For example, the front-end custom hardware subsystem can be implemented today using COTS field programmable gate arrays (e.g., Xilinx Virtex XCV1000) (Martinez, Moeller, and Teitelbaum 2001). Another advancement is the use of a PowerPC (0.5 GOPS/W per Power PC) from Motorola for the back-end programmable signal processor, resulting in higher power density. The interconnection hardware has also evolved to a widely accepted set of standards, such as the commercially available InfiniBand and rapid I/O. The trend for these systems has been to try approximating a throughput density in size and power commensurate with Moore's Law (about a factor of 10 every five years). However, the expectations are that this rate of advancement will likely slow down as programmable hardware continues to demand very high power levels. For example, the programmable Cell processor manufactured by IBM has an advertised peak of over 200 GFLOPS in 30 to 60 watts or, equivalently, a throughput power density between 7–14 GFLOPS/W at peak throughput. Obviously, these numbers must be de-rated based on the actual delivered throughput density for each specific application.

Over the 10-year period since this example HPEC system was developed, from the AD21060 DSP chip to today's Cell processor, the advancement in throughput power density has experienced a factor of about 88× (AD21060 at ~38 MFLOPS/W and Cell processor at ~3333 MFLOPS/W), which is just a bit lower than the factor of 100 increase expected at the exponential growth of Moore's Law. The computation has increased but so has the power. These ratios are further diluted when one takes into account all the other peripheral and support control hardware that make up a full HPEC system as previously described.

3.6 SUMMARY

This chapter discussed the generic topology of a complex HPEC system to highlight the key building components. These critical components—CPUs, memory, and interconnection network—must be very carefully balanced during the architecture of the system. Two implementation examples emphasized the design choices made during the implementation of the system. The first example was a front-end custom hardware system built using VLSI processing chips; the second was a COTS programmable signal processor performing back-end processing.

In addition to mapping algorithm requirements to different classes of hardware, the system architect must also ensure that the overall system meets the available size, weight, and power constraints. The overall system will draw significantly more power and will occupy a larger volume once all support hardware is taken into account, resulting in a much lower throughput per unit volume and per unit power than predicted by the performance of the microprocessors alone. Over several decades, the microprocessors, both custom VLSI and programmable chips, have had dramatic exponential growth in performance, commensurate with Moore's Law. However, the trend is slowing down, and predicted to slow down more, because of the significant increase in power consumption by these microprocessors. Therefore, HPEC architects and designers must pay careful attention to the overall system capabilities with all critical building components included in the expected system performance and final characteristics.

REFERENCES

Duato, J., S. Yalamanchili, and L. Ni. 1997. *Interconnection Networks: An Engineering Approach.* Los Alamitos, Calif.: IEEE Computer Society Press.

Greco, J. June 1996. A1000 Critical Design Review. Lexington, Mass.: MIT Lincoln Laboratory.

Hennessy, J. and D. Patterson. 1990. *Computer Architecture: A Quantitative Approach.* San Francisco: Morgan Kaufmann.

Martinez, D.R., T.J. Moeller, and K. Teitelbaum. 2001. Application of reconfigurable computing to a high performance front-end radar signal processor. *Journal of VLSI Signal Processing Systems* 28(1–2): 63–83.

Oppenheim, A.V. and R.W. Schafer. 1989. *Discrete-Time Signal Processing.* Englewood Cliffs, N.J.: Prentice Hall.

Patterson, D. 2001. Computer Architecture Course. University of California, Berkeley.

Teitelbaum, K. 1991. A flexible processor for a digital adaptive array radar. *Proceedings of the IEEE National Radar Conference*: 103–107.

4 High Performance Embedded Computers: Development Process and Management Perspectives

Robert A. Bond, MIT Lincoln Laboratory

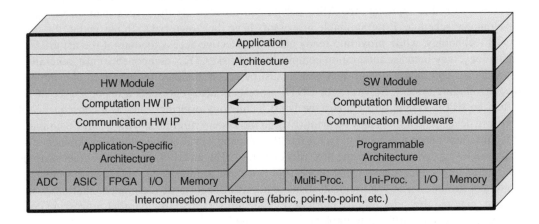

This chapter briefly reviews the HPEC development process and presents a detailed case study that illustrates the development and management techniques typically applied to HPEC developments. The chapter closes with a discussion of recent development/management trends and emerging challenges.

4.1 INTRODUCTION

This chapter presents a typical development process for high performance embedded computing (HPEC) systems, focusing on those aspects that distinguish HPEC development from the mainstream of embedded system development. HPEC system developments are influenced by requirement, plans, and implementation decisions of the systems in which they are embedded. In an advanced radar system, for example, it may be advantageous to field the analog radio frequency (RF) subsystems (antenna, transmitters, receivers, and control) prior to the availability of the high performance signal and data processors. This practice allows field data collections that will then lead to refinement of the algorithms targeted for the HPEC systems. However, to perform data collections, HPEC front-end subsystems are needed to perform analog-to-digital conversion, signal demodulation, and low-pass filtering. A high throughput, high-capacity recording system, which is itself an HPEC system, is also required. To accommodate the overall radar development plan, the

HPEC system is architected and implemented as three subsystems: front-end, data recorder, and back-end. These subsystems are then developed and integrated with the overall radar to support the phased, rapid deployment of the antenna. This, in fact, is the plan employed in the case study presented later in this chapter.

In addition to such external factors, the need to control technical, cost, and schedule risks also strongly influences HPEC development and management. A typical HPEC system will be a hybrid consisting of both custom hardware and programmable processors. It may consist of field programmable gate arrays (FPGAs), custom application-specific integrated circuits (ASICs) implemented in very-large-scale integration (VLSI) technology, custom boards, and a commercial off-the-shelf (COTS) programmable multicomputer [with several digital signal processors (DSPs) or microprocessor nodes]. The management of the development must be tailored to meet the particular technology choices since each technology has its own development cycle, cost, technical limitations, and risks. For example, developing a custom ASIC will tend to slow down the implementation; sometimes, especially if the chip is complex, it will be prudent to plan on two fabrication runs. It might make sense to mitigate this schedule risk by developing a lower-performing standard-cell or FPGA solution with a plan to retrofit the custom ASIC in a later iteration. With this approach, the ASIC design can proceed off the critical path of the system development schedule. Programmable processors have their own risks. In particular, the use of COTS technology, while providing ready and cost-effective access to state-of-the-art processor technology, may increase integration complexity and risk. COTS components must generally be used "as is," and, hence, integration issues must be accommodated by increased complexity in the surrounding system. Software for COTS processors also represents a major source of risk. Often, an HPEC system will employ the latest technology offerings, and as a consequence the software development tools may be primitive and the system runtime software may be immature. It is important for management to factor into the cost and schedule the added development effort that must be expended to overcome these deficiencies. Also, an HPEC system must meet throughput and latency requirements using processors that must accommodate severe size, weight, and power constraints. These requirements demand highly efficient software implementations, which, in turn, may require substantially more development effort. Usually, parallel processing codes, which are especially difficult to implement, are needed.

In this chapter, the HPEC development process is briefly reviewed, followed by a detailed case study that illustrates the development and management techniques typically applied to HPEC developments. The reader is referred to Chapter 3 for a more detailed technical discussion of the processor described in the case study. The chapter closes with a discussion on recent development and management trends, as well as emerging challenges.

4.2   DEVELOPMENT PROCESS

Conventional HPEC system development can be understood as an adaptation of the spiral development process, as defined by Boehm (1988). Figure 4-1 shows the basic process. Although the model was originally designed for software development, it can also be applied more generally to system development. The basic idea behind the spiral model is that it is most effective to manage the system development in a series of prototyping and development cycles that explicitly address risks. The iterative nature of this risk-driven methodology acknowledges that requirements are not expected to be fully known at the outset and must be explored through a series of prototypes.

As Boehm explains in his original article (1988):

> Each cycle of the spiral begins with the identification of the objective of the portion of the product being elaborated (performance, functionality, ability to accommodate change, etc.); the alternate means of implementing this portion of the product (design A, design B, reuse, buy, etc.); and the constraints imposed on the application of the alternatives (cost, schedule, interface, etc.).

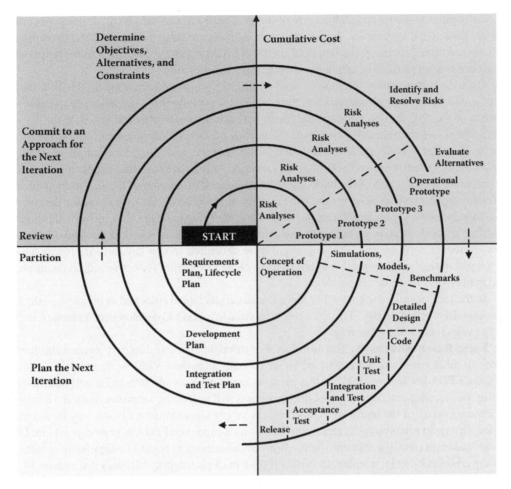

FIGURE 4-1   The spiral development process consists of a series of iterations that are viewed as building spirally to develop the overall system. Each cycle is intended to address a set of major risks in the development. A cycle may involve the development of a prototype, with each subsequent prototype addressing and removing risks and adding functionality. At some point, a transition from rapid prototyping to a production development approach (for example, a waterfall model) occurs, and subsequent spirals continue to add functionality iteratively. The complexity and duration of a spiral may be as short or long as the program pace and system evolution dictate.

In an HPEC development, the objectives are specified in terms of the overall system into which the computer is to be embedded. For example, consider a synthetic aperture radar (SAR) embedded in an unmanned aerial vehicle (UAV). The operational and performance goals of the UAV missions will determine radar performance objectives. The UAV form-factor constraints and the radar performance requirements translate into requirements for the SAR signal and data processors. The first cut at SAR performance would include cross-range and along-range resolution, sensitivity, modes of operation (e.g., scan and spot mode), false-alarm rates, etc. Candidate algorithms would be developed for each mode, which, along with operational parameters such as bandwidth and dynamic range, determines the throughput requirements of the processor. The UAV payload budget places constraints on the size, weight, power, cooling, and ruggedization for the HPEC system. Alternative HPEC architectures are developed to implement the modes and algorithms within specified form-factor constraints. Quite often, MATLAB codes of the algorithms will be developed and evaluated against real data or simulations. The MATLAB programs are then used to specify processing requirements. As illustrated more fully later in this chapter, the MATLAB programs can be restructured to mirror the architecture of the processor and can then be used to support processor integration and verification. In this approach, datasets are input to both the MATLAB program and the real-time processor; the output of the MATLAB program is compared to the output of the processor at each processing step to verify processor functionality.

Once the requirements, alternatives, and constraints for the processor are established, the next steps are to perform a risk analysis and then to define a prototyping spiral designed to retire some portion of the risk. At this point, management will determine the cost and schedule for the iteration and will also extrapolate a cost and schedule for the overall development. To size the processor development, a basic architecture design will be needed, including, for example, allocation of functionality to hardware and software; specification of the hardware that needs to be developed or procured; specification of the basic algorithm suite; and estimation of the size of the embedded software program.* The development time constraint imposed by the overall sensor development also influences the approach and the cost. For example, trying to compress a software development schedule beyond a certain point leads to an exponential increase in schedule risk and development cost. To reduce software risk, the program may be structured so that key software functionality is developed in the earlier spirals even though it may lead to software rework as requirements change and mature later in the development.

In the early spirals, the goal is to reduce the largest technical risks and to initiate lengthy tasks that may drive schedule risk. The most common risks in an HPEC development addressed in these early cycles include the following:

Form-factor constraints: The form-factor or throughput constraints may warrant the development of either custom hardware, based either on ASICs (custom VLSI or standard-cell chips) or efficient FPGA implementations. At this point, a deeper analysis and simulation will be required in which the precision and complexity of the algorithms will need to be weighed against the implementation complexity of the hardware. The efficiency of key algorithms on FPGAs may be a deciding factor, so a rapid prototyping of algorithm kernels on a candidate FPGA may be carried out. Often, the co-design of the algorithm and the hardware architecture can result in a significant reduction in system complexity and implementation risk. If the FPGA prototyping indicates that custom VLSI is needed, then the VLSI design can be initiated at an early stage in the program.

Algorithm and functional uncertainty: Typically, the most suitable algorithm techniques are generally not well understood at the outset and will continue to evolve throughout the program. The complexity of the algorithm, however, drives the throughput requirements of the processor and, when coupled with the form-factor constraints, will dictate implementation technologies. The particular algorithmic techniques may or may not have efficient implementations in hardware. Often, a simulation of the operational environment is used to help refine and focus algorithm alternatives. Usually, data collections in representative environments are carried out. Using real data has the advantage of helping to validate assumptions that may be evident in the simulations. An effective risk-mitigation strategy at this stage is having the processor architects work with the algorithm developers to determine the fundamental algorithm kernels and the likely range of operation of these kernels. Then, an analysis of alternatives and benchmarks of prototype implementations can be performed. In any case, it is again important that algorithms and architectures are developed in a co-design approach. Often, simple changes from an algorithmic perspective, such as using single precision versus double precision or integer arithmetic versus floating-point arithmetic, can have dramatically simplifying effects on hardware without much loss in performance.

---

* This is an incomplete list of the important requirements that the system designer will need to consider when assessing program cost and risks. For example, testing requirements, including integration test beds, test datasets, field testing, etc., are significant cost and schedule drivers. Mission assurance and system availability requirements will also affect cost and schedule. For systems requiring fault tolerance, such as mission-critical command-and-control modules in spacecraft, redundant equipment will be needed, driving up costs and testing requirements. Maintenance and upgrade requirements will influence up-front design costs. All of these and more must be considered in the overall HPEC development.

Synchronization and control: In HPEC applications, it is important to be able to exploit the regularity of data flow in the computation. This leads to simpler, lower risk implementations. Current trends toward multifunction and multimodal sensors and communication systems, however, are significantly increasing the need for context changes, buffering, and complex communication and synchronization requirements. In the early stages of the development, it is important to take explicit aim at simplifying the control complexity of the application through a co-exploration of system alternatives and processor architectures. System-level simulations, concept-of-operations (CONOPS) studies, and up-front processor architecture simulations will be employed to uncover the control requirements. Often, for the sake of reducing implementation complexity, the flexibility of the processor will be constrained, in turn constraining the CONOPS of the overall system.

Software complexity: Software programs in HPEC systems from two decades ago would typically have had a few thousand lines of algorithm code. Much of the HPEC architecture was implemented in custom hardware, either in custom boards using COTS hardware components or custom chips. Today, however, HPEC systems with 100,000 to 500,000 source lines of code are common. The development of large software codes for HPEC systems can be controlled through the use of early prototyping of key software modules, the reuse of software codes, the use of software frameworks previously developed for the particular class of application, and the use of middleware to raise the level of development abstraction, which has the effect of reducing overall development complexity.

Commercial off-the-shelf processor technology integration: COTS hardware is procured "off-the-shelf," thereby avoiding costly custom design and development. However, it brings its own set of risks that need to be addressed early in the development. For example, a single-board computer will need to be evaluated for application suitability. The board should be benchmarked using the key algorithm kernels at the scales that will be needed by the application. Preferably, a simplified application example, often referred to as a mini-application or compact application, should be implemented and benchmarked on the board. If the memory, input/output (I/O), or computation resources of a single board will not meet the application throughput and latency requirements, then additional boards will be needed. For most HPEC applications, multiboard, multicomputer COTS implementations are needed to meet performance requirements. Sometimes, it is not feasible or cost-effective to evaluate all candidate solutions at full scale; therefore, extrapolation to the larger system must be made. The sooner representative hardware can be benchmarked, the better. Moreover, the performance is not the only risk dimension. COTS programmable systems generally consume more power and require more space than ASIC counterparts. Thus, form-factor risks must be assessed. Furthermore, the development tools, runtime tools, scalability, and maturity of the product are all important risk factors.

Custom ASIC designs: Custom processor designs can deliver excellent performance with efficient size, weight, and power, but they are costly and time-consuming. They are inherently less flexible than are programmable systems. Therefore, for programs that require rapid development, the decision to use ASICs will require early commitment to an algorithm technique, with limited ability to change the algorithm later on. If the early algorithm variants are lower performing or, worse, inappropriate, the overall objectives of the system will be at risk. On the other hand, increasing the complexity of the hardware to accommodate potential algorithm changes increases the complexity of the custom designs, thereby leading to technical risks.

At the outset of each spiral, management assesses risk areas and decides on mitigation approaches. In the earlier cycles, "throwaway" prototypes and models may be used. The prototypes at this phase are focused on retiring specific risks. For example, a prototype FPGA implementation of a key algorithm kernel may be undertaken, or an efficient parallel code for a key kernel may be developed. In a subsequent spiral, it might be determined that the prototype is mature enough that it can serve as an operational baseline. For example, if the FPGA implementation turns out to be very efficient, it can then become the basis for a portion of the final architecture. Moreover, if the development has reached the point where requirements have stabilized for some portion of the system, then that portion can undergo a more traditional "waterfall approach" while other, less stable, components can continue in the prototyping cycle. An essential aspect of the spiral model is the review process that occurs at the completion of each cycle. This review is the opportunity for management and technical personnel to assess the success of the cycle and to establish the plans for the next cycle. It is also the opportunity to refine cost and schedule estimates and, if necessary, to adjust the scope of the program.

4.3 CASE STUDY: AIRBORNE RADAR HPEC SYSTEM

The RAPTOR* space-time adaptive-processing (STAP) radar signal processor project provides an excellent example of the HPEC spiral development process. At the time that this system was implemented (circa 1989), it represented the state of the art in HPEC systems. RAPTOR, shown in Figure 4-2, consists of three major subsystems:

1. *Custom front-end processor (FEP)*: The front-end requirement could only be addressed through the use of high performance custom VLSI chips. The VLSI chips were the heart of the digital in-phase and quadrature sampling (DIQ) subsystem, which demodulated the received radar signals and applied low-pass filtering. This subsystem processes 48 receiver channels and operates at a throughput of 100 billion operations per second (GOPS).
2. *Tape recorder subsystem (TRS)*: The high-capacity tape recorder system that operates at 30 Mbytes/s is capable of recording 30% of the range extent after DIQ low-pass filtering.
3. *Programmable signal processor (PSP)*: To perform digital filtering, jammer nulling, clutter nulling, detection, and target parameter estimation, the PSP consists of nearly 1,000 processors, housed in four interconnected chassis, and is capable of performing 85 billion floating-point operations per second (GFLOPS). The software program consisted of over 180,000 source lines of C language code, designed to implement highly optimized radar signal processing.

This case study focuses on the programmable signal processor of RAPTOR, although aspects of the overall RAPTOR development are briefly reviewed to provide context. During the first spiral, the overall requirements for the processor were set. It was determined that RAPTOR would process 48 narrowband receiver channels from an airborne radar phased-array antenna. The received RF signals would be downconverted to an intermediate frequency and digitally sampled. The digital signals would then be demodulated to baseband and low-pass filtered. After that, airborne moving-target indication (MTI), employing STAP, would be performed. After constant false-alarm (CFAR) detection, the target reports would be passed to a tracker and displayed to the system operators. The required airborne MTI processing chain comprised a set of sophisticated and computationally demanding signal processing algorithms. The transmitted radar waveform is referred to as a coherent processing interval (CPI). A CPI consists of a series of pulses transmitted at a regular interval referred to as the pulse repetition interval (PRI). The dataset received during a CPI can be thought of as a data cube. Each return signal can be labeled in the data cube by a {range gate, channel, PRI} triplet. The major functions performed in the RAPTOR airborne MTI (AMTI) digital processor are as follows [the reader is referred to Ward (1994) for more information on STAP MTI processing]: First, the receive signals are match filtered along the range dimension with replicas of the transmitted waveform to improve the SNR and localize the target returns in range. This stage is referred to as pulse compression. Then, the signals are filtered along the PRI dimension into Doppler frequency bins, each of which corresponds to a specific range rate of the return signal relative to the radar platform velocity vector. These returns are then adaptively combined along the channel dimension in an adaptive beamformer to produce 48 beams. The beamformer adaptive weights are computed by sampling the environment for jammer energy and then solving a least-squares optimization prob-

---

* Reconfigurable Adaptive Processing Test bed for Onboard Radars (RAPTOR)

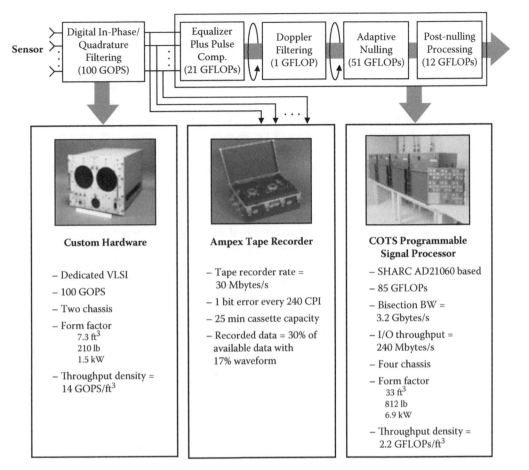

FIGURE 4-2 Airborne signal processor. (From Martinez, D.R., T.J. Moeller, and K. Teitelbaum. Application of reconfigurable computing to a high performance front-end radar signal processor. *Journal of VLSI Signal Processing Systems* 28(1–2): 67, Figure 3 © 2001 Kluwer Academic Publishing, with kind permission of Springer Science and Business Media.)

lem to determine a set of beamforming weights that remove (null) the jamming while maintaining high gain in the beam pointing directions. Subsequently, these beams are adaptively recombined to mitigate clutter interference. The clutter mitigation stage uses a space-time adaptive beamformer, producing another 48 beams that are adaptively determined to minimize clutter while preserving the signals of interest. The samples in these beams are tested for detections. State vectors consisting of range, azimuth, elevation, range rate, and radar cross section are estimated for all target detections. Finally, target reports consisting of target numbers and state vectors are sent to the back-end tracking and display subsystems. The required end-to-end algorithm turns out to be very complicated, involving over 100 signal processing steps. At the outset of the first development spiral, only a simplified version of the algorithm existed. The throughput requirement was initially estimated at between 100 GOPS and 200 GOPS, including both the front-end and PSP processing.

In the first spiral, an initial HPEC system architecture was established, and the system was factored into its three main subsystems. The following major risks were identified:

1. The throughput in the front-end of the system, where digital IQ processing is performed, would exceed 100 GOPS, although the operations were very regular and could be carried out in each of the 48 receive channels in parallel. Analysis and simulation determined that

18 bits (complex) of precision would be needed. Programmable processor technology at the time could not handle the throughput requirement, and so special-purpose hardware would be required.

2. The STAP and detection algorithms were immature and would require extensive development. Data collected from the radar antenna and receivers in a representative environment would be needed to validate the algorithm techniques. The late specification of these algorithms, the need to evolve the techniques as more insight into the operational environment was garnered, and the sophistication of the algorithms all pointed to the need for a programmable signal processor. It was likely that this PSP would need to be rated at over 50 GFLOPS. A programmable adaptive radar processor of this scale had never before been implemented.

To address the front-end processing challenge, a set of three spirals, shown in Figure 4-3, was carried out. In the first spiral, a survey of architectural and technology alternatives was conducted. The INMOS A100 chip and others were evaluated, and it soon became evident that there were no commercial chips that could simultaneously meet the throughput and precision requirements of the digital IQ processing. At this point, a second spiral was initiated. The spiral consisted of an analysis of alternatives for VLSI designs for a digital IQ chip, a matched-filter chip, a Doppler-filter chip, and a beamforming chip. The analysis determined that the digital IQ requirements could be met with a standard-cell custom design using 18-bit integer arithmetic. Full-custom VLSI was also investigated, but the schedule risk was assessed as too high, although the design would be able to deliver higher performance. During this spiral, the preliminary front-end processing architecture was refined. Standard-cell custom top-level designs for the matched filtering, Doppler filtering, and beamforming were developed that would meet the requirements of these functions. Due to the overall complexity of the timing, control, synchronization, and data movement in the front-end processing architecture, coupled with the large number of unique custom chip designs, the front-end processor implementation was assessed as being very high risk. As a result, an alternative architecture, based on programmable signal processing, was developed. In this architecture, the digital IQ function was allocated to a standard-cell custom design, but the remaining signal processing stages downstream of this function were allocated to programmable components. This approach was also

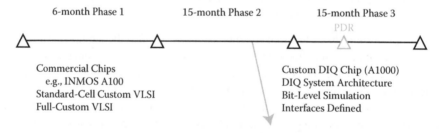

| 6-month Phase 1 | 15-month Phase 2 | 15-month Phase 3 |

Commercial Chips
e.g., INMOS A100
Standard-Cell Custom VLSI
Full-Custom VLSI

Custom DIQ Chip (A1000)
DIQ System Architecture
Bit-Level Simulation
Interfaces Defined

• Unacceptable risk of custom hardware for digital filtering and jammer nulling
• Risk of programmable alternative deemed acceptable
• Risk of custom DIQ acceptable (no programmable alternative)

FIGURE 4-3　Custom-hardware risk-mitigation timeline. The front-end signal processing requirements represented a significant technical risk. The risk mitigation approach involved three spirals (phases). The first spiral entailed a requirements analysis and a survey of commercial chips and technology options. The second spiral took the findings of the first spiral, which recommended a VLSI technology approach, and developed a set of alternative VLSI-based front-end architectures. This spiral resulted in an architecture that relegated the digital-filtering and jammer-nulling functions to programmable hardware and allocated the digital in-phase and quadrature sampling function to custom VLSI. The final spiral developed a detailed design that followed the architecture recommended in the second spiral.

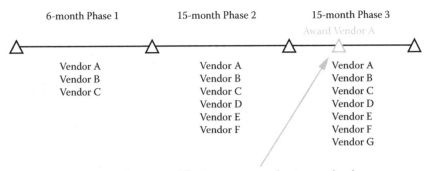

Contract modification to accommodate increased scale
(from 50 GFLOPS to 85 GFLOPS to add digital filtering and jammer nulling)

FIGURE 4-4   PSP spiral risk-mitigation timeline. The programmable signal processor hardware was a significant risk factor in the program. To mitigate this risk, a series of risk-mitigation spirals (phases) were carried out. The first spiral was a preliminary feasibility study. The second spiral was a detailed benchmarking phase. The final spiral was the processor procurement. During the procurement phase, the size and functional scope of the PSP were increased to take on some of the signal processing initially assigned to the front-end custom hardware.

recognized as a risky proposition, but pending the results of the benchmarking of the programmable processors, the risk was considered more manageable.

The risks associated with the programmable implementation of the STAP and detection algorithms were considered to be very high. Programmable signal processor technology had never been applied to a STAP problem of this size, and the need for a ruggedized, airborne form-factored computer further increased the challenge. Hence, an early investigation was conducted to assess this approach. The first step was to survey the state of the art in PSP technology. It became evident that while programmable supercomputing systems would be capable of handling the throughput, they were neither ruggedized nor of an appropriate form factor (size, weight, and power). On the other hand, embedded multicomputer technologies based on i860s, SHARC DSPs, and PowerPCs were emerging as potential candidates. Their performance on STAP algorithms, however, was unproven. At this point, management decided to address the risk through a comprehensive, nationwide study and benchmarking initiative. Three risk-mitigation cycles, which culminated in the procurement of a PSP hardware system, were carried out. The timeline and goals of the spirals are shown in Figure 4-4.

In the first spiral, a set of benchmarks and processor form-factor constraints were specified. The benchmarks consisted of several variants of STAP algorithms, specified for radars of various operational configurations. A more general-purpose linear algebra benchmark was also included. A wide range of radar and algorithm options was needed since it was still unclear at this early stage what the final configuration for the target radar would be and what algorithm variant would be chosen. A Request for Proposal (RFP) was issued to major radar prime contractors and computer vendors. There were three respondents, each of which was awarded a study contract. The results of the studies were favorable, indicating that modern multicomputer technology would be able to meet the throughput, latency, and form-factor requirements.

A second spiral was initiated with another RFP, this time requesting the implementation of the benchmarks on a scaled-down version of a proposed system and an architecture design for the target PSP system. Six vendors participated, including the original three. The results were once again quite promising, although the challenge of developing efficient parallel versions of the benchmarks was also evident. This spiral demonstrated that STAP algorithms could achieve real-time throughput on PSPs. The architectures were scaled to extrapolate to the full-scale STAP algorithm. The proposed architectures required in the range of 100 to over 400 processors. Computational efficiencies from 25% to 30% were predicted.

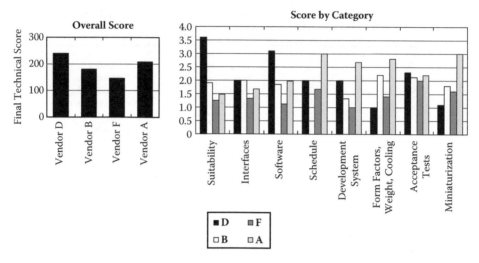

FIGURE 4-5 PSP procurement: proposal evaluation. The PSP proposals were evaluated in several categories. A cumulative threshold of 200 was established. Vendors with scores below this value (Vendors B and F) were, therefore, eliminated. Vendor D had a higher cumulative score than did Vendor A, but it was determined that the form-factor risks associated with Vendor D were too risky. Thus, Vendor A's proposal was chosen as the winning proposal.

The third spiral was initiated by a third RFP for the procurement of a full-scale, ruggedized, airborne multicomputer to perform STAP. The RFP was issued to the six participants and also to the vendor community in general. The responses were evaluated against a matrix of required capabilities (refer to Figure 4-5):

1. Suitability—an assessment of the architecture design, including predicted throughput and latency on the benchmarks
2. External interfaces—an assessment of the I/O capability
3. Software—an assessment of the runtime software suite
4. Schedule—an assessment of the processor fabrication and delivery time
5. Development system—an assessment of the availability and quality of a development system that would be delivered prior to the final system to allow early software development
6. Form factor—an assessment of the size, weight, power, and cooling metrics
7. Acceptance tests—an assessment of the quality of the proposed acceptance test procedures
8. Miniaturization—an assessment of the viability for the miniaturization in a future variant

The six vendors from the benchmarking phase and three other vendors were solicited for proposals. Of the original six, one decided not to bid. One of the newly selected vendors also did not bid. Of the remaining vendors, one was determined to be nonresponsive since many of the selection categories were not addressed. The remaining vendors (denoted vendors A, B, D, and F in the figure) were scored against the criteria listed above. Two of the vendors fell below the acceptance threshold. The comparison between the two remaining vendors was quite interesting. Vendor D had the highest overall score, but had a particularly low score in form-factor assessment. Vendor A had moderate scores across the board, but the overall suitability of the architecture was only marginal. After careful consideration, the evaluation committee decided that the form-factor risks associated with Vendor D were more likely to occur and be of potentially greater impact than the architectural risks associated with the Vendor A processor. The costs for the two processors were comparable. Thus, Vendor A was awarded the contract.

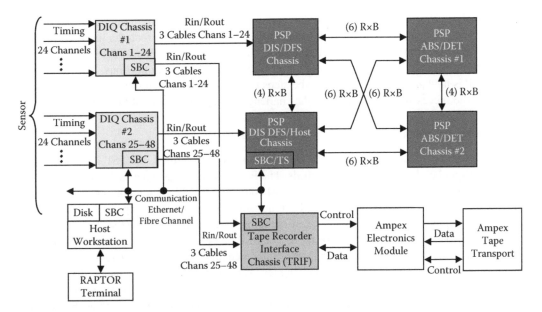

FIGURE 4-6 RAPTOR chassis-level block diagram.

At the same time that this evaluation was taking place, management was struggling with the realization that a full-custom front-end of the scope that was originally envisioned was too risky. The chief architect, who was a participant in both the front-end and the back-end spirals, developed a new architectural proposal that allocated the digital IQ functionality to custom hardware and subsequent stages to a scaled version of the Vendor A proposed processor. Program management, the chief architect, and the team leads for the front- and back-end systems conferred to weigh the relative risks and options. Out of this meeting, the critical decision was reached that the PSP would be scaled upward to encompass all of the processing after the DIQ task. Vendor A was notified and asked for a cost, schedule, and technical proposal (contract modification) to scale the processor. They sent a favorable response quickly. The proposed processor would be scaled from two chassis to four chassis, its throughput would be increased by about 40% to handle the additional processing, and the front-end I/O system would be scaled up by a factor of two to handle the greater input data rate. Although commercial hardware and operating-system software components were to be used, a system of the required size and complexity was not yet commercially available. The vendor proposed (as per contract requirements) to deliver the processor in nine months. A fully working STAP acceptance test benchmark and acceptance test would be carried out first at the factory and then at the radar system integration laboratory (SIL).

The sequence of risk analyses, studies, and architectural designs just described illustrates the complex and interdependent nature of HPEC system design. By using a risk-driven spiral development model, program management was able to navigate the early stages of the project through a series of important analyses of alternatives, informed by early benchmarking, simulation, and modeling. The early trade-offs between the custom and COTS hardware, between programmable and hardwired functionality, and between full-custom and standard-cell VLSI were crucial to the success of the development. In the absence of these trade-offs, management might have committed to a design whose critical shortcomings would only have become evident during a later cycle in the development, when the impact would have been much more severe.

Even with these early trade-offs, the project still had several risk areas, in particular those associated with the scale and complexity of the PSP. The architecture that emerged from these spirals is shown in Figure 4-6. The entire system was controlled by a host workstation connected over an Ethernet. Two 9U-VME chassis were allocated to the digital IQ (DIQ) subsystem. Each chassis handled 24 receiver channels. The outputs from the DIQ subsystem were duplicated so that data could be simultaneously sent to the PSP and the tape recorder. The tape recorder had its own chassis and special mount for the tape-recorder assembly. The PSP consisted of four 9U-VME chassis housing nearly 1,000 processors having special high-speed interchassis connections. The development of the PSP is now presented to illustrate the steps taken by management to address these risks and to ensure the successful development of the challenging PSP component of this HPEC system.

4.3.1 PROGRAMMABLE SIGNAL PROCESSOR DEVELOPMENT

Once the top-level architecture of the HPEC system was determined, the PSP development progressed as another dedicated series of spirals. The PSP hardware was not scheduled to be delivered for nine months, and it would be one of the most complex embedded multicomputers of its kind in the nation. It consisted of nearly 1,000 DSPs, requiring four chassis with a specially designed interchassis communication subsystem and a new operating system (that spanned the chassis, essentially treating the four chassis as a single system). The PSP had the maximum amount of memory that could be fit on each board. It had six high-speed input channels to accommodate the 48 channels of digital IQ from the front-end system. The long lead time in the PSP delivery meant that integration with the front-end system would have to begin almost immediately after acceptance of the PSP. There would be little time for further software development, as much of the time would be needed to integrate the radar system end-to-end and then integrate it into a test-bed aircraft.

Risk in the development was mitigated by taking the following measures:

1. To get a head start on software development, an early, off-the-shelf, development system was procured. The development system, while similar to the final system, had significant differences. It occupied only one chassis, it used i860 processors at each node instead of the triple-SHARC node (still being developed) of the final system, and it used an earlier version of the operating system. While not ideal, the system allowed many software components to be developed and tested early on in the development.
2. As a bridge between the full-scale system and the development system, a 1/4-scale, single-chassis PSP was procured that would arrive six months after contract award. This processor provided early access to a parallel (but reduced-scale) signal processor. It had the same type of processor boards, I/O hardware, and development software as the full-scale PSP and, hence, served as an excellent development and test surrogate.
3. A series of spirals was designed to incrementally add algorithm capability and to scale to the full, four-chassis PSP system.

As shown in Figure 4-7, the overall PSP software system factors into seven subsystems. The allocation of the subsystems within the overall RAPTOR chassis configuration is shown in Figure 4-6. All of the subsystems map onto multiple signal processing nodes within these chassis, with the exception of the task scheduler (TS) subsystem. The TS receives radar mode commands and configuration information from the radar, describing the waveform and other operational parameters. Each command arrives a few processing intervals in advance, giving the PSP time to set up for the new mode before the data are input into the PSP. The scheduler calculates internal parameters and then transmits a mode description to the master node in each parallel PSP subsystem. Each master node then disseminates the description to all of the nodes that compose its subsystem. The nodes in a subsystem then extract the parameters they need and set up in anticipation of the arrival of new mode data.

The parallel processing subsystems perform signal processing and data reorganization tasks. The data input subsystem (DIS) receives data from the front-end digital IQ hardware each pulse repetition interval. The DIS uses six input ports, each of which handles eight receiver channels and is controlled by a PowerPC. The DIS buffers the data and demultiplexes them into a format that

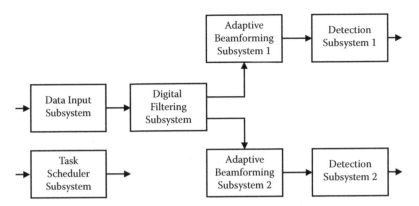

FIGURE 4-7   PSP subsystems. The programmable processors comprised seven subsystems. Five subsystems carried out complex signal processing operations: the digital filtering subsystem carried out pulse compression and Doppler filtering; the two adaptive beamforming subsystems (ABF1 and ABF2) performed jammer nulling, interference rejection, and beamforming; the two detection subsystems (DET1 and DET2) performed detection and target parameter-estimations. The task scheduler controlled the mode of operation of the other subsystems, and the data input subsystem performed high-speed data input and demultiplexing.

can be efficiently processed by the digital filtering subsystem (DFS). The DFS performs the pulse compression and Doppler filtering tasks. It exploits the parallelism in the receiver channels so that each triple SHARC node receives and processes all of the range gates for one of the 48 receiver channels. After filtering, the DFS must also perform data corner-turning, during which it reorganizes the data and transmits them to the downstream adaptive beamforming subsystem (ABS). The corner turn reorganizes the data so that the range gates for each Doppler bin are stored on a separate triple-SHARC processing node in the ABS. There are a total of 96 Doppler bins, and these are mapped one to each of 96 triple-SHARC nodes. The ABS performs jammer nulling and space-time adaptive processing, both of which are beamforming operations. The ABS is the most demanding subsystem, requiring over 30 GFLOPS. Fortunately, due to the early risk-mitigation studies and benchmarks, the throughput expectations of the SHARC nodes on the ABS computations had been assessed beforehand. It was (correctly) predicted that the 96 triple-SHARC nodes would not be able to meet the throughput requirements. To solve this, a ping-pong configuration was used, in which every odd-numbered CPI is forwarded to one set of 96 nodes and every even interval is forwarded to another set of nodes. Although this doubles the latency in the ABS, it allows the hardware to meet the throughput requirement. After the ABS, the beamformed data for three Dopplers were grouped and forwarded to a detection subsystem (DET). Odd processing intervals were forwarded to one DET subsystem and even intervals were forwarded to another subsystem. Each DET used 32 triple-SHARC nodes, with each SHARC in a node handling detections for a single Doppler bin. (However, to perform clustering properly, data needed to be shared between neighboring bins, thus complicating the synchronization and communication logic in this subsystem considerably.) Once detections were grouped and their state vectors (azimuth, elevation, range, range rate, and radar cross section) were estimated, the DET created a target report message that it forwarded to the back-end radar processor for tracking and display.

Each software development spiral followed the basic waterfall sequence shown in Figure 4-8. The sequence starts with a refinement of the requirements and a statement of the specific objectives for the spiral. MATLAB code is used as an executable specification of the algorithm requirements. The code is augmented by documentation that describes the theory and the algorithm design trades. MATLAB code is double-precision floating-point, and the PSP by default uses single-precision floating. Hence, a precision analysis needs to be conducted for any portion of the algorithm that may be expected to have precision issues. This analysis requires coordination between the algorithm designer and the real-time program software engineer. For example, during the early develop-

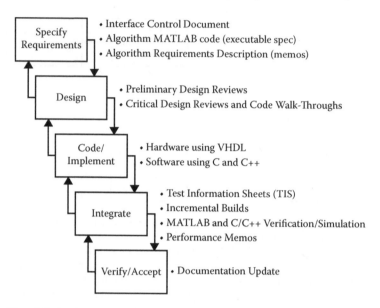

FIGURE 4-8 Waterfall development cycle. For each spiral increment that involved software or hardware development, a waterfall development model was followed. The figure also shows the major products of each phase in the waterfall.

ment of the QR decomposition for the jamming nulling stage, it was determined through numerical simulation that although the single-precision variants of the algorithm produced slightly different results from the MATLAB double-precision code, the end-to-end real-time system still operated within performance tolerances. Double-precision arithmetic requires roughly six times the number of operations as single-precision arithmetic; therefore, being able to stay with single precision was a significant computational benefit. To handle double precision, a QR co-processor or dedicated QR processing nodes would have been required, thereby increasing the size and complexity of the subsystem. Such algorithm architecture trade-offs were generally mediated between the verification and architecture teams, who had the joint responsibility to certify the completion of each spiral. When the spiral involved interfacing requirements to other subsystems in the embedded system, the interface control documents were updated and used as design references.

Figure 4-9 shows the nominal allocation of implementation effort and schedule to the development phases. It is important to realize that the actual effort for a cycle depends on the size of the overall system, not just the size of the increment, since adding new software functionality to an existing body of software requires more work than just creating the new functions; there is an increase in the number of interfaces (leading to additional design and code), the amount of regression testing, and the integration effort.

Figure 4-10 shows the planned sequence of development spirals. Management and technical leadership recognized early on that the task of scaling the PSP to its full size would be very challenging. Because of the late availability of the full-scale PSP, the software needed to be developed initially on a smaller system and ultimately scaled to the full-scale system. The 1/4-scale PSP system was a steppingstone on the way to the full-scale PSP and also served as the permanent development system once the PSP was deployed to the platform. To be able to run the software on three different systems of different parallel scales, with different node processors and runtime systems, a layered software architecture was developed, as shown in Figure 4-11. The architecture was based on a novel parallel middleware library called the STAP library (STAPL), developed using the C programming language.

STAPL provided an object-based library of scalable matrix algebra and signal processing components. Each object could be scaled from one to hundreds of nodes by changing configuration

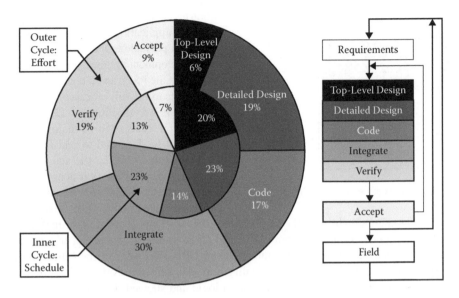

FIGURE 4-9   Software development cycle.

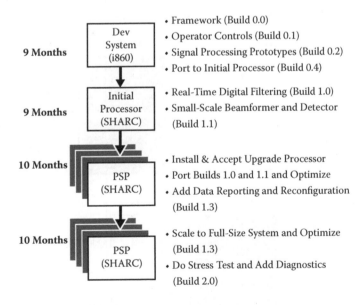

FIGURE 4-10   PSP development spirals.

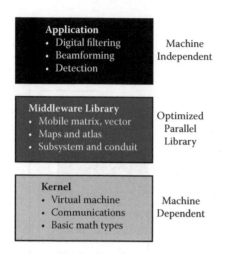

FIGURE 4-11   Layered software architecture.

parameters that were read into the program at start-up and communicated to the objects at initialization. STAPL objects used the underlying lower-level communication and computation libraries supported by the vendor system. In this manner, application software that used STAPL was both portable to the three vendor platforms (the development system, the 1/4-scale PSP, and the full-scale PSP), and it could be scaled from a few nodes to the full processor configuration.

As shown in Figure 4-12, the PSP was scaled in a series of three major spirals. It was important to show end-to-end functionality as part of each spiral, so the scaling steps were chosen to demonstrate a processing thread that exercised the major functionality of all the software subsystems. The first instantiation processed eight channels through the DFS, followed by 24 Doppler beams through the adaptive beamforming and detection subsystems. The odd CPIs were sent to one set of adaptive beamforming (ABF) and DET subsystems, and the even CPIs were sent to the other set. This first scaling step verified multichassis functionality and also allowed every major processing step to be exercised. It verified basic synchronization logic and showed that the task scheduler communication logic was correct and could meet real-time latencies. The next major scaling step increased the DFS to the full-scale, 48-channel configuration and increased the ABF and DET subsystems to half scale—48 Doppler configurations. Most significantly, this configuration was integrated into the full radar system to support system-level demonstrations. Several real-time performance issues that required further optimization were uncovered. Integration issues were also identified at all levels: operating system, middleware, and application. Although the middleware was designed to allow the application to scale, there were still several challenges in moving

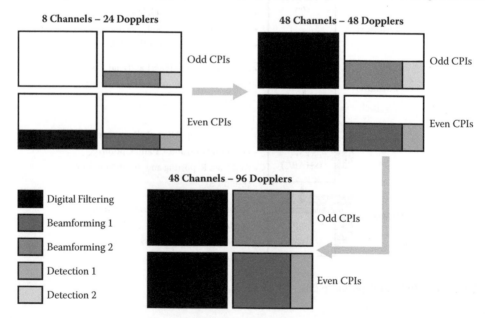

FIGURE 4-12   PSP scaling. The PSP was scaled in three major steps. The allocation of subsystems to chassis is shown for each step. In each step, the end-to-end algorithm was verified at the required scale.

from a 32-processor development system to a 1,000-processor, full-scale system. During system acceptance, the performance of the full-scale processor had been demonstrated using benchmark codes. The full-scale application, however, was significantly more complex, both in total lines of code and total throughput. As the application was scaled, the operating system resource utilization proved to be a major stress area. The full-scale application required the allocation of 1,000s of communication objects, causing the start-up time of the system to become unacceptably long. Loading and initializing the application took over two hours initially. This slowed the debug and test cycle to a crawl. The initial middleware design naively assumed that the operating system would be able to accommodate this load without deleterious consequences. Fortunately, although the solution involved numerous optimizations, the changes were mainly isolated to the middleware initialization; changes to the application were minimal. The final scaling step increased the ABF and DET subsystems to the full 96 Dopplers. The overall scaling development cycle took ten months, during which time the start-up time was further reduced in a series of optimizations, bringing the final, full-scale start-up time down to about 17 minutes—long, but acceptable. The PSP was integrated with the rest of the radar, and numerous performance optimizations were carried out on the ABS and DET subsystems to bring them to within the real-time allocation.

4.3.2   SOFTWARE ESTIMATION, MONITORING, AND CONFIGURATION CONTROL

One of the most important management tasks for an HPEC project is to develop accurate estimates of the project development effort, cost, and schedule. For programmable HPEC systems, this involves an estimate of the complexity of the delivered software product. One way to get an idea of the complexity of a software program is to estimate the size of the code in terms of non-blank, non-comment source lines of code (SLOC). Table 4-1 shows the initial estimate of the SLOC for the PSP application code. Similar estimates were developed for the communication, middleware, and mathematical kernels developed for the project. The estimates were derived after the first set of risk-reduction spirals (during the builds prior to Build 1.0, as shown in Figure 4-10). The figure also shows the measured lines of code at the end of the first full delivery (Build 2.0 shown in Figure 4-10) for each subsystem. The estimate turned out to be reasonably good (anything within 20% of the initial estimate is considered good). It was generated by an experienced team working in a familiar application area and was developed after the software algorithm requirements were relatively stable. However, the middleware library (not shown) ended up being about 52,000 SLOC, twice the original estimate. The increased complexity of the middleware was due to the extra coding required to accommodate parallel processing routines and scalability from 10s to 100s of nodes.

TABLE 4-1
PSP Code Estimates

Component	Measured Lines of Code	Estimated Lines of Code	Notes
Task Scheduler (TS)	2.6K	1.5K	
Data Input Subsystem (DIS)	3.2K	1.8K	
Digital Filtering Subsystem (DFS)	8.3K	8.8K	
Adaptive Beamforming (ABF)	9.9K	10.1K	Scope reduced
Detection (DET)	9.0K	10.1K	
System Control & System Loading (SC/SL)	4.3K	4.3K	Already coded
Host (CC/CD)	33.5K	27.9K	Display and control
Total Code	**70.8K***	**64.5K**	

* Does not include OS, Communications, Middleware Library, and Kernel code

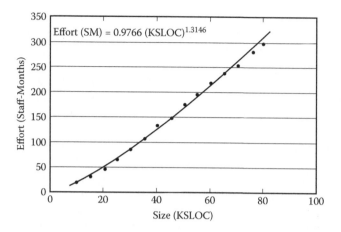

$$\text{Effort (SM)} = 0.9766 \, (\text{KSLOC})^{1.3146}$$

FIGURE 4-13 Software productivity. As the size of the program code grows, the effort (shown above) and development time (not shown) grow nonlinearly. Often, a power-law curve can be used to model the relationship.

There are many pitfalls to using SLOC as a means to estimate cost and schedule. Different software domains, different projects, and different organizations all influence the relationships between SLOC, software cost, and development schedule. Figure 4-13 shows the relationship used in this case study. It was based on metrics from previous projects; historical data predict that as the size of the software product grows, the effort grows with a power-law coefficient. This is intuitively explained by noting that the interactions between software components can grow combinatorially. Adding a line to an existing body of code requires testing the interaction of the new code with preexisting code (referred to as regression testing), as well as testing of the new functionality. The larger the body of previous code, the more likely that the new code will have interactions, and the more extensive the interactions are likely to be. Many other factors besides code size affect the overall cost and schedule. For example, the cost and schedule of the processor in the case study were strongly affected by the delayed availability of the processor, the system deployment schedule, the scale of the processor, the immaturity in the back-end detection and estimation algorithms, and the need to deploy the front-end processor prior to the programmable back-end processor. The series of risk-reduction spirals was also a strong determinant of development plan, cost, and schedule. It turned out that for the case study, the curve shown predicted the cost for unit-tested functionality to within about 10%, but it underestimated the integration cost by nearly 30% and project duration by several months. There was a 50% underestimation in the integration portion of the schedule, due principally to the protracted time spent scaling the processor, an unprecedented task that turned out to be much more difficult than anticipated. Thus, while software complexity estimation is a useful management tool, it must be emphasized that the predictions thus generated serve only as guidelines. Moreover, progress monitoring, configuration control, risk assessment, and project replanning are required throughout the entire development. The whole point of the spiral process is to address the technical, cost, and schedule risks in a way that allows management to iteratively rebalance priorities (cost, schedule, and technical) while meeting evolving and emerging requirements.

Figure 4-14 gives an example of software development monitoring. The progress represents a roll-up of the code integration status of the software planned for a particular spiral build. By monitoring at this level of detail, it is possible to predict the completion date of the integration phase of the spiral. Insight into the impact of external events, such as the introduction of a new operating system, can also be gleaned. Such indicators can aid in risk mitigation by pointing out where additional effort or attention is required.

Tables 4-2 and 4-3 show typical software issue-tracking reports. During each development cycle, it is important to track issues, develop solutions, and then schedule the incorporation of the

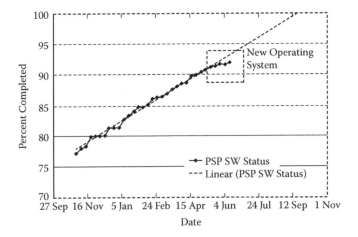

FIGURE 4-14 PSP integration progress. The figure shows the progress of code integration and verification over an eight-month period in the PSP development. The boxed region in the June time frame shows the slow-down in integration when a new operating system was installed in the PSP.

TABLE 4-2
Build 3.0 Issues List

Item	Subsystem	Description	Completed
128	ABS	Optimize ABS tag-files	X
85	ABS	110 msec glitch in ABS	X
123	ABS	alpha back is not being applied correctly	X
122	ABS	Threshold done "prior" to summing but this may be OK	X
121	ABS	Jammer nulling start range in CPI info not being used	X
110	ABS	Clutter training beam (Nc vs. Nctr)	X
120	ABS	Om Jns_selectClutterSet, initialize IC based on Doppler index	X
124	DET	Detecting targets from beam 19 twice	X
87	DET	Clutter editor agreement	X
108	DET	Increase # DET reports (again)	
89	DET	Grouper agreement	X
90	DET	SendToCache optimizations	
116	DFS	Corner Turn Conduit output defer until CPI interval ends	X
115	DFS	Avoid Corner Turn Conduit data copy for 96 Dopplers	X
94	DFS	Fix CPI timer	X
95	DFS	Tune SURV Tag files	
125	DFS	All data products fail to provide actual CPI number	X
127	DFS	There seems to be an extra word (at end of?) each of the messages	X
69	DFS	Integrate DFS data collection	X
114	TS	Debug Cal Data message timing between RCP, TS, DFS	X
64	ABS/DET	Send subset Cholesky data from ABS-> DET (36 × 36 not 36 × 180)	

Key: No shading = algorithm; light shading = performance; medium shading = integration;
X = completed.

TABLE 4-3
Build 3.1 Issues List

Item	Subsystem	Description	Completed
18	ABS	Move tgate_ndx and bgate_ndx to the AbsData structure	
19	ABS	Cns_computeCorrelations() performs a validity check	
13	ABS	Change objects to Basic types B, idxTset, baz, chtaper, spatialTaper	
15	ABS	destMats() method for MobMatrixCplxFltConduit	
106	ABS	Channel masking	X
20	ABS	Cns_apply SHoleWeights() calls SHARC assembly functions	
34	ALL	Error Handling (ability to handle bad CPIs)	X
91	DET	Improve logic for limiting target reports	
86	DET	Send "Tracker Aux Info" as part of DetCommRspMsg	
8	DET	Fix MobTargetReportList::copy Shutdown freeze up – use a non-blocking send	
126	DFS	EquPc raw A-scope should contain 2 adjacent channels but only contains one	
83	DFS	Time reverse and conjugate PC coefs at start-up. Change .I files	
117	DFS	CalSigValid: Real time for the per-channel part	
65	DFS	EQU mode needs to operate in real time (tag files required)	
118	TS/DFS/ABS	Implement Sniff mode	X
97	DIS	Send timing data to Fran's tool	X
100	SYS	Merge hosts	X
102	TS	Send timing data to Fran's tool	X
63	TS/DFS	Transition to SysTest on any CPI	X
130	DET	Implement MLE-based estimation algorithms	

Key: Heavy shading = algorithm; medium shading = performance; light shading = integration; no shading = robustness/maintainability; X = completed.

solutions into the configuration-controlled code. These tables show excerpts of the issues for PSP Builds 3.0 and 3.1, respectively. Both builds occurred after acceptance of the full-scale system, during the first few months of system operation. A configuration control board (CCB) consisting of the major program stakeholders was responsible for assessing, prioritizing, and scheduling work to be performed to resolve issues. The program manager was the CCB chairman; other participants included the integration lead, the verification lead, the PSP development manager, and the application lead. The issue descriptions are actual excerpts from the development and, as such, they are rather cryptic. They are shown here to give the reader an idea of the detailed types of software issues that arise in a complicated system development. The table entries are shaded to indicate issue categories. The main focus for Build 3.0 was algorithm enhancements. During Build 3.1, the focus was on robustness. Not shown in the tables is the supporting material, which included testing requirements, estimated implementation effort and time, and an impact assessment. The issues were signed off by the issue originator, the implementer, and a verification team member. The last column in each table codes (with an X) the issues completed when this snapshot was taken. The full tracking includes a list of the affected software modules (configuration controlled), as well as sign-off dates.

4.3.3 PSP Software Integration, Optimization, and Verification

The integration of the PSP was carried out first in a smaller system integration laboratory using the 1/4-scale PSP and a copy of the DIQ front-end subsystem. Integration continued in a full-scale SIL

TABLE 4-4

Performance on Key Kernels

Kernel	Per SHARC FLOP Count (millions)	Measured Execution Time (ms)	Computational Efficiency (%)
Equalization and Pulse Compression	7.73	110.8	87.2
QR Factorization of Jammer-Nulling Training Matrix	0.71	26.6	33.5
Orthogonalization of Jammer-Nulling Adaptive Weights	0.11	3.4	41.3
Application of Jammer-Nulling Adaptive Weights	4.02	62.7	80.3
QR Factorization of Clutter-Nulling Training Matrix	1.74	25.4	85.9
Application of Clutter-Nulling Adaptive Weights	2.49	42.5	73.2
Whitening of Clutter-Nulled Data	0.80	28.0	35.8

80 MFLOP/s peak SHARC 21060 DSP

using the full-scale PSP and another copy of the DIQ subsystem. Finally, the PSP was moved to the platform and integrated end-to-end with the radar. During platform integration, small-scale SIL was kept operational to handle initial testing of new software. The final codes, once supplied with scaled values for the radar configuration, were transferred to the full-scale SIL. To handle this code scaling, a full set of verification tests with both reduced-scale and full-scale parameters was developed. When a code base passed verification in the small-scale SIL, it was moved to the full-scale lab or (during radar demonstrations) directly to the platform, where the code was tested at full scale using tailored verification tests. Any changes made to the code base during the full-scale testing were sent back to the small-scale SIL, where the code was "patched" and re-verified. In this way, the two code bases were kept synchronized.

Code optimization was an important part of the program. At the outset, computationally intensive kernels were identified. These kernels were coded and hand-optimized during the early spirals. Table 4-4 shows the key kernels and the ultimate efficiencies achieved on the full-scale PSP. These kernels were first coded and tested on the development system. They were ported to the 1/4-scale system, where they were hand-optimized to take advantage of the SHARC instruction set and the SHARC on-chip memory. Parallel versions of the QR factorization routine were developed and the optimum number of processor nodes (i.e., the number of nodes giving the maximum speedup) was determined. During the subsequent builds, performance optimizations continued, with the most intense optimization efforts being carried out during the scaling spirals.

Earlier in this discussion, for the sake of clarity, the scaling spirals were depicted as occurring sequentially. In fact, to help expedite the integration schedule, these spirals were overlapped in time and integrated code was then transitioned to the radar for further integration and verification. The overlap is depicted in Figure 4-15. The first phase of each scaling spiral consisted of the initial scaling step, in which the system was booted, the application was downloaded, and the basic set of start-up tests was run; in the second phase, the scaled system was verified for correct algorithm functionality; during the final phase real-time performance was the major focus. It was during this last phase that extensive code modifications were carried out at several levels. For example, the movement of data and code into and out of internal SHARC memory was reviewed and carefully orchestrated to minimize external memory accesses; source-level codes were optimized by techniques such as code unrolling; data were rearranged for efficient inner loop access; and handcrafted assembly-level vector-vector multiply routines were created. After optimization in the full-scale SIL, the code base was delivered to the platform, where it underwent platform-level integration and verification. The SIL had a comprehensive set of external hardware that allowed the processor to be interfaced with other platform components and then tested.

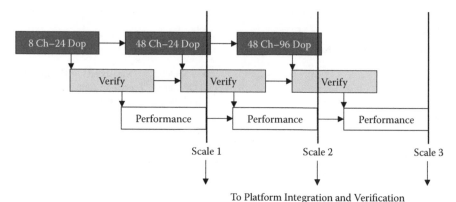

FIGURE 4-15   System integration laboratory scaling phases.

With such a hectic and complex development, tight configuration management was a necessity. The configuration management system tracked code updates and allowed the integration team to develop and manage multiple system configurations. Each subsystem program was tracked through incremental milestones. The major milestones were designated as

(a) *developmental*, at which point the code compiled successfully and was in the process of being tested at the unit level;
(b) *unit-tested*, at which point the code had passed unit-test criteria and was in the process of being integrated with other subsystems;
(c) *integrated*, at which point the code had completed verification; and
(d) *accepted*, at which point the overall subsystem code base had been accepted by an independent application verification team and had been signed off by the program manager and the integration lead.

A version-control system was used to keep track of the different code bases, and regression testing was carried out each day so that problems were detected early and were readily correlated with the particular phase and subphases in progress.

During the optimization phases, the performance of each subsystem was measured on canonical datasets, performance issues were identified, and the code was scrutinized for optimization opportunities. Some of the techniques applied included code unrolling (in which an inner loop is replicated a number of times to reduce the overhead associated with checking loop variables), program cache management (in which, for example, a set of routines was frozen in program cache when it was observed the code would be re-executed in order), and data cache management (in which, for example, data that were used sequentially by a set of routines were kept in cache until the last routine completed, only then allowing the system to copy the results back to main memory). Figure 4-16 shows a snapshot of the optimization performance figures for the 48 Doppler and the 96 Doppler (full-scale) systems, compared to the real-time requirement. At this point in the development, the adaptive beamforming and detection subsystems were still slower than real time, so these two subsystems became the focus of subsequent optimizations.

For example, once the ABS was scaled to the full 96 Doppler configuration (in the final scaling spiral), a disturbing communication bottleneck was identified, as shown in Figure 4-17. The full-scale data cube processed by the ABS consisted of 792 range gates for each of 48 channels for 96 Doppler bins. When the data were transported from the DFS to the ABS, the cube had to be corner-turned so that each ABS node received all of the range gates for all of the channels for a single Doppler bin. This communication step severely loaded the interchassis communication system. At first, the CPI data cubes were streamed through the system at slower than the full real-time rate. Full real-time perfor-

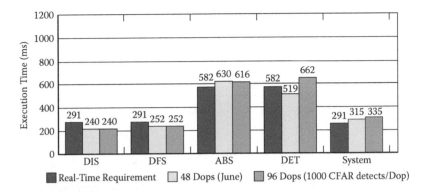

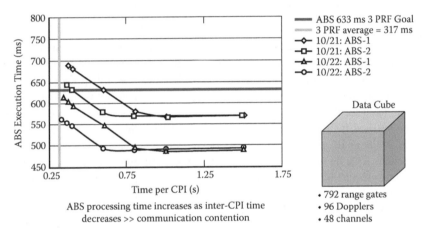

FIGURE 4-16   Performance measurements. Achieving real-time performance in the PSP was a significant challenge. The chart shows one of the many real-time performance assessments that were used during the development to focus optimization efforts. In this example, the DIS and DFS subsystems were measured at better than real-time performance for the reduced-scale (48 Doppler) system and the full-scale (96 Doppler) system. The ABS and the DET were still not at the real-time threshold at this point in the development.

FIGURE 4-17   ABS real-time performance.

mance required the transmission of a data cube every 317 milliseconds (on average). As the inter-CPI time was reduced, a point was reached where the communication system became overloaded and the end-to-end computation time in the ABS began to rise linearly with decreasing inter-CPI time. Thus, although the ABS version that was tested on 10/21 met the real-time goal (shown for the nominal schedule of three CPIs in a repeating sequence) when operated in isolation, when it was operated in the full-scale system with CPIs arriving at the full rate, the execution time climbed to an unacceptable level. Fortunately, the optimization team had been working on a new version of the adaptive weight computation that was significantly faster. The optimized version made more efficient use of internal SHARC memory, thereby saving significant time in off-chip memory accesses. Within one day, the new version was installed and the full-rate CPI case met the end-to-end performance goal with a few milliseconds to spare. The variance in these measurements was determined to be acceptably small by running several 24-hour stress tests. (The stress test also uncovered some memory leaks*, which took time to track down but were fixed prior to the completion of the build.)

---

* A memory leak is a coding bug that causes the software program to overwrite the end of an array or structure, so that, in a sense, the data "leaks" into memory locations that it should not occupy. The leak can be largely innocuous until a critical data or control item is overwritten, and then the result can be catastrophic. Sometimes, hours or even days of execution are required to uncover these sorts of bugs.

TABLE 4-5

PS Integration Test Matrix (16 and 48 Channels; 48 Dopplers)

Test Item	16-Channel Data Cube	48-Channel Data Cube	Stress Test (full size)	Real Time (3 of 6 CPIs)
DIS (Internal test driver)	done	done	done	done
DIS (External interface)	done	done	done	done
TS	done	done	done	done
DFS (Surveillance)	done	done	done	done
DFS (Equalization)	done	done	done	partial
DFS (Diagnostics)	done	done	partial	partial
Jammer Nulling (part of ABF)	done	done	done	done
Clutter Nulling (part of ABF)	done	done	done	done
DET	done	done	done	partial
Overall System	done	done	partial	partial

Key: Heavy shading = partial; light shading = done. 96% completed: 9/17/99.

The verification of the PSP for each build spiral was tracked for each subsystem and for the end-to-end system. Table 4-5 shows a typical snapshot of the top-level verification test matrix for the build that scaled the processor to 48 channels and 48 Dopplers. At this point in the build, the test items had been completed on the canonical set of input data cubes, but the 24-hour stress tests had not been completed for the DFS diagnostic node, and the DFS equalization and diagnostic modes had not been verified for full real-time operation. The DFS diagnostic mode was used to test the performance of the analog components in the radar front-end. The DFS equalization mode computed coefficients for a linear filter that was used to match the end-to-end transfer function of each receiver channel prior to adaptive beamforming. Equalization was needed to achieve the maximum benefit from the jammer-nulling processing that was carried out in the ABF subsystem. The tests verifying these modes were completed in the next month.

The PSP functional verification procedure involved a thorough set of tests that compared the output of the PSP to the output of a MATLAB executable specification. Figure 4-18 shows the general approach. Datasets were read by the MATLAB code and processed through a series of steps. The same datasets were input into the DIS and processed through the PSP subsystems. The results at each processing step for both the MATLAB code and the real-time embedded code were written out to results files. The results were compared and the relative errors between the MATLAB and real-time codes were computed. Note that the relative error at a particular step reflected the accumulated effect of all previous steps. The verification teams analyzed the relative errors to determine if the real-time computation was correct and within tolerances. The MATLAB code executed in double precision, so, in general, the differences were expected to be on the order of the full precision of a single-precision floating-point word. However, since the errors accumulated, it was important to evaluate the signal processing in the context of the overall processing chain to verify the correct functionality and acceptable precision of the end-to-end real-time algorithm. Figure 4-19 shows the plotted contents of an example PSP results file. In this example, the output is the lower triangular matrix (L-matrix) in the LQ decomposition of the training matrix used in the jammer-nulling adaptive-weight computation. The relative difference between this computation and the equivalent computation performed by the MATLAB specification code is shown in Figure 4-20. The errors are on the order of 10^{-6}, which is commensurate with the precision of the single-precision computation. The end-to-end real-time algorithm verification consisted of over 100 such verification tests for several datasets for each radar mode.

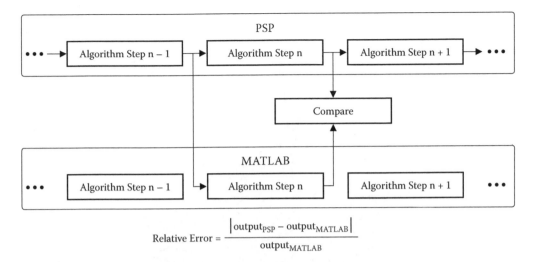

$$\text{Relative Error} = \frac{\left|\text{output}_{\text{PSP}} - \text{output}_{\text{MATLAB}}\right|}{\text{output}_{\text{MATLAB}}}$$

FIGURE 4-18 Verification test procedure. The same dataset is input into the PSP and the MATLAB code. The output of the PSP at each algorithmic step is compared to the equivalent output of the MATLAB. The relative error between the two computations is calculated and evaluated by the verification engineer.

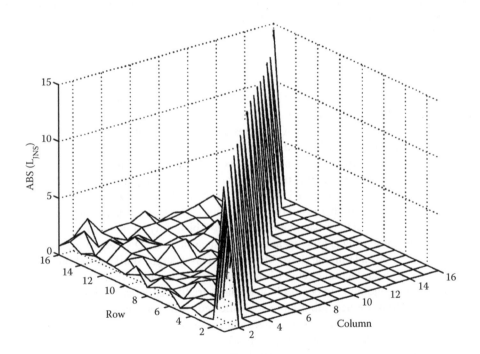

ABS Step 2 Parameters

Parameter	Value
Number of Channels	16
Number of Jammer-Nulling Training Samples	171
Jammer-Nulling Diagonal Loading Level	13 dB

FIGURE 4-19 Example verification computation. The ABS step 2 computation is the calculation of an LQ matrix factorization of a training dataset. Shown here is the magnitude of the entries in the lower diagonal matrix (L-matrix) computed by the PSP code on a controlled test dataset.

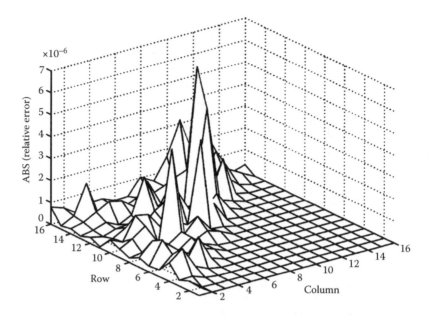

FIGURE 4-20 Example verification result. The L-matrix computed in the PSP step 2 computations is compared point-wise to the equivalent computation in MATLAB code (using the same input dataset). The relative error is plotted here. Note that the errors are on the order of the single-precision arithmetic performed in the PSP.

4.4 TRENDS

HPEC developers today have an ever-increasing repertoire of technology options. Full-custom VLSI ASICs, standard-cell ASICs, FPGAs, DSPs, microcontrollers, and multicore microprocessor units (MPUs) all have roles to play in HPEC systems. Form-factor constraints are becoming increasingly stressful as powerful HPEC systems are being embedded in small platforms such as satellites, unmanned aerial vehicles, and portable communications systems. At the same time, sensing and communication hardware continue to increase in bandwidth, dynamic range, and number of elements. Radars, for example, are beginning to employ active electronic scanning arrays (AESAs) with 1000s of elements. Radar waveforms with instantaneous bandwidths in the 100s of MHz regime are finding use in higher-resolution applications. Designs for analog-to-digital converters that will support over 5 GHz of bandwidth with over 8 bits of dynamic range are on the drawing table. Electro-optical sensing is growing at least as fast, as large CCD video arrays and increasingly capable laser detection and ranging (LADAR) applications emerge. Similar trends are evident in the communication industry. Digital circuitry is replacing analog circuitry in the receiver chain, so that HPEC systems are continuing to encroach on domains traditionally reserved for the analog system designers. Algorithms are becoming more sophisticated, with knowledge-based processing being integrated with more traditional signal, image, and communication processing. Sensors integrated with decision-support and data-fusion applications, using both wireline and wireless communications, are becoming a major research and development area.

As the capability and complexity of HPEC systems continue to grow, development methods and management methods are evolving to keep pace. The complex interplay of form-factor, algorithm, and processor technology choices is motivating the use of more tightly integrated co-design methods. Algorithm-hardware co-design techniques emphasize the rapid navigation of the trade space between algorithm designs and custom (or FPGA) hardware implementations. For example, tools that allow rapid explorations of suitable arithmetic systems (floating point, fixed point, block floating point, etc.) and arithmetic precision and that also predict implementation costs (chip real-estate, power consumption, throughput, etc.) are being developed. Techniques that map from high-level

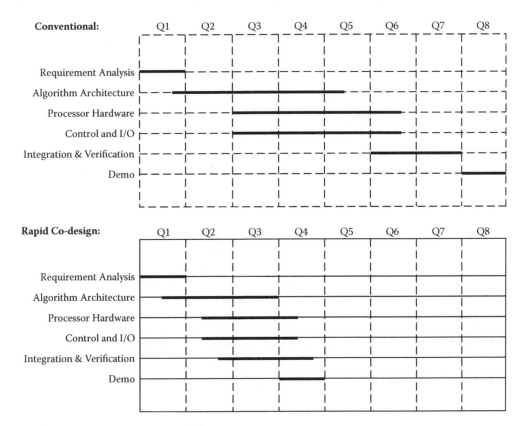

FIGURE 4-21   Rapid co-design of HPEC hardware.

prototyping languages (such as Simulink) to FPGA implementations are emerging. Recent trends are the development of fully integrated hardware design and algorithm exploration development environments. The benefits expected as these tools mature are shown in Figure 4-21 for a hypothetical HPEC system development. As the figure shows, co-design can allow the overlap of the algorithm and architecture design phases. Integration can begin sooner and will take less time since many of the integration issues are addressed earlier in the process.

For hybrid HPEC systems, hardware-software co-design techniques are emerging. These co-design approaches start by developing an executable model of the HPEC system. The mapping or binding of functionality to hardware and software is carried out using the model, permitting the design trade-offs to be explored early and the commitment to particular hardware to be made later in the development cycle. This approach leads to more effective architectures, but just as importantly, the integration of the hardware with the software proceeds more smoothly since the interfaces have already been verified via simulation. If hardware models are available, different technology choices can be explored via simulation to find an optimal assignment. Since much of the design space can be explored more rapidly (when proper models exist), this approach has the potential to dramatically shorten the overall development time through each development spiral.

Another, related, trend is the use of model-driven architectures (MDAs) and graphical modeling languages such as the Universal Modeling Language (UML) to specify software designs. The UML model can be executed to verify correct function; the overall architecture and interaction of the components can be explored; and then the actual code (for example, C++) can be generated and integrated into the HPEC system. Although this approach has not been widely adopted in HPEC designs, partly due to their complexity and the need to produce highly efficient codes, model-driven architecture design is expected to find application as it continues to mature.

Software middleware technologies, pioneered in the late 1990s (for example, STAPL discussed in this chapter), are becoming indispensable infrastructure components for modern HPEC systems. Most processor vendors provide their own middleware, and the standard Vector Signal Image Processing Library (VSIPL) is now widely available. In 2005, a parallel, C++ variant of the VSIPL standard (VSIPL++) was developed and made available to the HPEC community [http://www.vsipl. org]. Middleware libraries that support parallelism are particularly important since programmable HPEC systems are invariably parallel computers. The parallel VSIPL++ library gives HPEC developers a parallel, scalable, and efficient set of signal and image processing objects and functions that support the rapid development of embedded sensors and communication systems. Using libraries of this sort can reduce the size of an application code significantly. For example, a prototype variant of VSIPL++ called the Parallel Vector Library has been used to reduce application code by as much as a factor of three for radar and sonar systems.

Open system architectures (OSAs) are also being developed for HPEC product lines. The advantages of OSAs include ease of technology refresh; interoperable, plug-and-play components; modular reuse; easier insertion of intellectual property (e.g., advanced algorithms and components); and the ability to foster competition. For example, the Radar Open System Architecture (ROSA) was developed circa 2000 at MIT Lincoln Laboratory to provide a reference architecture, modular systems, common hardware, and reusable and configurable real-time software for ground-based radars. ROSA has had a revolutionary impact on the development and maintenance of ground-based radars. For example, the entire radar suite at the Kwajalein Missile Range was upgraded with ROSA technology. Five radars that previously were implemented with custom technology were refurbished using common processing and control hardware. Common hardware, nearly 90% of which was commercial off-the-shelf componentry, and a configurable common software base were used across the five systems. Currently, ROSA is undergoing an upgrade to apply it to phased-array radars, both ground-based and airborne.

Integrated development environments (IDEs) are becoming increasingly powerful aids in developing software. IDEs provide programmers with a suite of interoperable tools, typically including source code editors, compilers, build-automation tools, debuggers, and test automation tools. Newer IDEs integrate with version-control systems and also provide tools to track test coverage. For object-oriented languages, such as C++ and Java, a modern IDE will also include class browser, an object inspector, and a class-hierarchy diagram. At the same time, the environment in which programmers develop their software has evolved from command-line tools, to powerful graphical editors, to IDEs. Modern IDEs include Eclipse, Netbeans, IntelliJ, and Visual Studio. Although these IDEs have been developed for network software development, extensions and plug-ins for embedded software development are becoming available.

In conclusion, Figure 4-22 depicts the future envisioned for HPEC development, shown in the context of a high performance radar signal processor such as the one covered in the case study in this chapter. Future HPEC development is anticipated as a refinement of the classical spiral model to include a much more tightly integrated hardware-software co-design methodology and toolset. A high-level design specification that covers both the hardware and software components will be used. The allocation to hardware or software will be carried out using analysis and simulation tools. Once this allocation has been achieved, a more detailed design will be carried out and code will be generated, either from a modeling language such as UML or a combination of UML and traditional coding practices. The code will use a high-level library such as VSIPL++ and domain-specific reusable components. VSIPL++ components will rely on kernels that have been optimized for the target programmable hardware.

On the hardware side, the high-level design will be used in conjunction with parameter-based hardware module-generators to carry out the detailed design of the hardware. The hardware modules will be chosen from existing intellectual property (IP) cores where possible, and these IP cores will be automatically generated for the parameter range required by the application. Other components will still require the traditional design and generation, but once they have been developed,

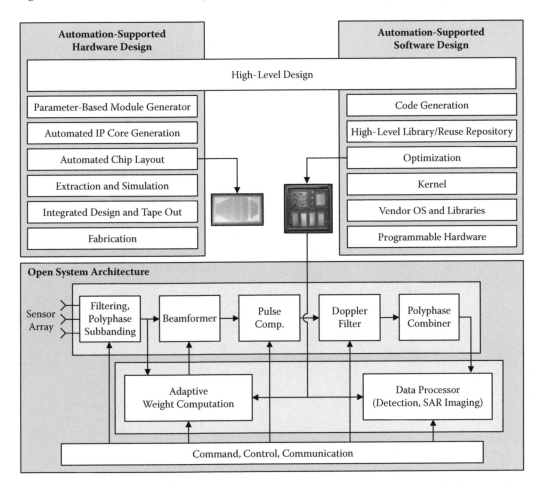

FIGURE 4-22 Vision of future development of embedded architectures. A high-level co-design environment is envisioned that will allow the system design to be allocated between hardware and software in a seamless and reconfigurable manner. Below the high-level design specification are toolsets that help to automate the detailed design, fabrication (or compilation), and testing of the HPEC system. The architecture interfaces are specified in the high-level design to enforce an open-architecture design approach that allows components to be upgraded in a modular, isolated manner that reduces system-level impact.

they can also be included in the overall IP library for future reuse. The next steps in the process will involve greater support for automatic layout of the chips and boards, extraction and detailed simulation, and finally tape out, fabrication, and integration. With a more integrated and automated process and tools as depicted, the spiral development and management process can be applied as before. However, the process will encourage reuse of both hardware and software components, as well as the creation of new reusable components. Domain-specific reuse repositories for both hardware (IP cores) and software (domain-specific libraries) will thereby be developed and optimized, permitting much more rapid system development, and will significantly mitigate cost, schedule, and technical risks in future, challenging HPEC developments.

REFERENCES

Boehm, B.W. 1988. A spiral model of software development and enhancement. *IEEE Computer* 21(5): 61–72.
Ward, J. 1994. *Space-Time Adaptive Processing for Airborne Radar Submitter.* MIT Lincoln Laboratory Technical Report 1015, Revision 1. Lexington, Mass.: MIT Lincoln Laboratory.

Section II

Computational Nature of High Performance Embedded Systems

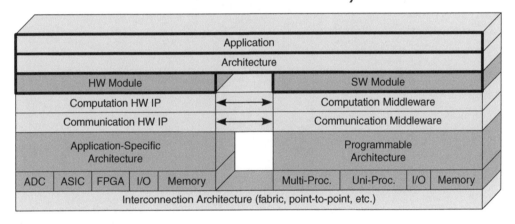

Chapter 5   Computational Characteristics of High Performance Embedded Algorithms and Applications

Masahiro Arakawa and Robert A. Bond, MIT Lincoln Laboratory

This chapter presents HPEC algorithms and applications from a computational perspective, focusing on algorithm structure, computational complexity, and algorithm decomposition approaches. Key signal and image processing kernels are identified and analyzed. Communications requirements, which often prove to be the performance limiters in stream algorithms, are also considered. The chapter ends with a brief discussion of future application and algorithm trends.

Chapter 6   Radar Signal Processing: An Example of High Performance Embedded Computing

Robert A. Bond and Albert I. Reuther, MIT Lincoln Laboratory

This chapter further illustrates the concepts discussed in the previous chapter by presenting a surface moving-target indication (SMTI) surveillance radar application. This example has been drawn from an actual end-to-end system-level design and reveals some of the trade-offs that go into designing an HPEC system. The focus is on the front-end processing, but salient aspects of the back-end tracking system are also discussed.

5 Computational Characteristics of High Performance Embedded Algorithms and Applications

Masahiro Arakawa and
Robert A. Bond, MIT Lincoln Laboratory

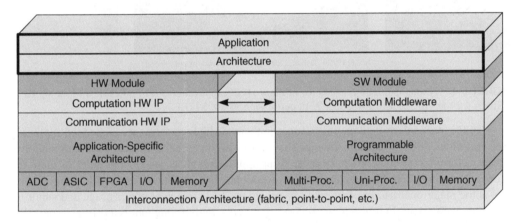

This chapter presents HPEC algorithms and applications from a computational perspective, focusing on algorithm structure, computational complexity, and algorithm decomposition approaches. Key signal and image processing kernels are identified and analyzed. Communications requirements, which often prove to be the performance limiters in stream algorithms, are also considered. The chapter ends with a brief discussion of future application and algorithm trends.

5.1 INTRODUCTION

One of the major goals of high performance embedded computing (HPEC) is to deliver ever greater levels of functionality to embedded signal and image processing (SIP) applications. An appreciation of the computational characteristics of SIP algorithms, therefore, is essential to an understanding of HPEC system design. Modern, high performance SIP algorithms demand throughputs ranging from 100s of millions of operations per second (MOPS) to trillions of OPS (TOPS). Figure 5-1 shows throughput requirements for several recent military embedded applications plotted against calendar year of introduction. Many of these applications are in the prototyping stage, but they clearly show the trend toward TOPS of computational performance that will be needed in fielded applications in the next ten years. Medical, communication, automotive, and avionics applications show similar trends, making HPEC one of the most exciting and challenging fields in engineering today.

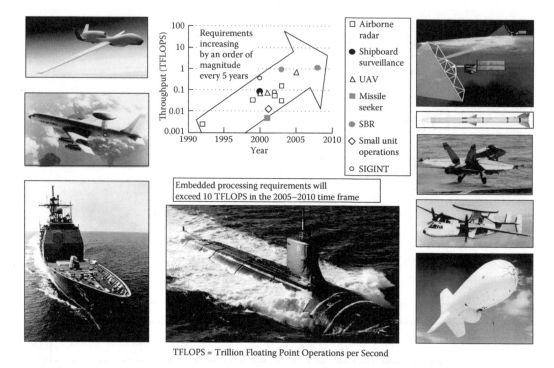

TFLOPS = Trillion Floating Point Operations per Second

FIGURE 5-1 Throughput requirements for military HPEC applications. (From Lebak, J.M. et al., Parallel VSIPL++, *Proc. IEEE* 93(2): 314, 2005. With permission. © 2005 IEEE.)

HPEC is particularly challenging because not only do high performance embedded applications face enormous throughout requirements, they must also meet real-time deadlines and conform to stringent form-factor constraints. In most other application areas, one or the other of these three principal considerations tends to dominate. For example, in scientific supercomputing, time-to-solution is the major concern. In commodity handheld products, efficient power usage and small form factor are paramount. In transaction processing systems, high throughput is a primary goal. HPEC is a juggling act that must deal with all three challenges at once.

HPEC latencies can range from milliseconds for high pulse repetition frequency (PRF) tracking radars and missile infrared (IR) seekers, to a few hundred milliseconds in surveillance radars, to minutes for sonar systems. The best designs will satisfy both latency and throughput requirements while minimizing hardware resources, software complexity, time to market, form factor, and other design goals. HPEC systems must fit into spaces ranging from less than a cubic foot to a few tens of cubic feet, and must operate on power budgets of a few watts to a few kilowatts. With these size and power constraints, achievable computational power efficiency, measured in operations per second per unit power (OPS/watt), and computational density, measured in operations per second per unit volume (OPS/cubic foot), often determine the overall technology choice.

Many HPEC systems require special ruggedization so that they can be embedded in mobile platforms. For example, airborne processors must meet stringent shock, vibration, and temperature robustness specifications and space-based processors have the additional requirement to be radiation tolerant. Cooling can be critical, especially since putting high performance circuitry, which is generally higher power, into constrained spaces produces a lot of heat. Figure 5-2 illustrates the span of various technologies across different regimes of power efficiency and computational density. Applications such as space radars, in which payload volume and power are at a premium, often call for application-specific integrated circuits (ASICs). The less severe constraints of an airborne sensor that may be embedded in an unmanned aerial vehicle, for example, may warrant the use of field programmable gate arrays (FPGAs) or programmable digital signal processors (DSPs).

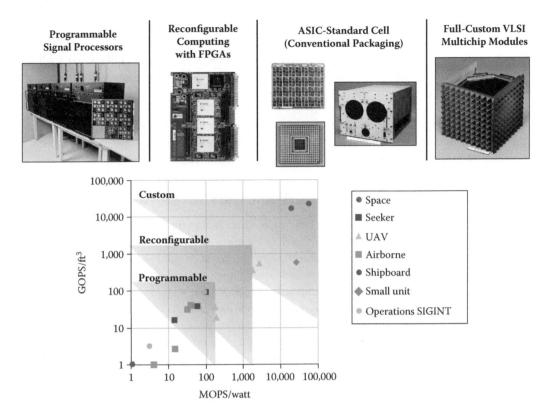

FIGURE 5-2   Computational form-factor requirements for modern HPEC applications.

In general, the preferred technology is the one that can meet form, fit, and function requirements while providing the most cost-effective solution, in which development costs and schedule, as well as production, maintenance, and upgrade costs, must all be factored into the evaluation.

HPEC requires a co-design approach, in which algorithm performance is traded off against processor implementation options, and custom hardware performance is weighed against programmable component flexibility and ease of development. Whatever technology is chosen, though, the computational structure and the complexity of the signal and image processing algorithms figure prominently in the HPEC design task. To effectively design high performance processors, computer system architects require a thorough understanding of the computational aspects of the signal and image processing algorithms involved. The design process requires the architect to first decompose the candidate algorithms into constituent stages, exposing computational parallelism, communication patterns, and key computational kernels. After decomposition, the algorithm components are mapped to processing hardware, which may be a combination of application-specific circuitry and more general-purpose programmable components. Often, the design proceeds iteratively. The algorithm may be modified to reduce computational complexity, to increase parallelism, or to accommodate hardware and software options (e.g., a specialized processing chip or an optimized library routine). Application performance must then be reassessed in light of the modified algorithm, so that a balance between performance and complexity can be reached. Quite often, different but equivalent algorithm variants—for example, time domain versus frequency domain filter implementations—are possible. Each variant affects computational complexity, communication patterns, word length, storage, and control flow; hence, each has an influence on the computer architecture or, conversely, is more or less suited to a particular architecture. As one can glean from this brief discussion, the complicated interaction between algorithm design and processor design is at the heart of HPEC system design.

This chapter discusses the computational characteristics of embedded signal and image processing algorithms, focusing on algorithm structure, computational complexity, and algorithm decomposition approaches. Key signal and image processing kernels are identified and analyzed. Communication requirements, which often prove to be the performance limiters in stream algorithms, are also considered. The chapter ends with a brief discussion of future application and algorithm trends.

In Chapter 6, the companion to this chapter, an example of high performance signal processors for ground moving-target indication (GMTI) is presented to further illustrate various aspects of mapping signal processing algorithms to high performance embedded signal processors.

5.2 GENERAL COMPUTATIONAL CHARACTERISTICS OF HPEC

High performance embedded computing can best be described in the context of the top-level structure of a canonical HPEC application, as shown in Figure 5-3.

HPEC processing divides into two general stages: a front-end signal and image processing stage and a back-end data processing stage. Back-end processing distinguishes itself from front-end computations in several important respects summarized in Table 5-1.

Generally, the goal of the front-end signal and image processing is to extract information from a large volume of input data. Typical functions include the removal of noise and interference from signals and images, detection of targets, and extraction of feature information from signals and images. The goal of the back-end data processing is to further refine the information so that an operator, the system itself, or another system can then act on the information to accomplish a system-level goal. Typical functions include parameter estimation, target tracking, fusion of multiple features into objects, object classification and identification, other knowledge-based processing tasks, display processing, and interfacing with other systems (potentially in a network-centric architecture).

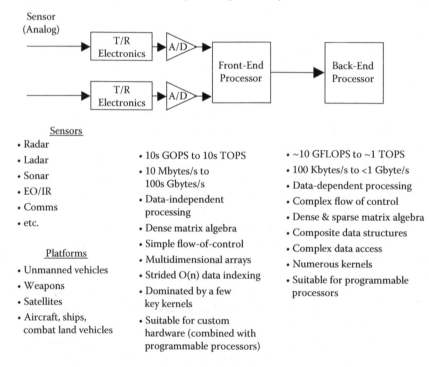

FIGURE 5-3 Canonical front-end and back-end architecture of an HPEC application.

TABLE 5-1
Comparison of Front-End Processing and Back-End Processing

Characteristic	Front-End Signal Processing	Back-End Data Processing
Processing Objective	Removal of noise and interference from signal and images; detection of targets and extraction of features information from signals and images.	Refinement of information and target parameters received from front-end; target tracking; fusion of multiple features into objects; object classification and identification; other knowledge-based processing tasks; display processing; interfacing with other systems.
Throughput and Latency	Short latencies (microseconds to seconds). High throughput (MFLOPS to TFLOPS).	Longer latencies (milliseconds to minutes). Moderate throughput (MFLOPS to GFLOPS).
Computations	(Multidimensional) signal and image processing. Multiply accumulate (MAC) operations dominate. Generally single-precision fixed or floating point, but occasionally double-precision floating point required.	Feature and target parameter estimation; tracking and data fusion. Discrimination, classification, and identification. Planning, scheduling, and threat analysis. Display processing.
Data Types	Signals (multidimensional) and images (hyperspectral). Data are organized into regular structures such as vectors, dense multidimensional arrays (3D data cubes), or tensors.	Composite data-structure (objects) such as track files, models of external world objects (e.g., complex targets with kinematics parameters as well as identification parameters), maps, linked lists (e.g., graph representations), etc. A mix of dense matrices and sparse matrices, with complex links between data structures.
Data Volume	Large volume of data per unit time. Increased sensor capability increasing data volume to gigabytes.	Moderate number of distinct objects (typically several orders of magnitude fewer objects than data elements from associated front-end).
Input	Single-precision fixed-point numbers in continuous or regular streams. Tens to hundreds of channels for radar and sonar; millions of pixels or voxels (volume or three-dimensional pixels) for optical systems. Often, the data are converted to complex format early in the processing stream. Systems have aggregate requirement in the Mbyte/s to Gbyte/s regime. Emerging high-end applications will exceed several Tbytes/s.	Features, detections, state vectors, and images. Usually single-precision fixed or floating point. High-end systems have aggregate bandwidth in the 10s kbyte/s to 10s Mbyte/s ranges.
Flow of Control	Linear flow of control, multiple computational stages, each with a set of nested loops that traverse the multiple dimensions in the datasets. Few data-dependent control paths and mode-dependent but data-independent processing loads.	Complex flow of control. Complex memory access patterns into composite data structures. Nested if-then-else decisions based on knowledge discovery. Search-like operations. Decision trees and traversal of sparse matrix or graph structures. Data-dependent control paths and processing loads.
Data Movement	Data are acted on only a few times and then transformed to a new data type for processing by the next stage in the computation. Corner turns are needed to align data in memory to efficiently traverse each dimension in the (typically) multidimensional dataset.	Data-objects are instantiated and tend to persist. Internal state of objects computed maintained (e.g., track state). Data access patterns are irregular with sometimes unpredictable mix of global and local references. Occasional all-to-all or multicast communications.

Continued

TABLE 5-1 (Continued)

Comparison of Front-End Processing and Back-End Processing

Characteristic	Front-End Signal Processing	Back-End Data Processing
Program Complexity	Typically around 20,000 to 50,000 source lines of code (SLOC), with some applications having as many as 250,000 SLOC. Program complexity[a] is generally low. Control structures are generally simple. Efficient synchronization mechanisms, independent data and instruction accesses, and instruction caches are all important. Many of the key kernels must be implemented in assembly to provide maximum efficiency. Optimized signal and image processing libraries and standards are becoming commonplace. End-to-end efficiency of ~25% typical on modern DSP, with some routines achieving much higher efficiency on specialized processors [e.g., fast Fourier transform (FFT) on SHARC].	Typically greater than 100,000 source lines of code. May be over 1,000,000 SLOC. Program complexity may be high, with dynamic flow of control and numerous conditional statements. Control structures and data structures may also be complex. Efficiency on modern DSPs may range from 2% to 15%, depending on the mix of arithmetic to data access and control code.

[a] Program complexity is not the same as computational complexity. The complexity of a program is a measure of how difficult the program is to create, debug, integrate, and test. Greater program complexity implies some combination of more possible paths through the program execution, larger composite data structures, and more data-dependent control flows. The total number of source lines of code (SLOC) is roughly related to these factors, and therefore provides a crude estimate of complexity. Cyclomatic complexity (McCabe 1976), which measures the number of independent paths through a program, is another way to measure program complexity.

The technology choices for front-end processors run the gamut from full-custom very-large-scale integration (VLSI), standard-cell ASICs, FPGAs, to programmable digital signal processors (DSPs) and microprocessor units (MPUs). Often, the best solutions are hybrid designs that incorporate a combination of these technologies.* The technology chosen for the back-end data processing is often a programmable multicomputer composed of DSPs or MPUs, or a shared memory multiprocessor.

Typically, front-end processing requires significantly greater computational throughput, whereas back-end processing has greater program complexity. As noted, throughput is usually measured in terms of OPS. In the HPEC domain, typical front-end algorithms can require anywhere from a few billion OPS to as many as a few trillion OPS. Back-end algorithms tend to require an order of magnitude or two fewer OPS.

HPEC latency requirements vary greatly from one application to the next. Latency is defined as the time interval from when a data item is input to the system to when a corresponding result is computed and output from the system. Sonar systems may have latencies on the order of seconds or minutes. Surveillance radars typically have latencies on the order of a second or less. Optical tracking devices may require latencies on the order of microseconds or a few milliseconds. The front-end and back-end latency allowances are determined by the overall latency requirement and must be traded off against each other. For example, tracking radars may have to close the tracking loop in a few milliseconds. Therefore, the amount of time the radar has to detect the target, update its track,

* Many times, the front-end signal processor is itself divided into a custom front-end and a programmable back-end. In these cases, the front-end will be performing the simpler though higher throughput operations and the programmable back-end will be performing the more sophisticated signal or image processing such as adaptive weight computations and parameter estimation.

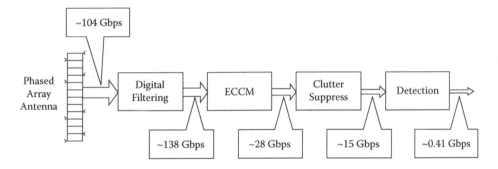

1. The input is from a 48-channel radar phased array sampled at 480 Msps with 12-bit real data per sample.

2. The detector assumes a maximum of 20,000 targets, with each target report being 256 bytes.

FIGURE 5-4   Data rates in a phased-array radar front-end processor.

generate a new position prediction, and then direct its antenna to point in a new direction (to keep the target in the radar beam) must not exceed a few milliseconds. This update rate is driven by the dynamics of the target being tracked. For a slow-moving target, a few hundred milliseconds or even a few seconds may be appropriate. For a highly maneuverable target such as a fighter aircraft, fire-control radars may need to operate with millisecond latencies.

Front-end computations perform operations that transform raw data into information. Computations proceed through a series of calculations carried out on continuous sequences, or streams, of data words that have been digitized and input to the front-end from a set of sensor channels. The analog-to-digital converters (ADCs) are themselves highly sophisticated components that set limits on the precision and bandwidth of the signals that can be digitally processed. Front-end SIP algorithms remove noise and interference, and extract higher level information, such as target detections, target or scene imagery, or communication symbols, from a complex environment being sensed by a multidimensional signal, image, or communication sensor. In phased-array radars, for example, signals with data rates in the 100s of millions of samples per second (MSPS) arrive at the front-end of the digital signal processor from tens of receiver channels. This results in an aggregate sample rate of billions of samples per seconds (GSPS). Data reduction may occur in several front-end stages. For example, Figure 5-4 shows an example of the reductions in data rate at each major stage in the front-end of a phased-array radar application. In this example, the ECCM (electronic countercountermeasures) stage contains a beamforming operation in which the channels are combined to form a small number of beams.* A major reduction occurs at the detection stage, in which a constant-false-alarm-rate algorithm rejects noise and identifies the range gates that contain targets. The number of targets and false alarms is significantly less than the total number of range gates that are tested, so that at the output of the front-end the data rate is typically a small fraction (less than 1/2%) of the input data rate.

Most of the computations performed in the front-end require regular and repetitive data-indexing patterns. The data accesses tend to be data-independent, meaning that the locations of the data in memory can be determined beforehand and do not depend on the values of the data. This is extremely important for custom hardware implementations that rely on regular, predictable memory accesses to optimize control logic and data flow. The operation mix applied at the front-end strongly favors arithmetic operations, such as multiplication, addition, and division, with a relative dearth of logic operations (e.g., Boolean operations, comparison operations) and data manipulations (e.g.,

---

* In some cases, the total number of beams may equal the number of channels, in which case there is no effective data reduction.

bit swaps and word swaps, complex indexing modes). Most high performance front-end algorithms consist of kernel operations from matrix or vector mathematics, with some special computational constructs, such as butterflies [q.v] that support fast Fourier transform (FFT) computations. Typical examples of front-end kernels are given in Table 5-2, along with their salient computation elements and example applications.

Most of the kernels identified in Table 5-2 require nested loops in which vector and matrix operations dominate. Occasionally, square roots, trigonometric operations, or exponential operations are needed. Although the examples given in the table are among the most common and the most computationally intensive, they are merely a small subset of the rich collection of kernels that are used in high performance signal and image processing algorithm. The reader is referred to the MATLAB language and the C/C++ language Vector, Signal, and Image Processing Library (VSIPL) [http://www.vsipl.org/] for more comprehensive treatments. *Matrix Computations* by Golub and Van Loan (1996) is the definitive text on the matrix mathematics that serve as the underpinning for today's high performance SIP algorithms.

Data precision and dynamic range issues are very much a concern in front-end processing, where analog signals are detected by sensors and then converted to digital values. The ADC must be chosen to match or exceed the precision, dynamic range, and processing rate (throughput) requirements of the application. By *dynamic range* we mean the difference between the largest and the smallest data values that can be represented. By *precision* we mean the value of the least significant bit. As the data are transformed by FFTs, finite impulse response (FIR) or infinite impulse response (IIR) filters, inner products, and other operations at successive stages in the front-end algorithm, the dynamic range of the data products increases. For example, in a beamforming operation, the dynamic range increases proportionally to the number of channels being combined. A 64-channel beamformer requires 64 times more dynamic range than that of the input channels; this translates into six more bits per data word. Operations in the power domain, in which the data are squared, require twice the number of bits. Operations on these longer words take more time and/or more transistors than their shorter counterparts. Thus, HPEC processors will use various arithmetic systems designed to give sufficient precision and dynamic range while optimizing throughput and minimizing other parameters, such as hardware utilization, power consumption, size, and weight. At the input to the front-end, ADCs generate fixed-point numbers. Immediately following the digitization phase, simple filtering operations such as baseband-quadrature sampling and low-pass filtering are common. Signal precision fixed-point arithmetic usually suffices for these computations and lends itself to efficient implementations using FPGAs, custom hardware, or programmable DSPs. Floating-point hardware may be needed for downstream operations with larger dynamic range requirements, such as QR decomposition, which is often used in radar and sonar adaptive beamforming. Block floating point, in which an exponent is shared for a block of words, is often employed as an intermediate between fixed and float representations, especially in VLSI implementations. In an FPGA implementation, either fixed or floating point may be employed, although the fixed-point variants are easier to implement and use less hardware real estate (fewer transistors and interconnects). If the precision of the arithmetic is chosen to closely match the application requirement, implementation efficiency can be improved significantly (Leeser, Conti, and Wang 2004). Signal processing applications in which the data are combined coherently, such as radar or sonar, require complex arithmetic. Some processors will perform better if a vector is represented as a list of real components followed by a list of imaginary components. For these processors an extra data reorganization step is needed to split apart a set of {real, imaginary} complex number pairs into two separate lists. This adds inefficiency in moving between scalar and vector quantities, and so the performance must be quantified in order to make the best trade-off. In all cases, the choice of data representation, both in dynamic range and precision, is intrinsically linked to the performance requirements of the application. Thus, a thorough analysis of the computational finite precision effects on system performance is an important component of any front-end system design.

TABLE 5-2

Common HPEC Front-End Computation Kernels

Front-End Kernels	Computation Elements	Applications
Time domain filters: FIR and IIR filters	$a = a + bc$, also called a multiply-and-accumulate or MAC operation. For efficiency, processors load registers b and c, perform the multiplication, add the product to the contents of register a, and store (accumulate) the result back in a. Complex filters require complex MAC operations (CMACs).	Time domain convolution, correlation, time domain frequency filtering (high, low, bandpass).
Fourier transforms: typically implemented with FFT and IFFT	The FFT and inverse FFT (IFFT) use a sequence of butterfly computations to efficiently compute the discrete Fourier transform (DFT).	Frequency domain filters and spectral analysis. For example, radar Doppler filtering can be accomplished by applying the FFT across a sequence of pulses. In optics, diffraction patterns can be computed using FFTs.
Vector and matrix multiplications and inner products; dense matrices are most common	One of the most common basic vector operations is called the *axpy*, which is $\mathbf{y} = \alpha\mathbf{x} + \mathbf{y}$, in which $\mathbf{y}$ and $\mathbf{x}$ are vectors and α is a scalar. Note that the dot product $c = \mathbf{y}^H\mathbf{x}$ is also the basis for matrix multiplication and can be accomplished with (C)MAC operations. Vector machines can load $\mathbf{y}$ and $\mathbf{x}$ into vector registers, each with a single instruction. The point-wise multiplication of two vectors or matrices is also common.	Numerous signal and image processing stages; in particular, beamforming for radar and sonar.
Matrix QR or LQ decomposition	The QR decomposition requires a succession of matrix-matrix multiplications, which factor a matrix A into the product of a unitary matrix, Q, and an upper triangular matrix, R. The LQ decomposition is similar except the matrix is factored into Q and a lower triangular matrix, L. The Householder and Givens algorithms compute reflections and rotation operation matrices, respectively. To compute these matrices, inner products and divisions are needed. The Givens rotation requires a square root computation (some variants exist that avoid this). The matrix A is multiplied by a succession of these matrices to zero the elements below the diagonal and thereby generate R. The product sequence of the operator matrices (reflections or rotations) forms Q.	The QR and LQ decompositions are basic constituents of many adaptive weight calculations, in which they are used to factor a sample matrix as part of the Weiner-Hopf optimization equation (Martin 1969).

Continued

TABLE 5-2 (Continued)

Common HPEC Front-End Computation Kernels

Front-End Kernels	Computation Elements	Applications
Matrix singular value decomposition (SVD)	A common variant of the SVD algorithm (used when the number of rows exceeds the number of columns) first performs a QR decomposition to reduce the matrix to a bidiagonal form and then iteratively solves for the singular values in the bidiagonal matrix. See Feng and Liu (1998) for a discussion of parallel variants of the SVD algorithm.	Solving systems of linear equations, performing spectral analysis, computing statistical correlations.

While regular memory accesses, dense-matrix mathematics, and simple program flow control dominate front-end processing, back-end processing has elements of more general-purpose computing. The front-end processor reduces the input signals or images to a set of higher-level objects such as detections or features; these objects are passed to the back-end processor for further processing. Thus, whereas front-end processing is mostly stream-based, back-end processing predominantly manipulates higher-level, composite objects. These more complex objects and the programs that use them distinguish back-end processing from front-end computations.

Some back-end tasks are similar to front-end tasks in their computational structure. For example, the back-end processing may include a refinement of the target or feature parameters that are computed in the front-end. Parameter estimation algorithms, such as maximum-likelihood estimation, employ vector or matrix mathematics over a grid of estimation points. Correlations, convolutions, threshold computations, normalizations and other scaling operations, distance computations, and rotation operations are common back-end operations that support such back-end tasks as automatic target recognition, discrimination, and object identification. All of these operations use dense matrix kernels similar to those used in the front-end. (Dense matrices are matrices in which most of the entries are nonzero.) Back-end tracking techniques, such as Kalman filtering and its variants, utilize state-space representations that lend themselves to dense matrix mathematics. Although many back-end computations depend on matrix kernels, these kernels are applied to many fewer (albeit more complicated) objects than in the case of front-end processing. Thus, the computational load in the back-end is not usually dominated by dense matrix mathematics. Instead, computations are characterized by a large percentage of logic and data manipulation (versus arithmetic) operations, highly irregular memory accesses, and complex program flow-of-control. Examples of tasks dominated by these computational elements include scheduling, planning, user interfaces, networked communications, databases, reasoning, and decision support tasks.

Many of these back-end tasks employ graph-based algorithms, such as Bayesian belief networks, Markov decision processes, associative databases, and decision-tree techniques. Graph-based algorithms can also be expressed as sparse-matrix algorithms. (Sparse matrices are matrices in which most of the entries are zeros.) Thus, one of the distinctions between front-end and back-end processing is that front-end processing is characterized by dense matrix tasks and back-end processing (at least in principle) by sparse matrix tasks.

Front-end stream data items have relatively simple internal structure. Typically, streams are simple sequences of real or complex integer or floating-point numbers. Often, these data are aggregated into vectors or arrays. Vectors and arrays have simple indexing computations: a vector reference $v(i)$ can be computed by simply multiplying the basic element size (for example, four bytes for a long integer) by $(i - 1)$ and adding the result to the initial location (assuming sequential storage with indexing that starts at 1). Frequently, vector or array computations involve accessing each adjacent element or every *n*th element in turn; this sort of indexing is readily accommodated by dedicated instructions and special hardware. Thus, index calculations can often be computed in a single instruction cycle. Furthermore, many DSP architectures have dedicated hardware for loop control, so that the regular access patterns for array, vector, and matrix operations can be programmed with fewer instructions and can be accomplished in fewer clock cycles. Often, the same operation is applied to each vector element (for example, when two vectors are added together or a vector is multiplied by a scalar). Special vector registers and vector arithmetic

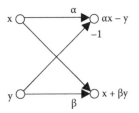

FIGURE 5-5 A variant of the basic butterfly operation used in the FFT algorithm. Implementation of the butterfly in hardware can significantly accelerate FFT computations. For example, the SHARC DSP has a special butterfly instruction that allows the FFT to be computed at an effective rate that is 50% greater than the nominal multiply-accumulate rate of the processor.

units are often used to speed up these operations with very little indexing overhead. For example, the PowerPC Altivec and the Pentium SSE instruction sets support these sorts of operations directly in hardware. Some matrix operations can require more complex indexing, but even so, the computations are regular and readily computed. Some DSPs have instructions especially designed for signal processing. For example, the FFT requires a butterfly operation, as shown in Figure 5-5, which consists of a special sequence of multiplies, sign changes, and additions. Analog Devices' ADSP2105 supports the butterfly operation as a special instruction implemented directly in hardware.

Back-end data objects typically contain complex, composite data structures with numerous fields. The fields themselves may be vectors, matrices, other composite data items, or pointers to any of these structures. For example, a simple target track, as shown in Figure 5-6, may consist of

```
struct track_history
{
char status[];              //string containing status
double x_pos;               //x coordinate of estimated position of track
double y_pos;               //y coordinate of estimated position of track
double x_vel;               //estimated x direction velocity of track
double y_vel;               //estimated y direction velocity of track
double snr;                 //snr from last associated target
double time;                //time from last target
struct VelWindow vw[5];     //represents different possible actual doppler
                            // values, used in resolving doppler ambiguity
struct CVHFilter cvhf;      //Abstraction from original
                            // system, may be removed.
struct target_report Msave[5];
                            //saves old targets for
                            // New/Novice tracks until
                            // they become established
char Hypothesis[];          //string containing
                            // hypothesis (enum should
                            // be used in compiled
                            // implementation)
double Q[4][4];             //process noise covariance matrix
double Pm[4][4];            //initial extrapolation covariance matrix
double Pp[4][4];            //updated extrapolation covariance matrix
double K1[4][2];            //Kalman gain stage 1 matrix
double K2[4][1];            //Kalman gain stage 2 matrix
}
```

FIGURE 5-6 Track and track file data structures.

dozens of entries. Moreover, in more complex structures, the entries may be pointers to other entries or to other objects outside of the data item. For example, a track-file object may link each target and track estimate and may also link the time-sequence of targets. High-level languages provide powerful referencing semantics and elegant syntax for iterating through both simple stream data items as well as complex data objects. However, if the references are dynamic, meaning that they may depend on data values or program state generated at runtime, the compiler will not be able to precompute a constant offset into the data object, and complex assembly-level code will be required. Thus, object accesses in back-end processing may be significantly more time-consuming than data accesses in front-end processing.

The caching characteristics of front-end and back-end codes also differ. Back-end data objects tend to be persistent, meaning that they last for many program iterations and have internal state that evolves over time. A track, for example, can exist for seconds or hours. Hence, a back-end program that operates on a set of tracks can benefit greatly from data caching. This benefit may be diminished, however, if the number of tracks that can be cached is less than the number of tracks in the application.

The data in front-end streams, on the other hand, contain transient data, meaning that they are accessed only a few times before being discarded; typically, the stream data are transformed to new data items, and the original data are no longer needed. This access pattern means that data caching often does not work well for streams; each data item is only reused a few times and, hence, there is little advantage in storing it in cache memory. For more complex operations such as matrix multiplication, however, there is modest data reuse since elements in the operand matrices are accessed multiple times. For in-place operations, in which the results are stored in the input data object, caching helps since the input locations are also accessed as output locations. Examples of such operations include in-place FFT, in which the original vector is replaced by the transformed vector, and QR decomposition in which the original matrix is replaced by the R matrix.

Dataset size influences performance for both front-end and back-end computations. For example, Figure 5-7 shows the throughput performance of a time-domain (convolution) implementation of a FIR filter as a function of input vector length. Notice how the computational throughput levels off once the data needed for the computation can no longer fit into the Motorola PowerPC G4 microprocessor level 1 cache, and then drops precipitously as level 2 cache capacity is exceeded.

Instruction caching can be very effective whenever a section of code is small enough and is frequently executed. Front-end codes tend to have nested for loops, in which the inner loops are small

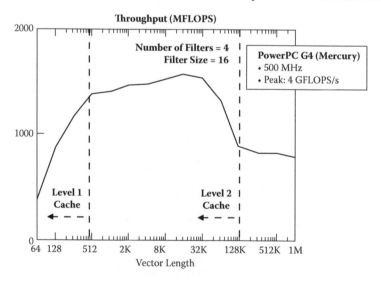

FIGURE 5-7 Performance of PowerPC Microprocessor Unit (MPU) on the convolution (time domain FIR filter). Notice how the performance is affected by the dataset size relative to the cache size.

and executed numerous times as they range over a set of data. Examples include FIR filters, inner products, and matrix multiplications. Thus, instruction caches can be very effective for these codes. Moreover, these inner loops often represent a substantial amount of the overall processing in the front-end algorithm, so that the end-to-end algorithm execution time can be dramatically improved by instruction caching. Although back-end processing may also have small, nested for loops, the complexity of these loops (amount of time spent in the loops compared to the rest of the program) is smaller than in front-end systems, and, hence, the caching advantage is less.

Back-end processing control flow is often influenced by the values of the data on which the computations act. This is referred to as *data-dependent* computation. For the most part, front-end processing control flow is *data-independent* and is only sensitive to exception conditions, such as word overflows and divide-by-zeros. In the front-end, the same operations are applied again and again on continual streams of data input to the processor. The data values are largely irrelevant to the processing applied to them. More complex data dependences can result in complicated programs, with many different execution paths. For example, in a target-identification algorithm, once a hostile target is identified, an entirely different set of actions is appropriate compared to the friendly target case. As another example, user interfaces must respond to user input and are, therefore, fundamentally data-dependent, the *data* in this case being the commands and queries from the users.

Data-dependent processing is generally ill-suited to custom ASIC hardware implementations, and so custom ASIC back-end processors are rare. Programmable DSPs or MPUs are the preferred solutions. The trade-off is between custom hardware for performance and programmable hardware for flexibility. FPGAs are emerging as a middle-ground technology solution. They can be programmed, but not as readily as a DSP or MPU, and they can provide high performance, but not as high as custom circuits. FPGAs perform well on regular operations that have data parallelism and simple control flow. They are also viable for simpler data-dependent operations such as searching and sorting. FPGAs have also been used for more complicated processing. For example, Linderman, Linderman, and Lin (2004) describe an FPGA implementation of a finite state machine that parses metadata for an embedded publish-and-subscribe communication system. Until very recently, FPGAs have not been flexible enough to handle general back-end processing. This limitation is being addressed in emerging FPGA offerings, however, that include both embedded soft-core and hard-core MPUs. A soft-core MPU implementation is one in which a portion of the FPGA is programmed to emulate a microprocessor core. A hard-core MPU implementation is one in which the actual MPU hardware core is embedded in the FPGA. Embedding MPU capabilities into FPGA chips make them, in effect, hybrid computing devices that may find use in both front-end and back-end processors.

Not only is back-end flow of control data-dependent, but the back-end workload is also data-dependent in two ways. First and most obvious, since the flow of control is data-dependent, and there is no guarantee that each control path has the same amount of computation, the workload naturally varies as the program execution unfolds. However, even when the flow of control remains the same, the back-end workload typically varies nondeterministically because the data are generated (in the front-end) by statistical processes that depend on the external environment. In the front-end, the workload is more deterministic up until the point of detection processing because it depends mainly on preset sensor parameters such as sample rate, dynamic range, sensor elements, etc. For example, an FFT performed on an input vector has the same computational complexity irrespective of the values contained in the vector; similarly, a FIR filter, a QR decomposition, or a matrix-vector product applied to front-end data streams does an amount of work that only depends on the size of the data streams and not on values contained in the data.* The workload in the back-end, on the

---

* Note, for a multimodal front-end system, such as a radar that performs both GMTI and SAR processing or one whose operational parameters (e.g., PRF, bandwidth, CPI) can change, the front-end workload will, indeed, vary as a function of the mode or parameter settings. However, the workload for any mode or set of parameters is still largely independent of the data.

other hand, depends on the number of data objects input to it from the front-end. This number is generally not deterministic and varies over time, causing the workload in the back-end to vary. For example, in a sonar or radar, the back-end may perform tracking operations on targets detected in the front-end. The workload will depend on the number of targets detected and being tracked, which in turn depends on the number of targets in the environment, the operating characteristics of the radar receiver, the front-end signal processing algorithms, and the environmental clutter or interference. Typically, the back-end processor needs to be sized for the worst case that needs to be supported, meaning that much of the time the processor may be underutilized.

Metrics to quantify program complexity can be a controversial subject, but it is safe to say that front-end codes are generally simpler than back-end codes. One way to measure code complexity is to simply count the number of (non-blank, non-comment) source lines of code (SLOC) in a program. Although this approach is open to much criticism, the size of the program measured in SLOC is a proven predictor of both the effort required to develop the program and the memory size of the executable program. Front-end programs for large HPEC applications may be as small as a few hundred SLOC and may occupy only a few Kbytes of core memory. Some front-end HPEC programs, however, may be over 100,000 SLOC and will occupy a few Mbytes of memory. Back-end programs are even larger. They may be more than 500,000 lines of code and may occupy several Mbytes of memory.

For front-end applications, stringent form-factor constraints and high throughout requirements will require highly efficient and small-sized codes. Typically, the code will need to be profiled for execution time and memory usage, certain sections or routines may need to be hand-optimized by the programmer to take advantage of specific processor features, and high-level language routines may need to be selectively abandoned in favor of more efficient assembly language codes. In the case in which very limited memory is available, the programmer will have to simultaneously optimize the program for both code size and code performance. As time (computational latency) and space (computer memory) resources become scarcer, a point is reached at which programming effort starts to grow exponentially. Thus, a line of code in the front-end may require substantially more effort than a line of code in the back-end. In many applications, the HPEC processor is required to provide spare processing capacity and memory in order to mitigate the development effort and to allow room for future modifications.

Another way to measure code complexity is to look at the internal structure of the code. For example, cyclomatic complexity measures the number of linearly independent paths through a program (McCabe 1976). Modules with high cyclomatic complexity are difficult to code and maintain. Function points are another way to characterize complexity (Matson, Barrett, and Mellichamp 1994). Whether based on cyclomatic complexity, lines of code, or function points, back-end processing programs tend to be larger, with more internal complexity, than front-end codes. Trackers, graphical user interfaces, decision support systems, schedulers, real-time control programs, and other back-end codes are considerably more complex than beamformers, filters, detectors, and other front-end computations. The amount of effort spent optimizing each line of code in a back-end algorithm may be less than what is needed in the front-end, but the size of the back-end program and its code complexity make up for this. Thus, it cannot be said that one domain is easier to handle than the other; they are just different.

Input, output, and data reorganization for front-end algorithms differ strikingly from those of back-end algorithms. In the front-end, the input data rate depends on the bandwidth of the signals and the number of channels. Data rates may range from a few Ksps or less for sonars to 100s of Gsps for wideband multichannel radars and charge-coupled device (CCD)-array optical systems. The back-end, on the other hand, receives a relatively small set of detections or image features. Referring back to Figure 5-4, one sees that the overall data rate in a typical radar is significantly reduced (in this case, by a factor of 254) from the front-end processor input to its output. Of course, the output serves as the input into the back-end processing. Similarly, in image processing systems, the front-end will either extract features or compress the images to reduce the data rate.

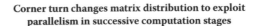

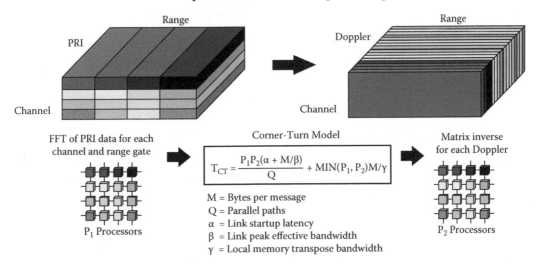

M = Bytes per message
Q = Parallel paths
α = Link startup latency
β = Link peak effective bandwidth
γ = Local memory transpose bandwidth

Many-to-many communication where each of P_1 processors
sends a message of size M to each of P_2 processors.
Total data cube size is $P_1 P_2 M$.

FIGURE 5-8   The figure shows a corner turn of a three-dimensional radar data cube. In this example, a Doppler filtering operation is first carried out across the pulse repetition intervals (PRIs, each of which is on a separate processor) converting the third dimensional of the cube to Doppler filter bins. The data are then corner-turned so that all of the data for a single Doppler bin are gathered on one node. At this point, beamforming, which combines the channels for all range gates, can be accomplished in each Doppler bin.

Back-end processing may exhibit irregular and complicated data-access patterns, but in one regard front-end processors must deal with an even more demanding data-access task. Many high performance front-end algorithms operate on multidimensional datasets. Phased-array radars and sonars, CCD optical sensors, two- and three-dimensional ladars, and hyperspectral imagers are a few examples. Invariably, the algorithms that process the sensor data need to operate across each dimension in the dataset. For computational efficiency, the data are organized in memory so that adjacent samples in the dataset along the dimension being sampled are in contiguous memory locations. This simplifies index calculations and provides good data locality. When the algorithm proceeds to the next dimension, however, it must first reorganize the data so the new dimension is now in contiguous memory locations. This data reorganization is referred to as a corner turn.

In a single processor, the two-dimensional corner-turn operation can be viewed as a transpose of the data in memory. The data need to be explicitly reorganized, or else they need to be accessed in transpose order; the choice depends on which approach is most efficient for a particular processor. Some DSPs have indexing instructions that permit efficient striding through memory, so the explicit reorganization is not needed. For data cubes and higher-dimensional datasets, stride permutations or sequences of two-dimensional corner turns can be used to reorganize or access the data. In a parallel processor, the data may be spread over a grid of processors. In this case, the corner turn requires a many-to-many communication pattern between processing nodes, as shown in Figure 5-8.

For higher-dimensional datasets, multiple corner turns may be needed. In phased-array radars, for example, returns from each channel (or from each RF subarray when analog beamforming is used) are received, digitally sampled in time, and input to the front-end processor. Sampling at regular time intervals is equivalent to sampling at regular range intervals, so that the data are naturally ordered as an array of range gates by receiver channel. In moving-target indication (MTI) or pulse-Doppler radar, multiple transmit waveforms are transmitted at a regular interval referred to as the pulse repetition interval (PRI). The same waveform is transmitted for multiple intervals, so that the returns across these intervals can be coherently combined. The total interval is referred to as the coherent processing interval (CPI). The pulses in the CPI allow the radar to use the Doppler shift principle to detect targets that have moved from one interval to the next. Thus, each sample is multidimensional, representing a return for a particular pulse, channel, and range gate. To process this "data cube," an algorithm must perform two corner turns: once to arrange the data along the pulse dimension to perform Doppler processing and once to arrange the data along the channel dimension to perform beamforming. Figure 5-8 shows the corner-turn operation that reorganizes a radar dataset after Doppler filtering, so that the data can be efficiently accessed to support a beamforming operation. In the figure, the data are spread over 16 processors. The samples have already been organized so that each processor contains all of the PRIs for a limited number of channels and ranges. Each processor performs Doppler filtering across the entire PRI set, for its subset of channel and range data. Then, the Doppler-filtered data are redistributed in an all-to-all communication step, so that each processor receives all the range gates for all of the channels for a subset of the Dopplers. At this point, each processor is able to beamform the data for the Dopplers assigned to it. Corner turns are particularly stressful for the communication system since all processors must simultaneously partake in multiple global communications. Unless there are dedicated point-to-point links, there will be contention for the communication hardware.

5.3 COMPLEXITY OF HPEC ALGORITHMS

To build an HPEC system with the right amount of computing power, one must first determine the computational workload that needs to be supported. For signal and image processing algorithms, this requires a derivation of computation time complexity in terms of arithmetic operations, parameterized by dataset dimensions. Certain data communication kernels, which do not do arithmetic operations but instead move data around, are also important HPEC kernels; those are dealt with later in this chapter. In the literature today, one can readily find the complexity expressions for most of the purely real SIP kernels, for example, in Golub and Van Loan (1996). However, expressions for complex-data kernels, which are important components of many SIP algorithms, are not that common. Table 5-3 gives computational complexity expressions for several kernels for both real and complex inputs. Although it is beyond the scope of this chapter to cover all or even most of the kernels used in SIP applications, we show how to derive the complexity of a few important ones. For a more comprehensive treatment, the reader is referred to Arakawa (2003).

The elements of the product matrix $C \in R^{mxn}$ of real matrices $A \in R^{mxp}$ and $B \in R^{pxn}$ are given by

$$C = AB \Rightarrow c_{ij} = \sum_{k=1}^{p} a_{ik}b_{kj} \ .$$

The summation over p is carried out for each of the mn elements in C. We perform p multiplications and p additions, giving an operations count of $2mnp$ for the matrix multiplication of real matrices. The derivation for complex matrices is likewise quite straightforward. The difference, of course, is that complex additions and multiplications are required. For each complex addition, the real parts must be added together and the complex parts added together, requiring, therefore, two addition operations. The complex product of two complex scalars x = a + jb and y = c + jd (in which j is $\sqrt{-1}$) is z = (ac – bd) + j(ad + cb). From this, it is apparent that a complex product requires six operations: four multiplications and two additions. For each of the mn elements in matrix C,

TABLE 5-3

Computational Complexity for Common Signal Processing Kernels

Signal Processing Kernel	Computational Complexity	
	Real Input	Complex Input
Matrix-matrix multiplication	$2mnp$	$8mnp$
Fast Fourier transform	$\frac{5}{2}n\log_2 n$	$5n\log_2 n$
Householder QR decomposition	$2n^2\left(m-\frac{n}{3}\right)$	$8n^2\left(m-\frac{n}{3}\right)$
Forward or back substitution	n^2	$4n^2$
Eigenvalue decomposition: eigenvalues only	$\frac{4}{3}n^3$	$\frac{16}{3}n^3$
Eigenvalue decomposition: eigenvalues and eigenvectors	$9n^3$	$23n^3$
Singular-value decomposition: singular values only	$4mn^2-\frac{4}{3}n^3$	$16mn^2-\frac{16}{3}n^3$
Singular-value decomposition: singular values and left singular vectors	$4m^2n+12mn^2$	$16m^2n+24mn^2$
Singular-value decomposition: singular values and right singular vectors	$4mn^2+12n^3$	$16mn^2+24n^3$
Singular-value decomposition: singular values, left and right singular vectors	$4m^2n+12mn^2+13n^3$	$16m^2n+24mn^2+29n^3$

In this table, the complexity is in terms of real multiplications and additions. For the matrix multiplication, the matrices are of dimensions $m \times n$ and $n \times p$. For the FFT, the vector size is n. The triangular system for forward and back substitutions is size n. The matrix operated on by the decomposition kernels is $m \times n$.

a total of eight p operations (four additions and four multiplications for each sum over p) must be performed, leading to a total operation count for complex matrix multiplication of $8mnp$.

The Fourier transform, one of the most important and fundamental operations in signal processing, is a more complicated example. The Fourier transform takes a signal defined in the time domain and converts it into a frequency domain signal. (The dual operation, which takes a frequency domain signal and transforms it to the time domain, is referred to as the inverse Fourier transform; it has the same computational structure as the Fourier transform and differs only in the conjugation of the coefficients and a scaling constant.) To compute the Fourier transform of an arbitrary signal using a digital computer, the signal must be sampled at discrete intervals to create a sequence of digitized samples. The discrete Fourier transform (DFT) is then applied to the sequence. The DFT is found at the heart of many HPEC applications, where it is used to analyze the frequency content of signals or images and to perform filtering operations such as circular convolution. Excellent discussions of the DFT appear in several textbooks, for example, Van Loan (1992). The direct expression for the computation of the DFT is

$$X[k] = \sum_{n=0}^{N-1} x[n]W_N^{kn}, k = 0, 1, ..., N-1,$$

in which $W_N^{kn} = e^{-j\frac{2\pi kn}{N}}$ is called the *twiddle factor*. The set of twiddle factors can be computed ahead of time if the size of the transform is known. However, as written, the transform requires N^2 complex

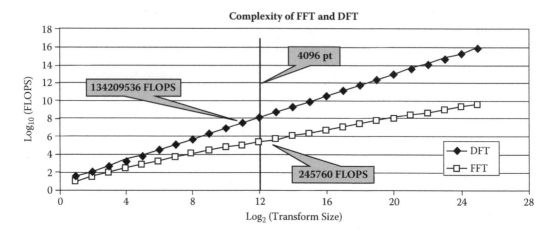

FIGURE 5-9 Complexity of FFT and DFT.

multiplications and $N(N-1)$ complex additions. Recall that a complex multiplication requires six real operations and a complex addition requires two real operations. Thus, the DFT requires $8N^2 - 2N$ operations. The FFT, on the other hand, can compute the DFT in $O(n\log_2 n)$ operations. As shown below, a radix-2 complex FFT can be computed in $5N\log_2(N)$ operations. For large values of N, this provides a substantial savings in operations and, hence, computation time. There are several FFT variants, but all exploit symmetries in the DFT equation to remove redundant operations. To illustrate the magnitude of the savings, Figure 5-9 compares the operation count for the DFT and the FTT for a range of transform sizes. The savings become very large very quickly. Note that, for example, a 4096-point complex DFT requires nearly 550 times more computation than does the FFT.

Below, the complexity of the basic decimation-in-time complex FFT is derived. The derivation of the complexity of a real FFT is a bit trickier and is given in Arakawa (2003). The decimation-in-time algorithm FFT decomposes the DFT into successively smaller DFTs. Consider the case of a transform size N that is a power of two. The DFT expression can be written by factoring it into a summation of the odd-numbered points and a summation of the even-numbered points. If $n = 2r$ is substituted for the even sum and $n = 2r + 1$ for the odd sum,

$$X[k] = \sum_{r=0}^{(N/2)-1} x[2r]W_N^{2rk} + \sum_{r=0}^{(N/2)-1} x[2r+1]W_N^{(2r+1)k}.$$

W_N^k can be factored out of the odd summation half to get

$$X[k] = \sum_{r=0}^{(N/2)-1} x[2r]W_N^{2rk} + W_N^k \sum_{r=0}^{(N/2)-1} x[2r+1]\left(W_N^2\right)^{rk}.$$

Since $W_N^2 = e^{-2j(2\pi/N)} = e^{-j2\pi/(N/2)} = W_{N/2}$, the above equation can be rewritten as

$$X[k] = \sum_{r=0}^{(N/2)-1} x[2r]W_{N/2}^{rk} + W_N^k \sum_{r=0}^{(N/2)-1} x[2r+1]W_{N/2}^{rk}.$$

Notice that the two summations on the right-hand side of this equation are DFTs of length $N/2$. Thus, a DFT of length N has been factored into two DFTs of length $N/2$, with a constant multiplier in front of the second DFT. Since the original DFT length was a power of two, these two DFTs have

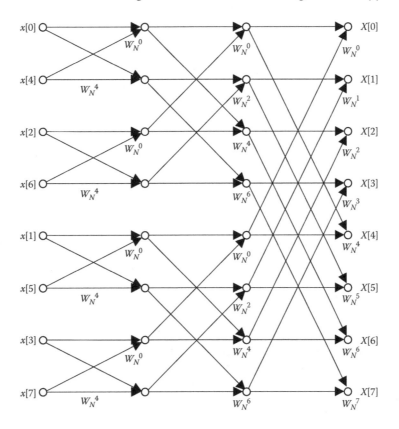

FIGURE 5-10   Data flow for an 8-point complex FFT.

lengths that are powers of two. Hence, the same factorization can be applied to each of these smaller DFTs, and so on, until the original DFT has been factored into a series of $v = \log_2(N)$ DFTs, each of length 2. This factorization can be represented using flow-graph notation, as shown in Figure 5-10 for an 8-point FFT.

In this figure, the branches (arrows) entering a node (circle) denote the addition of the quantities from which the branches originated, and a coefficient next to the head of a branch denotes a scaling of the quantity by the coefficient. If no coefficient is indicated, the scaling factor is assumed to be unity by default. The flow graph in the figure is composed of a basic computational unit which (because of the way it looks) is called a butterfly. Figure 5-11 shows the basic butterfly operation followed by two simplifications that can be used for FFT processing. The simplifications are accomplished by noting that

$$W_N^{N/2} = e^{-2j(2\pi/N)(N/2)} = e^{-j\pi} = -1 \text{ so that } W_N^{r+N/2} = W_N^{N/2}W_N^r = -W_N^r.$$

These specialized butterflies can be used to redraw the 8-point FFT as shown in Figure 5-12. At this point, the overall computational complexity of the FFT can be readily determined. Each butterfly requires one complex multiplication, one complex addition, and one complex subtraction, for a total of $6 + 2 + 2 = 10$ real FLOPs. There are $v = \log_2(N)$ butterfly computation stages, and each stage consists of $N/2$ butterflies. Hence, for a complex, decimation-in-time FFT, the complexity formula arrived at is

$$C_{FFT} = 10 * N / 2 * \log_2\left(N\right) \text{ FLOPs} = 5N \log_2\left(N\right) \text{ FLOPs}.$$

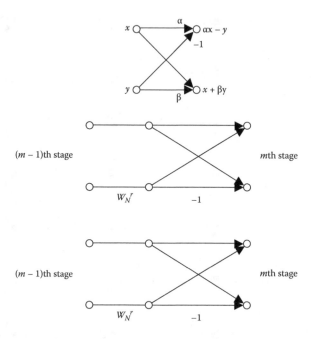

FIGURE 5-11   The basic butterfly operation can be specialized in the FFT by exploiting mathematical properties of the FFT twiddle factors. These specializations reduce the number of multiplications needed in an FFT.

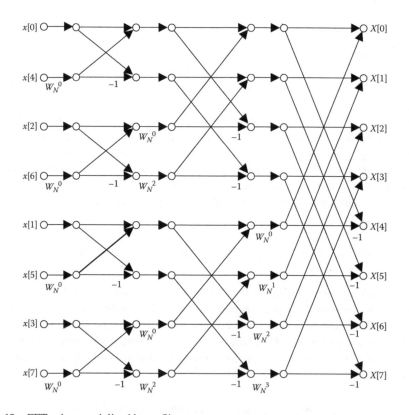

FIGURE 5-12   FFT using specialized butterflies.

The FFT has been studied extensively and several variants exist. The above analysis was performed by successively dividing the FFT in half to ultimately express the DFT in terms of 2-point DFTs. This is referred to as a radix-2 FFT. It is also possible to factor the DFT by larger radix DFTs. For example, radix-4 FFTs are commonly used in optimized DSP implementation. Split-radix FFTs, in which parts of the algorithm are carried out at different radices, also exist. The reader can find many excellent texts that deal with FFT techniques, such as Nussbaumer (1982) and Van Loan (1992).

The QR decomposition (or equivalently the LQ) is used extensively in adaptive signal and image processing as part of the weight computation algorithm. The complexity derivations for the complex and real cases are similar. The complex case is examined below. A complex-data matrix $A \in C^{mxn}$ can be factored into an unitary matrix $Q \in C^{mxn}$ and an upper triangular matrix $R \in C^{nxn}$ so that $A = QR$. The QR decomposition can be accomplished using three algorithms: Modified Gram-Schmidt, Givens, and Householder. The complexity of the Householder algorithm is derived, but all of the algorithms are based on the idea of iteratively constructing the upper triangular matrix R by applying a series of unitary transformation matrices to A:

$$P_n * ... * P_1 * A = R .$$

The product of two unitary matrices is unitary, and so is the Hermitian transpose of a unitary matrix. Thus, setting $P_n * ... * P_1 = Q^H$, the above equation becomes

$$Q^H A = R .$$

Premultiplying both sides of this equation by Q,

$$QQ^H A = Q\ R .$$

The definition of a unitary gives us

$$QQ^H = Q^H Q = I .$$

So we get

$$QQ^H A = IA = A ,$$

and, hence,

$$A = Q\ R .$$

The major difference between the different QR factorization algorithms is the choice of the matrices P_i. The Householder algorithm uses a series of reflection matrices; the Givens algorithm uses a series of rotation matrices. The MGS algorithm constructs projection matrices. Reflections, rotations, and projections are unitary operations. The discussion continues with the Householder algorithm. The Householder reflection matrix is defined as

$$H = I - \beta v v^H.$$

The ith reflection vector v is chosen to zero out the elements in the ith column that are below the ith row. In the above, $\beta = -2 / (v^H v)$ is a normalization scalar. Each Householder reflection H is applied to the submatrix, as shown in Figure 5-13, that has elements starting at the ith column

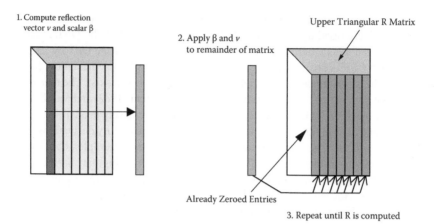

FIGURE 5-13   Householder QR decomposition.

and ith row through to the end of the matrix (Mth row and Nth column). Figure 5-14 shows the basic Householder algorithm. The algorithm does not explicitly compute H; it is more efficient to use v directly, as follows:

$$w = \beta A^H v$$

$$\left(I - \beta vv^H\right)A = A - \beta vv^H A$$

$$= A - v\left(\beta v^H A\right)$$

$$= A - vw^H.$$

For the ith iteration, computing $w = \beta A^H v$ requires $2(m-i) + 8(m-i)(n-i)$ FLOPS. The $A - vw^H$ computation requires $6(m-i)(n-i)$ FLOPS for the outer product and $2(m-i)(n-i)$ FLOPS for the subtraction. Thus, ignoring the second-order term, the ith iteration requires $16(m-i)(n-i)$ FLOPS. For the entire matrix, then, the complexity for the Householder QR decomposition (ignoring lower-order terms) is the sum over the n columns:

$$C_{QR(complex)}^{House} = \sum_{i=1}^{n} 16(m-i)(n-i)$$

$$= 16 \sum_{i=1}^{n} \left(mn - (m+n) - i^2\right)$$

$$= 16mn^2 - 16(m+n)\sum_{i-1}^{n} i + 16 \sum_{i-1}^{n} i^2$$

$$= 16mn^2 - 16(m+n)\frac{n(n+1)}{2} + 16\frac{n(n+1)(2n+1)}{2}$$

$$= 16mn^2 - 16(m+n)\frac{n(n+1)}{2} + 16\frac{n(n+1)(2n+1)}{6}$$

$$= 16mn^2 - 8(m+n)(n^2+n) + \frac{8}{3}(2n^3 + 3n^2 + n).$$

```
[num_rows, num_cols] = size(A);
for col = 1:num_cols
  % Compute the Householder vector v
  [v, beta] = house(A(col:num_rows, col));

  % Apply the Householder vector to the remainder of the matrix
  w = beta + v' + A(col:num_rows, col:num_cols);
  A(col:num_rows, col:num_cols)= A(col:num_rows, col:num_cols) - v + w;

  % Zero out the remainder of the column.
  A((col + 1):num_rows, col) = 0;
end % for col
```
Householder QR Decomposition

```
if (isreal(x))
  n = length(x);
  sigma = x(2:n)' * x(2:n);
  v = [1; x(2:n)];
  if sigma == 0
    beta = 0;
  else
    mu = sqrt(x(1)^2 + sigma);
    if x(1) <= 0
      v(1) = x(1) - mu;
    else
      v(1) = -sigma / (x(1) + mu);
    end % if
    beta = 2 * v(1) ^ 2 / (sigma + (v(1) ^ 2));
    v = v / v (1);
  end % if
else
  v = x;
  nx = norm(x);
  v(1) = x(1) + nx;
  beta = 1 / (nx * (nx + x(1)));
end % if
```
Householder QR Reflection Vector Computation

FIGURE 5-14   Pseudocode for Householder QR reflection vector computation.

Keeping only the third-order terms, the complexity of the complex Householder QR is

$$C_{QR(complex)}^{House} \approx 8mn^2 - \frac{8}{3}n^3.$$

The difference between the complexity of the real and complex QR works out to be a factor of four fewer operations [see Arakawa (2003) for details].

$$C_{QR(real)}^{House} \approx 2mn^2 - \frac{2}{3}n^3.$$

In analyzing the complexity of an algorithm, it is important to consider the complexity of the data movement. Multidimensional datasets generally require corner-turn operations so that computations can in turn access each dimension from sequential memory. In a parallel processor, this will require a many-to-many communication between sets of processors. If the same set is being used to process along each dimension, then the communication is all-to-all, as shown in Figure 5-8. The complexity of the communication depends both on the size of the dataset and the partitioning of the dataset (i.e., on the size of the processor array). It is easy to estimate the communication complexity of the corner turn of a dataset of size S from P_1 processors to P_2 processors. Each source processor must send a message (if we use a distributed memory system) to each destination processor. In all,

$N_{msg} = P_1 * P_2$ messages are needed. The size of the message is the amount of data needed (plus the message overhead, O_{msg}, in bytes) by the destination. Each source will send $1/P_2$ fraction of its data to each source processor. Each source processor has S/P_1 data in all. Thus, a message is of size $M = S / P_1 * P_2 + O_{msg}$. The complexity in terms of bytes communicated is simply

$$C_{ct}^{P1xP2} = P_1 * P_2 * M \ \ bytes = S + P_1 * P_2 * O_{msg} \ \ bytes .$$

If a destination processor is also a source processor, it does not need to send a message to itself, so in this case there are fewer messages. However, instead of sending a message, each of these processors must corner-turn (transpose) its own data in memory, so the overall number of bytes accessed remains the same, except that some of the bytes are accessed for local movement and others for global communication. In HPEC systems with high bandwidth data streams or large volumes of data, the amount of time spent doing corner turns can be significant. Often, processor designs are employed that can "hide" the communication latency by using pipelining techniques, in which data are transmitted, and while they are in flight the source processors are able to continue with other computations. The communication time (while the data are "in flight") affects end-to-end latency, but the throughput of the processor is increased since it can do work during this time. Often, the performance of the communication network (and, hence, its complexity) is increased to provide higher link bandwidth and more parallel paths between processors. Many multicomputer designs, such as those from Mercury Computer Systems, have focused on providing high performance communication systems to enable HPEC applications.

5.4 PARALLELISM IN HPEC ALGORITHMS AND ARCHITECTURES

High performance embedded computing algorithms contain computations that can take place concurrently. Computer architectures attempt to exploit this concurrency with hardware execution units that operate in parallel. In fact, virtually all HPEC implementations employ parallel processing in order to meet throughput and latency requirements. Parallelism is exploited at all levels of granularity, from bit-level parallelism within basic operations to entire program replication, as shown in Figure 5-15. Custom VLSI processors embed fine-grained parallelism directly into their computational structures. Standard-cell ASICs, structured ASICs, and FPGAs likewise explicitly capture the algorithm decomposition and parallelism. The basic internal building blocks become progressively larger (coarser grained), and internal communication paths become more general-purpose as the technology moves from standard cells to structured ASICs to FPGAs all the way through to programmable DSPs and MPUs. The basic trade-off is the high degree of computational efficiency and power density that custom implementations can deliver, weighed against the ease of design and the flexibility that programmable solutions afford. Standard-cell, structured ASIC, and FPGA technologies occupy the regime between the fully programmable and the fully custom technologies. In all cases, a detailed understanding of algorithm structure and parallelism is required to develop effective implementations. In high performance applications, multiple chips must be employed, and hybrid solutions are developed that exploit the advantages of each of these technologies.

There are two fundamental types of parallelism: *task* or *functional parallelism*, and *data parallelism*. In data parallelism, a data object is decomposed into subobjects, each of which is similarly operated on by a computation. For example, matrices can be partitioned into submatrices. Division of a matrix by a constant and point-wise multiplication of two partitioned matrices are examples of data parallelism, as shown in Figure 5-16.

It is easy to see that, with data decomposition, work can be partitioned between the multiple processors in a parallel computer. The same principles hold for an FPGA or VLSI chip in which the data can be distributed over the chip and independent sets of gates can be dedicated to each data item. These operations are often referred to as embarrassingly parallel when there is no need to

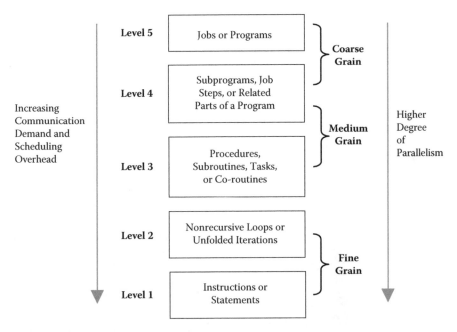

FIGURE 5-15   Levels of parallelism. (From Hwang, K., *Advanced Computer Architecture: Parallelism, Scalability, Programmability*, © 1993 McGraw-Hill. With permission.)

An algorithm contains data parallelism if the same operation
can be applied concurrently over a set of data objects

• For example: the matrix operation A = B + C;

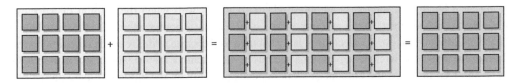

• The degree of parallelism (DoP) is $3 \times 4 = 12$

• In this example, the operation can be applied
  independently to the individual data elements
  Embarrassingly parallel !

• Other partitionings possible, as shown below

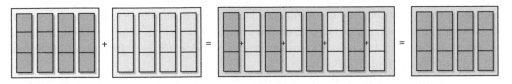

• DoP for column vector partitioning is $1 \times 4 = 4$

FIGURE 5-16   "Embarrassingly parallel" data parallelism.

Parallel, matrix multiplication is an example of a data parallel algorithm that requires *synchronization* and *communication* between the parallel units.

The partitioning of the algorithm and mapping of the data attempt to balance processor load with communication overhead.

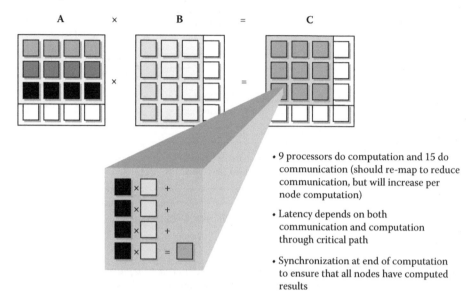

- 9 processors do computation and 15 do communication (should re-map to reduce communication, but will increase per node computation)
- Latency depends on both communication and computation through critical path
- Synchronization at end of computation to ensure that all nodes have computed results

FIGURE 5-17 Data parallelism in matrix multiplication.

communicate data between the parallel partitions and synchronization is only needed to signal the completion of the composite operation. Matrix-matrix multiplication, as shown in Figure 5-17, is an example of a more complex data-parallel operation.

There are a few standard ways to partition arrays and distribute them over compute nodes to accommodate data-parallel algorithms. Figure 5-18 shows the most common partitioning strategies. The most basic approach is to partition the data (assumed in the figure to be a two-dimensional array, but the same concepts apply to vectors and higher-dimensional arrays) into blocks and to assign each block to a separate compute node. This works well when the amount of work to be performed is proportional to the amount of data on the node. Another approach is to distribute the data cyclically, so that every nth column is distributed to the same node. A cyclic partitioning more evenly distributes the work for operations such as parallel matrix factorizations (for example, QR, SVD, and LU) in which the amount of work depends not only on the amount of data on the node, but also on the location of the data within the matrix. Block and cyclic distributions can be seen as special cases of the general block-cyclic arrangement of data. In the general block-cyclic distribution, blocks of size m are distributed cyclically so that every nth block is assigned to the same node. The choice of block size and the number of processors to use depend on both load balancing and the communication overhead of the mapping. Sometimes it is important to share data "at the edges" of each partition, in which case a block (or block-cyclic) partitioning with data overlap is used. Often it is important to redistribute a dataset from one block-cyclic organization to another (a corner turn can be seen as a special case of this). To do this efficiently at runtime requires a complex calculation that is supported by a technique known as Processor Indexed Tagged Family of Line Segments (PITFALLS) (Chung, Hsu, and Bai 1998).

As the overall computation proceeds through each stage, it may be necessary to redistribute the data to achieve the best parallelism for the next computation. Important data reorganizations in HPEC parallel processors (apart from the corner turn discussed already) are shown in Figure 5-19. In a broadcast operation, the same data are sent to all processors. In a gather operation, data from a

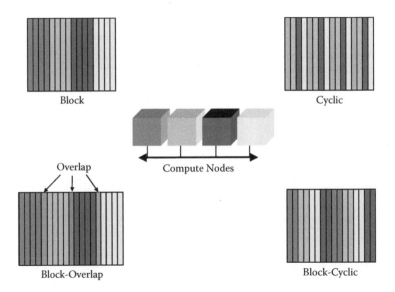

FIGURE 5-18   Common partitioning strategies.

Communication Routines

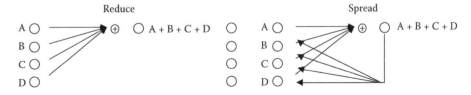

FIGURE 5-19   Important data reorganizations in HPEC parallel processors.

set of processors are gathered onto a node. In a scatter operation, a set of data are distributed onto a set of processors. Computations may be carried out in concert with communication. The general reduction operation takes a set of data, combines the data according to a defined operation (for example, addition), and stores the result on a processor. The general spread (or expansion) operation performs the reduction, takes the result of the reduction, and sends it to a set of processors.

Functional or task parallelism, on the other hand, looks for independence between tasks so that these tasks can be executed concurrently. Figure 5-20 gives an example of task parallelism in an airborne phased-array radar jammer-nulling system (JNS). We will not go into the computational details of each step since the purpose here is to illustrate the parallelism of the major steps, but a brief description of the system follows. The nulling technique has two major phases, a setup phase and a beamforming phase. Within the setup phase, the JNS performs three computations: (a) it computes the position of a clutter ridge, where strong radar returns from clutter are expected; (b) it receives the sensor data and reorganizes it in beam-parallel order; and (c) it receives pointing directions specify-

An algorithm contains task parallelism if different operations
can be applied concurrently over a set of data objects

• An example: A Jammer Nulling Subsystem

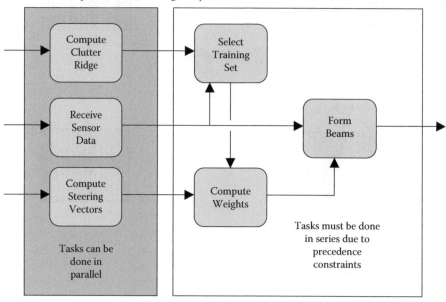

FIGURE 5-20   Example task parallelism in airborne phased-array jammer-nulling system.

ing the surveillance directions for the radar. Each of these computations is independent of the others and, hence, can be performed in parallel. In the beamforming stage, the JNS performs three more computations: (a) it uses the clutter ridge computation to direct its selection of training signals which *are not* due to the clutter ridge (thereby avoiding clutter signals, which are caused by the radar itself and not due to jammers); (b) it computes a set of adaptive weights that is constrained to provide high gain in the steering vector directions and trained using the training weights; and (c) it computes a set of jammer-nulled beams by combining the sensor channel data using the adaptive weights. In this phase, the tasks are not task parallel: the beamforming task must await the computation of the adaptive weights, and the adaptive weight computation must await the selection of the dataset.

There are two other forms of parallelism, referred to as round-robin and pipeline scheduling, which are often employed in front-end processors. Both of these techniques exploit the repetitive nature of the incoming data streams, and each trades computational latency for throughput. Suppose, for example, that a series of datasets are input to the JNS set-up phase in the front-end processor. In a round-robin, as shown in Figure 5-21, each successive dataset is dealt out to a free processor (or set of processors). As such, round-robins can be viewed as data-parallel constructs. As long as there are no dependencies from one dataset to the next, the processors can work concurrently on their respective datasets. If the round-robin is going to keep up with the input data rate, there must always be an idle processor prepared to accept a newly arrived dataset. The trade-off, of course, is that the total time to produce a result is longer than the interval between dataset arrivals (if one set of processors could keep up, there would be no need to round-robin). If N, the number of processor sets in the round-robin, is balanced with the input rate of the datasets, then the latency equals the dataset interval multiplied by N. In the JNS example, therefore, the latency for the results from processing the first data set is $N*T_D$, in which T_D is the interarrival time of sensor datasets (in seconds). After the first computation emerges from the round-robin, subsequent solutions are generated every T_D seconds.

Pipelining, as shown in Figure 5-22, is another form of parallelism. It is particularly suited to front-end processing. Often, a front-end algorithm can be divided into a set of processing stages

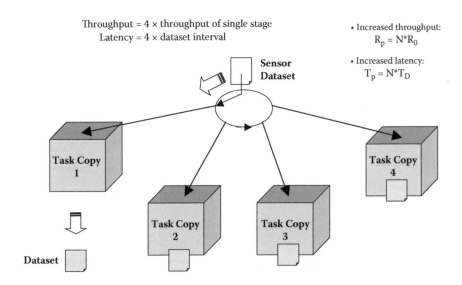

Throughput = 4 × throughput of single stage
Latency = 4 × dataset interval

• Increased throughput:
$$R_p = N*R_0$$

• Increased latency:
$$T_p = N*T_D$$

• Once pipeline is full, a processed dataset is produced each dataset interval

• Each stage must be prepared to accept data at 1/N rate of input, where N is the depth of the round-robin

FIGURE 5-21 Example of round-robin parallelism.

• Throughput of pipeline = 4 × throughput of single stage
  Latency of pipeline = 4 × latency of single stage

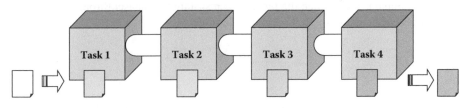

• A processed dataset is produced each dataset input interval (once pipeline is full)

• Increased throughput:
$$R_p = SUM(R_j), j = 1, ...N$$

• Increased latency:
$$T_p = SUM(T_j), j = 1, ...N$$

Independent datasets can be processed concurrently through pipelining

FIGURE 5-22 Example of pipeline parallelism.

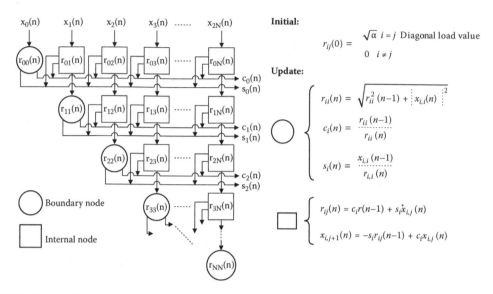

FIGURE 5-23 Givens QR decomposition on a systolic array with node indexing showing the order of processing.

in which the output of one stage is the input to the next and so on. In this case, there are definite dependencies between tasks since each stage typically depends on results computed in the previous stages, but once again the multiple independent datasets can be exploited, although each dataset is a transformation of the previous data. Since each pipeline stage performs a distinct task, pipelining can be viewed as a special type of task parallelism.

Let's suppose a series of radar datasets is arriving at the input to the front-end processor. Each dataset comprises the data for a CPI, which is the basic processing interval for an MTI radar. In the case of an N-stage pipeline, as shown in Figure 5-22, the first dataset enters the first processing stage. Now, if the amount of work allocated to the first stage is completed before the next CPI arrives, and the results are forwarded to the second processing stage, then the first stage is now free to handle the next dataset. As long as each stage can finish its work within the allotted time, which for the radar example is a CPI, then the pipeline can sustain the dataset throughput. The results of the first fully processed dataset will be available in time:

$$T_{pipeline} = \sum_{all\ tasks} t_{task} \ .$$

If each task is precisely balanced to the CPI, so that $t_{task} = CPI$ for all tasks, then $T_{pipeline} = N * CPI$. Results from subsequent datasets arrive every CPI thereafter.

Systolic processing is another important form of parallelism. It is akin to pipelining in that the data are moved from one processing unit to the next, but it also exploits both the distribution of data across an array of processors as well as the efficiency of neighbor-to-neighbor communications. Figure 5-23 gives an example of a high performance systolic algorithm. The algorithm carries out the Givens variant of the A = QR matrix decomposition.

Figure 5-24 shows the details of the mapping of the Givens QR decomposition for an FPGA implementation (Nguyen et al. 2005). The data for matrix A are clocked into the systolic array, one row at a time, from the top of the triangular array. As shown in the figure, the Givens rotations are computed on the array diagonal (shown as circles) and applied (in the nodes drawn as boxes) to the matrix values as they are clocked synchronously down the triangle rows. The two extra columns (shown as hexagons and diamonds) on the right are for computing adaptive weight vectors through

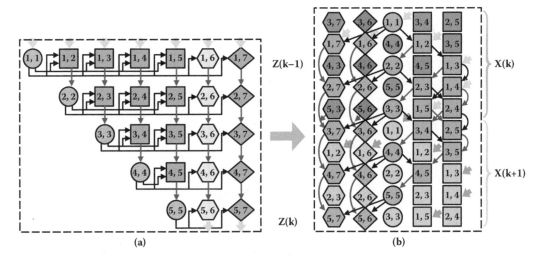

FIGURE 5-24 **(Color figure follows page 278.)** FPGA mapping for a systolic Givens QR decomposition: (a) the logical data flow and operations in the systolic array; and (b) the folded array, in which the execution of the green datagram is overlapped with the brown from the previous time frame, and orange from the next time frame.

two back-substitution steps. The FPGA design uses a mapping proposed by Walke (2002) that folds the array such that two decompositions are overlapped at one time, allowing the idle processors from the first decomposition to operate on the second decomposition while the first is still being processed. In this way, FPGA processing units [referred to as configurable logic blocks (CLBs)] that perform the decomposition are load-balanced and remain fully utilized for the entire computation.

For software programmable parallel systems such as multicomputers, *parallel speedup* and *parallel efficiency* are two important metrics for evaluating parallel algorithm implementations. Speedup for a parallel algorithm distributed over N processors is

$$S_N = T_S \, / \, T_P \, ,$$

in which T_S is the execution time of the serial algorithm and T_P is the execution time of the parallel algorithm. Ideally, the speedup achieved is equal to the number of processors. Parallel efficiency for an algorithm distributed over N processors is defined as

$$E_N = S_N \, / \, N \, .$$

The ideal efficiency is 100% and the ideal speedup is N. Actual speedup is usually less since the parallel components need to spend time to communicate data and synchronize between themselves. This time is overhead that cannot be used to do computation.* As the algorithm is distributed over more and more nodes, the overhead may exceed the benefit of using the additional computing power and the speedup may begin to decrease. Figure 5-25 shows this effect for a parallel implementation of the Householder QR decomposition algorithm. In this example, the speedup peaks at around 8 to 12 nodes.

For the case in which the computation can be divided into components that require the same amount of computation and do not need to communicate with each other, the speedup can be linear

---

* Often, it is possible to overlap the data communication with computation. For example, if a stream of data is being processed, it may be possible to compute a result for a portion of the stream, transmit the result to the next computation, and while the communication is taking place, start on the next computation. This technique reduces the communication overhead but usually does not eliminate it altogether.

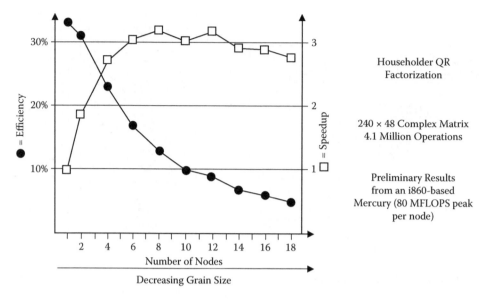

- Efficiency decreases with number of nodes

- Ultimate limits to speedup

FIGURE 5-25 Speedup and efficiency for Householder QR decomposition algorithm on a parallel programmable processor. Notice how the speedup levels off and then decreases as more processors are used. This performance degradation is also evident in the decreasing efficiency and is due to communication and synchronization overheads in the parallel algorithm.

with the number of processors, as long as each component can be given to an unoccupied processor (or set of processors). Surprisingly, sometimes the speedup on N processors can actually exceed N. This phenomenon, known as superlinear speedup, can occur when the single-processor algorithm does not make efficient use of the single-node memory hierarchy, for example, when the data cannot fit into the cache on a single processor. When the algorithm is parallelized and the data are then split up over several nodes, each node may have a dataset that is small enough to completely fit into cache. Thus, the parallel algorithm is able to avoid the performance overhead of cache misses. If this overall performance gain offsets parallelization overheads, then the algorithm will exhibit superlinear speedup.

Consider the following example. Suppose an algorithm is specified to first sum the rows of a matrix and then to find and return the largest sum. One way to do this is to partition the matrix into row vectors and then sum each row vector. The next step is to do a comparison to determine the largest summand. If each row of data is assigned (mapped) to an independent processor to perform the row summation, then the speedup for this stage of the algorithm would be equal to the number of rows, R. However, the summation must be followed by the comparison stage to find the largest sum. Thus, the speedup for this partition will be slightly less than R; the comparison stage requires communication (of sums) between processors. One way to do the compare in parallel is as follows: row R processor sends its sum to the row $R - 1$ processor, while at the same time, row $R - 2$ processor sends its sum to row $R - 3$, and so on. At the end of this first stage, $R/2$ communications and compares have been carried out. This comparison process is repeated for the $R/2$ processors that contain the results of the comparisons, etc., until one processor has the final comparison. This is known as a parallel binary compare. $R - 1$ comparisons are done in $\log_2(R)$ stages.

If the matrix is R rows and C columns, the total number of additions for a serial algorithm is $(C - 1)*R$. The total number of comparisons is $(R - 1)$. Thus, the total complexity of the algorithm is $(C - 1)*R + R - 1 = RC - 1$. For the data-parallel algorithm, each of $N = R$ processors does $C - 1$

additions. Then, the binary comparison begins. At least one processor ends up doing $\log_2(R)$ comparisons (while the binary comparison progresses, more and more of the processors become idle, thus wasting potential computing cycles, so one can see that this algorithm is not 100% parallel efficient). So, if (unrealistically) communication takes zero time and additions take as long to compute as comparisons, the speedup for a parallel system with $N = R$ processors would be

$$S_N = \frac{T_S}{T_P} = \frac{RC - 1}{C - 1 + Log_2 R} .$$

The efficiency of this parallel implementation is

$$E_N = \frac{S_N}{N} = \frac{RC - 1}{N(C - 1 + Log_2 R)} .$$

As C approaches infinity, the complexity of the parallel binary comparison becomes relatively insignificant compared to the number of additions along each row, and Sn and En approach ideal values:

$$E_N = 1 \text{ and } S_N = N(= R) .$$

On the other hand, as R approaches infinity (so that the machine size $N = R$ also approaches infinity), the speedup approaches infinity and the efficiency approaches unity. Although infinite-sized problems and processors are not particularly realistic, the trends revealed by such analyses are useful for exploring various mapping approaches.

Another way of looking at the parallel algorithm is to consider the problem size as a constant and to seek finer and finer grained parallelism mapped to a parallel architecture. Consider the objective of finding the maximum parallelism in the algorithm examined above. First observe that the summation along each row could also be accomplished as a parallel binary add. Thus, if there are C columns in the matrix, we could add pairs of elements, and then add these sums in pairs, and so on until the entire row has been summed. This parallel variant still requires $C - 1$ summations for each row, but the row can be summed in $\log_2 C$ stages. The longest sequential path through the parallel algorithm must do at least $\log_2 C$ addition stages followed by at least $\log_2 R$ comparison stages (using the same parallel binary comparison algorithm). Thus, the minimum parallel algorithm execution time run is at least as long as it takes this sequence of $\log_2 C + \log_2 R$ operations to execute. Thus, the speedup is limited by the longest sequential portion of the algorithm. This observation generalizes to what is known as Amdahl's Law.

More formally, letting δ represent the fraction of operations in a parallel algorithm that must be performed sequentially (so that $1 - \delta$ is the fraction that can be completely parallelized), then the maximum speedup that can be achieved by a parallel processor with N processors is bounded by

$$S_N \leq \frac{1}{\delta + \dfrac{(1 - \delta)}{N}} .$$

As the processor size, N, increases, the speedup will approach $1/\delta$. For example, if half of the algorithm cannot be parallelized, the greatest achievable speedup is two, no matter how many processors are used.

Although algorithms in back-end processing exhibit the types of task and data parallelism discussed previously, they also have parallelism over the major data objects that are processed. For

Image Processing Pipeline

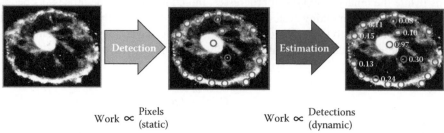

$$\text{Work} \propto \frac{\text{Pixels}}{\text{(static)}} \qquad \text{Work} \propto \frac{\text{Detections}}{\text{(dynamic)}}$$

Static Parallel Implementation

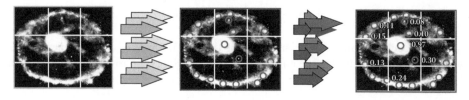

Load: balanced   Load: unbalanced

• Static parallelism implementations lead to unbalanced loads

FIGURE 5-26 **(Color figure follows page 278.)** Parallelism in an image processing algorithm. The algorithm first detects celestial objects in the image and then performs parameter-estimation tasks (location, size, luminance, spectral content) on the detections. The algorithm-to-architecture mapping first exploits the data parallelism in the image to perform detection. If the detections are not distributed evenly amongst the processors, a load imbalance will occur in the estimation phase.

example, consider the image processing example shown in Figure 5-26. In this example, an image of a nebula is first processed to determine a set of detections. The detection task, as with typical front-end processing, proceeds in a data-parallel manner. Each processor is given a subset of the data (with some overlap to handle the edges) and determines how many celestial objects are in its dataset. This is followed by an estimation task that operates on each *detection* using surrounding data. Since the detections are distributed unevenly throughout the image and the image was spread out evenly across the processors to do detection, the estimation task workload is unevenly distributed across processors. In order to achieve an equal workload for all processors, the detections (and related data) must be redistributed evenly across the processor set.

Typically, HPEC applications exploit multiple decomposition strategies. Consider the phased-array radar processor example shown in Figure 5-27. The figure shows a high performance embedded processor (circa 2000) that performs airborne radar space-time adaptive processing. The system supports up to 48 receive channels from a narrowband phased array. The processor architecture consists of a custom VLSI digital I/Q subsystem for performing signal downconversion and low-pass filtering, a high-data-rate recorder, and a four-chassis Mercury Computer Systems multicomputer containing nearly 1000 processors (SHARC DSPs and Motorola PowerPCs). The signal processing chain implemented by the processor is shown in Figure 5-28.

The processing chain implemented in the C programming language on the processor consists of three major programmable signal processing subsystems:

1. Digital filtering subsystem (DFS): The DFS receives 48 channels of downconverted streams from the digital I/Q subsystem and equalizes each stream to reduce any channel

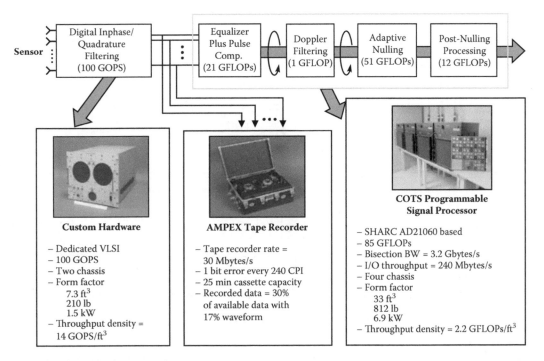

FIGURE 5-27 High performance embedded processor that performs airborne radar space-time adaptive processing using 48 channels of narrowband radar data from a phased array.

transfer function mismatches. Then, DFS performs waveform pulse compression and Doppler filtering.

2. The adaptive beamforming subsystem (ABS): The ABS adaptively removes jamming and clutter in two beamforming steps.

3. A detection and estimation subsystem (DET). The DET performs constant-false-alarm-rate (CFAR) detection, removes clutter discretes, clusters multiple detections that represent the same target, and then estimates the kinematics state vector of each target (range, azimuth, range-rate).

The processor also hosts a radar signal processor scheduler subsystem, which receives mode information from the radar back-end processor (not shown), derives the detailed mode information needed by each subsystem, and transmits the information to the subsystems to synchronize the overall processing chain.

Figure 5-29 shows the mapping of the radar signal processing chain onto the 1000-processor massively parallel signal processor. Pipelining, data parallelism, task parallelism, and round-robin scheduling are all employed. First, the data need to be buffered at the processor input, where they are demultiplexed* before digital filtering can be employed. At the same time, the scheduler, acting in parallel to the data input, receives control input, converts it to subsystem-specific commands, and transmits (dotted lines) the commands to the signal processing subsystems. To support the throughput requirement of the radar, the mapping pipelines the dataset for each coherent processing interval (a CPI is the time needed to acquire a radar data cube consisting at the input of returns for each range gate, for each channel, for each pulse repetition interval) through each of the digital filtering, adaptive beamforming, and detection subsystems. The beamformer and detection subsystems are

---

* The digital I/Q subsystem outputs the channel data in sets of eight, so that the first range for eight channels is transmitted to the input subsystem, then the second range gate for eight channels, and so on. The input subsystem needs to reorganize the data so that 48 digital filters can operate along the range dimension of each channel in parallel.

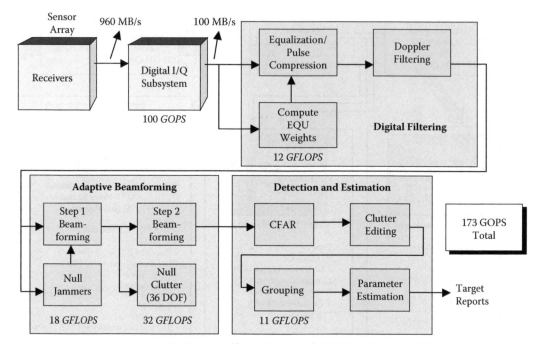

FIGURE 5-28 Signal processing chain implemented by the high performance embedded radar processor shown in Figure 5-27.

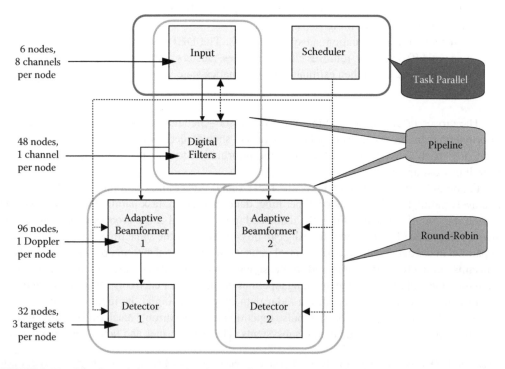

FIGURE 5-29 Mapping of the radar signal processing chain onto the 1000-processor massively parallel signal processor.

paired into pipelines, and the pairs are handed data in a round-robin schedule. The overall latency for the processor is five CPIs, which meets the two-second track-loop constraint set by the back-end processor. Within each stage, data parallel partitioning is used to provide sufficient computation power. In the digital filtering subsystem, the data are processed on a per-channel basis. Forty-eight channels are processed, with each channel assigned to a node (each node has three DSPs). Within each beamformer, the data are partitioned by Doppler bin. There are 96 Doppler bins and each is assigned to a separate node. The odd-numbered CPIs are sent to the first adaptive beamformer, and the even-numbered CPIs are sent to the second adaptive beamformer. Ninety-six nodes are used for each beamformer, so that the overall ABS uses 182 nodes. Each detector paired to a beamformer consists of 32 nodes, so that the overall DET subsystem consists of 64 nodes. Each node handles the detections for the beams from three Doppler bins. This leads to load imbalance in the detectors since detections in general are not evenly distributed in Doppler space. However, this approach was chosen for simplicity and was motivated by the relatively large communication overhead between the nodes. The processing chain sustains a throughput of about 73 GFLOPS and is capable of handling a variety of radar waveforms and CPIs. The remaining 12 GFLOPS of throughput (the programmable processor had a peak of 85 GFLOPS) were dedicated to overall control, monitoring, status, and spare processing capacity.

This particular implementation achieved an impressive overall efficiency of 30%. The DFS was able to achieve 52% efficiency since its filtering and FFT operations were well matched to the DSP processor architecture. The ABS achieved 24% efficiency, with much of the overhead located in the adaptive weight computation. The DET subsystem achieved an average of 12% efficiency. This lower efficiency was due to a combination of load imbalancing (the node with the most targets takes the most time, and the other nodes finish their work and must then idle), the numerous comparison operations (in the CFAR, for example), and the data-indexing complexity of the algorithms.

5.5 FUTURE TRENDS

Trends in HPEC algorithms, architectures, and applications are evident in the systems being prototyped today in government and industry research laboratories. One of the most notable trends is that signal and image processing algorithms have begun to incorporate knowledge databases, knowledge-based computations, and feedback from higher-level computations into front-end sensor processing. This has led to a blurring of the traditional boundary between front-end and back-end computations. It has also led to a decrease in the efficiency of algorithm implementations on conventional processors, thus leading to research into new computing architectures.

At the Defense Advanced Research Projects Agency (DARPA), the Knowledge-Aided Sensor Signal Processing and Expert Reasoning (KASSPER) Program has investigated knowledge-aided signal processing for radar systems. Figure 5-30 shows an example of a KASSPER processing architecture for an airborne ground moving-target indicator (GMTI) radar. This architecture has been demonstrated in real time on a 480 GFLOPS, ruggedized, Mercury Computer Systems multicomputer (Schrader 2004). The algorithm uses prior knowledge of the environment (for example, the locations of roads, terrain contours, types of ground cover, etc.) to significantly improve the adaptive processing stage, thereby improving overall system performance. The novel components of the front-end architecture are the knowledge database, knowledge cache, and the look-ahead scheduler. The knowledge database contains terrain information that has been formatted for efficient real-time retrieval and use. During system operation, the knowledge cache holds the portion of the knowledge store that is within the radar's field of view. The knowledge preprocessing step performs real-time operations such as transformations between geographic coordinates and radar system coordinates. The application of the knowledge is performed locally as the aircraft flies over the terrain. This requires that the signal processing be done at a granularity that is finer than usual for a GMTI adaptive algorithm. This, in turn, reduces the efficiency of the end-to-end code to about 5%, whereas efficiencies in the range of 25% to 30% are typically achievable for more conventional

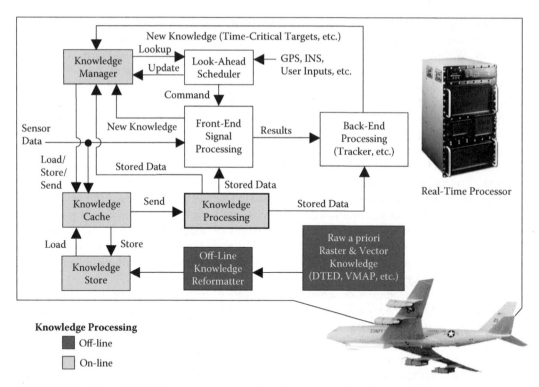

FIGURE 5-30　The KASSPER program is researching knowledge-aided airborne radar signal processing.

GMTI algorithms. Thus, as knowledge-based algorithms migrate to the front-end signal processor, new HPEC architecture will be needed to deliver the requisite computational power with acceptable processing efficiency.

Several recent research and development initiatives are addressing these computational challenges. For example, the DARPA Polymorphous Computing Architectures (PCA) program has initiated the development of processors that can be configured or "morphed" to efficiently handle applications that are a mix of both signal processing and knowledge processing algorithms. These new processors, such as the Tera-ops Reliable Intelligently-adaptive Processing System (TRIPS) processor developed at the University of Texas (Sankaralingam et al. 2003), the RAW processor developed at MIT (Taylor et al. 2002), and the MONARCH processor developed jointly by Raytheon and the University of Southern California (Vahey et al. 2006), are tile-based processors containing several processing units on a single chip. These architectures are better suited to finer-grained parallelism than are conventional MPUs and CPUs. The gaming industry is also motivating the development of processors capable of handling finer-grained computations, which may make the processors more suitable to a mix of signal and knowledge processing. For example, the IBM/Sony Cell processor packages nine processing units on a single chip: one master processor and eight synergistic processing elements designed to handle finer-grained computations (Pham et al. 2006). Graphics processing units (GPUs) such as the GeForce 6800 are also emerging as programmable alternatives (Montrym and Moreton 2005). Parallel variants of SIP and knowledge-based algorithms that are matched to these new architectures are currently major research interests. At present, the level of expertise required to develop efficient codes for these processors raises important questions of programmability, portability, and productivity that will need to be addressed if these technologies are going to achieve wide-scale acceptance.

As algorithms and processors continue to grow in sophistication, sensors continue to increase in bandwidth, sensitivity, dynamic range, and the number of sensing elements. Moreover, digital technology continues to encroach on the traditional analog domain as analog receivers give way to

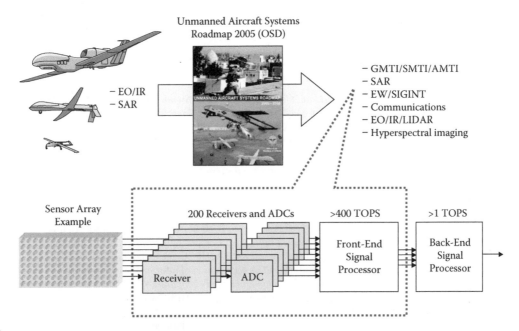

FIGURE 5-31   Sensor array signal and image processing for future UAV applications may require TOPS of computational throughput in small form factors.

digital counterparts and digital receivers become tightly integrated with sensor front-ends. Thus, the sheer volume of sensor data that must be processed is increasing dramatically, placing ever greater demands on conventional digital stream processing algorithms. For example, element-level digital array technology for next-generation phased-array radars is being developed that will require several TOPS of computing power. Figure 5-31 shows an example of a wideband phased-array radar embedded in a small form-factor unmanned aerial vehicle (UAV). In this future system, the phased array receiver inputs are digitized at the element level (so that there is no need for analog subarrays). Element-level digitization affords the maximum flexibility in the number and placement of beams, and also provides the maximum dynamic range. The reliability of digital hardware, ease of calibration, and the decreasing cost per unit performance are increasing the cost-effectiveness of digital receiver systems of this sort. The computational requirement placed on the digital hardware, however, is on the order of 100s of TOPS for front-end processing. On the order of TFLOPS of throughput will be required in the back-end. If knowledge-based algorithms such as those being developed in the KASSPER program are used, the computational demands will increase even more.

The emergence of sensor networks and the growing importance of integrated sensing and decision support systems are both influencing HPEC sensor systems and extending the need for high performance into the distributed computation domain. For example, distributed processing architectures and algorithms aimed at optimally scheduling communication and computation resources across entire networks are areas of active research. The interfaces between in-sensor processing and network distributed processing, the advent of new data fusion and tracking algorithms, advances in network data-mining algorithms, new wireless technologies, and ad hoc networks are all beginning to impact HPEC system designs. In particular, the advent of service-oriented architectures is motivating the need for HPEC systems that are explicitly designed for "plug-and-play" capabilities, so that they can be readily incorporated into larger, networked systems-of-systems.

It is readily apparent that with algorithms becoming both more sophisticated and computationally more demanding and varied, with sensors becoming more capable, with the boundary between front-end and back-end computations beginning to blur as knowledge processing moves farther forward, and with multiple HPEC systems being combined to form larger, networked systems, the future of HPEC promises to be both challenging and exciting.

REFERENCES

Arakawa, M. 2003. *Computational Workloads for Commonly Used Signal Processing Kernels.* MIT Lincoln Laboratory Project Report SPR-9. 28 May 2003; reissued 30 November 2006.

Chung, Y.-C., C.-H. Hsu, and S.-W. Bai. 1998. A basic-cycle calculation technique for efficient dynamic data redistribution. *IEEE Transactions on Parallel and Distributed Systems* 9(4): 359–377.

Feng, G. and Z. Liu. 1998. Parallel computation of SVD for high resolution DOA estimation. *Proceedings of the IEEE International Symposium on Circuits and Systems* 5: 25–28.

Golub, G.H. and C. Van Loan. 1996. *Matrix Computations.* Baltimore: Johns Hopkins University Press.

Hwang, K. 1993. *Advanced Computer Architecture: Parallelism, Scalability, Programmability.* New York: McGraw-Hill.

Leeser, M., A. Conti, and X. Wang. 2004. Variable precision floating point division and square root. *Proceedings of the Eighth Annual High Performance Embedded Computing Workshop.* MIT Lincoln Laboratory, Lexington, Mass. Available online at http://www.ll.mit.edu/HPEC/agenda04.htm.

Linderman R., M. Linderman, and C.-S. Lin. 2004. FPGA acceleration of information management services. *Proceedings of the Eighth Annual High Performance Embedded Computing Workshop.* MIT Lincoln Laboratory, Lexington, Mass. Available online at http://www.ll.mit.edu/HPEC/agenda04.htm.

Martin, J.C. 1969. A simple development of the Wiener-Hopf equation and the derived Kalman filter. *IEEE Transactions on Aerospace and Electronic Systems* AES-5(6): 980–983.

Matson, J.E., B.E. Barrett, and J.M. Mellichamp. 1994. Software development cost estimation using function points. *IEEE Transactions on Software Engineering* 20(4): 275–287.

McCabe, T.J. 1976. A complexity measure. *IEEE Transactions on Software Engineering* SE-2(4): 308–320.

Montrym, J. and H. Moreton. 2005. The GeForce 6800. *IEEE Micro* 25(2): 41–51.

Nguyen, H., J. Haupt, M. Eskowitz, B. Bekirov, J. Scalera, T. Anderson, M. Vai, and K. Teitelbaum. 2005. High-performance FPGA-based QR decomposition. *Proceedings of the Ninth Annual High Performance Embedded Computing Workshop.* MIT Lincoln Laboratory, Lexington, Mass. Available online at http://www.ll.mit.edu/HPEC/agendas/proc05/agenda.html.

Nussbaumer, H.J. 1982. *Fast Fourier Transform and Convolution Algorithms. 2nd corrected and updated edition.* Springer Series in Information Sciences, vol. 2. New York: Springer-Verlag.

Pham, D.C., T. Aipperspach, D. Boerstler, M. Bolliger, R. Chaudhry, D. Cox, P. Harvey, P.M. Harvey, H.P. Hofstee, C. Johns, J. Kahle, A. Kameyama, J. Keaty, Y. Masubuchi, M. Pham, J. Pille, S. Posluszny, M. Riley, D.L. Stasiak, M. Suzuoki, O. Takahashi, J. Warnock, S. Weitzel, D. Wendel, and K. Yazawa. 2006. Overview of the architecture, circuit design, and physical implementation of a first-generation cell processor. *IEEE Journal of Solid-State Circuits* 41(1): 179–196.

Sankaralingam, K., R. Nagarajan, H. Liu, C. Kim, J. Huh, D. Burger, S.W. Keckler, and C.R. Moore. 2003. Exploiting ILP, TLP, and DLP with the polymorphous trips architecture. *IEEE Micro* 23(6): 46–51.

Schrader, G. 2004. A KASSPER real-time signal processor testbed. *Proceedings of the Eighth Annual High Performance Embedded Computing Workshop.* MIT Lincoln Laboratory, Lexington, Mass. Available online at http://www.ll.mit.edu/HPEC/agenda04.htm.

Taylor, M.B., J. Kim, J. Miller, D. Wentzlaff, F. Ghodrat, B. Greenwald, H. Hoffman, P. Johnson, J.-W. Lee, W. Lee, A. Ma, A. Saraf, M. Seneski, N. Shnidman, V. Strumpen, M. Frank, S. Amarasinghe, and A. Agarwal. 2002. The Raw microprocessor: a computational fabric for software circuits and general-purpose programs. *IEEE Micro* 22(2): 25–35.

Vahey, M., J. Granacki, L. Lewins, D. Davidoff, G. Groves, K. Prager, C. Channell, M. Kramer, J. Draper, J. LaCoss, C. Steele, and J. Kulp. 2006. MONARCH: A first generation polymorphic computing processor. *Proceedings of the Tenth Annual High Performance Embedded Computing Workshop.* MIT Lincoln Laboratory, Lexington, Mass. Available online at http://www.ll.mit.edu/HPEC/agendas/proc06/agenda.html.

Van Loan, C. 1992. *Computational Frameworks for the Fast Fourier Transform* (Frontiers in Applied Mathematics series, no. 10). Philadelphia: Society for Industrial and Applied Math.

Walke, R. 2002. Adaptive beamforming using QR in FPGA. *Proceedings of the Seventh Annual High Performance Embedded Computing Workshop.* MIT Lincoln Laboratory, Lexington, Mass. Available online at http://www.ll.mit.edu/HPEC/agendas/agenda02.html.

6 Radar Signal Processing: An Example of High Performance Embedded Computing

Robert A. Bond and Albert I. Reuther, MIT Lincoln Laboratory

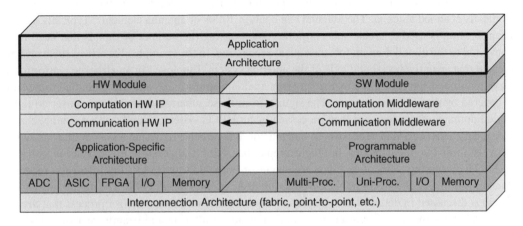

This chapter further illustrates the concepts discussed in the previous chapter by presenting a surface moving-target indication (SMTI) surveillance radar application. This example has been drawn from an actual end-to-end system-level design and reveals some of the trade-offs that go into designing an HPEC system. The focus is on the front-end processing, but salient aspects of the back-end tracking system are also discussed.

6.1 INTRODUCTION

The last chapter described the computational aspects of modern high performance embedded computing (HPEC) applications, focusing on computational complexity, algorithm decomposition, and mapping of algorithms to architectures. A canonical HPEC processing taxonomy—with a front-end component that performs stream-based signal and image processing and a back-end component that performs information and knowledge-based processing—was presented. In this chapter, the concepts discussed in the previous chapter are illustrated further by presenting a surface moving-target indication (SMTI) surveillance radar application. This example, although simplified, has been drawn from an actual end-to-end system-level design and reveals some of the trade-offs that go into designing an HPEC system. Instead of presenting a definitive processor design, this chapter covers the major considerations that go into such designs. The focus is on the front-end processing, but salient aspects of the back-end tracking system are also discussed.

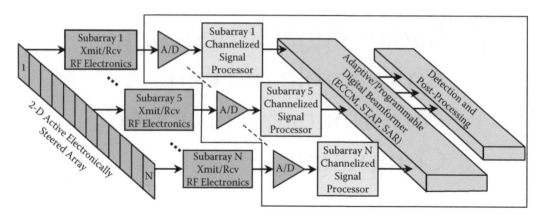

FIGURE 6-1   Wideband airborne radar processing architecture.

A canonical wideband airborne radar processing architecture is shown in Figure 6-1. The figure shows the basic elements of an airborne phased-array radar with a two-dimensional active electronically scanned antenna. The antenna can contain 1000s of antenna elements. Typically, analog beamforming is used to create subarrays, thereby reducing the number of signals that need to be converted to the digital domain for subsequent processing. In the example, 20 vertical subarrays are created that span the horizontal axis of the antenna system. Employed in an airborne platform, the elevation dimension is covered by the subarray analog beamformers, and the azimuthal dimension is covered by digital beamformers. The signals from these subarray channels are converted to the digital domain, where they are then processed by an HPEC system. Usually, either synthetic aperture radar (SAR) or SMTI processing is carried out. It is possible to design a system that can switch between these two radar modes, but the example that will be explored in this chapter is restricted to SMTI processing. For an excellent treatment of both MTI and SAR, the interested reader is encouraged to read Stimson's book, *Introduction to Airborne Radar* (1998). The digital processing, which is discussed in detail in this chapter, divides roughly into a channelizer process that divides the wideband signal into narrower frequency subbands; a filtering and beamformer front-end that mitigates jamming and clutter interference, and localizes return signals into range, Doppler, and azimuth bins; a constant-false-alarm-rate (CFAR) detector (after the subbands have been recombined); and a post-processing stage that performs such tasks as target tracking and classification.

SMTI radars can require over one trillion operations per second (TOPS) of computation for wideband systems. Hence, these radars serve as excellent examples of high performance embedded computing applications. The adaptive beamforming performed in SMTI radars is one of the major computational complexity drivers. Ward (1994) provides an excellent treatment of adaptive beamforming fundamentals. The particular SMTI algorithm used in this example, shown in Figure 6-2, is based on Reuther (2002). It has a computational complexity of just over 1 TOPS (counting all operations after analog-to-digital conversion up to and including the detector) for the parameter sets shown in Table 6-1. SMTI radars are used to detect and track targets moving on the earth's surface. The division between the onboard and ground-based processing is determined by considering the amount of processing that can be handled on board the platform and the capacity of the communication system that transmits the processed data down to the ground computing facility for further processing. For many SMTI radars, the natural dividing point between onboard and off-board processing is after the detector stage. At this point, the enormous volume of sensor data has been reduced by several orders of magnitude to (at most) a few thousand target reports. The principal challenge in the airborne front-end processors is to provide extremely high performance that can fit into a highly constrained space, operate using low power, and be air-vehicle qualified. This chapter focuses on the computational complexity of the front-end of the SMTI radar application. Parallel

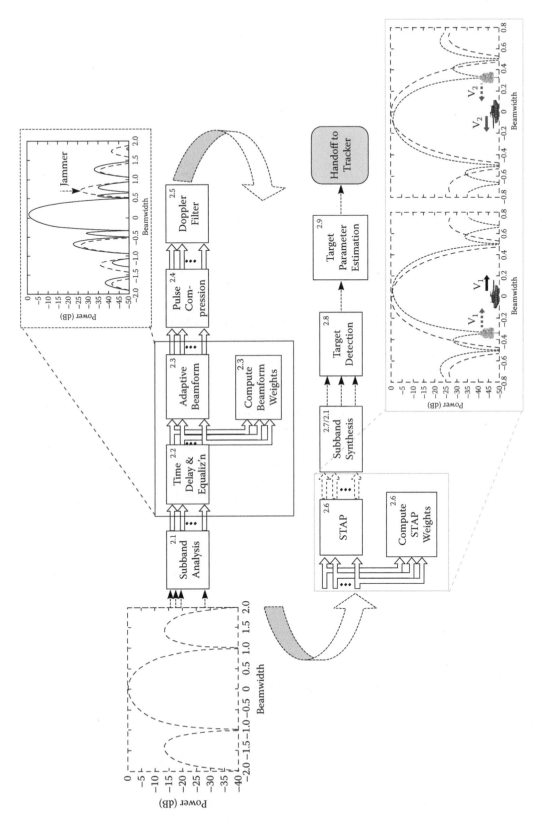

FIGURE 6-2 Example SMTI algorithm. (From Lebak, J.M. et al., Parallel VSIPL++, *Proc. IEEE* 93(2): 314, 2005. With permission. © 2005 IEEE.)

TABLE 6-1
Key Operational Parameters for the SMTI Radar

SMTI Radar Parameters		
Nch	20	Number of channels
f_samp	480,000,000	Sampling frequency (Hz)
dec_ratio	96	Decimation ratio
Nsubband	48	Number of subbands
Nppf	128	Number of polyphase filters
Nppf_taps_dn	12	Number of polyphase filter taps (analysis)
PRF	2,000	PRF (Hz)
Npri	200	Number of PRIs per CPI
Nstag	2	Number of PRI staggers
duty	10%	Transmit duty factor
Ntide_taps	0	Number of taps for time delay & EQ
Nbm_abf	4	Number of beams formed in ABF*
Nbm_stap	3	Number of beams formed in STAP**
Nppf_taps_up	15	Number of polyphase filter taps (synthesis)
Nequ_taps	1	Number of equ taps to use in the ABF*
bytes_per_complex	4	Number of bytes per complex data sample
num_target_per_dop	100	Number of targets per Doppler bin
target_report_size	256	Number of bytes per target report
flops_per_byte	12	Number of flops per byte for general computation
SMTI Radar-Derived Parameters		
f_samp_dec	3,000,000	Decimated sampling frequency (Hz)
Nrg_dec	2,250	Number of decimated range gates per PRI
Nfft_tde	4,096	FFT size for time delay & EQ
Ndof_abf	20	Number of degrees of freedom for ABF
Ntraining_abf	120	Number of training samples for ABF
Nfft_pc	4,096	FFT size for pulse compression
Ndop	199	Number of Doppler bins per stagger
Ndop_stap	8	Number of degrees of freedom for STAP
Ntraining_stap	48	Number of training samples for STAP
Nrg	216,000	Number of range gates per PRI into subband analysis
Nrg_syn	81,000	Number of range gates per PRI out of subband synthesis

* ABF — adaptive beamforming
** STAP — space-time adaptive processing

decomposition strategies and implementation alternatives are developed. The salient computational characteristics of a typical back-end processor tracking algorithm are also discussed.

6.2   A CANONICAL HPEC RADAR ALGORITHM

The SMTI radar algorithm described, shown in Figure 6-2, is a modern design using wideband signals for improved range resolution. Refer to Reuther (2002) for more details. Before the radar data are received, a radar signal consisting of a series of pulses from a coherent processing interval (CPI) is transmitted. The pulse repetition interval (PRI) determines the time interval between transmitted pulses. Multiple pulses are transmitted to permit moving-target detection, as will be described later on. The pulsed signals reflect off targets, the earth's surface (water and land), and man-made structures such as buildings, bridges, etc.; a fraction of reflected energy is received by the radar antenna. The goal of the SMTI radar is to process the received signals to detect targets (and estimate their positions, range rates, and other parameters) while rejecting clutter returns and noise. The radar must also mitigate interference from unintentional sources such as RF systems transmitting in the same band and from jammers that may be intentionally trying to mask targets. As mentioned above, the radar antenna typically consists of a two-dimensional array of antenna elements. The signals from these elements are combined in a set of analog beamformers to produce subarray receive channels. The channel signals subsequently proceed through a set of analog receivers that perform downconversion and band-pass filtering. The signals are then digitized by analog-to-digital converters (ADCs) and input to the high performance digital front-end. The ADCs must operate at a sufficiently fast sampling rate to preserve the range resolution provided by the waveform. The radar in our example has been designed to achieve about one-meter range resolution; on transmit, a 180 MHz linear FM waveform is used. The ADCs sample at 480 Msps, which amounts to oversampling of the signal by a factor of 4/3 over the Nyquist rate. Key radar parameters are shown in Table 6-1.

Digitized data cubes, as shown in Figure 6-3, are input to the SMTI processing chain continuously during each 100 ms CPI. The input data cubes consist of one spatial dimension, the channel dimension, and two temporal dimensions: the fast time dimension, which corresponds to the ADC sampling interval, and the slow-time dimension, which corresponds to the PRI. The fast-time dimension is used to isolate a target to a time bin, which is equivalent to a range gate (that is, the target's slant range distance from the radar). The slow-time dimension is used, after Doppler processing, to isolate the target to a Doppler frequency bin, which is equivalent to the target range-rate (the along-range speed of the target with respect to the radar).

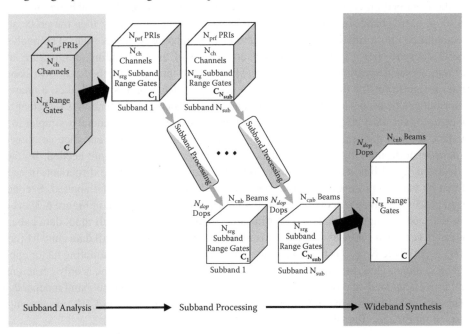

FIGURE 6-3 This figure shows an overview of the subband processing in terms of the data that are processed. The data are organized in three-dimensional "data cubes." The figure shows the data cube input to the subband analysis phase, the resultant set of subband data cubes into the subband processing, the data cubes out of the subband processing, and the reconstructed data cube out of the subband synthesis phase. The dimensions of the data cube entering the processor are range-gates × receiver channels × PRIs. At the output of the processor, the data cube has dimensions range-gates × beams × Doppler bins.

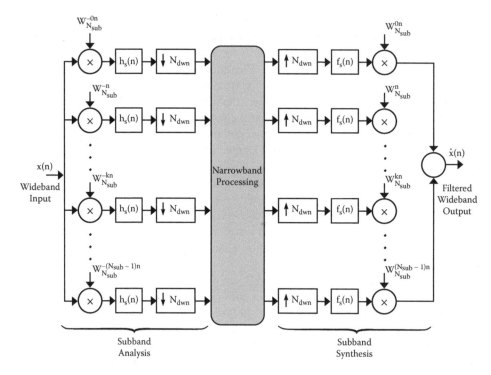

FIGURE 6-4   Subband filtering.

In the front-end processor, the wideband returns are decomposed by a subband analysis stage into a set of narrowband signals and processed as narrowband data cubes; then the processed data are recombined into a full wideband data cube in the subband synthesis stage. The analysis and synthesis steps are shown in Figure 6-4. The advantages of the subbanding architecture are two-fold. First, signal dispersion across the face of the array increases as the waveform bandwidth becomes a significant fraction of the carrier frequency. If dispersion is small, then a phase-based beamformer can be used. However, in a wideband system where the dispersion is too great, a time-delay beamformer must be used. Phase-based digital beamformers are generally simpler to implement since channel data can be multiplied by a simple complex number and then combined. In a time-delay beamformer, more sophisticated, tap-delay-line architectures are needed. By factoring the wideband signal into a set of narrower subband signals, the effective dispersion in each subband can be made small enough so that phase-based beamforming can be applied. Second, each subband processing chain can operate in its data independently. As shown in Figure 6-3, the overall size of the subband input data cube is significantly smaller than the overall input data cube. In the example chosen, each subband data cube has fewer samples than the overall data cube by the factor $(4/3)(N_{subband})$, where $N_{subband}$ is the number of subbands and 4/3 is an oversampling factor applied by the subband filters. Thus, the amount of processing required in each subband is only a fraction of the overall processing load, and dedicated processors can be applied to each subband independently.

The full SMTI processing algorithm proceeds as shown in Figure 6-2. The processing chain consists of nine stages (five of which are carried out within each subband):

1. subband analysis
2. time delay and equalization
3. adaptive beamforming
4. pulse compression
5. Doppler filtering

6. space-time adaptive processing (STAP)
7. subband synthesis (recombining)
8. detection
9. estimation

This signal processing chain then reports the resulting targets to the tracker.

Within a subband, time-delay and equalization processing compensate for differences in the transfer function between subarray channels. The adaptive beamforming stage transforms the subbanded data into the beam-space domain, creating a set of focused beams that enhance detection of target signals coming from a particular set of directions of interest while filtering out spatially localized interference. The pulse compression stage filters the data to concentrate the signal energy of a relatively long transmitted radar pulse into a short pulse response. The Doppler filter stage applies a fast Fourier transform (FFT) across the PRIs so that the radial velocity of targets relative to the platform can be determined. The STAP stage is a second beamforming stage, designed to adaptively combine the beams from the first beamformer stage to remove ground clutter interference. The subband synthesis stage recombines the processed subbands to recoup the full bandwidth signal. The detection stage uses constant false-alarm rate (CFAR) detection to determine whether a target is present. The estimation stage computes the target state vector, which consists of range rate, range, azimuth, elevation, and signal-to-noise ratio (SNR). Often, the estimation task is considered a back-end processing task since it is performed on a per-target basis; however, it is included here with the front-end tasks since in many existing architectures it is performed in the front-end where the signal data used in the estimation process are most readily available.

The target state vectors are sent in target-report messages to the back-end tracker subsystem. The tracker, shown in Figure 6-5, employs a basic kinematics Kalman filter that estimates target position and velocity (Eubank 2006). A track is simply a target that persists over a time interval (multiple CPIs). By using information from a sequence of target reports, the tracker can develop a more accurate state vector estimate. The Kalman filter provides an estimate of the current track velocity and position. These estimates are optimal if the statistics of the measurements (the target reports) are Gaussian and the target dynamics are linear. The linearity condition is rarely strictly met in practice, so extensions to the basic filter to account for nonlinear dynamics are often used (Zarchan 2005). For non-Gaussian statistics, more general approaches such as Bayesian trackers and particle tracking filters can be employed (Ristic, Arulampalam, and Gordon 2004). For any tracker, each target detected during the most recent CPI is either associated with an existing track, or else it is used to initiate a new track. Tracks that do not associate with any targets for a predetermined number of CPIs are dropped.

In feature-aided-tracking (FAT), shown in Figure 6-6, the kinematics data used for association are augmented by target features. This can improve the target-to-track association process, especially in dense target environments where tracks can cross each other or where multiple tracks may be kinematically close to the same target. The variant of FAT employed here is referred to as signature-aided tracking (SAT). In SAT, the high range resolution provided by a wideband SMTI radar is used to determine the

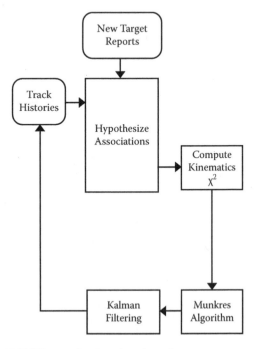

FIGURE 6-5 Back-end tracker subsystem.

range profile or *signature* of a target. The previous target-track associations are used to compute a representative track signature, and as new targets are detected, candidate targets are associated by considering both their kinematics correlation to the track and the similarity of their radar signature to the track signature. This process will be discussed further when some of the computational aspects of this approach are covered later in the chapter.

Table 6-2 shows the computational complexity and throughput of each of these front-end stages along with the totals for the front-end. Overall, the front-end processing requires 1007 GOPS (giga, or billion, operations per second), not counting the equalization stage, which is included in the table for completeness but is not actually needed in the chosen example.*

6.2.1 Subband Analysis and Synthesis

The computational aspects of the SMTI processing stages are discussed next. The subband analysis and synthesis phases, shown in Figure 6-4, are complements of each other. The subband analysis can be implemented by demodulating the incoming signal into a series of subband signals, low-pass filtering each of these subband signals, and downsampling the subsequent filtered signals. The filter inputs are the range vectors of the input radar data cube, as shown in Figure 6-3. Demodulation is performed by mixing or multiplying each sample by a complex downshifting value. The low-pass filtering ensures that signal aliasing is minimized in the downsampling step. Downsampling is conducted by simply extracting every Nth sample. When these operations are completed, the output is a set of subband data cubes, each with the same number of channels and PRIs but $1/N$th the number of range gates. Conversely, subband synthesis entails upsampling the subband data, low-pass filtering the upsampled subbands (using a 15-tap filter in our example), modulating the subband signals back up to their original frequencies, and recombining the subbands by superposition addition. Upsampling is performed by inserting zero-valued samples between each sample. Again, low-pass filtering is used, this time to minimize signal aliasing in the upsampling step. The modulation is performed by multiplying each sample by a frequency upshifting value.

The subband analysis can be accomplished with a time-domain polyphase filter to perform low-pass filtering and downsampling, followed by a fast-time FFT to decompose the signals into frequency bins. The computational complexity of the subband analysis computation can be determined as follows: first of all, a 12-tap finite impulse response (FIR) filter is applied to each decimated range gate, for each polyphase filter, for each channel. The FIR filter requires 12 real multiply-accumulate operations, for a total of 24 operations (counting a multiply and add as two operations,

---

* Depending on the number of taps required, the equalizer can add as many as another 1000 GOPS to the workload. The equalizer has the task of matching the transfer functions of each receiver channel to a reference channel. The degree to which the channels are matched determines the maximum null depth that the adaptive beamformer can achieve. One of the principal goals of the beamformer is to place nulls in the directions of jammer sources. This, in turn, determines the largest jammers that can be handled by the radar without significant performance degradation. The equalizer is implemented as an N-tap finite impulse response (FIR) filter that performs a convolution on the data from each receiver channel. The number of taps is determined by the required null depth, the channel transfer functions, and the intrinsic mismatch between channels. The overall channel mismatch depends strongly on the mismatch in the analog filters chosen for the receivers. In the chosen example, the receiver design allows a single complex coefficient tap to match the channels. For computational efficiency, this coefficient is folded into the downstream beamformer weights, so that the equalizer stage can be removed altogether. If an equalizer were needed, it would be implemented either as a time-domain FIR filter for a small number of taps or a frequency domain FIR filter if the required number of taps were large. The 1000 GOPS number given above is for the case where a full frequency-domain filter is required. In this case, each PRI is transformed by an FFT, a point-wise multiply with the filter frequency coefficients, and the result is converted back to the time domain by an inverse FFT (IFFT). This is carried out on every PRI for every channel. The main point in discussing the subtleties of the equalizer stage is that a careful co-design of the end-to-end system, including both the analog and the digital hardware, can pay large dividends in reducing overall system complexity. The designer of the receiver might not pick the appropriate analog filters without a proper appreciation of the impact on the digital processor. Furthermore, jamming mitigation requirements may be too severe; margin may have been added "just in case," thereby forcing the digital hardware to become unreasonably complex. A few extra dB of jammer margin and a relaxation on the analog filter specifications, and the size and complexity of the front-end processor could potentially double.

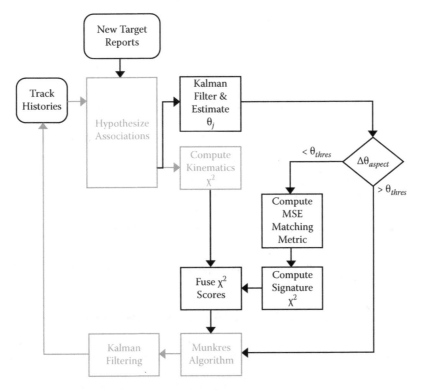

FIGURE 6-6   Feature-aided tracking using signature data.

TABLE 6-2

Computational Throughput of the Example SMTI Radar, Based on the Configuration Parameters from Table 6-1

	SMTI Radar Throughput per Stage		
Stage	Fixed Point (GOPS)	Floating Point (GFLOPS)	Aggregate (GOPS)
Subband Analysis	478		478
Time Delay & EQ	[up to 1000]		478
Adaptive Beamforming	139	0.20	617
Pulse Compression	198		816
Doppler Filtering	66		881
STAP	41	2.36	925
Subband Synthesis	76		1,001
Detection	0.00	5.37	1,007
Total	999	7.93	1,007

Note that the equalizer is not needed for the example radar system since the receivers are well-enough matched that it is sufficient to use a single phase and amplitude correction that is combined into the beamformer weights. A full equalizer filter might require a long FIR filter. The complexity of a 25-tap convolution FIR filter implemented in the frequency domain is shown.

of course). The total complexity for the subbanding polyphase filtering step is, therefore (using the parameter values in Table 6.1),

$$C_{sb\_ppf} = 24 * N_{ppf} * N_{dec\_rg} * N_{ch} = 138.24 \text{ MOPs}.$$

The polyphase filter is applied every PRI. Thus, the total throughput is

$$F_{sb\_ppf} = C_{sb\_ppf} / PRI = 138.24 \text{ MOPS} / 0.5 \text{ ms} = 276.48 \text{ GOPS}.$$

The FFT is applied across the polyphase filter output. Using the complexity formula for a real FFT (see Chapter 5) shows that each FFT requires

$$C_{ppf\_fft} = 5 / 2N_{ppf} \log(N_{ppf}) = 2240 \text{ FLOPs}.$$

The FFT is repeated for every decimated range gate, for each PRI, and for each channel, for a total subbanding FFT complexity:

$$C_{sb\_fft} = 2240 * N_{ppf} * N_{dec\_rg} * N_{ch} = 100.8 \text{ MOPs}.$$

Again, this work must be accomplished in a PRI, so that the total throughput for the FFT stage is

$$F_{sb\_fft} = C_{sb\_fft} / PRI = 100.8 \text{ MOPS} / 0.5 \text{ ms} = 201.6 \text{ GOPS}.$$

Thus, the total throughput of this stage is

$$F_{sb} = F_{sb\_fft} + F_{sb\_ppf} = 478.1 \text{ GOPS}.$$

In a similar manner, one can derive the throughput requirement for the subband synthesis stage. The synthesizer uses a longer convolver, but acts on beams instead of channels; computationally, the net result is that the synthesis phase has fewer operations, requiring about 76 GOPS throughput. Given the high throughput and the regular computation structure of subband analysis and synthesis, these operations are ideally suited to application-specific integrated circuit (ASIC) or field programmable gate array (FPGA) implementations. In a highly power-constrained platform such as an unmanned aerial vehicle (UAV), a custom very-large-scale integration (VLSI) approach may be the best solution. Figure 6-7 shows a custom subband analysis chip set consisting of a polyphase filter chip and an FFT chip. This chipset was developed circa 2000 by Song et al. (2000). It uses 0.25-micron fabrication technology and achieves 31 GOPS (44 GOPS per watt) and 23 GOPS (38 GOPS per watt) of throughput for the polyphase filter and FFT chips, respectively. The internal architecture uses a bit-level systolic processing technique that can be scaled to smaller process geometries, so that future versions with higher throughput and lower power are readily implementable.

6.2.2 ADAPTIVE BEAMFORMING

Once the wideband signals have been decomposed into subbands, an adaptive beamformer combines the channels to produce a set of beams for each subband. The purpose of this computation stage is to remove interference (for example, jammers) while preserving high gains in beam-steering directions. There are numerous techniques for adaptively determining a set of beamformer weights, but they all boil down to the solution of a constrained least-squares problem. Two principal design choices are the

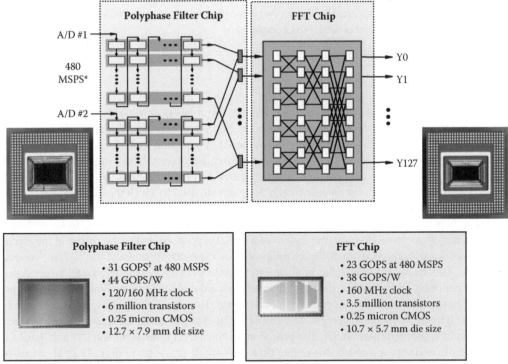

FIGURE 6-7   Custom subband analysis chip set.

manner in which the environment is sampled for interference and the constraints used to formulate the adaptive optimization equation. A straightforward approach, which nevertheless has the basic computational aspects present in more advanced techniques, is employed here. The first step is to capture a statistical picture of the interference environment. To do this, the first PRI in each CPI is designated as a receive-only PRI; that is, the radar does not transmit during this interval. Hence, samples collected during this PRI contain only interference signals. To provide a statistically significant representation of the environment, several samples must be collected for each subband. The number of samples needed is related to the N_{dof}, the number of degrees of freedom in the adaptive problem. Ward (1994) shows that, for effective performance, $2\,N_{dof}$ to $5\,N_{dof}$ samples are required. In the algorithm presented here, for each subband $5{*}N_{ch}$ samples (note that $N_{dof} = N_{ch} - 1$) are collected to form a sample matrix, $A^{(s)}$, of dimension $N_{ch} * 5N_{ch}$. Then, an extra N_{ch} samples diagonal loading matrix is appended (bringing the overall matrix size to $N_{ch} * 6N_{ch}$). Diagonal loading is used to desensitize the adaptive weight computation to perturbations due to lower-level noise.

The adaptive weights are the solution to the classical least-squares Weiner-Hopf equation (Haykin 2002):

$$W_{bm}^{(s)} = \frac{\left(\hat{R}^{(s)}\right)^{-1} V_{bm}}{V_{bm}^{H}\left(\hat{R}^{(s)}\right)^{-1} V_{bm}} .$$

In the above equation, $W_{bm}^{(s)}$ is a matrix of beamforming weight vectors, one for each of bm beams. The superscript (s) denotes that a separate matrix is computed for each subband. The matrix

$\hat{R}^{(s)}$ is an estimate of the covariance matrix (with the diagonal loading included), one for each subband. The estimated covariance matrix is computed as

$$\hat{R}^{(s)} = A^{(s)} * \left(A^{(s)} \right)^{H},$$

where, for notational simplicity, the diagonal loading factor has been absorbed into $A^{(s)}$. The term V_{bm} is a matrix of column steering vectors. Each steering vector serves to define a look direction for one of the radar receive beams. Although the Weiner-Hopf equation looks formidable, the denominator evaluates to a scalar and simply serves as a normalizing factor. Thus, the real challenge is to compute the inverse of the estimated covariance matrix. In practice, one usually does not form the covariance matrix. Instead, the sample matrix is dealt with directly (with the diagonal loading factors appended). By doing so, multiplication of $A^{(s)}$ by its Hermitian transpose is avoided. The real problem, however, is that after forming $\hat{R}^{(s)}$, the weight computation would have to be performed in the radar signal power domain (the square of a signal is proportional to the power in the signal). The power-domain quantities in $\hat{R}^{(s)}$ have roughly twice the dynamic range of the "voltage" domain signals in $A^{(s)}$. Because SMTI radar applications attempt to extract extremely small target signals in harsh jamming environments, they require a large signal dynamic range. Typically, a single-precision, floating-point number can contain the full signal. However, when the signal is converted to the power domain, subsequent computations must use double-precision, floating-point arithmetic to avoid loss of signal information. Double-precision arithmetic requires significantly more hardware and typically executes at a much slower rate than does single precision in digital signal processing (DSP) hardware. By staying in the voltage signal domain and using the sample matrix, one can avoid this increase in dynamic range and continue to work in single precision.*

The approach is as follows: substituting for $\hat{R}^{(s)}$ into the Weiner-Hopf equation and dropping both the normalization constant and the subband superscript for notational convenience, one gets

$$W_{bm} = \left(\hat{R} \right)^{-1} V_{bm} = \left(A * A^{H} \right)^{-1} V_{bm}.$$

This equation can be solved using LQ decomposition and back-substitutions as follows: first, rewrite the above equation as

$$\left(A * A^{H} \right) * W_{bm} = V_{bm}.$$

Using the LQ decomposition of matrix A, substitute the L and Q factors in the above equation and simplify it by observing that $Q*Q^{H} = I$, since Q is unitary:

$$\left(LQ * \left(LQ \right)^{H} \right) * W_{bm} = V_{bm}$$

$$\left(L(Q * Q^{H})L^{H} \right) * W_{bm} = V_{bm}$$

---

* An additional important benefit is that the single-precision sample matrix is more numerically stable (has a lower condition number) than its (estimated) covariance matrix counterpart.

$$\left(LL^H\right) * W_{bm} = V_{bm}$$

$$L\left(L^H W_{bm}\right) = V_{bm} .$$

Since L is a lower triangular matrix, the above equation can be readily solved for W_{bm} with two matrix backsolves. First, set $Z = L^H W_{bm}$ and then solve for Z through back-substitution using

$$LZ = V_{bm} .$$

Then, solve for W_{bn} through another back-substitution using

$$L^H W_{bm} = Z .$$

While the weights are being computed and downloaded to the beamformer, the channel range-gate data must be buffered. In highly dynamic jamming environments where the interference statistics correlated with direction of arrival are changing rapidly, it is important to capture a more timely representation of the interference. This is accomplished by computing a set of weights at the front-end of a CPI and another set at the back-end (at the front-end of the next CPI, actually). After the back-end set of weights is computed, the two sets are interpolated to produce weight sets that roughly capture the changing interference environment over the duration of the CPI, time-aligned to the middle of each PRI. These weights are then applied to the buffered CPI data, one set of weights for each PRI. This is an example of a co-design trade-off between computational simplicity, latency, memory, and algorithm performance. The hardware would be simpler and less memory would be required if the interpolation step were removed. To evaluate the design trade, one must understand the cost in algorithm performance, which will require, at a minimum, a faithful simulation of the environment and may require the collection of real-world data.

If a full CPI is allocated to compute a weight set and then another CPI is allocated for the next weight set to become available, the design incurs two CPIs of latency waiting before the weights are available. The processor must, therefore, buffer the next CPI as well as the current one. Then, while the weights are being applied, the processor needs to store the input from yet another CPI. The net storage requirement in the beamformer at the input is, therefore, three CPIs worth of data, or about 5.2 Gbytes of data (assuming 4 bytes integer data for each complex sample). More sophisticated control schemes can reduce the memory requirement slightly, but the cost of the extra memory is generally well worth the implementation simplicity.

The weight computation operations described above can be carried out in custom ASICs, although often programmable DSPs are chosen for their flexibility. DSPs enable the use of more sophisticated training strategies and accommodate algorithm upgrades and tuning. Moreover, due to numerical stability, the adaptive weight computation usually requires floating-point arithmetic, which is generally poorly suited to custom ASICs but finds efficient support in high-end DSPs. (FPGAs are also an option.) For the SMTI application being explored here, note that the adaptive weight computation requires a throughput of about 200 MFLOPs. This throughput is computed using the complexity expressions for the LQ decomposition and backsolve operations presented in Chapter 5. The calculation is straightforward:

$$F_{abf\_weights} = (C_{lq} + 2C_{backsolve} * N_{beams} + C_{in} * N_{pri}) * N_{subband} / CPI .$$

- **2.7 GFLOPS peak (2 TMS320C6713s)**
- 256 Mbytes SDRAM
- 4 Mbytes Flash
- JTAG & 16 Kgate FPGA
- 5W typ.
- 6 × 6 × 0.6 mm
- 424 pins
- 10 krad (Si) Flash die
- Latchup mitigation via 2 off-board electronic circuit breakers

Processor Node ca. 2003
(grid-array packages)

FIGURE 6-8 Design of a 2.7 GFLOPS dual-DSP processing node based on the Texas Instruments TMS320C6713.

The complexity of the LQ decomposition is

$$C_{lq} = 8 * N^2_{dof\_abf} * (N_{training\_abf} - (N_{dof\_abf} / 3)) \text{ FLOPs}$$

$$= 8 * 20^2 * (120 - (20 / 3) \text{ FLOPs} = 362,667 \text{ FLOPs}.$$

The total complexity of the backsolve stage (2 backsolves are done for each beam) is

$$2C_{backsolve} * N_{beams} = 2 * 4 * N^2_{dof\_abf} * N_{beams} = 2 * 4 * 20^2 * 4 \text{ FLOPs} = 12,800 \text{ FLOPs}.$$

The interpolation complexity, which is performed for each pair of weights for each PRI, is computed as

$$C_{in} * N_{pri} = 8 * N_{ch} * N_{pri} \text{ FLOPs} = 8 * 20 * 200 \text{ FLOPs} = 32,000 \text{ FLOPs}.$$

Substituting into the throughput equation and using the coherent processing interval time of 100 ms, one gets

$$F_{abf\_weights} = (362,667 + 12,800 + 32,000) * 48 / 0.1 \text{ FLOPS} = 195.6 \text{ MFLOPS}.$$

Figure 6-8 shows a design of a 2.7 GFLOPS dual-DSP processing node based on the Texas Instruments TMS320C6713. If one DSP is allocated to performing the adaptive weight computation, then the efficiency of the code would have to be (0.1956/1.35)*100 = 14.5%. This efficiency is well within the expected range for a DSP performing LQ decompositions and backsolves.* A benchmark should be developed to determine the performance on candidate DSP boards, and efficient assembly codes could be developed for these kernels, if needed. The LQ decomposition and backsolve kernels are reused in the STAP beamforming stage that is performed later in the processing chain,

---

* The processor discussed in Chapter 4, for example, achieved an efficiency of 33.5% on a QR decomposition on the SHARC DSP.

so hand-optimized kernels parameterized for a range of matrix sizes would be beneficial. Often, vendors develop their own hand-optimized libraries that can be used to construct more complicated routines such as the LQ factorization. More sophisticated standard middleware libraries, such as the Vector, Signal, and Image Processing Library (VSIPL), have implementations optimized for specific platforms and processors. VSIPL, for example, provides a QR computation object (which can be used in place of the LQ if the data matrix is transposed). If it proved difficult to achieve the required serial-code efficiency, the computation could be split between the two DSPs in the node, thereby performing the calculations in parallel. Breaking the matrix up into two blocks would not provide good load balancing because the LQ computation proceeds across the matrix from left to right on successively smaller matrices. Eventually, the DSP with the first (left-most) block would have no computations to perform while it waited for the second DSP to complete the factorization on the second block. However, by cyclically striping the matrix (odd columns to one DSP and even columns to the other), both DSPs can remain busy throughout the computation, thereby providing much better load balancing.

The next computation stage is the beamformer itself. The computation consists of a series of inner products in each subband:

$$Y_{pri}^{(s)} = \left(W_{bm}^{(s)} \right)^{H} X_{pri}^{(s)} .$$

The resulting matrix $Y_{pri}^{(s)}$ is organized such that each row is a beam and each column is a range gate. This matrix-matrix product must be computed for each PRI in each subband. This produces a data cube in each subband, as shown in Figure 6-9, that has dimensions $N_{pri} * N_{rg\_dec} * N_{bm\_abf}$. The overall throughput requirement for the beamforming operation is the complexity of a complex matrix multiplication of the basic beamformer multiplied by the number of times the beamformer is applied, divided by the length of a CPI:

$$F_{abf} = (8 * N_{dof\_abf} * N_{rg\_dec} * N_{bm\_abf}) * N_{pri} * N_{subband} / CPI .$$

The net throughput for the parameters in our example (see Table 6-1) is about 140 GOPS. The operations can be performed using single-precision floating point. There are a few options for an appropriate computational architecture. Figure 6-9 shows the dimensions of data parallelism for each transform within a subband. The highlighted dimension in each cube shows the principal processing vector for the subsequent transform. As such, this dimension is the innermost processing loop, and for efficiency it is best to organize the elements of these vectors in contiguous memory locations. The other two dimensions provide opportunities for data-parallel decomposition. The arrival of the data from the previous stage and the timing of the data also influence the algorithm decomposition strategy. The beamformer input data cube has dimensions $N_{pri} \times N_{rg\_dec} \times N_{ch}$. As mentioned, the adaptive beamformer performs an inner product between the vector of range gates for each channel and the beamforming weights. If multiple beams are being formed, then these vectors need to be multiplied by a matrix, where each column in the matrix is a set of weights for a separate beam. Data for a range gate arrive simultaneously for each channel. The next fastest dimension is the range dimension, appropriately known as the fast-time dimension; the PRI dimension is the slow-time dimension. The range dimension is, therefore, the next parallel dimension.

One way to carry out the computation is to use a systolic beamforming approach. Figure 6-10 shows a candidate hardware architecture, designed to accept 20 channels by 48 subbands at the input and to generate 5 beams and 48 subbands at the output. In this architecture, 20 custom beamformer chips are used. Shown ahead of the beamformers are (starting from the front) (1) polyphase filter and FFTs that form the subband analysis stage; (2) a sample matrix inversion (SMI) collector

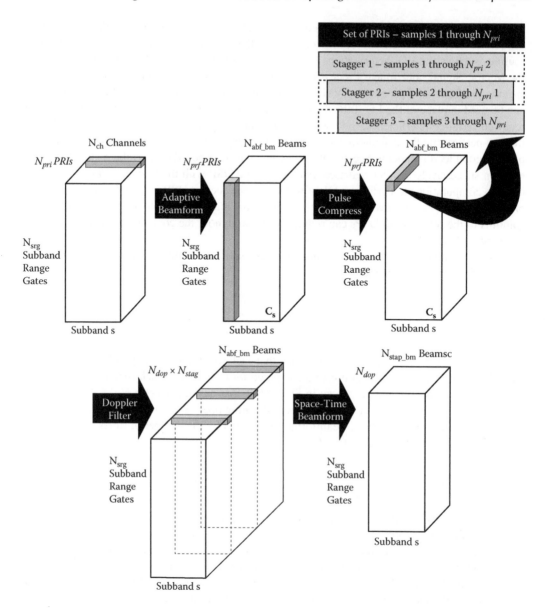

FIGURE 6-9 Data cube transformation within a subband. The figure shows a series of data cubes and the operations that transform each data cube to the next in the series. The dimension that is operated on is shown within the data cube as a shaded bar. For example, the first data cube is transformed by adaptive beamforming to the second data cube. Thus, the "channel" dimension is transformed into the "beam" dimension. There is ample data parallelism that the beamforming operation can exploit: each PRI can be beamformed in parallel and, within a PRI, each range gate can be beamformed in parallel.

function that selects training data for the adaptive beamformer and sends the data to an adaptive weight computer (not shown); and (3) a set of memory buffers that holds the channel data while the weights are being computed and downloaded to the beamformers. Each beamformer chip combines the samples from four channels for each of 12 subbands into partial beams. Four of these beamformer chips are clustered to generate partial beams for 48 subbands. The partial beams are communicated vertically in the diagram so that they can be combined with the partial beams from the other clusters. Each cluster outputs one complete beam for the 48 subbands. Note that the dynamic range of each beam may be up to 20 times the dynamic range of the input, but, on the other hand,

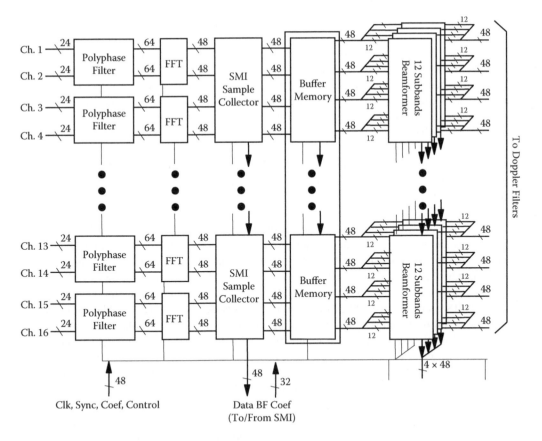

FIGURE 6-10   Wideband beamformer.

20 channels are combined into one output (for each beam), so 1/20 the data rate is needed. Thus, these two factors counterbalance so that the same number of output lines as input lines suffices, although the output lines must multiplex the output data. The interchip communication path must accommodate the movement of partial sums for five beams for each subband, thereby presenting a design challenge. For the application discussed here, only four beams are required from the adaptive beamformer (the ability to compute a fifth beam can be viewed as a future growth option), so the interchip communication load would be less. These beams are sent to the next processing stage (pulse compression), where the data must be demultiplexed and corner-turned.

The total throughput requirement of the beamformer in our example is 140 GOPS; each of the 20 beamformer chips must handle 1/20th of this load, leading to a requirement for 7 GOPS per chip. This is readily within the capability of modern FPGAs and standard cell ASICs. For example, the Xilinx Vertex 2 6000 FPGA can deliver up to 15 GOPS of throughput. Using custom ASIC technology, it would be feasible to aggregate the 20 chips into a single, high performance VLSI implementation. One challenge in using a single chip, though, is the need for very-high-speed I/O to marshal data into and out of the chip using a relatively small number of I/O pins. If the beamformer is implemented with programmable DSPs, the mapping onto a parallel processor needs to be considered. In this case, one might choose to beamform each subband in a separate DSP-based multicomputer. This would require an interconnection network upstream of the beamformer to gather the data for each subband from each receiver channel and store it in a buffer. From this point on, the signal processing could proceed independently for each subband. Consider the circa-2003 flight-qualified dual CPU (central processing unit) node discussed previously and shown in Figure 6-8. Each subband requires 140 GOPS/48 subbands = 2.9 GOPS. Thus, if the implementation can achieve an efficiency of 100*2.9 GOPS/(2.7 GOPS/node *2 nodes) = 54%, two nodes per subband would be sufficient. Beamforming is fundamentally an inner product, which uses multiply-accumulate (MAC) operations. DSPs can generally achieve an efficiency of over 50% on tasks that are dominated by MAC operations, so the 54% efficiency requirement is a reasonable goal. To further refine the design, one would need to benchmark the candidate DSP to get a more accurate efficiency estimate. The power consumption of each node is estimated at about 5 W. Thus, each subband beamformer consisting of two boards would consume 10 W. The processing could be partitioned in several ways. For example, each of the four processors in the two-node configuration could beamform four channels in the 20-channel system. The processing in this case would proceed in a systolic fashion. The first DSP would form four partial beams across five channels and would then pass these partial beams to the next processor, which would form its own four partial beams, add them to the partial beams from the first DSP, and so on until the last DSP adds the input partial beams to its partial beams and outputs five fully beamformed signal streams. Alternatively, a data-parallel approach could be used in which each processor could be dedicated to producing a single beam, for a total of four beams. This arrangement is likely to be more efficient than the systolic approach since inter-DSP communication is not required. However, the systolic approach has the advantage that it is more scalable. For example, if the computational efficiency falls short of the required 54%, an extra DSP can be allocated to the subband for a total of five DSPs. The same systolic architecture would apply, but with five DSPs each DSP would beamform four channels. An even more scalable approach would be to allocate 1/4 of the range extent to each processor. In this case, it would be easy to divide the range extent into five segments and use an extra DSP if four DSPs could not deliver the required throughout. For all of these configurations, processing 48 subbands would require 96 nodes and would consume about 480 W. This is significantly more power than would be required for an FPGA implementation, which could achieve about 3 GOPS/W using present-day technology, for a total of around 47 W. A custom VLSI beamformer chip with 200 GOPS throughput would consume on the order of 1 W using the 0.9 nm lithography process. The need to buffer three CPIs in about 5.2 Gbytes of synchronous dynamic random access memory (SDRAM) leads to a relatively large power consumption, and this memory is needed whether VLSI, FPGA, or DSP technology is used for the computation. Each 128 Mbyte SDRAM module, when being accessed, will consume about 2 W. The total power consumption if all memory is being accessed could be as high as 82 W.

The main advantage of the programmable approach is that it is flexible; it is easy to adjust the number of channels, beams, subbands, and PRIs since these are programmable software parameters. Thus, the programmable solution is often pursued for SMTI radars embedded in manned airborne platforms, where power and size are less constraining than in UAV configurations. In the example presented here, the circa-2003 processor board shown in Figure 6.8 has been used. This processor is significantly less capable than the latest processors. For example, modern processors such as the IBM Cell, introduced in 2005, and the Raytheon MONARCH (Prager 2007) are capable of delivering impressive programmable performance. The Cell processor is rated at over 200 GOPS while consuming an estimated 100–200 W (about 1–2 GOPS/W). The MONARCH processor promises to deliver up to 64 GOPS while consuming between 3 W and 6 W (about 10 to 20 GOPS/W).* In practice, flight-qualified boards typically lag their commercial counterparts by one or two generations, and as new processors become available, it takes a few additional years before they appear in ruggedized systems. The MONARCH processor is a case of the technology being developed expressly for

---

* In the same timeframe (circa 2006) beamformer VLSI chips capable of greater than 1 TOPS performance are being prototyped at MIT Lincoln Laboratory using the 90 nm complementary metal oxide semiconductor (CMOS) process. These processors will consume about 1 W. It is not the intent of this chapter to pass judgment on one type of technology over another. If pure performance were the only consideration, custom VLSI would always outperform programmable processors by orders of magnitude in power efficiency and throughput per unit volume. However, development cost, system complexity, time-to-market, technology refresh, flexibility, and scalability are all important factors that will figure into the overall design evaluation. The goal of this chapter is to illustrate the design trades that must be considered in architecting a high performance embedded processor of a chosen technology type.

Department of Defense applications, so the transition to an air-qualified system may be very rapid. The technology used in the example presented here has been chosen principally because it is a good example of a form-factored embedded processor card and serves well to illustrate the mapping and parallelization techniques that apply to all programmable HPEC system designs.

6.2.3 PULSE COMPRESSION

Pulse compression follows the adaptive beamformer and can be carried out independently in each subband. Pulse compression is a matched-filter operation. The return signal is correlated with a conjugate replica of the transmitted waveform. The transmit waveform typically extends for several range gates, thereby permitting a larger amount of power to impinge on the target. By match filtering the return signal, the returned waveform can be compressed into the range gate where the target resides. An efficient implementation of the pulse compressor often uses a fast convolver implementation, in which the signal is converted by an FFT to the frequency domain, point-wise multiplied with the waveform conjugate replica, and then converted back to the time domain by an inverse FFT (IFFT). For longer convolutions, this frequency-domain, or fast convolver, approach is more computationally efficient than the direct, time-domain convolution. As shown in Figure 6-9, the pulse compression operation is carried out in each subband across the samples in a PRI. For the fast convolver, the length of the FFT (and IFFT), N_{fft-pc}, is chosen to be the first power of two greater than the number of samples in the PRI. For the parameters in this example, $N_{fft-pc} = 4096$. The computational throughput of the frequency domain pulse compressor is computed as

$$F_{pc} = (2 * C_{fft-pc} + C_{mult} * N_{fft-pc}) * N_{subband} / PRI ,$$

where C_{fft-pc} is the complexity of the FFT (and inverse FFT) applied to 4096 complex samples (the PRI samples need to be zero-padded out to the correct number of samples), and C_{mult} is the complexity of a point-wise complex multiply of a conjugate waveform replica sample with the transformed PRI sample. The above complexity evaluates, for the parameters in the design example, as

$$F_{pc} = (2 * (5 * 4096 * \log_2(4096)) + 6 * 4096) * 48 / 0.0005 = 198.18 \text{ GFLOPS} .$$

Given the high throughput requirement, VLSI or FPGA implementations for this stage of the processing would be appropriate, but programmable technology can also be used for larger platforms. The application of programmable DSPs is discussed below. First of all, it makes sense to allocate a pulse compressor for each subband, so that the total throughput requirement of each subband pulse compressor is $F_{pc}^{(s)} = 198.18 / N_{subband} = 4.13$ GFLOPS for each of the 48 subbands. If one uses the same 2.7 GFLOP/s dual-processor DSP node (Figure 6-8) discussed earlier and assumes a nominal efficiency of about 50%, then each subband would require $N_{nodes}^{(s)} = 4.13 / (2.7 * 0.50)$ or about three DSP nodes. Dividing the computation between three nodes would result in a reduction in the computational efficiency due to the movement of data between six DSPs and the required synchronization overhead. Another approach would be to round-robin PRIs between nodes, so that each processor would receive every sixth PRI. This would incur an additional latency in the computation of five PRIs (2.5 milliseconds), a reasonable trade-off for improved efficiency. Since the Doppler filter (discussed next) is the next downstream computation, it makes sense to output the pulse-compressed data in corner-turn order, so that each range gate is stored in memory adjacent to range gates with the same index (i.e., range gates at the same range are placed adjacent to each other in memory).

Using 5 W per node, the total power requirement for each subband pulse compressor would be about 15 W. The total power consumption for the pulse compression stage across all subbands would be 720 W. Since the required number of nodes is dominated by the efficiency of an FFT, one would expect the system implementers to optimize the FFT code at the assembly level. Many

DSPs can achieve close to 100% efficiency on the FFT. For example, the SHARC DSP has a special butterfly hardware instruction to provide maximum FFT efficiency. Thus, for a highly optimized programmable pulse compressor, a greater than 50% efficiency is quite reasonable. Since the FFT is a common benchmark for DSP performance, vendors usually provide detailed information on FFT performance as a function of size, FFT arithmetic type (complex or real), and whether the FFT is performed in place (meaning that input is overwritten by the output) or out of place. It is a good idea, however, to verify all vendor-quoted performance figures within the context of the application code in order to measure overheads not present in single-kernel benchmarks.

6.2.4 Doppler Filtering

Doppler filtering follows the pulse compressor. It is also performed independently in each subband. The Doppler filter operation is a Fourier transform across the PRI dimension. The STAP computation downstream of the Doppler filter requires at least two Doppler filter operations, with one Doppler filter operating on data that are staggered in time compared to the data from the other. Thus, the length of the vector input into the Doppler filters, as shown in Figure 6-9, is nominally the number of PRIs minus the number of staggered windows needed by the STAP computation. The figure shows three staggers for illustrative purposes, but in this example two are chosen, the minimum required by the STAP computation. The first Doppler filter uses a vector composed of PRIs 1 to 199; the second uses a vector of PRIs 2 to 200. Since two staggers are used, the net result is that two Doppler-filtered cubes are generated, each of which captures the Doppler content across slightly different time windows. This temporal diversity is exploited by the STAP computation to steer beams in Doppler space that can null out clutter that competes with small targets in specific Doppler bins. The overall computational throughput of the Doppler filter is computed as

$$F_{dop} = (5N_{dop} \log_2(N_{dop}) * N_{rg\_dec} * N_{bm\_abf}) * N_{stag} * N_{subband} / CPI .$$

The values for the parameters in this equation are given in Table 6-1. The equation is interpreted simply as the complexity of the Doppler FFT acting on the input data cube, repeated for each stagger and subband. The computation must be performed with the latency of a CPI, giving the overall throughput requirement of 66.65 GOPS. The subband throughput requirement is 66.65/48 = 1.39 GOPS. If one assumes about 50% efficiency for this computation, about 2.78 GOPS per subband are required, so allocating one 2.7 GOPS node per subband is feasible.* The power consumption of the Doppler filter implementation would, therefore, be the power consumption of 48 nodes, or 240 W.

6.2.5 Space-Time Adaptive Processing

Space-time adaptive processing is the final step before subband synthesis. The goal of STAP is to reduce clutter that competes with small targets. Movement of the platform allows clutter (unwanted returns) from beam sidelobes to obscure smaller mainbeam targets. The relative range rate between

---

* In the computational allocations so far, no margin has been added for requirements growth; moreover, a fairly aggressive optimization phase has been assumed. In typical military system acquisitions, on the order of 50% spare processing and 100% spare memory are specified for programmable systems to allow for future growth. Providing ample resources for the software development also helps to reduce the cost and effort of the initial development. Generally, when an algorithm implementation reaches about 80% of the maximum capacity of the computation, communication, or storage resources, the effort to develop the implementation begins to grow dramatically. Therefore, while the mappings developed here are quite feasible, it is important to refine the design by conducting benchmarking that provides good estimates of the expected utilization of processing, memory, and communication. The cost of software development must be weighed against the hardware cost. If the production volumes are expected to be high, the optimization effort to fit the software into a resource-constrained processor may be economically justified. In this case, although the software cost is high, it can be amortized over the large number of units that are sold; at the same time, reduced hardware cost will translate into lower production costs.

the radar and the ground along the line of sight of the sidelobe may be the same as range rate of the target detected in the mainbeam. In this case, both ground clutter and target returns occupy the same range bin even after Doppler processing has been used to resolve the signals into Doppler bins. If the target cross section* is very small, its returned energy will be masked by the sidelobe clutter. To remove this clutter, the STAP algorithm used here constructs a two-dimensional filter that uses two Doppler-staggered data cubes to provide the temporal diversity needed to remove the competing clutter. By adaptively combining the returns in the two Doppler data cubes, a spatio-temporal null can be steered to remove clutter that impinges at the same Doppler but from a different azimuth, while preserving small mainlobe signals. The topic of STAP is treated in detail in several texts. The reader is referred to *Fundamentals of Radar Signal Processing* by M.A. Richards (2005) for an excellent treatment. The STAP computation is similar to the adaptive beamforming stage except that a two-dimensional (space and time) adaptive filter is constructed instead of a purely spatial filter (beamformer). Once again, the optimal filter weights are computed using the Weiner-Hopf equation. The same computational technique is applied here, namely, the computation of the sample covariance matrix is avoided in favor of the "voltage domain" approach, in which the sample matrix is factored into L and Q and two backsolves are employed to compute the adaptive weights. The sample matrix is a matrix of dimensions: $N_{dof\_stap} \times N_{training\_stap}$. N_{dof_stap} is the number of degrees of freedom exploited in the STAP computation. It is equal to the number of beams in a Doppler data cube multiplied by the number of staggered Doppler cubes. So, in the example presented here, $N_{dof\_stap} = 8$. $N_{training\_stap}$ is the number of training samples collected from each of the eight beams; thus, $N_{training\_stap} = 6 * N_{dof\_stap} = 48$. The length of the STAP steering vector is the number of beams that must be combined, i.e., $N_{dof\_stap} = 8$. The steering vector is composed of two spatial steering vectors stacked one on the other, with the second vector being a replica of the first, except that all of its entries are Doppler-shifted by one bin. In the example presented here, three STAP weight vectors, one for each desired look direction, are computed. The overall computational throughput of the STAP weight computation is calculated in a similar manner as the calculation of the throughput requirement for the adaptive beamformer weight computation. The major difference is that in this case the weights are not interpolated across PRIs (there are no PRIs at this point). Instead, a separate weight matrix is computed for each Doppler bin. The overall complexity of the weight computation is given by

$$F_{stap\_weights} = (C_{lq} + 2C_{backsolve} * N_{bm\_stap}) * N_{subband} * N_{dop} / CPI .$$

The complexity of the LQ decomposition is

$$C_{lq} = 8 * N_{dof\_stap}^2 * (N_{training\_stap} - (N_{dof\_stap} / 3)) \text{ FLOPs}$$

$$= 8 * 8^2 * (48 - (8 / 3)) \text{ FLOPs} = 23,211 \text{ FLOPs}.$$

The complexity of the backsolves stage is the complexity of two backsolves for each STAP beam vector:

$$2C_{backsolve} * N_{bm\_stap} = 2 * 4 * N_{dof\_stap}^2 * N_{bm\_stap} = 2 * 4 * 8^2 * 3 \text{ FLOPs} = 1,536 \text{ FLOPs}.$$

---

* The radar cross section of a target is a measure of the power reflected by the target, which generally varies with the aspect of the target with respect to the incident energy. Cross section is generally normalized to the reflected power of a unit-area surface that isotropically reflects all of the incident energy. Refer to Richards (2005) for a more precise definition.

Substituting into the throughput equation and using the coherent processing interval time of 100 ms yields

$$F_{stap\_weights} = (23,211 + 1,536) * 48 * 199 / 0.1 \text{ FLOPS} = 2,364 \text{ MFLOPS}.$$

This STAP weight computation is often performed in single-precision floating point using programmable hardware. The processing node discussed previously has a peak throughput of 2.7 GFLOPS. About 25% efficiency can be expected on this computation on a single processor, so (2.364/2.7)/0.25 ~ 3.5 nodes are required to compute all the weight matrices for all subbands and Doppler bins. One way to proceed would be to spread the computation of a single weight matrix over multiple nodes and compute each of the 48*199 = 9552 weight matrices in turn in this fashion. Each node contains two processors, so this computation needs to be mapped onto at least seven processors. However, spreading the LQ computation over seven processors will lead to a significant reduction in parallel efficiency due to the fine-grained nature of the computation and the overheads incurred in communicating between nodes. Figure 6-11 shows the overall efficiency for a parallel QR (or equivalently, an LQ) factorization as a function of the number of processors for a prototype weight computer developed at MIT Lincoln Laboratory. Although different communication systems and processors will yield different quantitative results, and the size of the matrix will also influence the achieved performance, the overall trend in the curve will be similar for any conventional programmable processor. In the figure, a cyclic distribution of the matrix is used, as this provides the best load balancing for the chosen Householder decomposition technique. Notice that the computational efficiency is over 30% for a single node, but falls off rapidly as the computation is spread over more processors. In fact, the maximum speedup achieved for this implementation is just over three and occurs when eight nodes are used. It is striking to observe

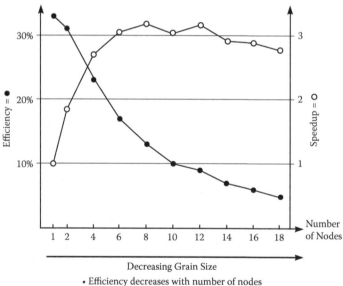

- Efficiency decreases with number of nodes
- Ultimate limits to speedup

Householder QR Factorization

240 × 48 complex matrix

4.1 million operations

Preliminary results from an i860-based Mercury (80 MFLOPS peak per node)

FIGURE 6-11   Overall efficiency for a parallel QR factorization as a function of the number of processors.

that this implementation could never achieve the weight computation throughput required here, no matter how many nodes were employed.

Fortunately, other parallel partitionings exist that are much more efficient. For example, a separate STAP weight computer subsystem can be allocated for each subband. Each subband requires 2.364/48 GFLOPS ~ 50 MFLOPS. The 2.7 GFLOPS node would require an efficiency of only 100* 50/2700 ~ 2%. However, this implementation would then require 48 nodes and would consume 48 nodes * 5 W/node = 240 W. To reduce power (and space) consumption, multiple subbands can be allocated to each node. Assuming about 25% efficiency by exploiting the subband data parallelism (which avoids spreading the LQ computation between processors) 1/4 of the 48 subbands (i.e., 12 subbands) can be allocated to a node. Each processor on the node would handle six subbands. The overall throughput requirement of the node would be 2.364/4 GFLOPS ~ 591 MFLOPS. This would require a very reasonable efficiency of 100*(591/2700) ~ 22%. Using four nodes, the STAP weight computation subsystem would therefore consume four nodes * 5 W/node = 20 W.

Once again, this discussion underscores the need to perform benchmarking on the candidate processor system, especially if form-factor constraints are severe, as would be the case for an SMTI radar in a small UAV. It also points to an important design guideline: if parallelism is needed in a programmable system, one should avoid fine-grained parallelism and exploit coarse-grained parallelism whenever possible. Fine-grained implementations will have significant overheads for most embedded multicomputers, so that parallel efficiency and speedup will drop off rather quickly.

Once the weights are computed, the STAP beamformer operation uses them to combine the eight input beams (four beams in each of two Doppler cubes) in the concatenated data cube, $X^{(s)}_{concat\_dop}$, to produce a new set of three output beams. The resultant STAP beams adaptively minimize the clutter energy in each Doppler bin while focusing in the directions specified by the steering vectors. The beamforming operation consists of a series of inner products in each subband in each Doppler bin:

$$Y^{(s)}_{dop} = \left(W^{(s)}_{bm\_stap} \right)^H X^{(s)}_{concat\_dop} .$$

The resulting matrix $Y^{(s)}_{dop}$ is organized such that each row is a STAP beam and each column is a range gate. The matrix-matrix product must be computed for each Doppler in each subband. This produces a data cube in each subband, as shown in Figure 6-9, that has dimensions $N_{dop} * N_{rg\_dec} * N_{bm\_stap}$. The overall throughput requirement for the beamforming operation is the complexity of a complex matrix multiplication of the basic beamformer multiplied by the number of Doppler bins and subbands, divided by the length of a CPI:

$$F_{stap} = (8 * N_{dof\_stap} * N_{rg\_dec} * N_{bm\_stap}) * N_{dop} * N_{subband} / CPI .$$

The net throughput, using the parameters in Table 6-1, is about 42.3 GOPS. Once again using the subband dimension of data parallelism, the computation can be divided amongst 48 subband processors, each of which will need to support 42.3 GOPS/48 = 881 MFLOPS. The beamformer performs multiply-accumulate operations, which have dedicated hardware support on DSPs, so an efficiency of 50% or more is to be expected. The required efficiency if each 2.7 GOPS node handles a separate subband is 100*(881/2700) = 33%, well within the 50% range. Since each node consists of two processors, the computation will have to be divided. There are two ready alternatives. Each processor can either handle half of the range gates or half of the Doppler bins. If 48 nodes are used, then at 5 W per node, the STAP beamformer would consume 240 W. If a lower-power solution is required, hand-optimization of the beamformer code, using various techniques such as loop unrolling, would be required. Benchmarking would be used to develop an accurate estimate of the achievable efficiency. If the measured efficiency is, say, 66% or greater, two subbands can be squeezed into a single node, with each node processor dedicated to a separate subband. Since DSPs are designed to perform efficient inner products, a goal of 66% efficiency is reasonable. However, if benchmarking shows that the achievable efficiency is less than 66%, say 50%, then it may be necessary to distribute three subbands across two nodes, using the range or Doppler dimension as the partitioning dimension. In this way, 2*2.7 GOPS = 5.4 GOPS of processing power can be applied to an aggregate three-subband beamformer requiring 3*0.881 = 2.643 GOPS. The net efficiency required with this partitioning is, therefore, 100*(2.634/5.4) = 49%. This configuration would need 2/3*48 = 32 nodes, thereby reducing power consumption to 160 W.

6.2.6   SUBBAND SYNTHESIS REVISITED

The subband synthesis phase, as shown in Table 6-2, has a throughput requirement of 76 GOPS. ASICs or FPGAs would be suitable choices for this computation. For example, noting that the synthesis process is the complement of the analysis process, one could use similar chip technology to that discussed for subband analysis. From a computational perspective, the major difference between the synthesis and analysis processes is that in analysis there is a fanning out of the data stream into multiple subbands, and in synthesis special attention must be paid to the fan-in of the multiple subbands. From the algorithmic perspective, in subband analysis the filtering operation occurs first followed by the FFT, whereas in synthesis these two steps occur in the reverse order. A 15-tap low-pass filter (versus the 12 taps used in the analysis phase) is used, which has been shown to provide excellent performance after the subband processing.

6.2.7   CFAR DETECTION

Detection is done after subband synthesis. The detection stage serves as the interface between the front-end signal processing and the back-end data processing. A basic constant false-alarm rate detector is employed that tests for a target in every range gate. The CFAR threshold computation, as shown in Figure 6-12, computes an estimate of the noise background around the cell (range gate) under test (CUT). This computation is accomplished by averaging the power in the cells around the CUT. Once the estimate is formed, the power in the CUT is computed and the ratio of these two numbers gives an estimate of the SNR of the CUT. A threshold value is compared to the SNR of the CUT and if the SNR is greater, a target detection is declared. The process is called a constant false-alarm rate process because, under assumed noise statistics (Gaussian), the probability that the energy in CUT will exceed the threshold in the absence of a target is a constant. Thus, the rate at which false alarms will occur will be a constant that depends on the threshold setting. The radar designer typically picks an acceptable noise-only false-alarm rate and then tries to design a system that maximizes the probability of detection.

The computational complexity of the CFAR is easy to compute. For each CUT, an average of the power in the surrounding cells is computed. This can be done with a moving average, where a previously computed average is updated with the energy in the next cell. Usually, an excised section is placed around the CUT to guard against the possibility that part of the target power might appear in the adjacent cells, thereby biasing the noise estimate. First, the power in every CUT is computed. Updating the moving average consists of simply subtracting the trailing-edge cell powers and adding the leading-edge cell powers. Since cells have been excised around the CUT, there are two sections in the moving average and hence two trailing and two leading edges. Thus, four floating-point operations are required for each CUT to compute the noise average. To compute the power in the CUT, the real and imaginary parts of the signal are squared and added together, for a total of three operations. Then, the ratio of the noise power to the signal power is computed, which requires a floating-point divide. Hence, to compute the SNR in each cell requires eight operations. The comparison operation is not counted in the throughput calculation since it is not an add, a multiple, or a divide, although arguably, since it is an operation that must be performed to complete the

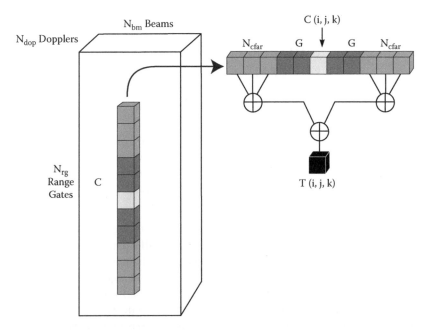

FIGURE 6-12 The CFAR threshold computations. For each cell under test (labeled C), N_{cfar} range gates to the left and right of the guard cells (labeled G) are averaged to compute a threshold T.

calculation, it should be taken into account. Note that this is the first instance in the processing flow in which a computationally significant non-arithmetic operation emerges. This is, in effect, an indication that the processing flow is moving from the front-end processing regime to the back-end processing regime, where computations tend to involve a larger number of non-arithmetic operations.

The detector is also the last stage in the processing stream in which a large volume of data is operated on by a relatively simple set of computations. The data cube processed by the detector contains $N_{dop}*N_{rg\_syn}*N_{bm\_stap} = 199*81,000*3 = 48,357,000$ cells (range gates). After the detector, the processing operates on targets and tracks. The reduction in the volume of data elements is from about 50 million range gates into the detector to 20,000 target reports or less output from the detector. (One hundred target reports have been budgeted in each Doppler, for a total of 19,900 reports. This estimate includes false alarms and provides margin for load imbalances due to target clustering.) The back-end processing operations that are carried out on the targets and tracks, however, are much more complicated than processing carried out in the subband processing stages and the detector. Thus, the detector serves as the interface between the front-end stream processing and the back-end data processing stages of the computing system. The reader should refer to Chapter 5 for a more in-depth discussion comparing back-end and front-end computational characteristics.

The overall detector throughput of the CFAR stage is eight operations per range gate (cell) in the synthesized data cube divided by a CPI:

$$F_{det} = (8 * 48,537,000)/0.1 = 5.37 \text{ GFLOPS}.$$

Given the large amount of data parallelism in this computation and the relative simplicity of the calculation, the detector could be efficiently mapped to a set of FPGAs. However, the detector is often mapped to the same processors that handle the estimation of target azimuth, range, range rate, and SNR, since the estimation algorithms make use of the data in the vicinity of the detections. Parameter estimation can be quite complex and is better suited to software-programmable devices. Thus, detection is often handled in programmable hardware. Hybrid architectures that use FPGAs for the basic detector and DSPs or microprocessor units (MPUs) for the estimation processing are also feasible.

TABLE 6-3

SMTI Memory Requirements (Mbytes)

Subband Analysis	0
Adaptive Beamforming	5,218
Pulse Compression	346
Doppler Filtering	346
STAP	889
Subband Synthesis	0
Detection	240
Total	7,038

Several data-parallel dimensions can be exploited to map the detector onto a set of processors. For example, the Doppler dimension readily affords 199 parallel data partitions, one for each Doppler bin. With this partitioning, the along-range averaging (the noise calculation) can operate on range gates arranged contiguously in memory. Each Doppler bin requires 5.37/199 = 27 MFLOPS. Assuming a conservative estimate of 20% efficiency for a DSP performing detection processing and using the 2.7 GFLOPS nodes, 2700*.2/27 = 20 Dopplers per node can be processed. Thus, to process the 199 Dopplers, 10 nodes are needed. The total power consumption would be 10 nodes*5 W/node = 50 W.

6.3 EXAMPLE ARCHITECTURE OF THE FRONT-END PROCESSOR

So far, the computational aspects of the entire front-end sensor processing algorithm, from the subband analysis stage through to the detection stage, have been explored, and various mapping strategies and processing hardware options have been discussed. A small UAV system is now considered as the host platform for the SMTI radar. The platform imposes form-factor constraints that help narrow the trade space: the radar processor must weigh less than 1 kg, it must consume less than 1 kW prime power, and it must occupy less than 4 cubic feet.* The processor must also be aircraft flight-qualified.

The adaptive weight computations must be done in floating point to accommodate dynamic range requirements, and although a baseline algorithm has been chosen, there is a strong desire to provide programmable capabilities so that the algorithm can be enhanced over time as new or refined techniques emerge.

The focus in this chapter has been the mapping of a front-end algorithm to a high performance computing system, but there are several other important design considerations that need to be factored into the design. For example, the memory requirements for each stage of the computation need to be considered. Table 6-3 shows the estimated memory requirements for the processor for each processing stage. Also, given the high bandwidth of the sensor data and the need to perform several complex data reorganization steps through the course of the processing chain, a complete design would need to carefully consider the interconnection network. In fact, for wideband applications, the movement of data is every bit as important as the processing. Table 6-4 gives an estimate of the communication bandwidth between the computation stages for the application considered here. Both the raw data rates and the data reorganizations that must occur between stages are important design consideration. For example, the subband synthesis phase must collect data from all subbands, requiring careful design of the fan-in interconnects system. Between the beamformer and the Doppler filter stages, the data must be buffered for a complete CPI (so that all PRIs are collected), and then the data are operated on in corner-turn order. This requires a communication

---

* These form-factor numbers represent a portion of the payload budget in a typical Predator-class UAV.

TABLE 6-4

Communication Data Rate per SMTI Stage

Stage	Input (Gbytes/s)	Input (Gbits/s)
Subband Analysis	12.96	103.7
Adaptive Beamforming	17.28	138.2
Pulse Compression	3.46	27.6
Doppler Filtering	3.46	27.6
STAP	6.88	55.0
Subband Synthesis	2.58	20.6
Detection	1.93	15.5
Downlink	0.05	0.4

system with high bisection bandwidth. The circuitry required for the memory and interconnects can consume a significant amount of power, volume, and weight, and therefore needs careful consideration in a full design.

Figure 6-13 shows a first-cut assignment of processing hardware for each processing stage in the UAV processor front-end. Table 6-5 presents a first cut at the allocation at the module level: the throughput per module and per subsystem and the estimated power requirements are given. Custom ASICs based on 0.25 CMOS technology have been chosen for the high-throughput signal processing chain. Standard cell ASICs would also be a viable design in place of custom VLSI. This would reduce the nonrecurring engineering costs significantly, but would incur about a factor of 10 reduction in chip-level power efficiency (operation per watt). Programmable DSPs have been chosen for the weight computations (referred to as SMI or sample matrix inversion in the figures) since (a) floating point is required and (b) techniques for sample matrix data selection and the weight-computation algorithms are complex and generally subject to change as improvements are discovered. A software implementation accommodates both of these considerations. The detector is also implemented with DSPs, which have the programmability to handle both the CFAR computation and the more complex target parameter-estimation algorithms. These processors must also control the interface to the communication system that will downlink the reports to a ground station. For the programmable DSPs, the flight-qualified, 4.7 GFLOPS, 5 W nodes described previously have been selected.

The overall HPEC system as configured consumes about 220 W and provides an aggregate computation capability rated at 1053 GOPS. It is interesting that more than half of the power is used

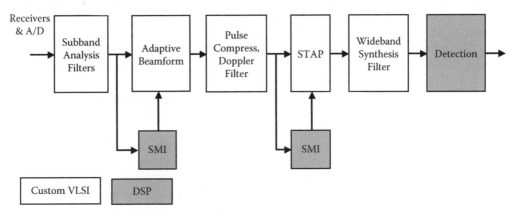

FIGURE 6-13 First-cut assignment of processing hardware for each processing stage in the front-end.

TABLE 6-5

First Cut at the Allocation at the Module Level

Stage	Throughput (GOPS)	Units	Total (GOPS)	Power (Watts)	Memory (Mbytes)	Memory Power (Watts)	Total Power (Watts)
Analysis-PPF	28	10	280	6	–	–	6
Analysis-FFT	20	10	200	5	–	–	5
ABF	7	20	140	4	5248	82	86
ABF-SMI*	2.7	1	2.7	5	–	–	5
PC	4.2	48	202	5	384	6	11
DF	1.4	48	67	2	364	6	8
STAP	4.2	10	42	1	880	14	15
STAP-SMI*	2.7	4	10.8	20	–	–	20
Synthesis-PPF	2.3	24	55	2	–	–	2
Synthesis-FFT	1.1	24	26	2	–	–	2
DET*	2.7	10	27	50	512	8	58
Total			1053	102		116	218

*denotes DSP; all others are custom VLSI

to access memory. The memory access power consumption is actually an upper-bound estimate since it assumes that all of the memory is being accessed at one time, which is clearly unrealistic, but the size, weight, and power budgets of memory modules are nevertheless very important aspects of HPEC design. Also, note that the custom VLSI is very power efficient and consumes only 27 W of the 102 W budgeted for computation. The remaining 75 W are consumed by the DSP boards although they only have a peak throughput of 38.5 GFLOPS, which is less than 4% of the overall processor load (albeit the computations are floating point). The power-consumption estimates account for memory, but do not include the input and output (I/O) circuitry. Based on a rule of thumb that I/O is generally 1/3 of the overall power budget, the I/O subsystems could increase the power budget by 110 W, more or less. A detailed analysis would be required during detailed design, in order to determine a more accurate number.

6.3.1   A Discussion of the Back-End Processing

This chapter concludes with a brief discussion of the back-end data processing which, in the example chosen here, consists of a kinematics Kalman filter tracker augmented by feature-aided tracking algorithms. The signature-aided tracking variant of FAT has been chosen. SAT and another FAT variant, known as classification-aided tracking (CAT), can be found described in Nguyen et al. (2002). Both SAT and CAT aid in the track-to-target association process, with CAT also leading to target classification, important information for the radar operator. For example, a target may be classified as a hostile entity such as a tank, and classification to a particular tank type would add further knowledge to be exploited in a kill prosecution. The feature-aided tracker improves association accuracy; its greatest value is in dense target environments, especially for cases in which target paths cross or come into close proximity to one another.

In front-end processing, operations are applied to SMTI radar data streams. Relatively few operations are applied to each datum, and the data are transformed as they move through the processing chain. This is evident in Figure 6-9, which shows how channel data are transformed to beam data, and PRI data are transformed to Doppler bin data, etc. The throughput of the computations is dominated by the sheer size of the data cubes involved, and the computations stay the same regardless of the values of the data. In other words, the computations are data invariant.

This allows us to use an HPEC processing architecture that exploits workloads that can be determined in advance, several dimensions of data parallelism, natural pipelining, and static data-flow interconnectivity.

By contrast, the back-end processing computations occur on a per-target or per-track basis. Since the number of targets in the environment is not known *a priori* and the number of false alarms is at best statistically predictable, the workload will vary nondeterministically over time. The number of operations applied to a target or track is on the order of 100s or 1000s times more than the number of operations applied to each data element in a front-end data cube. Also, tracks persist over time (as long as the radar is operated and the real-world targets stay within detection range), and hence they must be stored, accessed, and updated over time periods that are long compared to the rather ephemeral front-end data streams, which represent snapshots of the world on the order of 100 ms (a CPI).

Since the majority of computations involve track objects and their data, it is natural to process these objects in parallel. In Figure 6-5, the first stages of the tracking system are all concerned with associating targets to tracks. The first step serves to exclude target-track associations that are highly unlikely. Various criteria can be used, but one of the easiest is to disqualify a target-track pair if the range difference between the track and target is too great. The complexity of this step is O (nm), where n is the number of targets and m is the number of tracks. In a fully track-parallel configuration, the targets could be broadcast to the tracks and the comparison could be made independently to each track. In the most fine-grained mapping, each track is handled by a separate, parallel thread. The column vectors containing the chi-squared values of the target-track pairs are computed in parallel, one column per track. The chi-squared value is the Euclidean distance between the target and track positions, normalized by the standard deviation of the track-position estimate. Target-track pairs that have been excluded are set to a maximum value. The columns are consolidated into a matrix for use by the (serial) Munkres algorithm (Munkres 1957), which determines the optimal association. Once the final associations are made, the Kalman filters for all tracks are updated using the associated target data.

To get an idea of the computational complexity of this basic tracker, refer to Figure 6-14. If there are, say, 1000 targets and tracks, the throughput (for the 100 ms update interval) to perform track filtering and update is about 12 MFLOPS. This workload is nearly five orders of magnitude lower than the front-end processing workload. The workload varies linearly with the number of tracks, so even a tenfold increase in the number of targets would only result in a 120 MFLOPS workload. As a host computer, a symmetric multiprocessor architecture with a multithreading operating system could be used. If parallelism is required, each track could be processed by a separate thread (as described above), and the threads would be distributed evenly amongst the P processors. If one track per thread is too inefficient due to operating system overheads in managing a large number of threads, then multiple tracks could be assigned to a thread.

The signature-aided tracker has the task of improving the target-track association and can be especially effective in dense target environments where tracks may be converging or crossing. The net effect of the SAT algorithm is a new chi-squared matrix that is subsequently used by the Munkres algorithm. The new matrix has elements that are simply the chi-squared values from the kinematics tracker added to chi-squared values from the feature comparisons.

To improve the association performance, a significant computational price must be paid. The SAT processing requires the computation of the mean-squared error (MSE) between the stored high-range-resolution (HRR) signature of the track and the HRR profile of the target. (The HRR profile is simply the range gates in the vicinity of the target report. For high-enough bandwidth, the target will occupy multiple range gates, thus providing a profile for the target.) The MSE value is then used to compute a chi-squared value. Only targets that are within an azimuth tolerance of a track are considered. If the azimuthal difference is too great, the target signature will most likely have decorrelated too much from the stored track profile. Another twist is that the stored HRR profile will, in general, have to be scaled and shifted to make the comparison more valid. The stored profile is first scaled to the maximum value and then shifted incrementally. The MSE is computed at each shift and the smallest value is retained. In the worst case where no targets have been eliminated in the coarse association and azimuth checks, the number of MSE computations is SNM, where S is the number of increments, N is the number of targets, and M is the number of tracks. The MSE complexity for the ith target compared to the jth track is

$$MSE_i^j = S * \frac{\sum_{k=1}^{K} w_k \left(t_k^j - h_k^i \right)}{\sum_{k=1}^{K} w_k}.$$

In the above, w_k is a weight that gives relatively more importance to samples near the center of the target (the range gate of the detection); t_k^j is the kth HRR profile value (power level) for the jth target. h_k^i is the kth HRR profile value for the ith track. One can assume that the weights and their normalizing factor are computed ahead of time. If trucks or tanks are the targets of interest, then with the 180 MHz bandwidth SMTI radar, which affords a range resolution of about 1 ft, about 16 elements are expected for an HRR profile. Suppose the profiles may be misaligned by up to eight increments, so that $S = 8$; the complexity of a single MSE measurement is about

$$C_{mse} = S * 2K = 8 * 2 * 16 = 256 \text{ FLOPs}.$$

If there are $N = 1000$ targets and $M = 1000$ tracks, then for a CPI = 0.1 seconds, the overall throughput requirement of this computation is

$$F_{mse} = 256 * \text{NM/CPI} = 256 * 1000 * 1000 / 0.1 = 2.56 \text{ GFLOPS}.$$

Apart from the Munkres algorithm, the other tasks that constitute the association logic can be parallelized along the track dimension. Thus, the above throughput requirement could be met by spawning 1000 threads on a shared-memory multiprocessor. Of course, the speedup would not be 1000, but would be close to the number of processors that the system contains. To minimize thread overheads, one would divide the number of tracks amongst processors. For example, in a 16-processor shared multiprocessor, roughly 1000/16 = 63 tracks would be processed per thread, with one thread per processor. Some processors support multithreading or hyperthreading, so more threads could be used profitably in these systems. The Munkres algorithm, unfortunately, has a complexity that grows exponentially with the size of the association matrix, and efficient parallel variants have not been developed (to the knowledge of the authors). Thus, this computation can become a bottleneck limiting the scalability of the association phase. Figure 6-14 shows the exponential increase in Munkres computational complexity as a function of number of tracks and targets. In the figure, the algorithm complexity has been divided between comparison operations and arithmetic computations. When the tracking system contains about 1000 tracks and targets, the Munkres algorithm requires about an order of magnitude more computations than the track filters.

Many other computations may need to take place in the back-end processor. For example, the high range resolution can be used to do target classification. This would involve comparing the profiles to stored profiles that correspond to known target classes. An alignment, similar to what was done for the SAT, would be performed and then comparisons would be carried out for a range of profile replicas, with each replica being a rotated version of the profile. Other operations that might be performed would include radar mode scheduling, radar display processing, and interfacing to a sensor network.

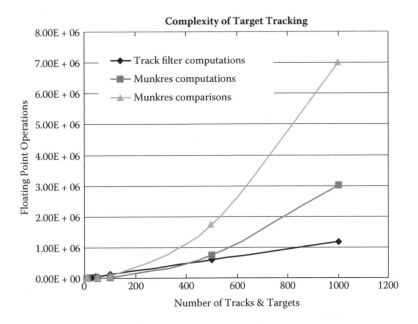

FIGURE 6-14   Computational complexity of basic tracker.

6.4   CONCLUSION

The computational aspects of a 1 TOPS throughput wideband SMTI radar application have been presented to illustrate the analysis and mapping of a challenging HPEC algorithm onto computational hardware. The dramatic form-factor difference between custom ASICs and software programmable processors has been very apparent. Power consumption was focused on as a key design metric, but it is also evident that significantly more hardware is required for a programmable solution. This translates directly into greater weight and volume. FPGAs have also been discussed as a middle-of-the-road alternative. If form factor and performance were the only issues, one might question why one should even consider the use of programmable processing in SMTI applications. Indeed, for cases in which form factor considerations dictate, ASICs and a judicious use of programmable processors are the dominant architectural ingredients. For many platforms, however, the form-factor constraints are less severe and, hence, more amenable to programmable technology. Programmable processors, when they can be used, are generally preferred to ASICs for HPEC for several important reasons. In particular, if an algorithm has a complicated flow of control or requires irregular memory accesses, a custom ASIC approach is often too difficult and costly to implement. Software programmable DSPs or MPUs, on the other hand, contain instruction sets explicitly designed to handle such algorithms. Moreover, DSPs and MPUs can be reprogrammed to accommodate algorithm changes. This is especially important in prototype systems or systems for which algorithm technology is expected to evolve. In an ASIC, a particular algorithm or algorithm kernel is "hard-wired" into the silicon, and a complete redesign may be needed to implement a new algorithm. Scalability, an aspect of flexibility that deserves special attention, is generally easier to accommodate in programmable systems. With sufficient design margin, programmable solutions can be scaled upward by changing soft-coded parameters. Programmable parallel processing systems encode the parallel partitioning in software. Thus, different mappings can be employed that best exploit a particular algorithm's parallelism as the application scales. For example, with careful design, one could change the number of Doppler bins assigned to a detector node. When the design of the detection algorithm is changed in a way that increases its complexity, fewer Doppler bins can be assigned to each node and either spare processing power can be exploited or more nodes can be added to the system. While it is feasible to parameterize ASIC-based designs in a similar manner, the number of reconfiguration options necessarily needs to be kept small to control the complexity of the hardware.

The nonrecurring engineering cost to develop an ASIC, especially a custom VLSI chip, must be accounted for in the overall cost. High performance embedded computing applications generally require the manufacture of only a handful to a few hundred systems, so the cost of the ASIC development can be amortized only over a few systems. Using commercially developed commodity DSPs or MPUs can, therefore, lead to a significant cost savings. On the other hand, in military applications, performance often overrides considerations of development cost, and VLSI can usually provide the best performance for a specific HPEC application.

If commercial processors are used, one can also exploit the Moore's Law rate of improvement of commercial technologies. After a few years, processors with twice the performance of the ones initially chosen can be used, providing proper attention has been paid to the use of standard commercial off-the-shelf (COTS) hardware components, so that new processor boards can be inserted into the system. The fast pace of Moore's Law, however, is a double-edged sword. Soon, components that have been used in the HPEC implementation may become obsolete. Either the design must explicitly accommodate the insertion of new technology, or else lifetime buys of spare parts will be required. One of the most important considerations is the cost of porting the software from the obsolete processor to the new one. This often involves a repartitioning of the parallel mapping. By isolating hardware details from the software through an intermediate level of software referred to as *middleware*, it is possible to develop portable software applications. Standard middleware libraries such as the Vector, Signal, and Image Processing Library (VSIPL) [http://www.vsipl.org/] can greatly improve system development time by providing highly optimized kernels and support for hardware refresh by isolating machine detail from the application code. Recently, parallel middleware libraries such parallel VSIPL (Lebak et al. 2005) have begun to emerge. These libraries provide mechanisms for mapping the algorithms onto parallel processors, thereby greatly reducing the amount of application code needed for programmable HPEC systems and providing highly optimized parallel signal and image processing algorithm kernels.

In conclusion, this chapter has illustrated how HPEC systems, such as the 1 TOPS SMTI radar presented here, place challenging form-factor, throughput, and latency demands on computing technologies and architectures. The design trade space is complicated and interdependent. Consequently, HPEC systems are usually hybrid computing systems in which ASICs, FPGAs, DSPs, and MPUs all have roles to play. Due to the intimate relationship between the algorithms and the HPEC design options, a co-design approach, in which algorithm variants and modifications are considered jointly with HPEC designs, invariably produces the best solution. Although a specific set of requirements and technologies were presented here for illustrative purposes, as technologies evolve and new algorithms and larger-scale systems emerge, the underlying mapping techniques and design trade-offs considered in this chapter will continue to apply to the challenging discipline of HPEC design.

REFERENCES

Eubank, R.L. 2006. *A Kalman Filter Primer.* Statistics: a series of textbooks and monographs, 186. Boca Raton, Fla.: Chapman & Hall/CRC.
Haykin, S. 2002. *Adaptive Filter Theory*, 4th edition. Upper Saddle River, N.J.: Prentice Hall.
Lebak, J., J. Kepner, H. Hoffmann, and E. Rutledge. 2005. Parallel VSIPL++: an open standard software library for high-performance parallel signal processing. *Proceedings of the IEEE* 93(2): 313–330.
Munkres, J. 1957. Algorithms for the assignment and transportation problems. *Journal of the Society for Industrial and Applied Mathematics (SIAM)* 5: 32–38.
Nguyen, D.H., J.H. Kay, B. Orchard, and R.H. Whiting. 2002. Feature aided tracking of moving ground vehicles. *Proceedings of SPIE* 4727: 234–245.
Prager, K., L. Lewins, G. Groves, and M. Vahey. 2007. World's first polymorphic computer—MONARCH. *Proceedings of the Eleventh Annual High Performance Embedded Computing Workshop*. MIT Lincoln Laboratory, Lexington, Mass. Will be available online at http://www.ll.mit.edu/HPEC/.

Reuther, A. 2002. *Preliminary Design Review: GMTI Processing for the PCA Integrated Radar-Tracker Application.* MIT Lincoln Laboratory Project Report PCA-IRT-2.

Richards, M.A. 2005. *Fundamentals of Radar Signal Processing.* New York: McGraw-Hill.

Ristic, B., S. Arulampalam, and N. Gordon. 2004. *Beyond the Kalman Filter: Particle Filters for Tracking Applications.* Boston: Artech House.

Song, W., A. Horst, H. Nguyen, D. Rabinkin, and M. Vai. 2000. A 225 billion operations per second polyphase channelizer processor for wideband channelized adaptive sensor array signal processing. *Proceedings of the Fourth Annual High Performance Embedded Computing Workshop.* MIT Lincoln Laboratory, Lexington, Mass.

Stimson, G.W. 1998. *Introduction to Airborne Radar,* 2nd edition. Mendham, N.J.: SciTech Publishing.

Ward, J. 1994. *Space-Time Adaptive Processing for Airborne Radar Submitter.* MIT Lincoln Laboratory Technical Report 1015. DTIC #ADA-293032.

Zarchan, P. 2005. *Fundamentals of Kalman Filtering: A Practical Approach,* 2nd edition. Reston, Va.: American Institute of Aeronautics and Astronautics.

Section III

Front-End Real-Time Processor Technologies

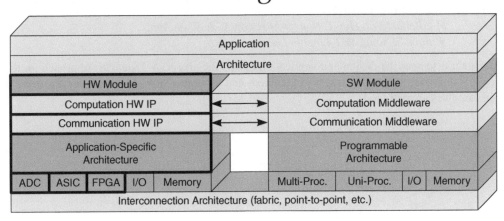

Chapter 7   Analog-to-Digital Conversion
James C. Anderson and Helen H. Kim, MIT Lincoln Laboratory

This chapter outlines the performance metrics commonly used by engineers to specify analog-to-digital conversion (ADC) requirements. An overview of the technological issues of high-end ADC architectures is also presented.

Chapter 8   Implementation Approaches of Front-End Processors
M. Michael Vai and Huy T. Nguyen, MIT Lincoln Laboratory

This chapter describes a general design process for high performance, application-specific embedded processors and presents an overview of digital signal processing technologies.

Chapter 9   Application-Specific Integrated Circuits
M. Michael Vai, William S. Song, and Brian M. Tyrrell, MIT Lincoln Laboratory

This chapter provides an overview of application-specific integrated circuit (ASIC) technology. Two approaches to ASIC design are described: full-custom and synthesis. The chapter concludes with a case study of two high performance ASICs designed at MIT Lincoln Laboratory.

Chapter 10 Field Programmable Gate Arrays
Miriam Leeser, Northeastern University

This chapter discusses the use of field programmable gate arrays (FPGAs) for high performance embedded computing. An overview of the basic hardware structures in an FPGA is provided. Available commercial tools for programming an FPGA are then discussed. The chapter concludes with a case study demonstrating the use of FPGAs in radar signal processing.

Chapter 11 Intellectual Property-Based Design
Wayne Wolf, Georgia Institute of Technology

This chapter surveys various types of intellectual property (IP) components and their design methodologies. The chapter closes with a consideration of standards-based and IP-based design.

Chapter 12 Systolic Array Processors
M. Michael Vai, Huy T. Nguyen, Preston A. Jackson, and
William S. Song, MIT Lincoln Laboratory

This chapter discusses the design and application of systolic arrays. A systematic approach for the design and analysis of systolic arrays is explained, and a number of high performance processor design examples are provided.

7 Analog-to-Digital Conversion

James C. Anderson and
Helen H. Kim, MIT Lincoln Laboratory

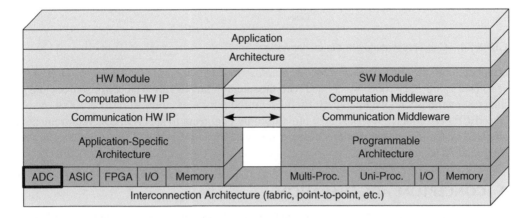

This chapter outlines the performance metrics commonly used by engineers to specify analog-to-digital conversion (ADC) requirements. An overview of the technological issues of high-end ADC architectures is also presented.

7.1 INTRODUCTION

An analog-to-digital converter (ADC) converts an analog signal into discrete digital numbers. In a sensor application, the ADC interfaces a front-end processor to an IF (intermediate frequency) circuit and converts the IF signal for digital processing. ADC characteristics (e.g., dynamic range, sampling rate, etc.) often determine the application design and performance. For example, channelization is commonly used to divide a wideband signal into narrow subbands and thus suppress the noise floor for detection of weak (below noise) signal targets. The performance of channelization is hinged on the ADC spurious-free dynamic range (SFDR). Without a sufficiently large SFDR, the target cannot be distinguished from the spurious noise.

High performance ADCs are critical in defense applications. However, in the last decade, the demand for high performance, low-cost ADCs has been driven largely by the need for rapidly improving embedded digital systems such as multifeatured cellular telephones and personal digital assistants (PDAs) to interface with a "real-world" environment. The wide range of ADC options available today requires that embedded systems designers make complex device-selection decisions based on a trade-off space with many variables, including performance, cost, and power consumption.

In July 1989, the draft form of the *IEEE Trial-Use Standard for Digitizing Waveform Recorders* (IEEE Std 1057) was issued with the intent of providing a set of measurement standards and test techniques for waveform recorders (IEEE 1989). Device characteristics measured as set forth in this standard were consistent and repeatable, so that performance comparisons could be made between devices provided by many different manufacturers (Crawley et al. 1992; 1994). Subsequently, in

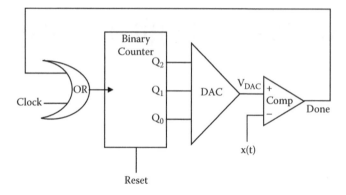

FIGURE 7-1   Conceptual analog-to-digital converter.

December 2000, the *IEEE Standard for Terminology and Test Methods for Analog-to-Digital Converters* (IEEE Std 1241-2000) was issued with the intent of identifying ADC error sources and providing test methods with which to perform the required error measurements (IEEE 2000). A subset of the IEEE performance metrics most often used by engineers to specify ADC requirements for embedded systems is outlined in this chapter, followed by an overview of the technological issues of high-end ADC architectures.

7.2   CONCEPTUAL ADC OPERATION

ADC parameters that impact performance can best be illustrated using the highly simplified, conceptual ADC design shown in Figure 7-1. Note that this conceptual ADC is not a practical architecture and is provided solely to illustrate the concept of analog-to-digital conversion.

This conceptual ADC has three bits of "resolution" (binary counter output Q_2, Q_1, and Q_0). In practice, a *sample-and-hold* (SAH) circuit prevents the ADC input, a time-varying analog voltage $x(t)$, from changing during the conversion time. For this discussion, assume that $x(t)$ does not change during the following operation. The counter stops counting when its output, which is converted into a voltage by the DAC (digital-to-analog converter), equals $x(t)$ as indicated by the comparator COMP. The counter output is then the ADC output code corresponding to $x(t)$.

Example DAC output voltages resulting from the changing binary counter values (i.e., the DAC's *transfer function*) are given in Table 7-1. In this conceptual design, the least significant bit (LSB) Q_0 corresponds to a *step size* of 0.25 volt. The resulting overall ADC transfer function (ADC digital output code vs. analog input voltage over the limited range of interest) is shown in Figure 7-2.

7.3   STATIC METRICS

The most basic ADC performance metrics deal with the static, DC (direct current, or zero-frequency constant input) performance of devices.

7.3.1   OFFSET ERROR

The offset error of an ADC, which is similar to the offset error of an amplifier, is defined as a deviation of the ADC output code transition points that is present across all output codes

TABLE 7-1

Digital-to-Analog Converter Transfer Function

ADC Output Code (Q2, Q1, Q0)	V_{DAC} (volts)
000	−0.75
001	−0.50
010	−0.25
011	0.00
100	0.25
101	0.50
110	0.75
111	1.00

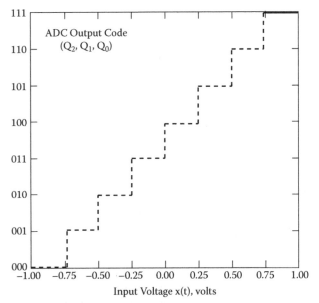

FIGURE 7-2   Conceptual ADC transfer function.

(Bowling 2000). This error has the effect of shifting, or translating, the ADC's actual transfer function away from the ideal transfer function shown in Figure 7-3. For example, by comparing the conceptual ADC's transfer function of Figure 7-2 with that of the ideal transfer function defined by Figure 7-3, it is apparent that the conceptual ADC's offset error is −1/2 LSB. In practice, the ideal transfer function may be defined by either Figure 7-2 (known as the *mid-riser* convention) or Figure 7-3 (known as the *mid-tread* convention) depending on the manufacturer's specifications for a particular ADC (IEEE 2000). Once the offset error has been characterized, it may be possible to compensate for this error source by adjusting a variable "trimming" resistor in the analog domain, or by adding (or subtracting) an appropriate offset value to the ADC output in the digital domain. Note that changes in the offset error as a function of time create a *dynamic offset error*, the effects of which may be mitigated through a variety of design techniques described later.

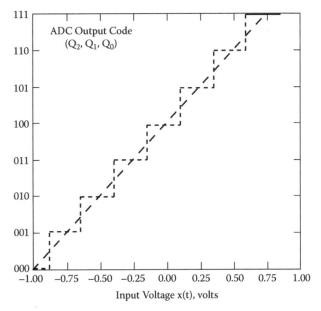

FIGURE 7-3   Ideal ADC transfer function.

7.3.2 Gain Error

The gain error of an ADC is similar to the gain error of an amplifier. Assuming that the ADC's offset error has been removed, the gain error then determines the amount of "rotational" deviation away from the ADC's ideal transfer function slope (i.e., the dashed diagonal line of Figure 7-3). Once the gain error has been characterized, it may be possible to compensate for this error source by adjusting a variable "trimming" resistor in the analog domain, or by multiplying (or dividing) the ADC output by an appropriate scaling factor in the digital domain. As with offset error, changes in the gain error as a function of time create a *dynamic gain error.*

7.3.3 Differential Nonlinearity

For an ideal ADC, the difference in the analog input voltage is constant from one output code transition point to the next. The differential nonlinearity (DNL) for a nonideal ADC (after compensating for any offset and gain errors) specifies the deviation of any code in the transfer function from the ideal code width of one LSB. DNL test data for an ADC may be provided in the form of a graph that shows DNL values (relative to an LSB) versus the digital code. If specifications indicate a minimum value of –1 for DNL, then *missing codes* will occur; i.e., specific output codes will not reliably be produced by the ADC in response to any analog input voltage. Changes in DNL as a function of time create a *dynamic DNL.*

7.3.4 Integral Nonlinearity

Integral nonlinearity (INL) is the result of cumulative DNL errors and specifies deviation of the overall transfer function from a linear response. The resulting linearity error may be measured as a deviation of the response as compared to a line that extends from the origin of the transfer function to the full scale point (*end-point* method) or, alternatively, a line may be found that provides a *best fit* to the transfer function and deviations are then measured from that line (best-fit method). Although the best-fit method produces lower INL values for a given ADC and provides a better measure of distortion for dynamic inputs, the end-point method provides a better measure of absolute worst-case error versus the ideal transfer function (Kester and Bryant 2000). Changes in INL as a function of time create a *dynamic INL.*

7.4 DYNAMIC METRICS

The static parameters described in the last section also have dynamic AC (alternating current, or time-varying input) counterparts that play an important role in the design of high-speed devices, as discussed below.

7.4.1 Resolution

ADC resolution, specified in bits, is a value that determines the number of distinct output codes the device is capable of producing. For example, an 8-bit ADC has 8 bits of resolution and $2^8 = 256$ different output codes. Resolution is one of the primary factors to consider in determining whether or not a signal can be captured over a required dynamic range with any degree of accuracy. For example, some humans can hear audio signals over a 120 dB dynamic range from 20 Hz to 20 kHz. Therefore, for certain high-fidelity audio applications, an ADC with at least a 20-bit resolution is required (i.e., $20\log_{10}2^{20} \approx 120$ dB) that can operate over this frequency range. Similar considerations apply when developing ADC requirements for communication systems that must simultaneously receive radio signals from distant aircraft as well as from aircraft nearby. As long as the signals in space (i.e., signals traveling through some transmission medium) can be captured, it may be possible to apply digital post-processing to overcome many ADC and other system-level deficiencies. Conversely, if the ADC resolution is not adequate to capture the signals in space over the necessary dynamic range, it is often the case that no amount of digital post-processing can compensate for the resulting loss of information, and the ADC may introduce unwanted distortion (e.g., *flat topping* or *peak clipping*) in the digitized output.

7.4.2 Monotonicity

An ADC is monotonic if an increasing (decreasing) analog input voltage generates increasing (decreasing) output code values, noting that an output code will remain constant until the corresponding input voltage threshold has been reached. In other words, a monotonic ADC is one having output codes that do not decrease (increase) for a uniformly increasing (decreasing) input signal in the absence of noise.

7.4.3 Equivalent Input-Referred Noise (Thermal Noise)

Assume that the input voltage to an ADC, $x(t)$, consists of a desired signal (in this case, a long-term average value that is not changing with time) that has been corrupted by additive white Gaussian noise (WGN). Although the source of this WGN may actually be wideband thermal noise from amplifiers inside the ADC (with the WGN being added to signals within the ADC), this noise is modeled as an equivalent noise present at the input for analysis purposes [i.e., equivalent input-referred noise (Kester and Bryant 2000)].

Qualitatively, if one were to take a series of M measurements (where M is an integer >0) of the value of $x(t)$ using, for example, an analog oscilloscope with unlimited vertical resolution, then add the resulting sample values and divide by M, one would expect the noise to "average out" and leave only the desired DC value of interest. Quantitatively, this amounts to forming the sum of M Gaussian random variables, each of which has the same mean μ (voltage appearing across a unit resistance) and variance σ^2 (noise power in the same unit resistance), to obtain a new Gaussian random variable with mean $M\mu$ (corresponding to a signal power of $M^2\mu^2$) and variance $M\sigma^2$ (Walpole and Myers 1972), then dividing the result by M to obtain the average. Whereas the signal-to-noise ratio (SNR) for any individual sample, expressed in dB, is $20\log_{10}(\mu/\sigma)$, the SNR for the sum (or average) of M samples is given (in dB) as

$$\text{SNR}_{\text{WGN}} = 10\log_{10}(M\mu^2/\sigma^2) = 20\log_{10}(\mu/\sigma) + 10\log_{10}M \,. \tag{7.1}$$

This result indicates that, in the presence of WGN, an SNR improvement (relative to a single sample) of up to $10\log_{10}M$ dB could be obtained by digitally processing M samples (e.g., processing four samples may provide up to a 6 dB SNR processing-gain improvement when dealing with WGN).

Although Equation (7.1) was motivated by a case in which the signal portion of the input waveform did not change with time, the same result applies for any synchronously sampled periodic signal (similar to an analog oscilloscope operated in triggered-sweep mode). More generally, once any input waveform has been digitized, narrowband signals can be separated, to a great extent, from broadband WGN in the frequency domain using a variety of digital signal processing techniques.

7.4.4 Quantization Error

In addition to other noise sources, each ADC digital output sample represents the sum of an analog input waveform value with a quantization error value. The time-domain sequence of quantization error values resulting from a series of samples is known as *quantization noise*, and this type of noise decreases with increasing ADC resolution. For this reason, embedded system designers often choose an ADC having the highest possible resolution for the frequency range of interest.

Unlike WGN, quantization noise is correlated with the input waveform. For example, if the input $x(t)$ is a constant voltage, then the quantization error and ADC output do not change from sample to sample. In this particular case, unlike WGN, no SNR improvement can be achieved by digitally processing multiple samples. Similarly, it can be shown that for many time-varying input waveforms, the quantization noise appears in the same frequency bands as the input waveform and at other frequencies as well (Bowling 2000; Kester and Bryant 2000).

A further characterization of the quantization noise is possible using probability theory (Drake 1967). Normalizing both the signal and quantization noise values to the ADC's full-scale value, the following result for quantization-noise-related SNR is valid for any number of resolution bits (N):

$$\text{SNR}_{\text{QUANT}} \approx (1.76 + 6.02N) \text{ dB} . \tag{7.2}$$

7.4.5 RATIO OF SIGNAL TO NOISE AND DISTORTION

The ratio of signal to noise and distortion (SINAD) is measured using a nearly full-scale sine wave input to the ADC, where the sine wave frequency is nearly half the sampling rate. A fast Fourier transform (FFT) analysis is performed on the output data, and the ratio of the root-mean-square (RMS) input signal level to the root-sum-square of all noise and distortion components (excluding any zero-frequency component) is computed. The SINAD value, therefore, includes the effects of all noise (e.g., thermal and quantization), distortion, harmonics (e.g., DNL effects), and sampling errors (e.g., aperture jitter in the sample-and-hold circuitry) that may be introduced by the ADC. SINAD for a given ADC typically decreases as the sampling rate increases. The SINAD ratio may also be abbreviated as SNDR (signal to noise-plus-distortion ratio).

7.4.6 EFFECTIVE NUMBER OF BITS

The effective number of bits (ENOB) for an ADC is computed by rearranging Equation (7.2) and using the measured SINAD instead of the theoretical $\text{SNR}_{\text{QUANT}}$:

$$\text{ENOB} = (\text{SINAD} - 1.76)/6.02 , \tag{7.3}$$

where SINAD is expressed in dB. The ENOB for a given ADC typically decreases as the sampling rate increases.

Equation (7.3) indicates that each 6 dB increase in SINAD corresponds to an improvement of one effective bit. Therefore, when WGN is the dominant factor limiting SINAD, up to a one-bit improvement can theoretically be obtained by digitally processing sets of four samples in accordance with Equation (7.1). For example, data from a 100 MSPS (million samples per second) ADC having ENOB = 10 may be processed to generate data similar to that from a 25 MSPS ADC having ENOB = 11.

7.4.7 SPURIOUS-FREE DYNAMIC RANGE

The SFDR is a frequency-domain measurement that determines the minimum signal level that can be distinguished from spurious components, and includes all spurious components over the full Nyquist band regardless of their origin (IEEE 2000; Kester and Bryant 2000). SFDR for a given ADC typically decreases as the sampling rate increases. SFDR values may be referenced either to a carrier level (dBc) or to an input level that is nearly full-scale (dBFS), as shown in Figure 7-4. SFDR is often caused by the worst of second or third harmonic distortion. Since it is one of the main factors limiting the effectiveness of digital signal processing techniques for signal enhancement, a number of technologies have been developed to improve SFDR performance (Batruni 2006;

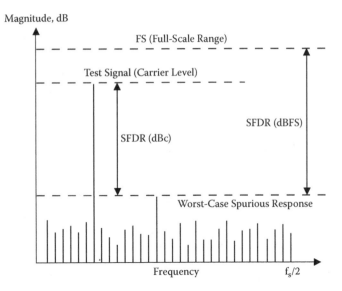

FIGURE 7-4   Spurious-free dynamic range measurement.

Lundin, Skoglund, and Handel 2005; Raz 2003; Velazquez and Velazquez 2002; White, Rica, and Massie 2003).

7.4.8   DITHER

When dealing with low-level signals (on the order of a few LSBs), it is often useful to add a user-controlled dither waveform to the input waveform, causing an increase in the number of times the ADC output changes value. For example, by adding a square wave with voltage that alternates between 0 and 1/2 LSB to the sample-and-hold output, assuming the square wave has half the period of the clock used for the sample-and-hold, it may be possible to obtain sample pairs from which a new LSB can be formed via digital post-processing. Such *deterministic* dither is not limited to square waves, and sinusoids or triangle waveforms are sometimes used for ADC testing purposes (Sheingold 1972).

A more popular approach is to add *random* dither such as WGN with 1/2 LSB RMS voltage (Callegari, Rovatti, and Setti 2005; Zozor and Amblard 2005). Since 1998, some ADC manufacturers have provided on-chip circuitry that adds *pseudorandom* dither to the input while subtracting a digital estimate of the dither value from the output. In such cases, dither amplitude that is 25% of the ADC's full-scale range may be used (IEEE 2000).

7.4.9   APERTURE UNCERTAINTY

For high-speed ADCs in particular, the dominant SNR limitation may come from jitter in the sampling clock or sample-and-hold circuitry, and other short-term timing instabilities such as phase noise (IEEE 1989). For a full-scale sine wave input with frequency f_{in} (in Hz) and total RMS aperture uncertainty τ (in seconds), the maximum SNR (in dB) is approximately (Kester and Bryant 2000)

$$\text{SNR}_{\text{APERTURE}} = -20\log_{10}(2\pi f_{in}\tau) . \tag{7.4}$$

For example, the aperture uncertainty of the conceptual ADC sampling at 1 Hz must be <16 msec RMS to achieve ENOB $\approx$ 3. Similarly, if aperture uncertainty is the dominant component of SINAD, then an ADC with $\tau = 1$ ps RMS sampling a 100 MHz sine wave will have ENOB < 10.3.

Note that, for a given aperture uncertainty, $\text{SNR}_{\text{APERTURE}}$ decreases by 6 dB (and ENOB decreases by 1 bit) for every doubling (i.e., octave) of f_{in}.

7.5 SYSTEM-LEVEL PERFORMANCE TRENDS AND LIMITATIONS

Historic ADC device-level performance improvement rates may not be sustainable as technical and economic limits are approached, and rates may also level off after requirements for key markets have been met. For example, by 2002, ADCs for the digital audio market had achieved 24-bit resolution (with ENOB = 18) over a 40 kHz analog input bandwidth and provided an audio quality exceeding the discrimination limits of human hearing (AES 2003; Aude 1998; Neesgaard 2001). Future performance improvements for these ADCs are likely to be limited by internal thermal noise considerations, but the bandwidths of these devices may continue to increase due to future improvements in semiconductor processes. Although devices with greater bandwidth may well find application in broadband sonar systems, the sonar market may be much smaller than the historical consumer digital audio market. Therefore, future improvements are more likely to lie in the areas of cost and power reduction or addition of special features (e.g., on-chip data buffers and anti-aliasing filters) rather than performance parameters.

In many cases, individual ADC chips can be interconnected to form small subsystems that fill a need not otherwise met at the component level. By 2003, for example, commercial off-the-shelf (COTS) ADC modules were available with 12-bit resolution (ENOB = 9.8) at 400 MSPS created from a pair of 12-bit ADCs (each with ENOB = 10.5 @ 210 MSPS) and onboard digital post-processing. This two-way time-interleaved, or "ping-pong," approach doubled the sampling rate while losing ~0.7 effective bit compared to the individual component ADCs. More recently, time-interleaved designs ranging from 16-way (Elbornsson et al. 2005) to 80-way on a single chip (Poulton et al. 2003) have been developed. Also in 2003, an effort to combine ADCs for higher resolution yielded COTS modules with 16-bit resolution (ENOB = 12.9) at 80 MSPS created from four 14-bit ADCs (each with ENOB = 12.0 @ 105 MSPS) and onboard digital post-processing. Note that, in the following, no distinction is made as to whether any given specification was achieved using a single monolithic ADC versus a multichip module.

For applications in which the input waveform has a small bandwidth with respect to its center frequency (e.g., as in radio receivers), it is often convenient to represent the digitized signal as a sequence of complex numbers in the time domain. This representation may be achieved by multiplying the narrowband analog input waveform with a cosine wave having the same center frequency as that of the input waveform, then low-pass filtering the result to form an in-phase, or I, waveform prior to conversion by an ADC (Shaw and Pohlig 1995). The narrowband waveform is simultaneously multiplied by a sine wave and filtered to form a quadrature, or Q, waveform prior to conversion by a separate ADC. Once digitized, the I samples are generally taken as the real part of a complex sample, while the Q samples form the imaginary part. Using this approach, it is possible to digitize the input signal using a pair of ADCs, each of which operates at a sampling rate (in samples per second) that numerically corresponds to the input signal bandwidth (in Hz), rather than using a single ADC operating at a rate that numerically corresponds to twice the input signal bandwidth and produces purely real samples. For such applications, it is desirable to use a pair of ADCs on the same chip to minimize I/Q channel mismatch, and many dual converters have on-chip circuitry to facilitate operation in either an I/Q or two-way time-interleaved mode.

7.5.1 TRENDS IN RESOLUTION

Figure 7-5 is intended to give a sense of improvements in resolution for state-of-the-art COTS ADCs over a 20-year time span. One particularly noteworthy trend is that, from 1992 through 2003, 12-bit devices (ENOB ≈ 10) roughly doubled in speed every three years.

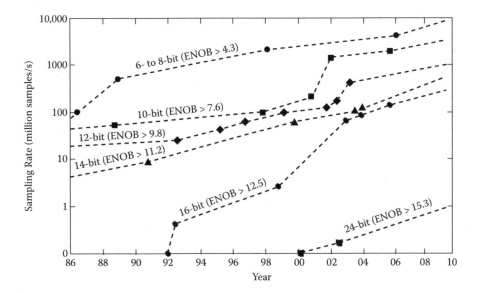

FIGURE 7-5   Resolution improvement timeline.

Trend lines at the right-hand side of Figure 7-5 give a general indication of possible near-term future resolution improvements. For example, as of 2004, a system had been demonstrated [incorporating the non-COTS ADC of Poulton et al. (2003)] that could digitize a 5 GHz sine wave (using a 10 GSPS effective sampling rate with data from a 20 GSPS data stream) with ENOB = 5.0. Therefore, there is no theoretical limitation that would preclude a COTS ADC from achieving such a performance level in the foreseeable future. Aperture uncertainty limits the near-term future performance of 10-bit ADCs (ENOB ≈ 8) to 3 GSPS, 12-bit ADCs (ENOB ≈ 10) to 1 GSPS, and 16-bit ADCs (ENOB ≈ 12) to 200 MSPS, while 14-bit devices are expected to lie between the 12- and 16-bit ADCs. Thermal noise is expected to limit the 24-bit devices (ENOB ≈ 16) to <1 MSPS.

7.5.2   TRENDS IN EFFECTIVE NUMBER OF BITS

From 1999 through 2005, the ENOB for 100 MSPS state-of-the-art COTS ADCs improved at a rate of approximately 0.28 bit per year, as shown in Figure 7-6. Over the same time period, the sampling

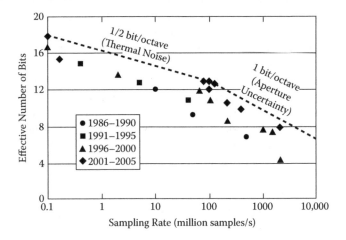

FIGURE 7-6   ENOB (effective number of bits) improvements.

rate of the fastest COTS ADCs (with ENOB ≈ 7.7) roughly doubled (i.e., the equivalent of up to 0.5 bit processing gain in 6.3 years).

Future ENOB improvements may depend on the ability of manufacturers to overcome certain design challenges which may or may not prove to be "hard limits." For example, a 34 ohm resistor produces 150 nV RMS thermal Johnson noise across a 40 kHz bandwidth (i.e., the mean-square voltage per Hz is $4RkT$, where R is in ohms, k is Boltzman's constant = 1.38×10^{-23} Joule per degree Kelvin and T is 298 degrees Kelvin at room temperature), which is one-half the step size of a 24-bit resolution ADC with 5 V full-scale range. If design of ADCs with such a low input resistance presents a significant challenge, then future ENOB improvements for low-speed, high-resolution ADCs are likely to be limited by thermal noise from their internal circuitry. Although such noise can be mitigated through the use of cryogenics and superconducting materials, it remains to be seen if such technologies are appropriate for COTS ADCs (Mukhanov et al. 2004). Similarly, although optical technologies are being investigated to mitigate the effects of aperture uncertainty (and for other reasons as well), such technologies have not historically been applicable to COTS ADCs (Juodawlkis et al. 2003; Miao et al. 2006).

As shown in Figure 7-6, a monolithic 16-bit (ENOB = 12.8) 100 MSPS COTS ADC was available by the end of 2005. This device lies near the intersection of two theoretical performance limiter lines (the dotted lines shown in Figure 7-6), one representing thermal noise corresponding to a 2000 Ω equivalent resistance and the other representing 0.2 ps aperture jitter (Le et al. 2005; Walden 1999). Note that these performance limiters are not intended to represent hard limits for COTS devices, as semiconductor process improvements continue to reduce aperture uncertainty. Rather, as shown in the following, these lines are used to illustrate an important design trade-off space.

When digitizing a signal that has frequency components distributed throughout a range from 0 to 12.5 MHz, for example, it is necessary to place a 12.5 MHz bandwidth low-pass anti-aliasing filter prior to the ADC (although some modern ADCs have such circuitry on chip). In this example, the ADC must sample the signal at or above the Nyquist rate of 25 MSPS. Figure 7-6 indicates that a suitable COTS ADC may not be readily available at this sampling rate, so instead consider the use of an ADC with ENOB = 12.8 operating at 100 MSPS to obtain a set of samples. The resulting samples may then be digitally processed to obtain a data stream approximating that from an ADC operating at 25 MSPS and having ENOB up to 13.8 (i.e., an improvement of 0.5 bit/octave). In this example, the anti-aliasing filter bandwidth remains at 12.5 MHz to eliminate any out-of-band signals that might otherwise dominate ADC resolution requirements. Note that the digital post-processing discussed here is directed toward reducing noise that has been introduced by the ADC, rather than reducing noise that may have corrupted a signal prior to its arrival at the ADC input (as discussed later). The 100 MSPS ADC with ENOB = 12.8 could similarly be used in lieu of ADCs with sampling rates down to approximately 100 KSPS (i.e., 10 octaves, with correspondingly lower anti-aliasing filter bandwidths), providing up to five additional effective bits. Below 100 KSPS, as shown in Figure 7-6, COTS ADCs with ENOB ≈ 17.8 are available that may offer lower cost and power versus the 100 MSPS ADC.

Figure 7-6 shows a rapid drop-off in performance (more than 1 bit/octave) above ~100 MSPS for COTS technology ca. 2005, primarily due to aperture uncertainty. As a result, it may not be desirable to use higher-speed ADCs with reduced ENOB in lieu of lower-speed devices with higher ENOB above ~100 MSPS. For example, if a signal is sampled at 2 GSPS using an ADC with ENOB = 7.8, then digital post-processing may only provide the equivalent of ENOB = 9.8 at 125 MSPS, whereas COTS ADCs are available that provide ENOB = 12.6 (i.e., an improvement of 2.8 effective bits) at this rate.

7.5.3 Trends in Spurious-Free Dynamic Range

Like ENOB, the SFDR data shown in Figure 7-7 fall off with increasing sampling rate and exhibit a steeper fall-off above 100 MSPS. Over the last 20 years, SFDR for state-of-the-art COTS ADCs has

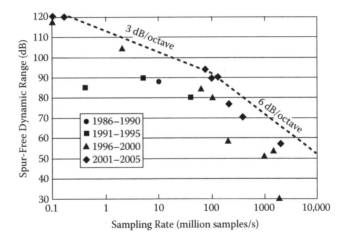

FIGURE 7-7   SFDR (spurious-free dynamic range) improvements.

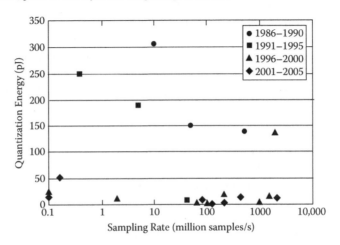

FIGURE 7-8   EQ (energy per effective quantization level).

generally improved at a rate consistent with improvements in ENOB. However, it must be cautioned that Figure 7-7 shows only approximate values due to variations in manufacturer's specifications (e.g., values may be in dB with unspecified measurement technique, dBFS or dBc).

7.5.4   TRENDS IN POWER CONSUMPTION

One useful power-performance figure of merit is the ADC energy required per effective quantization level, or EQ:

$$EQ = Power/[2^{ENOB} \times (Sampling\ rate)], \tag{7.5}$$

where power is in watts, sampling rate is in samples per second, and EQ is in Joules (noting that 1 pJ = 10^{-12} watt-sec). As shown in Figure 7-8, there are many examples of COTS ADCs in the 2001–2005 time frame with EQ < 5 pJ.

For sampling rates in excess of approximately 200 MSPS, an on-chip demultiplexer (DMUX) capability is often included to provide multiple ADC output data streams at rates compatible with low-cost, low-power processors and memories. The DMUX, whether on-chip or external, may be an important factor in system-level power consumption. For example, a 2 GSPS ADC in 2005 included an on-chip 1:4 DMUX and provided EQ = 15 pJ. This is a substantial improvement over 2002, when the same manufacturer provided a comparable ADC with EQ = 27 pJ, or EQ = 61 pJ when operating with the manufacturer's external DMUX.

7.5.5   ADC IMPACT ON PROCESSING GAIN

In the preceding discussion, care has been taken to distinguish noise introduced in the analog-to-digital conversion process from other types of noise (e.g., noise that corrupts a signal traveling through a transmission medium, or that enters a receiver chain prior to an ADC). However, it is often the case that digital signal processing techniques applied to an ADC output data stream attempt to compensate for many different noise sources simultaneously, regardless of the noise's origin (e.g., receiver "nonlinear equalization"). The degree to which such post-processing can succeed depends to some extent on ADC performance parameters, as outlined in the following.

First, consider the numerical representation of data from a 24-bit ADC having ENOB = 18. Assume that accuracy of the conversion process is limited by thermal noise coming from within the ADC, and it is not necessary to add dither. If each 24-bit ADC output sample is to be numerically represented as an "IEEE 754 format" 32-bit floating-point word, then care must be taken to map the ADC's most significant bit (MSB) into the sign bit of the floating-point word, and the ADC's 23 LSBs into the 23-bit mantissa of the floating-point word. A variety of digital signal processing techniques can then be applied that will automatically replace the sample bits with "effective" bits in the 23-bit mantissa (up to a limit of 5 bits from processing gain in addition to the original 18 effective ADC bits). In practice, the ADC's SFDR performance typically limits the processing gain to approximately 5 bits (i.e., an additional 30 dB) or less.

In applications that require separation of narrowband signals from broadband noise, an algorithm such as the FFT may be used to provide processing gain. In the complex radix-2 FFT, for example, each butterfly stage provides a factor of two bandwidth reduction, causing rejection of half the noise power and providing a 3 dB (0.5 bit) SNR improvement. A $2^{10} = 1024$-point complex FFT has 10 stages, with the output potentially providing up to 5 bits processing gain simultaneously within each of the 1024 distinct output frequency bins (assuming any desired signal lies entirely within one frequency bin). This result is comparable to the 30 dB maximum processing gain when 1024 samples are processed to remove WGN, as indicated in Equation (7.1). Since FFT processing is typically performed on samples of a broad bandwidth ADC input waveform that contains both signals and noise, ADC performance parameter variations across the input waveform's entire frequency range must be taken into consideration for post-processing purposes.

7.6   HIGH-SPEED ADC DESIGN

The increasingly closer placement of the ADC to the antenna in a radar processing flow has created strong demands of wideband and high dynamic range ADCs. For instance, a 0.3 m spot size of synthetic aperture radar (SAR) imaging requires 600 MHz of instantaneous bandwidth, which is currently limited by ADC performance. In addition, clutter cancellation and Doppler processing require high SNR, SINAD, and SFDR, which implies at least 8 to 10 ENOB. The high-speed ADCs discussed in this section are those that operate in the range from hundreds of MSPS to several GSPS and provide 8 to 10 ENOB.

The current semiconductor process has produced transistors with high transit frequencies (f_T, the frequency at which the transistor current gain drops to unity). For example, transistors fabricated in the 90 nm CMOS (complementary metal oxide semiconductor) and 130 nm SiGe BiCMOS (bipolar CMOS) processes have f_T's that approach 100 GHz and 200 GHz, respectively. A myth in ADC technology is that "device scaling alone will push f_T up and thus the sampling rate higher, and power dissipation lower." However, scaling alone has not provided the desired ADC performance. New architectures and circuit topologies are needed to take advantage of new devices. For example, the 8-bit Maxim MAX108 (1.5 GSPS), which was fabricated in 2000 using a bipolar process with 29 GHz f_T, dissipates approximately 5.3 W. In 2005, National Semiconductor introduced a dual 8-bit ADC (ADC08D1500), which was built using a 0.18 μm CMOS process with a 49 GHz f_T. This 1.5 GSPS dual ADC chip dissipates only ~1.8 W. This nearly 6× reduction in power dissipation was made possible by applying new circuit techniques and architectural solutions that take advantage of the faster process.

The effect of scaling on the performance of ADCs is a mixture of benefits and detriments. For example, the reduced power-supply voltage forced by the reduction of gate length (CMOS) or emitter width (bipolar transistors) has been a serious challenge for analog and ADC designers for some time. The scaling without new circuitry and architecture will not yield a dramatic improvement in performance. The development of high-speed ADCs requires a thorough understanding of the limiting factors of resolution, sampling rate, and power dissipation, as well as the impacts of architecture and circuit design techniques on them. Two of the most common architectures for medium resolution ADCs, flash and pipeline architectures, will be used to discuss these design issues.

7.6.1 Flash ADC

A parallel or flash architecture achieves the highest conversion rate at the cost of area and power consumption. As shown in Figure 7-9(a), an n-bit converter comprises $2^n - 1$ comparators and subsequent decoding stages. The comparators compare the analog input voltage V_{in} simultaneously with $2^n - 1$ equally spaced reference voltages V_{r1} through V_{rn}, quantizing the signal into a so-called thermometer code. The thermometer code, which consists of a list of "0's" followed by a list of "1's," is named after the reading style of a mercury thermometer. For example, in an 8-bit flash ADC, $n = 8$, so 255 comparators are used. If the full-scale reference voltage V_{ref} is 1 V, the resistor ladder divides V_{ref} into 256 equal segments (i.e., $V_{r1} = 1/256$ V, $V_{r2} = 2/256$ V, etc.). Figure 7-9(a) shows that the level of V_{in} triggers the lower five comparators to produce a thermometer code of 00...011111. The digital encoder converts the thermometer code to a binary number 00000101 (5 in decimal).

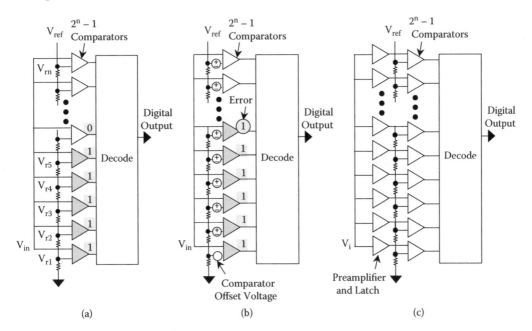

FIGURE 7-9 An n-bit flash ADC: (a) basic architecture; (b) model with comparator offset voltages; (c) the use of preamplifiers and latches to correct offset voltages.

The above explanation assumes that all comparators (255 in the example) are sensing the incoming voltage accurately. However, due to mismatch between transistors, comparators unavoidably have offset voltages, which are modeled by the voltage sources added to the reference voltages in Figure 7-9(b). Typical comparator offset voltages are 2 to 10 mV for silicon bipolar technology, 15 to 50 mV for CMOS technology (Bult and Buchwald 1997), and >50 mV for III-V (e.g., GaAs) technology. In the 8-bit ADC example, a 5 mV offset voltage (>1 LSB = 4 mV) could have caused comparators 1 to 6 (or 1 to 4) to be triggered and resulted in an error, as shown in Figure 7-9(b). Since offset voltages are caused by random, statistical events in the fabrication process, in order for the 8-bit ADC to have a 99% yield, the offset voltage should be <0.2 LSB (0.8 mV in the ongoing example) (Kinget and Steyaert 1996; Uyttenhove and Steyaert 2002).

Figure 7-9(c) shows that in order to minimize the impact of offset voltages, each comparator is equipped with a preamplifier (preamp) and a latch (see also Figure 7-11). The preamp gain reduces the offset voltage in the reference caused by the latch. The gain of a preamp implemented in a bipolar technology is typically 3× to 5×. Depending on the required amount of offset reduction, more than one stage of preamplification may be needed to achieve the necessary gain. Unfortunately, preamps also have offset voltages (0.1 to 2 mV for silicon bipolar technology, 3 to 10 mV for CMOS technology, and >20 mV for III-V technology), so a delicate design optimization is needed to achieve the desired accuracy.

As the gate length (or emitter width) scales down, the supply and reference voltages must also be scaled down. In Figure 7-9, if V_{ref} is 0.5 V, the LSB is reduced to merely 2 mV. The offset voltage must also be scaled down to maintain the same performance. However, it was shown that the mismatch parameters contributing to offset voltages had been scaling down up to, but not beyond, the 180 nm CMOS technology (Kinget and Steyaert 1996). The reason for the discontinuity of scaling is that there are two major mismatch parameters: variation in threshold voltage V_T, and variation in gain β. As the oxide thickness scales down with a finer gate-length technology, the mismatch due to V_T has been scaling proportionally. However, the mismatch due to β has not scaled as fast as that of V_T. Up to the 180 nm CMOS technology, the impact of V_T mismatch dominates. For 120 nm and a gate-over-drive voltage (i.e., gate voltage – V_T) of 0.2 V, the β mismatch becomes dominant (Uyttenhove and Steyaert 2002).

The slow scaling of the offset voltage is the fundamental limitation to accuracy in advanced ADC technology. Additional preamp stages and larger-size transistors and resistors can somewhat improve the situation at the expense of higher power dissipation since larger devices require more current to drive. The power-dissipation situation may worsen when the technology is scaled for higher f_T. Analog designers have developed architectural solutions to overcome this limitation. Techniques that alleviate the non-scaling effect of offset voltage are discussed below.

Instead of increasing the preamp device size and thus the power consumption, various averaging techniques have been successfully implemented (Bult and Buchwald 1997; Choi and Abidi 2001; Kattmann and Barrow 1991). First presented in Kattmann and Barrow (1991), the averaging technique has reduced the dynamic differential nonlinearity error by 3× and improved the 8-bit ADC yield by 5×.

The averaging technique presented in Kattmann and Barrow (1991) places series resistors between neighboring preamp output loads, as shown in Figure 7-10(a). Instead of taking an output voltage from an individual preamp, the averaging resistor network averages the preamp output value. Figure 7-10(b) shows a preamp bank with series resistors R_{A1} to R_{An-1} placed between R_{L1} and R_{Ln}. While Figure 7-10(b) shows an implementation of the preamps in bipolar technology, the same architecture is also applicable to the CMOS technology. Other examples of averaging technique to further improve offset voltages can be found in Bult and Buchwald (1997) and Choi and Abidi (2001).

Another factor limiting the ADC resolution is the input-referred noise of the preamp and latch combination. The signal at the latch is higher in amplitude so the noise of the preamp usually dominates the result. Figure 7-11 shows a simplified comparator consisting of a differential pair preamp

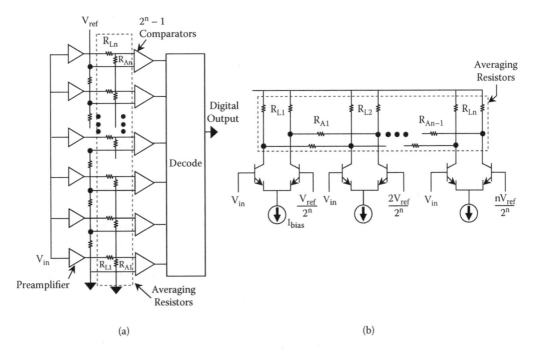

FIGURE 7-10 (a) A simplified flash ADC with averaging resistor network; (b) preamps with averaging resistor network.

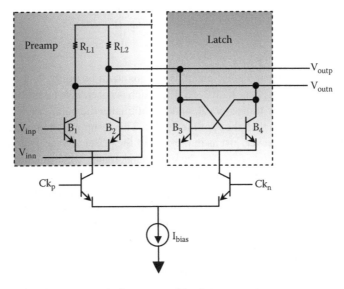

FIGURE 7-11 A comparator composed of preamp and latch.

and a latch implemented in bipolar technology. The input-referred noise of the preamp consists of thermal and shot noise.

The noise spectral density is the thermal and shot noise of transistors B_1 and B_2 and the thermal noise of the load resistors R_{L1} and R_{L2} (Gray et al. 2001). The spectral density of input-referred noise is

$$v_n^2/\Delta f = 4kT(r_{b1} + r_{b2} + r_{e1} + r_{e2}) + 4kT(1/2g_{m1} + 1/2g_{m2})$$
$$+ 4kT(1/g_{m1}^2 R_{L1} + 1/g_{m2}^2 R_{L2}),$$

(7.6)

where r_{b1} and r_{b2} are the base resistance of transistors B_1 and B_2, respectively; r_{e1} and r_{e2} are the emitter resistance of transistors B_1 and B_2, respectively; g_{m1} and g_{m2} are the transconductance of B_1 and B_2; R_{L1} and R_{L2} are the load resistors of the preamp. In Equation (7.6), the first and second terms are the thermal noise and shot noise of B_1 and B_2, respectively, and the third term is the thermal noise of R_{L1} and R_{L2} (see Section 7.5.2).

The following example is used to illustrate the limiting effect of the input-referred noise on the ADC resolution. Typical values are used in this example: the collector currents of B_1 and B_2 are ~1 mA; g_{m1} and g_{m2} are ~0.02 A/V; R_{L1} and R_{L2} are 200 Ω, which give ~400 mV of differential swing at the output; and r_b and r_e are ~50 Ω and ~35 Ω, respectively. Also, assume that the tail current of the comparator, I_{bias}, is 2 mA. The offset requirement usually determines the differential gain of the preamp. Using the typical values in this example, this gain is

$$A_v = R_{L1}/(1/g_{m1} + r_{e1}) = 200/[(1/0.0386) + 35] \approx 3.3 .$$

(7.7)

The spectral noise density is determined by Equation (7.6) as $v_n^2/\Delta f = 3.3 \times 10^{-18}$ V²/Hz. If the input signal bandwidth is 1 GHz, the input referred noise voltage is ~58 µV. For an 8-bit ADC, the total input-referred noise of 255 preamps is ~922 µV. The noise-induced SNR loss, which is referred to as $SNR_{penalty}$ and defined in Equation (7.8), should be limited to <3 dB or about 1/2 LSB (Vorenkamp and Roovers 1997).

$$SNR_{penalty}(\text{dB}) = 20 * \log \sqrt{1 + \frac{v_n^2}{\varepsilon_q^2}} ,$$

(7.8)

where v_n is the total input-referred noise and ε_q is the quantization noise.

In this example, the $SNR_{penalty}$ is 2.2 dB < 3 dB, which is acceptable (assuming the input swing is 1 V). The thermal noise of the transistors B_1 and B_2 dominates the noise, which is typical when the signal bandwidth is much less than the f_T of the device. The designer can increase the device size to reduce the thermal noise. However, just as in the case of the offset voltage, a larger device area does not only reduce r_b and r_e, but also increases the input capacitance and limits the bandwidth. This is a design trade-off that a designer must make.

The limited preamp bandwidth also affects the ADC resolution. As shown in Figure 7-12(b), the ADC input signals distributed to the multiple comparators do not arrive at the same time. This variable delay can cause an incorrect input signal level when the latch is triggered, leading to an error in the digital readout as indicated in the example of Figure 7-12(a). The propagation delay is caused by the parasitic nonlinear capacitance of the comparators, the routing capacitance, the varying input signal level, and the limited bias current. A combination of these effects causes a third-order harmonic distortion in the ADC (Dalton et al. 1998), which causes distortion and degrades SFDR and SINAD of the ADC (Peetz, Hamilton, and Kang 1986). As the preamp bandwidth is limited by the nonlinear capacitance, the relationship between the third-order harmonic distortion and the preamp bandwidth at different input voltage swings can be established (van de Plassache 1994). For instance,

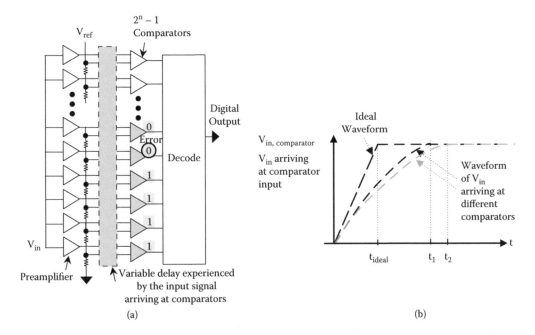

FIGURE 7-12   (a) Comparators experiencing an effective variable delay at the input; (b) waveforms of input signal arriving at comparator inputs at different times.

according to van de Plassche (1994), if we want to achieve −55 dBc of third-order distortion level for an input swing of 1 V, the preamp bandwidth should be 10× the input signal bandwidth.

Ideally, the unity gain bandwidth of a preamp is close to the f_T of the technology. However, in order to reduce offset and noise, the devices in an ADC are often designed to be larger than minimum size so their f_T's are typically less than the value quoted (for minimum size devices). Also, although the bias current is set high enough to drive the input swing without causing distortion, it is generally set at a level that is slightly lower than what is required to support the peak f_T. This bias current reduction is necessary to avoid the high-level injection that causes f_T to drop rapidly if the bias current exceeds the level associated with the peak f_T.

The bandwidth of the latches in a flash ADC is much higher than the preamps. The flash ADC sampling rate is thus constrained by the speed of its preamps. As an example, the unity gain frequency can be ~35 GHz for a preamp fabricated in 0.18 μm CMOS (with a peak f_T of 49 GHz). As the gain-bandwidth product is a constant, if a preamp gain of 10× is needed to reduce the offset voltage, the bandwidth of preamp is ~5 GHz. The input signal bandwidth for the ADC is then 0.5 GHz, which requires a Nyquist sampling rate of 1 GSPS.

As discussed and illustrated in Figure 7-12(b), the signals arriving at the comparators may experience variable delay. In theory, a flash ADC structure does not require the use of a sample-and-hold amplifier (SHA). However, the use of an SHA to feed the comparators alleviates the dispersion problem pronounced at higher sampling rates (e.g., >100 MSPS). Figure 7-13 shows an 8-bit flash ADC enhanced with an SHA. The SHA samples and holds the input value (at 22 mV in Figure 7-13) and the comparators compare the held value with equally spaced reference voltages.

7.6.2   Architectural Techniques for Power Saving

As mentioned earlier, the MAX108 (Maxim) and the ADC08D1500 (National Semiconductor) ADCs dissipate 5.3 W and 1.8 W, respectively. Besides a 2× improvement in f_T, (49 GHz versus 29 GHz), the significant power reduction (6×) is the result of the architectural techniques (e.g., folding, interpolation, interleaving, and calibration) used extensively in the ADC08D1500.

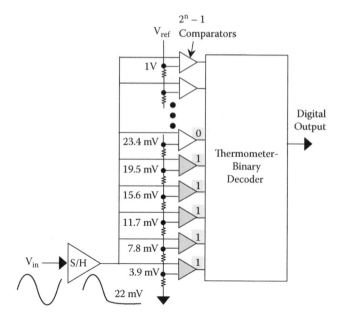

FIGURE 7-13 Eight-bit flash ADC with a sample-and-hold circuit.

In the example shown in Figure 7-9, each comparator dissipates 7.2 mW (assuming a 1.8 V power supply). For an 8-bit flash ADC, the comparators alone consume ~2 W (255 × 7.2 mW). A high-speed ADC requires the use of an SHA, which can consume 100 mW or more. Note that the total ADC power dissipation figure also has to account for the power consumption of the resistor ladder network, decoder, bias circuits, demultiplexer, and clock driver. For a flash architecture, the power dissipation depends on the resolution and the sampling rate. The power dissipation is proportional to 2^N, where N is the number of resolution bits. Also, for a given resolution, the power dissipation goes up with the sampling rate (Onodera 2000).

One of the power-saving techniques is to reduce the number of comparators by folding. The folding technique allows a single comparator to be associated with multiple preamps. Figure 7-14(a) shows a modified flash architecture with folding, which is similar to the two-step architecture of a coarse-fine approach shown in Figure 7.14(b). In the two-step architecture, a coarse flash ADC gen-

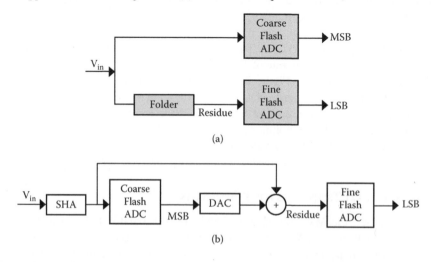

FIGURE 7-14 (a) Flash ADC architecture with a folder; (b) two-step ADC architecture.

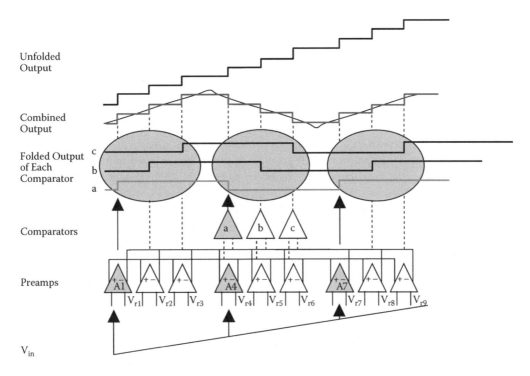

FIGURE 7-15 Folding example.

erates the MSBs of the result. A DAC converts the MSBs back into an analog signal. The residue (i.e., the difference between the original input signal and the DAC output) goes through a fine flash ADC to generate the LSBs of the result. The major difference between the folding and two-step architectures is that the former remains a parallel approach, in which a folder circuit produces the residue for the conversion of LSBs. In theory, the folding ADC does not require an SHA, but in practice one is often included to minimize the distortion, especially at high sample rates (e.g., >100 MHz).

The bandwidth of an op-amp-based, switched-cap SHA is roughly $f_T/6$ to $f_T/10$ (Cho 1995). As the bandwidth must be at least 10× the sampling rate to minimize distortion, the sampling rate of an ADC with sample-and-hold is limited to ~ $f_T/60 - f_T/100$.

The maximum input voltage level is divided into regions, each of which covers multiple reference voltages. The input signal is then folded into these regions. Figure 7-15 shows an example of folding by three (Bult and Buchwald 1997). Instead of assigning a single reference voltage to a comparator, in a folding architecture each comparator is associated with three separate preamps. For instance, comparator a is connected to preamps A1, A4, and A7.

For the sake of illustration, V_{in} is shown as a ramp signal in Figure 7-15. When V_{in} passes the threshold set by reference voltage $Vr1$, the output of preamp A1 switches from low to high. As V_{in} continues to increase, it exceeds $Vr4$ of the second preamp A2, and as this amplifier has reversed polarity, it will cause the comparator to go from high to low. At the moment the input signal passes the amplifier on the right, the comparator will again change from low to high. Comparator b will be connected to preamps A2, A5, and A8, producing the folded output curve b, and Comparator c will be connected with A3, A6, and A9, producing the folded output curve c. Figure 7-15 also shows the combined output and the unfolded output.

For an 8-bit ADC, the number of comparators can be reduced from 255 to 87 ($2^N/k + k - 1$, where k is the folding factor and assume $k = 3$). However, the power dissipation would not be reduced by a factor of three because the bias current of the preamps must increase to accommodate the folding. An interpolation technique can be used to eliminate some preamps and thus further reduce the power dissipation. For instance, instead of using 255 preamps for distinct reference volt-

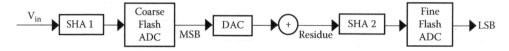

FIGURE 7-16   Two-step pipeline ADC.

age points, a design can use 64 preamps and interpolate the reference voltage points between them with resistive dividers. Because multiple folding amplifiers are connected to a comparator, the load resistor and parasitic capacitance at the output of the folding amplifiers can limit the bandwidth and cause distortion. Therefore, the folding amplifier should be designed to have a bandwidth at least 10× larger than the maximum input frequency to keep the SINAD degradation to less than 3 dB (Limotyrakis, Nam, and Wooley 2002). With fewer comparators, the large offsets become a significant problem that requires offset cancellation. This offset issue is especially significant in the CMOS technology, which is the reason why folding and interpolation techniques appear first in the bipolar process (van de Grift and van de Plassche 1984; van de Plassche and Baltus 1988), and subsequently migrate to the BiCMOS (Vorenkamp and Roovers 1997) and CMOS (Bult and Buchwald 1997; Taft et al. 2004) processes.

7.6.3   Pipeline ADC

Section 7.6.2 discussed the use of folding [Figure 7-14(a)] and two-step architectures [Figure 7-14(b)] to mitigate the problems of high power dissipation and large footprint area of a full flash architecture. In a two-step ADC, the entire conversion—which consists of the operations of an SHA, a coarse flash ADC, a DAC, a subtraction, and a fine flash ADC—must be completed in one sampling period. The sampling rate is thus limited by the settling time of these components. A folding architecture alleviates this issue by precomputing the residue. Alternatively, as shown in Figure 7-16, an SHA (SHA2) can be inserted before the fine flash ADC to form a two-step pipeline operation. While SHA2 holds a sample (i.e., the residue) for the conversion of the fine ADC, SHA1 holds the next sample for the coarse ADC, virtually doubling the throughput. The two-step architecture can be generalized into the k-stage pipeline shown in Figure 7-17.

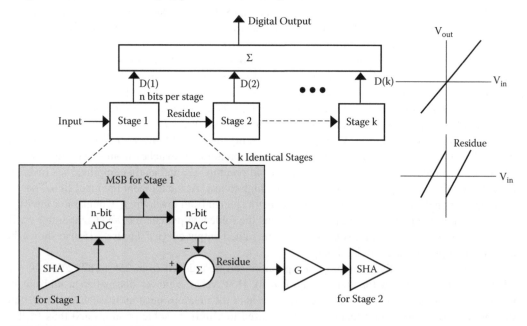

FIGURE 7-17   Pipeline ADC.

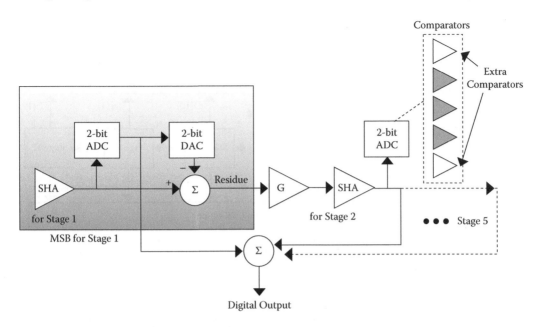

FIGURE 7-18 Ten-bit pipeline ADC with 2-bit-per-stage ADC and 1-bit redundancy.

Even though each pipeline stage only has to contribute *n* bits of the conversion result, its SHA must have enough dynamic range and bandwidth so that later stages can perform accurate conversions. Therefore, the SHA is often the limiting factor of a high-speed pipeline ADC. For example, a 10-bit pipeline ADC has five stages, each of which contributes two bits. The first stage samples and holds the analog input signal V_{in}, generates 2 bits of the digital output, converts the 2-bit digital signal into an analog signal, and subtracts it from the held output of the SHA. The difference (residue) is amplified and then processed by the next stage. In theory, amplification is unnecessary, although a larger signal level eases the subsequent operations. As the same process repeats for the subsequent stages, the resolution requirement is reduced by a factor of four for each stage. In the example, the first-stage SHA has to provide a resolution of 10 bits, the second-stage SHA only needs a resolution of 8 bits, etc.

The comparator offset in a pipeline can be corrected digitally. Assume that each stage is resolving 2 bits for a 10-bit pipeline ADC in Figure 7-18. In addition to the three comparators required for the conversion, two extra comparators can be added for each of the second to the fifth stages. The extra comparators detect any overflow level outside the nominal level set by the three main comparators. The ADC output can be digitally corrected by adding or subtracting (depending on the overflow direction) the detected error. In this example, a 1-bit redundancy is achieved by overlapping the ranges of neighboring (1st and 2nd, 2nd and 3rd) stages.

Instead of adding extra comparators as shown in Figure 7-18, another approach (see Figure 7-19) adds extra stages to perform digital calibration (Karanicolas, Lee, and Bacrania 1993). For instance, 12 stages can be used for a 10-bit pipeline ADC to provide two stages of redundancy.

With the offsets in comparators and preamps and the mismatches in capacitors, the residue deviates from ideal characteristics. The residue can be outside the nominal range, resulting in a missing decision level as shown in Figures 7-20(b) and (c). In order to ensure that the residue range is within the detection level, the gain (G) is reduced according to the amount of maximum mismatch and offset, as shown in Figure 7-20(d). The reduced gain is made up by the extra stages. The gain reduction prevents the missing decision levels but cannot deal with the missing codes. A calibration is performed digitally in each stage to recover the missing code (Karanicolas, Lee, and Bacrania 1993).

Another source of error is the nonlinearity of the SHA. In order to minimize the nonlinearity, a closed-loop op-amp is used in pipeline ADCs. However, the closed-loop amplifier has a higher

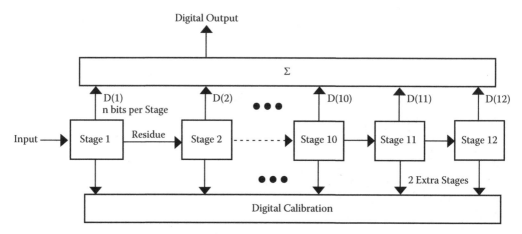

FIGURE 7-19   Ten-bit pipeline ADC with digital calibration.

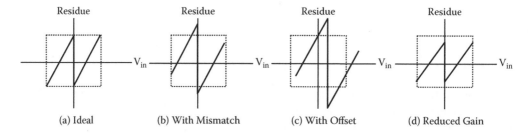

FIGURE 7-20   Residue characteristics.

noise and a smaller bandwidth, and consumes more power than does its open-loop counterpart. An open-loop amplifier can be used where error caused by the amplifier nonlinearity is corrected digitally (Murmann and Boser 2004). This technique is claimed to save a pipeline ADC power by 75% (Murmann and Boser 2004). Since this technique allows the use of an open-loop amplifier in a pipeline ADC, the sampling rate can also improve.

7.7   POWER DISSIPATION ISSUES IN HIGH-SPEED ADCS

It is very difficult to compare power-dissipation performance of high-speed ADCs. Power dissipation depends strongly on the architecture and device technology used. For a given architecture, the power dissipation increases linearly with a sampling rate (if $f_s \ll f_T$). In flash ADCs, as the number of resolution bits N increases, the required power increases at a faster rate than 2^N because the offset voltages must be reduced. In pipeline ADCs, the power increases nonlinearly depending on the number of bits per stage and the digital correction circuitry (Lewis 1992).

In CMOS, the sample-and-hold and gain stages can be combined into a single circuit to save power. Also, low-power CMOS digital circuits can correct for offsets, mismatches, gain errors, and nonlinearity. Such techniques cannot be readily applied to the III-V technologies, which must use very large devices to reduce the offset (since III-V transistors have much higher offset than Si and digital offset cancellation is not readily available). Clearly one must choose a device technology that can take advantage of power-reduction schemes (power-efficient architectures and digital correction circuits).

7.8   SUMMARY

This chapter introduced the performance metrics used to characterize ADCs. The development trends of ADC technology were discussed. The design issues and performance deciding factors of flash and pipeline ADCs were presented. Since digital circuit performance improves at a rate much faster than analog circuit performance, it is expected that increasingly more digital techniques will be applied to improve the performance of ADC circuits.

REFERENCES

Aude, A.J. 1998. Audio Quality Measurement Primer. Intersil Corp. Application Note AN9789.

Audio Engineering Society Standard Committee. 2003. AES17-1998 (r2004): AES standard method for digital audio engineering—Measurement of digital audio equipment (Revision of AES17-1991).

Batruni, R.G. 2006. Analog to digital converter with distortion correction. U.S. Patent No. 7,002,495. Filed Jan. 5, 2005; issued Feb. 21, 2006.

Bowling, S. 2000. Understanding A/D converter performance specifications. AN693, DS00693A. Chandler, Ariz.: Microchip Technology, Inc.

Bult, K. and A. Buchwald. 1997. An embedded 240 mW 10-b 50-MS/s CMOS ADC in 1-mm². *IEEE Journal of Solid-State Circuits* 32(12): 1887–1895.

Callegari, S., R. Rovatti, and G. Setti. 2005. Embeddable ADC-based true random number generator for cryptographic applications exploiting nonlinear signal processing and chaos. *IEEE Transactions on Signal Processing* 53(2): 793–805.

Cho, T. 1995. Low-Power Low Voltage A/D Conversion Techniques Using Pipelined Architectures. Ph.D. thesis, University of California, Berkeley.

Choi, M. and A.A. Abidi. 2001. A 6b 1.3-Gsample/s A/D converter in 0.35 μm CMOS. *IEEE Journal of Solid-State Circuits* 36(12): 1847–1858.

Crawley, H.B., R. McKay, W.T. Meyer, E.I. Rosenberg, and W.D. Thomas. 1992. Performance of flash ADCs in the 100 MHz range. *IEEE Transactions on Nuclear Science* 39(4): 780–783.

— 1994. Recent results from tests of fast analog-to-digital converters. *IEEE Transactions on Nuclear Science* 41(4): 1181–1186.

Dalton, D., G. Spalding, H. Reyhani, T. Murphy, K. Deevy, M. Walsh, and P. Griffin. 1998. A 200-MSPS 6-bit flash ADC in 0.6-μm CMOS. *IEEE Transactions on Circuits and Systems-II: Analog and Digital Signal Processing* 45(11): 1433–1444.

Drake, A.W. 1967. *Fundamentals of Applied Probability Theory.* New York: McGraw-Hill Book Company.

Elbornsson, J., F. Gustafsson, and J.-E. Ekund. 2005. Blind equalization of time errors in a time-interleaved ADC system. *IEEE Transactions on Signal Processing* 53(4): 1413–1424.

Gray, P., P. Hurst, S. Lewis, and R. Meyer. 2001. *Analysis and Design of Analog Integrated Circuits*, 4th edition. Hoboken, N.J.: John Wiley & Sons.

Institute of Electrical and Electronics Engineers. 1989. *IEEE Trial-Use Standard for Digitizing Waveform Recorders.* IEEE Std 1057 (issued July 1989 for trial use; see also IEEE Std 1057-1994), SH12740. New York: Institute of Electrical and Electronics Engineers, Inc.

Institute of Electrical and Electronics Engineers. 2000. *IEEE Standard for Terminology and Test Methods for Analog-to-Digital Converters*, IEEE Std 1241-2000, SH94902. New York: Institute of Electrical and Electronics Engineers, Inc.

Juodawlkis, P.W., J.J. Hargreaves, R.D. Younger, R.C. Williamson, G.E. Belts, and J.C. Twichell. 2003. Optical sampling for high-speed, high-resolution analog-to-digital converters. *Proceedings of the International Topical Meeting on Microwave Photonics* 75–80.

Karanicolas, A.N., H.-S. Lee, and K.L. Bacrania. 1993. A 15-b 1-Msample/s digitally self-calibrated pipeline ADC. A 10-bit 5-Msample/s CMOS two-step flash ADC. *IEEE Journal of Solid-State Circuits* 28(12): 1207–1215.

Kattmann, K. and J. Barrow. 1991. Technique for reducing differential nonlinearity errors in flash A/D converters. *Proceedings of the IEEE Solid State Circuits Conference* 170–171.

Kester, W. and J. Bryant. 2000. *Mixed-Signal and DSP Design Techniques.* Norwood, Mass.: Analog Devices, Inc.

Kinget, P. and M. Steyaert. 1996. Impact of transistor mismatch on the speed-accuracy-power trade-off of analog CMOS circuits. *Proceedings of IEEE Custom Integrated Circuits Conference* 333–336.

Le, B., T.W. Rondeau, J.H. Reed, and C.W. Bostian. 2005. Analog-to-digital converters. *IEEE Signal Processing Magazine* 22(6): 69–77.

Lewis, S.H. 1992. Optimizing the stage resolution in pipelined, multistage, analog-to-digital converters for video-rate applications. *IEEE Transactions on Circuits and Systems-II: Analog and Digital Signal Processing* 39(8): 516–523.

Limotyrakis, S., K.Y. Nam, and B. Wooley. 2002. Analysis and simulation of distortion in folding and inter-polating A/D converters. *IEEE Transactions on Circuits and Systems-II* 49(3): 161–169.

Lundin, H., M. Skoglund, and P. Handel. 2005. Optimal index-bit allocation for dynamic post-correction of analog-to-digital converters. *IEEE Transactions on Signal Processing* (53)2: 660–671.

Miao, B., C. Chen, A. Sharkway, S. Shi, and D.W. Prather. 2006. Two bit optical analog-to-digital converter based on photonic crystals. Optical Society of America, *Optics Express* 14(17): 7966–7973.

Mukhanov, O.A., D. Gupta, A.M. Kadin, and V.K. Semenov. 2004. Superconductor analog-to-digital convert-ers. *Proceedings of the IEEE* 92(10): 1564–1584.

Murmann, B. and B.E. Boser. 2004. *Digitally Assisted Pipeline ADCs Theory and Implementation.* Norwell, Mass.: Kluwer Academic Publishers.

Neesgaard, C. 2001. Digital Audio Measurements. Texas Instruments, Inc., Application Report SLAA114.

Onodera, K.K. 2000. Low-Power Techniques for High-Speed Wireless Baseband Applications. Ph.D. thesis, University of California, Berkeley.

Peetz, B., B.D. Hamilton, and J. Kang. 1986. An 8-bit 250 mega samples per second analog-to-digital con-verter without a sample-and-hold. *IEEE Journal of Solid-State Circuits* 21(6): 997–1002.

Poulton, K., R. Neff, B. Setterberg, B. Wuppermann, T. Kopley, R. Jewett, J. Pemillo, C. Tha, and A. Montijo. 2003. A 20 GS/s 8b ADC with a 1MB memory in 0.18 μm CMOS. *2003 IEEE International Solid-State Circuits Conference.* Session 18 (Nyquist A/D Converters), Paper 18.1.

Raz, G.M. 2003. Highly linear analog-to-digital conversion system and method thereof. U.S. Patent No. 6,639,537. Filed March 30, 2001; issued Oct. 28, 2003.

Shaw, G.A. and S.C. Pohlig. 1995. I/Q baseband demodulation in the RASSP SAR benchmark. MIT Lincoln Laboratory Technical Report RASSP-4, Lexington, Mass.

Sheingold, D.H., Ed. 1972. *Analog-Digital Conversion Handbook.* Norwood, Mass.: Analog Devices, Inc.

Taft, R.C., C.A. Menkus, M.R. Tursi, O. Hidri, and V. Pons. 2004. A 1.8V 1.6-GSample/s 8-b self-calibrat-ing folding ADC with 7.26 ENOB at Nyquist frequency. *IEEE Journal of Solid-State Circuits* 39(12): 2107–2115.

Uyttenhove, K. and M. Steyeart. 2002. Speed-power-accuracy trade-off in high speed CMOS ADCs. *IEEE Transactions on Circuits and Systems-II* 49(4): 280–287.

van de Grift, R.E.J. and R.J. van de Plassche. 1984. A monolithic 8-bit video A/D converter. *IEEE Journal of Solid-State Circuits* SC-19(3): 374–378.

van de Plassche, R. 1994. *Integrated Analog-To-Digital and Digital-To-Analog Converters.* Norwell, Mass: Kluwer Academic Publishers.

van de Plassche, R.J. and P. Baltus. 1988. An 8-bit 100-MHz Full-Nyquist analog-to-digital converter. *IEEE Journal of Solid-State Circuits* 23(6): 1334–1344.

Velazquez, S.R. and R.J. Velazquez. 2002. Adaptive parallel processing analog and digital converter. U.S. Patent No. 6,339,390. Filed June 20, 2001; issued Jan. 15, 2002.

Vorenkamp, P. and R. Roovers. 1997. A 12-b 60-MSample/s cascaded folding and interpolating ADC. *IEEE Journal of Solid-State Circuits* 32(12): 1876–1886.

Walden, R.H. 1999. Analog-to-digital converter survey and analysis. *IEEE Journal on Selected Areas in Com-munications* 17(4): 539–550.

Walpole, R.E. and R.H. Myers. 1972. *Probability and Statistics for Engineers and Scientists.* New York: The Macmillan Company.

White, L.B., F. Rica, and A. Massie. 2003. Spurious free dynamic range for a digitizing array. *IEEE Transac-tions on Signal Processing* 51(12): 3036–3042.

Zozor, S. and P.-O. Amblard. 2005. Noise-aided processing: revisiting dithering in a sigma-delta quantizer. *IEEE Transactions on Signal Processing* 53(8): 3202–3210.

8 Implementation Approaches of Front-End Processors

M. Michael Vai and Huy T. Nguyen, MIT Lincoln Laboratory

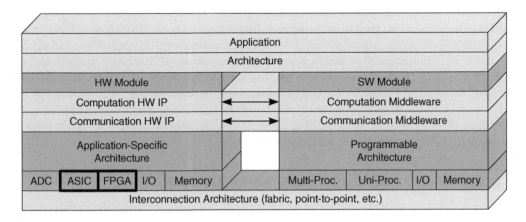

This chapter describes a general design process for high performance, application-specific embedded processors and presents an overview of digital signal processing technologies.

8.1 INTRODUCTION

The principal functions of front-end processors are signal and image processing algorithms, such as digital filtering, fast Fourier transforms (FFTs), matrix operations, beamforming, detections, and tracking. Besides improving the signal-to-noise (uncorrelated noise) ratio, the primary objective of a front-end processor is to transform the raw analog-to-digital converter (ADC) data into information useful to the user. An important outcome of this transformation is the reduction in data volume per unit time. This reduction allows data to be converted from hundreds of gigabytes per second at the output of the ADCs to a few megabytes per second. In this chapter, front-end processing is defined as the operations starting at the output of ADCs and ending when both the data rate and computation requirement are reduced low enough to be handled by a computer (e.g., a programmable digital signal processor or general-purpose workstation).

The design and implementation of front-end processors must overcome many challenges. At high data rates (e.g., several gigasamples per second), the signal and image processing algorithms have very demanding computational requirements. They may require tera-operations per second (TOPS) of computational throughput and terabytes of holding memory as buffers for incoming data and intermediate data products. Many applications demand hard real-time performance (e.g., a latency of 10 milliseconds or less) since the system must meet stringent system deadlines. Furthermore, these systems often have to meet the requirements of small form factor, light weight, low power, radiation tolerance, and low cost.

Military applications have some of the most demanding front-end performance requirements. They have to (1) deliver high performance at low cost, (2) allow a high degree of adaptability, flexibility, scalability, and reconfigurability, (3) provide high bandwidth, and (4) take advantage of massively parallel processing. Ironically, military applications usually neither demand the volume nor have the funding that leads to the development of special technologies. Instead, they have to rely on commercially developed computing technology to meet their requirements. Fortunately, commercial applications (e.g., games, communications, medical equipment, etc.), because of their potential high volume and high profit, develop computing technology very rapidly.

This chapter first describes a general design process of high performance, application-specific embedded processors. This is followed by an overview of the digital signal processing technologies: full-custom and synthesized application-specific integrated circuits (ASICs), field programmable gate arrays (FPGAs), and general programmable processors (GPPs). These technologies will be individually discussed in detail in later chapters. This chapter compares them from various aspects: computational efficiency, computational density, design complexity, and cost. Front-end processing, due to its performance requirement, typically requires the use of "hardware" technologies (i.e., ASIC and FPGA). The reader should note that hardware technologies offer higher levels of delivered performance at the expense of limited to no flexibility in programming. In contrast, GPPs offer flexibility in software programming but typically deliver much lower performance relative to their peak throughput capabilities. A couple of case studies that illustrate the implementation of front-end processors at both chip and system levels will be provided.

8.2 FRONT-END PROCESSOR DESIGN METHODOLOGY

Figure 8-1 shows a typical design flow of a high performance application-specific signal processor. The development of a processor has to go through a modeling phase followed by a physical design phase. In the modeling phase, a floating-point simulation model, typically written with a high-level programming language (e.g., C, MATLAB, etc.), is used for algorithm simulation and verification. Despite the apparent benefits of floating-point operations, their hardware implementations are several times more complicated and thus more expensive than their fixed-point counterparts. Often, the floating-point simulation model is converted into a fixed-point model in the precision analysis

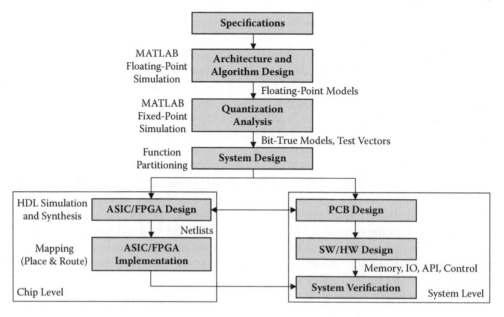

FIGURE 8-1 High performance front-end processor development flow.

step to determine implementation requirements (e.g., word size, etc.). The fixed-point simulation model is also referred to as a "bit-true" model since it can be used to develop binary test vectors and output vectors.

The algorithm is developed into a processor in the physical design phase, which is further divided into chip-level design and system-level design. The chips can be implemented with a number of different technologies, each of which has its own benefits and limitations. The rest of this section examines these technologies with the objective of enabling a proper technology selection for a high performance embedded signal processor. Note that the adaptation of an application-specific device in a system will require the support of a printed circuit board (PCB), power supply, input/output, memory, control, etc. ASICs almost always require the development of custom boards. On the other hand, commercial off-the-shelf (COTS) FPGA boards are available, even though their capability must be carefully investigated before adaptation.

The choice of an implementation technology depends on the constraints imposed by the embedded platform, which could be as large as an aircraft carrier or as small as a micro unmanned aerial vehicle (UAV). For example, UAV applications are constrained to very low power and weight. On the other hand, the size, weight, and power (SWAP) requirements are likely to be relaxed for ground-based applications. Therefore, there is no one best choice of computing technology.

8.3 FRONT-END SIGNAL PROCESSING TECHNOLOGIES

An embedded signal processor designer often has to balance various competing objectives: development cost, production cost, performance (operation speed and power consumption), time to market, and volume expectation. It is very difficult, if not impossible, to deliver a design that is optimized for all of these objectives. The constraints specific to an application should be used to select from available implementation technologies. In many cases, the optimal system is a hybrid system consisting of ASICs, FPGAs, and GPPs. The FPGA technology will provide the benefits of low nonrecurring expenses (NRE) and reconfigurability. The ASIC technology will be used in the most demanding portion of the system to keep the power consumption and form factor under control.

Figure 8-2 shows the concept of building an optimized heterogeneous signal processing system that includes GPPs, FPGAs, synthesized ASICs, and full-custom ASICs. Given an application,

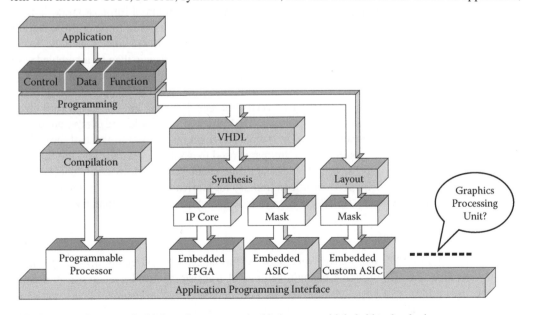

FIGURE 8-2 Concept of a high performance embedded system with hybrid technologies.

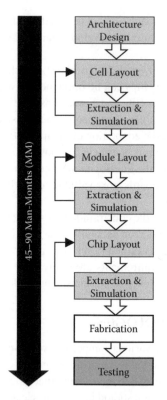

FIGURE 8-3 Full-custom ASIC design flow.

the design will separate its functions into the implementation, using different technologies to achieve an optimal mix-and-match processing system. Other innovative devices should also be allowed to participate in this computing paradigm. One example of such devices is the graphics processing unit (GPU), which has shown low-cost, high performance processing power for certain scientific computation applications (e.g., linear algebra) (GPGPU website 2007; GPGPU Workshop 2006). The advancement of GPUs has been driven by the insatiable appetite of the gaming industry.

The following sections provide general overviews of ASIC and FPGA technologies. More details of these technologies can be found in Chapters 9 and 10.

8.3.1 FULL-CUSTOM ASIC

The layout of a module or even a complete integrated circuit (IC) can be manually created by specifying individual transistor dimensions, locations, and interconnections. This approach is commonly referred to as the full-custom design. Potentially very high performance can be achieved in a full-custom design since the circuit is under the direct control of the designer. However, the performance benefit comes at the price of an extremely high design complexity. In the case of a high volume production, the non-recurring expenses cost can be divided among a large number of devices. Otherwise, this design approach is reserved for meeting extremely high performance requirements.

There are architectural properties that can be explored to facilitate a full-custom design. If a circuit has a high degree of regularity, then only a few types of cells need to be crafted. Another useful property is local communication between cells, which allows interconnections to be formed when cells are put together without the need for explicit routing. Cells can then be assembled using a process called *tessellation*, which is analogous to laying the tiles on a bathroom wall.

Figure 8-3 shows the typical design flow of a full-custom ASIC. The layouts of the cells (i.e., building blocks) are first created. In order to verify the cells, layouts are converted into circuit schematic diagrams in a process called *extraction*. The extracted schematics are compared to target schematics with a layout versus schematic (LVS) tool. The function and performance of the cells are determined by simulation. A tessellation technique is used to assemble the cells into the final chip layout. The entire circuit is verified by LVS, design rule check (DRC), electric rule check (ERC), and simulation before it is delivered to a silicon foundry for fabrication, which is commonly referred to as a *tape-out*. The ASICs are tested when they come back from the silicon foundry.

8.3.2 SYNTHESIZED ASIC

While full-custom ASICs can deliver very high performance, it is extremely difficult to manually lay out an entire IC consisting of multimillions of transistors. Another approach to creating an ASIC uses a library of basic building blocks, called standard cells. Standard-cell libraries are commercially available for specific fabrication processes. A library usually contains primary logic gates (e.g., inverter, NAND, NOR, etc.) as well as function blocks (e.g., adder, multiplier, memory, etc.). The quality of a library is a significant determining factor of the design performance, so choosing a library is critical to the success of a design.

A typical standard-cell ASIC design flow is shown in Figure 8-4. The device to be designed is described in a hardware description language (HDL) program, which can be simulated for verification and synthesized into a netlist of standard cells. Currently, the popular HDLs are VHDL (very-high-speed integrated circuit hardware description language) and Verilog, which are programming languages specially designed to allow the description of circuits. On the other hand, SystemC, a language based on the context of C++ and enhanced with a class library of hardware-oriented constructs, is also gaining popularity. Advocates of SystemC claim the use of an HDL that originated as a general programming language can facilitate the design and verification of a system from concept to implementation in hardware and software.

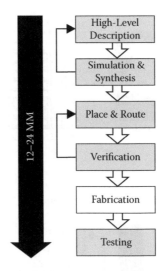

FIGURE 8-4   Synthesized ASIC design flow.

The standard cells are then arranged on the available chip area with all the associated signals connected. The task of arranging the cells on the layout, called the placement process, attempts to determine the best location for each cell. The routing process takes the result of a placement and automatically completes the interconnections. A chip-level verification (LVS, DRC, ERC, etc.) is performed before a tape-out.

8.3.3   FPGA Technology

While ASICs have the potential of achieving the highest performance offered by a semiconductor process, their high design complexity (especially in the case of a full-custom design) and high fabrication cost often make them cost prohibitive. FPGA-based designs are appropriate for applications that do not require ASIC-like performance and can thereby benefit from much lower total development costs. As shown in the conceptual architecture of Figure 8-5, an FPGA is a fully manufactured device that contains an array of configurable logic blocks, memory blocks, arithmetic blocks, and interconnects that are designer controllable. Advanced FPGAs may have microprocessor hard cores* embedded to implement functions that are more suitable for a programmable processor than for a dedicated circuit, such as a controller. Alternatively, microprocessor soft cores are also available.

The FPGA design flow is shown in Figure 8-6. Similar to the standard-cell ASIC design flow, an HDL program describing the circuit is created for simulation and synthesis. The resultant netlist is mapped, placed, and routed to create configuration data, which contain the information needed to configure the FPGA into an application-specific design. To promote sales, FPGA vendors often provide free and low-cost design tools customized for their devices.

There are three types of FPGAs. SRAM (static random access memory)-based FPGAs store configuration data in SRAMs so they are reprogrammable. However, the configuration is volatile since they are lost when the power is off. Anti-fuse-based FPGAs are one-time programmable and nonvolatile. Flash-based FPGAs use flash memories and are both reprogrammable and nonvolatile.

---

* Please see Section 8.4 for a definition of hard and soft cores.

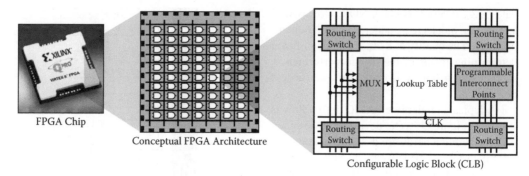

FPGA Chip

Conceptual FPGA Architecture

Configurable Logic Block (CLB)

FIGURE 8-5   FPGA architecture.

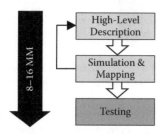

FIGURE 8-6   FPGA design flow.

The development trends of FPGA technology will have significant impact on embedded signal processing. The availability of platform FPGAs, which are optimized for specific application domains, enable a higher degree of customization, and thus speed and power. Traditionally, in many cases processor development is input/output (I/O) limited. FPGAs are now embedded with gigabit serial I/Os, providing very-high-speed interchip and intersystem interconnections. Also, high (algorithmic)-level design tools have been developed to attempt solving the problems of increasingly high design and verification costs and the scarcity of hardware design expertise (see Mentor Graphics' Catapult Synthesizer online at http://www.mentor.com/products/ and the Xilinx System Generator at http://xilinx.com).

Many people misunderstand the design complexity of FPGA design. In demanding applications, especially when a high percentage of FPGA resources have to be used, the design is actually harder than ASIC design due to the constraint of available resources. People often mistake FPGA as a programmable processor. In fact, an FPGA is a piece of hardware configurable for a specific application and its design requires no less understanding of digital circuit design technology. Also, while an FPGA is reconfigurable and it is possible to incorporate last minute "field" changes, it may also trigger a complete design cycle. The fact that FPGA manufacturers keep the configuration bitstream format proprietary makes the physical design tools proprietary. Therefore, low-cost FPGA design tools are often available through the FPGA vendors.

An FPGA is a good implementation platform for prototyping and low-volume production. When the volume of a product warrants, it may be desirable to take an FPGA design and convert it into an ASIC. The most common reason for FPGA-to-ASIC conversion is to get better speed and lower power consumption since FPGAs are significantly more power hungry than their ASIC counterparts. While theoretically the HDL program originally designed for an FPGA can be resynthesized into a standard-cell ASIC, in reality the design effort is often close to a new design cycle. The resynthesizing effort can be facilitated by targeting a gate-array style architecture designed to somewhat mimic an FPGA structure.

One approach for FPGA-to-ASIC conversion is to use a mask programmed device with the same architecture as the FPGA to be replaced. Improved performance is a direct result of replacing programmable routing connections with metal customization. The advantage of this approach is that the original FPGA designed for prototyping and early production can be extended into an ASIC without modifications. However, this is an FPGA vendor-supplied solution.

8.3.4 STRUCTURED ASIC

A new kind of semicustom mask programmable device, the structured ASIC, can be viewed as a technology located between FPGAs and ASICs. The structured ASIC appears to have the potential to achieve close to ASIC-level performance without high nonrecurring expenses, long design cycles, and high risk factors. Structured ASIC devices typically provide premanufactured logic components, routing fabric, and a number of commonly used functional blocks (e.g., high-speed I/O, memory, microprocessors, etc.). The devices are manufactured except for a few metal layers, which are used to customize the final devices for a specific design.

The advantages of a structured ASIC over an ASIC are lower design, manufacturing, testing, and integration costs. Compared with FPGAs, structured ASICs offer higher performance with lower power consumption and potentially a smaller footprint.

8.4 INTELLECTUAL PROPERTY

Design reuse contributes to both risk mitigation and the reduction of development time. The most direct use of this principle is the use of functions from a previous product. The intellectual property (IP) core approach extends the concept to the use of functions designed by another party. IP cores are available for a large variety of functions, ranging from adders to microprocessors. Details about the IP technology are provided in Chapter 11. The following discussion refers to IP cores used in ASIC/FPGA development.

There are three types of IP cores: soft, firm, and hard. A soft IP core is represented with HDL at the register transfer level (RTL). Users will be required to perform the typical steps (synthesis, place and route, and verification) in the design flow following HDL development. Since the IP representation is provided in HDL, it is relatively technology independent.

A firm IP core moves a bit closer to the physical side by delivering the IP design in a netlist of generic logic cells. No synthesis is required. The responsibility of the user is to perform the physical designs steps (place and route, verification). This approach depends on the assumption that generic logic cells are available in most cell libraries. A firm IP has a certain degree of technology dependence, but remains relatively generic.

On the extreme end of the spectrum, a hard IP core can be delivered as a layout in the form of a drop-in module. The user only needs to perform verification. Since the IP is in a physical form, it is technology dependent.

8.5 DEVELOPMENT COST

There are two costs involved with a front-end processor: nonrecurring expenses and unit production cost. The NRE is mainly the design cost (chip and system) and in the case of ASIC, the mask manufacturing cost.

ASIC has performance that is orders of magnitude better than that of software technology. However, it has a very high NRE, which can go to more than $10M. Figure 8-7 shows example ASIC NREs associated with various semiconductor technologies. Special design methodologies, such as systolic arrays (see Chapter 12), can be used to mitigate the design cost, which generally dominates the NRE.

The design of an ASIC typically needs a team of 9–16 engineers with 12–24 specialties. The mask cost for submicron technology (e.g., 90 nm) is $1M to $2M. There is a high infrastructure cost (millions) to acquire and maintain computer-aided design (CAD) tools. Also, the chance of first silicon failure is quite high; statistically about half of the designs required two or more spins. Therefore, ASIC technology is usually reserved for applications with requirements that cannot be met otherwise. Alternatively, the application must have a volume large enough to amortize the NRE.

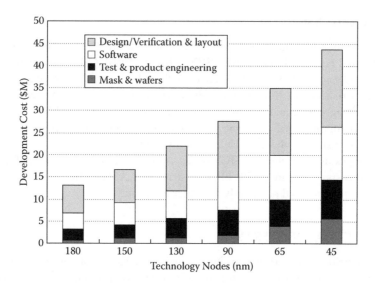

FIGURE 8-7 NRE cost for ASIC development.

FPGA is an application-specific hardware technology, but it is a COTS device and has all the benefits of COTS products. Probably because of the seemingly self-contradictory features (COTS and application-specific hardware), a lot of myths about FPGA technology persist. According to *EE Times* (June 2004), the "field programmable gate array" might be the most misnamed device in the electronic repertoire. It is important to have a clear understanding of what it is and what it isn't to develop a successful processor system.

The cost of using an FPGA has two facets. First, there is the cost of acquiring the FPGAs, which are COTS and have the NRE amortized. Therefore, they are significantly less expensive compared to low-volume ASICs. Large and powerful FPGA devices only cost a few thousand dollars. Second, there is the cost of actually customizing the FPGA to make it application specific.

Figure 8-8 shows the relationships between unit chip cost and production volume. The fact that FPGAs can be converted to structured ASICs or cell-based ASICs allows a system to take advantage of the FPGA benefits, such as fast time to market and low NRE, at the early phase of the life cycle, while taking advantage of better performance when the volume grows into mid to high volume.

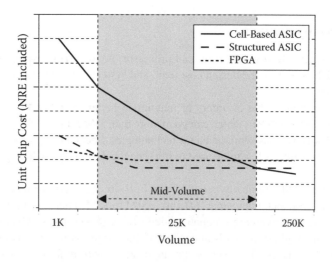

FIGURE 8-8 Volume versus unit chip cost.

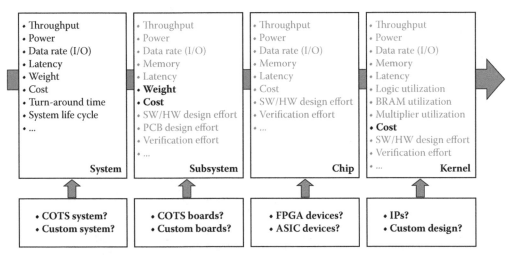

Application-specific system requirements.   Multiple-level performance parameters.

FIGURE 8-9   System design issues.

Figure 8-9 shows the design issues at different levels of a system design. The top design problem is the "here's the spec so build it" mindset. The current system design process consists of a number of sequential design operations (system, subsystem, chip, and kernel levels), each of which searches for a local optimum in its own domain.

These issues can only be mitigated by benchmarking performance parameters at multiple levels. However, realistic and scalable benchmarks are difficult to obtain. Vendor-touted performance is often obtained under idealistic conditions (i.e., theoretical performance). Furthermore, benchmarking must be performed at both chip and system levels to model behaviors across multiple chip/board behaviors (e.g., through a backplane connection or an Ethernet connection). Without an accurate benchmark at the system level, the same chip could produce very different performance when it is used in different boards.

Industry has long acknowledged the difficulty of benchmarking. Since it is virtually impossible for a vendor to measure and report on what's relevant to every design team considering a particular tool or technology for their project, vendors have always been reluctant to publish benchmark results, except with reference to their own previous generations. In addition, the most meaningful metrics are domain- or application-specific, so designing a fair and accurate test methodology that yields reasonable metrics is a daunting task. Finally, the benchmarking challenge goes beyond quantifying the performance of a device; the designer must be convinced that the device can function at speed on a real board.

The system design issues illustrated in Figure 8-9 include the design complexity of a system at different levels. This is important since the schedule and budget must be set correctly. For example, the verification time accounts for 33% of chip-level design and up to 75% of system design, which apparently has to be included in the schedule and budget.

Figure 8-10 illustrates the concept of a system development infrastructure. The application programming interface (API) provides a library of function calls to facilitate the software development on the host computer. Typical functions include chip (ASIC or FPGA) configuration and memory access. Note that COTS FPGA boards may have proprietary APIs so that designs are not portable between vendors. API development is often expensive as it may require the same level of effort used to develop the chips. The map allows the system to be easily verified against a bit-true model, a fixed-point model, or a floating-point model.

In addition to the cost of chips, system hardware and software development costs are also significant. The system will need one or more PCBs, other chips (e.g., memory), and hardware/soft-

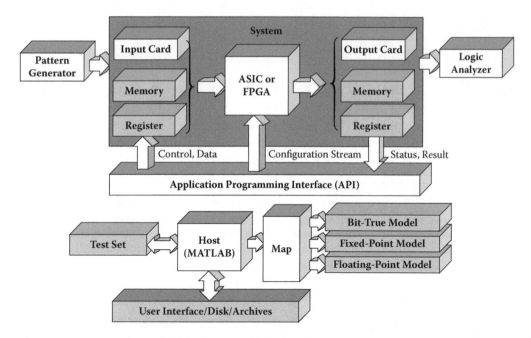

FIGURE 8-10 Conceptual system development infrastructure.

ware system interface with other devices. It is especially a challenge if we have to integrate FPGAs, ASICs, and high-speed I/O on a complex PCB. For example, a complex FPGA has more than 1000 pins, most of which are user-configurable. The large pin count on a package causes the PCB pin density to be significantly increased, resulting in routing congestion and a high number of PCB layers. Signal paths have to be precisely matched in length to enable high-speed operations. At this level of complexity, the conventional practice of throwing a system design over the wall for the PCB designer no longer works. This integration requires a lot of synergy between design team members. A chip/system co-design approach should be used to achieve optimization of the design at both system and chip levels.

A one-of-a-kind ASIC always needs a custom-designed PCB to operate. On the other hand, COTS FPGA boards are available and they may be preferable to custom boards. These COTS boards have standard form factors (e.g., VMS) and contain one or more FPGAs and supporting devices (e.g., memory). Also, the COTS board vendor may provide IP cores (e.g., memory controller) to simplify the design process. An appropriate selection of a COTS board, of course, will save a lot of design time. However, the claims given by the vendor have to be physically verified as their boards have been measured under ideal conditions. This is especially critical when the application is expected to push the technology envelope.

8.6 DESIGN SPACE

Figure 8-11 shows the signal processor design space defined by three computing technologies: ASIC, FPGA, and GPP. This chart summarizes the SWAP (size, weight, and power) characters of each technology using a two-dimensional space. The size and power characteristics are directly indicated on the diagram. The horizontal axis measures the computational efficiency in terms of giga-operations per second per watt, or GOPS/W. The vertical axis measures the computational density in giga-operations per liter (GOPS/L). In contrast, the weight characters have to be derived. Since a system comprises chassis, chips, circuit boards, power supplies, and other accessories, its weight will be proportional to its computational efficiency and density. The requirements of several example defense applications and their applicable implementation technologies are indicated in Figure 8-11.

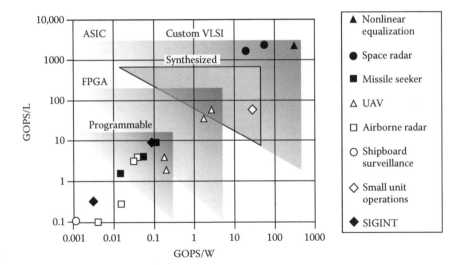

FIGURE 8-11 Digital signal processing system design space.

In this chapter, both microprocessors and digital signal processors (DSPs), special microprocessors optimized for digital signal processing, are referred to as GPPs. Chapters 13 and 26 of this book are dedicated to the development of microprocessor technology. The issues of software development and programming are explained in Chapters 15 through 19. GPPs have the benefit of being software controlled; however, their use in front-end processing is often limited by their performance, which is in the order of 0.1 GOPS/W and 10 GOPS/L. The readers should note that in the case of a GPP, its performance is determined by its sustained throughput, which typically is only a small percentage (e.g., 10–20%) of its peak throughput.

According to Moore's Law, to get a 10× improvement in performance will take five years at the current improvement rate of a 2× factor every 18 months. The rate of advances in semiconductor feature size reductions will affect the values in Figure 8-11. The International Technology Roadmap for Semiconductors keeps track of these technology advances (ITRS website 2007). The following prediction assumes a continuing advancement in semiconductor performance, as experienced for the last several decades. It will take approximately two to three years before an FPGA system is able to meet the platform system goals currently achievable with ASIC technology. A GPP system will take even longer to be a viable technology. This prediction is, in fact, rather optimistic and is not applicable to all cases, as Moore's Law does not apply to data communication and memory access.

Before we begin our case studies in the next section, some statistics of the design activities and design complexity are provided here. According to a 2006 survey conducted by *EE Times*, around 25% of the PCB designs have nine or more layers; an average system has 3.0 to 4.0 PCBs, 1.7 to 2.1 FPGAs, and 2.2 to 2.8 ASICs. The mean clock speed is 400 to 800 MHz.

8.7   DESIGN CASE STUDIES

Two high performance front-end processor design cases will be presented in this section. The first processor is a hybrid processor, which includes both ASICs and FPGAs. The second processor is based on the FPGA technology only.

8.7.1   CHANNELIZED ADAPTIVE BEAMFORMER PROCESSOR

Today's sensor array systems commonly use adaptive beamforming techniques to isolate a target from interference (e.g., clutter, jamming, etc.). The beamforming computational requirement increases with the number of sensors in the array, the number of beams to be formed, and the

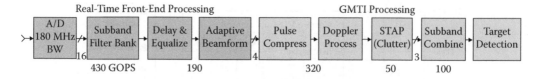

FIGURE 8-12   Processing chain of a 16-channel subband GMTI application.

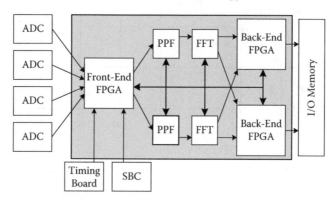

FIGURE 8-13   Four-channel adaptive beamforming module.

number of taps per beam. For wideband applications, a subband-based beamforming approach is several times more efficient than the traditional time-tap delay beamforming. Beamforming in the subband domain requires fewer time-taps to achieve the same performance. This reduction provides a significant computational advantage. Furthermore, adaptive weight computation and application in the subband domain can be performed at a lower sample rate across all subbands in parallel. These properties allow increased flexibility in the adaptive beamformer implementation.

Figure 8-12 shows the processing chain of a 16-channel subband ground moving-target indication (GMTI) application developed at MIT Lincoln Laboratory. The application's goal is to achieve a 1 m range resolution. An analysis of the processing chain indicates that the most demanding part is the real-time front-end processing, which takes the ADC output and performs subbanding, delay and equalization, and adaptive beamforming. The actual GMTI processing is within the capacity of general-purpose computing, such as a computer cluster.

The system takes a modular approach for implementation. The 16 channels are divided into four boards, each of which processes four channels. This modular design creates a significant data communication challenge. In the adaptive beamforming processing, signals from all channels have to be used jointly in the calculation of optimal beamforming weights. Since the 16 channels are distributed among four boards, this requirement creates a large amount of interboard traffic.

Figure 8-13 depicts the basic module of this system, a four-channel, four-beam, 48-subband adaptive beamforming processor. The two back-end X2V8000-5 FPGA chips deliver 30 GOPS to compute the beams, while a smaller front-end FPGA is used for system control and data formatting. The subbanding operation is performed by two sets of full-custom ASICs, each consisting of a polyphase filter (PPF) chip and an FFT chip and forming a 50 GOPS subband channelizer. Chapter 9 has more details about this channelization chip set. As a whole, the module performs 130 GOPS at a 480 MHz ADC sampling rate. A picture of this beamforming board is shown in Figure 8-14.

Each beamforming FPGA is responsible for forming four adaptive beams across 24 subbands, or 96 complex beams with 16-bit I (in-phase) and Q (quadrature-phase) data. The beamforming mechanism occupies 35% of the FPGA area. The 96 incoming serial I/Q interleaved 16-bit complex inputs had to be received and processed at 160 MHz to meet real-time system requirements.

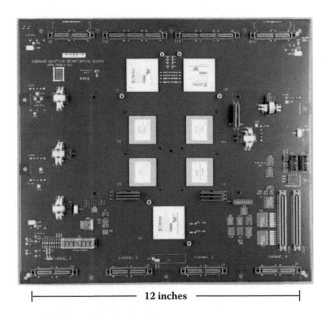

|———————— 12 inches ————————|

FIGURE 8-14 Four-channel adaptive beamforming processor board.

This application demonstrates that with careful design methodology FPGAs can achieve computational capability formally attainable exclusively with ASICs. Besides the challenge of meeting the processing requirements, FPGAs also face other issues, such as heat dissipation and radiation tolerance, when operating in harsh environments such as space. Mitigation approaches are available to overcome this environmental integration hurdle and need to be further explored by the embedded systems community.

The processor shown in Figure 8-14 delivers 130 GOPS of throughput at 25 watts. This power efficiency is over 100 times better than that of a programmable microprocessor. The custom chips perform the subbanding function and provide 100 GOPS. The two beamforming FPGAs account for the rest, 30 GOPS.

Several techniques were used to optimize the FPGA designs. First, the beamforming architecture was specifically optimized for an FPGA implementation, taking into consideration the nature of an FPGA. The result architecture was then manually placed on the FPGA to maximize density, minimize delay, and lower power consumption.

Beamforming is the process of computing the inner product of weight vector, W_i, and input vector, Y_i, indicated by the following equation:

$$Beam = \sum_{i=0}^{n} W_i Y_i,$$

where n is the number of channels.

A straightforward computation of a complex beam using this equation would require four multipliers and six adders for the complex multiplication. This implementation is unacceptable for a system that requires the computation of 96 complex beams or the equivalent of 384 multiplications, where each multiplication uses either one embedded multiplier or 176 slices (one X2V8000-5 contains 168 embedded multiplier blocks and 46,600 slices). Because of the obvious mismatch between the demand and availability of resources, a different, more feasible implementation using distributed arithmetic was adopted.

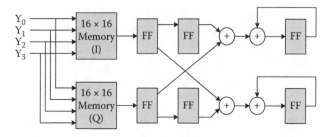

FIGURE 8-15   Beam computation using distributed arithmetic.

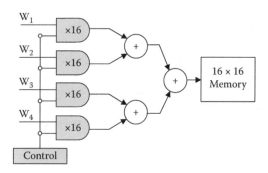

FIGURE 8-16   Updating of DA-encoded weight table.

Distributed arithmetic (DA) constitutes a rearrangement of the operations at the bit level to eliminate costly multiplications and to optimally utilize FPGA configurable resources (Andraka and Berkun 1999). The cost of the operation being done in serial is that the latency to compute a beam is dependent on the bit length of the words. In the MIT Lincoln Laboratory system, the 96 FPGA inputs produced by the PPF/FFT chip sets are in a 32-bit serial I/Q interleaved format, providing a perfect match with the distributed arithmetic scheme.

Figure 8-15 illustrates the DA hardware allocated to the computation of each of the 96 beams. These resources are able to do the equivalent of four multiplications and six additions using ~15% of the resources compared to the brute-force approach. All 96 realizations of this hardware consume ~18% of the chip's area, or ~8500 slices. In addition, a novel approach for dynamically encoding new weights and updating the DA memories as required in adaptive beamforming is implemented as shown in Figure 8-16. This mechanism determines the values to be stored in the 16 × 16 memories depicted in the DA hardware (Figure 8-15). Each instance of this component utilizes ~95 slices. Because it is not necessary to reload all of the weights in parallel to meet system timing requirements, only two instances are implemented for each subband, thus utilizing ~10% of the FPGA area, or 4550 slices.

Although the design described thus far consumes only 28% of the resources, it does not yet include I/O overhead and control structures. If the 96 beams of the design were implemented completely in parallel without buffering, 96 32-bit words would have to be output at the same time, therefore requiring 3072 output pads. Instead only 48 output pads were used. This was accomplished by time-multiplexing the output into three 16-bit double-data-rate buses. The multiplexing is accompanied by insertion of delays into DA inputs to stagger the beginning and ending of DA computations. Buffering and multiplexing the data accounts for 2–3% of the FPGA area, while the remaining 5–6% is due to control structures, routing resources, and buffering for timing. All design elements together produce a beamformer that consumes 35% of X2V8000-5, or ~16,300 slices.

Meeting the real-time processing clock rate of 160 MHz required several design iterations. The first iteration with automatic place-and-route showed the design was only capable of running at 80 MHz. Combinational logic minimizing and registering improved timing to 100 MHz. To meet the specified clock rates, hand-placement of the design was necessary. The high regularity and parallelism of the design (24 subbands, four beams each) reduced the complexity of this task significantly. A relationally placed macro (RPM) of a subband was created, which involved hand-placement of four beams and two weight loaders in a control-centered fashion. These RPMs were replicated 24 times and then arranged in the chip using area constraints. An RPM for the control structure was also constructed to ensure consistent signal distribution and timing between modules. By hand-placing the beamforming logic to minimize control and I/O routing, performance improved to 170 MHz, surpassing the desired 160 MHz system specification. Figure 8-17 shows the final routed design.

As shown in Figure 8-18, the cost of designing the system is comparable with that of the chip design. While chip design is undeniably a daunting task, the integration of chips, software components, and other devices (e.g., ADC, memory, etc.) into a system is at least equally challenging.

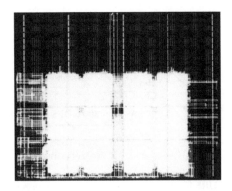

FIGURE 8-17 Routed beamformer design at 170 MHz.

8.7.2 RADAR PULSE COMPRESSION PROCESSOR

This case study describes the rapid prototyping of a high performance real-time processor (RTP) for a radar application. This prototyping project is significant for two reasons. First, the use of FPGA technology in the development of a high-bandwidth, high-throughput, real-time system was demonstrated. Also, after determining that the use of COTS FPGA boards would present a high risk to the project schedule, the custom FPGA-based processor board (see Section 8.7.1), previously developed at MIT Lincoln Laboratory for a different functionality, was quickly adopted as the prototyping platform for this RTP. Despite the need to go through a PCB modification and manufacturing cycle, the development of this RTP was successfully completed in one year.

The RTP input consists of two reference and two echo channels with a data rate of 1.2 GSPS per channel. The real input data are converted into a complex format and cross-correlated to produce four correlation outputs (i.e., range compression). A block diagram of this processing is shown in Figure 8-19.

An analysis was first performed to match up the computational and communication requirements with COTS FPGA boards. While COTS boards may have adequate computational resources (i.e., two to three large FPGAs), they are limited by insufficient data communication bandwidth and power supply. Although a design could have been created to circumvent the COTS limitations, MIT Lincoln Laboratory decided that the COTS approach was posting an unnecessarily high risk on the project schedule. In fact, the result of this analysis showed that a custom FPGA board was warranted to accommodate the high computational and communication requirements.

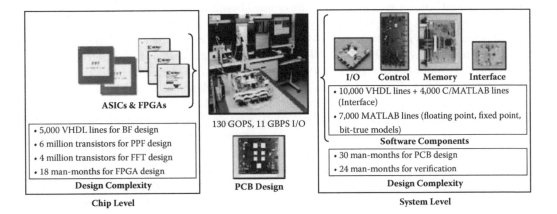

FIGURE 8-18 Design complexities at chip and system levels.

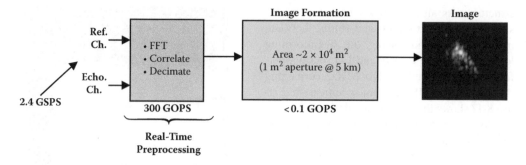

FIGURE 8-19   Real-time pulse compression processing chain.

FIGURE 8-20   Real-time sensor pulse compression processor.

The development of a custom board was ruled out due to schedule constraints. Instead, a different approach was pursued, which was to adopt an MIT Lincoln Laboratory FPGA-based wideband adaptive beamforming processor board, previously explained in Section 8.7.1, as the implementation platform for this RTP prototype.

Figure 8-20 shows a picture of the RTP prototype, which contains two beamforming boards integrated with four 1.2 GSPS ADC (Max 108) daughtercards to compute four real-time correlation outputs. The beamforming boards were enhanced by incorporating high-speed demultiplexer circuits to match the ADC output data rate (1.2 GSPS) with the FPGA interface (300 MSPS). Both FPGAs and custom ASIC chips were employed in the original beamforming board. In this RTP prototype, the ASIC chip sites were not populated since the entire application was implemented in the FPGAs. A total of six Xilinx Virtex II 8000 FPGAs clocked at 150 MHz were used to perform the cross correlations. The RTP has a throughput of 450 GOPS with a power efficiency of 3.5 GOPS/W.

Figure 8-21 illustrates the correlation processing between one reference channel and one echo channel. The objective of this architecture is to perform a correlation between 32K sample segments from these two channels, which would normally require the use of 32K-point FFTs. A segmented correlation scheme was devised to calculate only the required 4K lags from the correlation, thus reducing the size of the FFT from 32K to 8K. An additional benefit of this segmented correlation scheme is that it allows the use of different correlation lengths to improve signal-to-noise ratio (SNR). The implementation utilized block floating-point computations. The word sizes at different stages were chosen carefully to provide the precision required by the applications.

The 8K-point FFT in Figure 8-21 is an innovative systolic implementation, which uses first-in-first-out (FIFO) buffers and switches to manipulate the data flow so that the data are processed by the butterflies with the right coefficients at the right time. This design has a high degree of

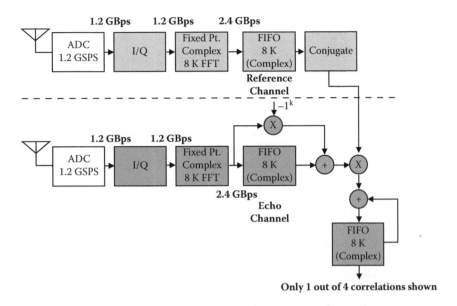

FIGURE 8-21   Cross-correlation between a reference channel and an echo channel.

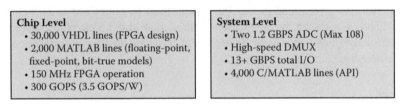

Chip Level	System Level
• 30,000 VHDL lines (FPGA design) • 2,000 MATLAB lines (floating-point, fixed-point, bit-true models) • 150 MHz FPGA operation • 300 GOPS (3.5 GOPS/W)	• Two 1.2 GBPS ADC (Max 108) • High-speed DMUX • 13+ GBPS total I/O • 4,000 C/MATLAB lines (API)

FIGURE 8-22   Design complexity at chip and system levels.

regularity and uses mostly local communication, both of which are desired features for an FPGA implementation. Another important feature of this FFT design is that it is self-contained within the FPGA and does not require any off-FPGA memory to support its operation. More details of this FFT design can be found in Chapter 10 of this book.

The adoption of a custom FPGA board into a rapid prototyping platform for a different functionality was demonstrated. Figure 8-22 shows the design efforts at both chip and system levels. The benefits of adopting existing processor boards for the RTP were numerous. The material cost was significantly reduced from a similar COTS implementation. The FPGA board had an airborne form factor and provided adequate power rating and communication bandwidth to meet RTP requirements. On the other hand, the adoption of an existing platform did not eliminate the system-level development effort. As shown in Figure 8-22, significant effort was still spent at the system-level design and optimization, mainly in the development of an API for the current system.

8.7.3   Co-Design Benefits

We would like to conclude this case study with an anecdote that demonstrates the benefits of co-design methodology. The co-design methodology was applied to prototype a real-time processor, resulting in improvements of 10× and 4× in size and power reduction, respectively, over an existing design. The initial baseline design was significantly overdesigned due to a lack of effective metrics and algorithm/architecture co-design. The baseline processor was originally designed to perform floating-point operations and carried three to five times the complexity. Furthermore, the baseline system was underdesigned using marketing performance and eventually ran into power density problems. On the other hand, the optimized system was created by algorithm/architecture co-design and chip/system co-optimization.

8.8 SUMMARY

This chapter introduces three technologies—ASIC, FPGA, and GPP—for the development of signal processors. ASICs and FPGAs are often required in front-end signal processors that demand high throughputs and small form factors. ASICs offer, at the expense of high NRE, the highest computation throughput at the lowest unit SWAP. FPGAs have less throughput per watt compared to ASICs, but they do not incur NRE (for chip fabrication). Therefore, ASICs are commonly reserved for applications demanding performance that cannot be met with other technologies.

FPGAs will continue to be an important computing technology for Department of Defense applications, allowing them to leverage advances developed in the commercial world. More and more FPGAs will be used in prototyping because the performance gap between FPGA and ASIC is narrowing and the ASIC design cost gets increasingly higher. FPGAs have been successfully used in ASIC prototyping, ASSP (application-specific standard part) replacement, and real-time system development and implementation. There are some FPGA developments that will further this trend. Instead of forcing designers to use one-configuration-fits-all FPGAs, vendors are providing platform FPGAs, which allow a higher level of customization by minimizing and matching elements for different approaches. The result is better performance.

A high performance front-end processing system must be optimized at both chip and system levels. This chapter has shown the application of these algorithm/architecture, software/hardware, chip/system co-design techniques to the development of two high performance processors with very high power efficiency.

REFERENCES

Andraka, R. and A. Berkun. 1999. FPGAs make a radar signal processor on a chip a reality. *Conference Record of the 33rd Asilomar Conference on Signals, Systems, and Computers* 1: 559–563.

General-Purpose Computation on GPUs (GPGPU) website. General-purpose computation using graphics hardware. 2007. Available online at http://www.gpgpu.org/. Accessed 23 May 2007.

General-Purpose GPU Computing: Practice and Experience. Supercomputing '06 Workshop, 13 November 2006, Tampa, Fla. Available online at http://www.gpgpu.org/sc2006/workshop.

The International Technology Roadmap for Semiconductors. 2007. Available online at http://www.itrs.net. Accessed 13 August 2007.

9 Application-Specific Integrated Circuits

M. Michael Vai, William S. Song, and Brian M. Tyrrell, MIT Lincoln Laboratory

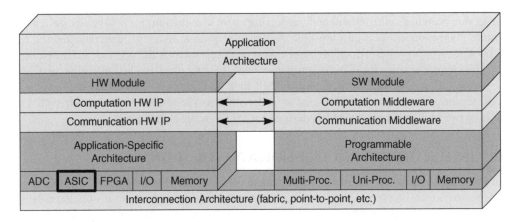

This chapter provides an overview of application-specific integrated circuit (ASIC) technology. Two approaches to ASIC design are described: full-custom and synthesis. The chapter concludes with a case study of two high performance ASICs designed at MIT Lincoln Laboratory.

9.1 INTRODUCTION

Three types of embedded computing devices were introduced in Chapter 8: application-specific integrated circuits (ASICs), field programmable gate arrays (FPGAs), and general programmable processors (GPPs). FPGAs and GPPs are user configurable/programmable commercial off-the-shelf (COTS) products. ASICs are *custom-fabricated* for specific functions in particular applications.

The performance of an ASIC could be 10–100 times better than that of its FPGA or GPP counterpart. However, ASIC technology is often associated with high cost and long development time, which is true in low-volume (e.g., < 250,000 units) applications. Interestingly, the cost of a silicon die is rather inexpensive. A 1 cm² silicon die, which can house millions of transistors, is only a few dollars in material cost. In contrast, the production of the first batch of chips (e.g., a few hundred chips) can run into millions of dollars. This is referred to as the nonrecurring expenses (NRE) of ASIC development. The costs of chip design and fabrication setup (e.g., photomask preparation) are NRE examples.

The high NRE of an ASIC device can be amortized over the entire population to be produced. Therefore, the unit cost of an ASIC should be determined using its current and future volume of production. For example, the automotive industry is currently the largest consumer of ASICs. A vehicle has up to 50 embedded ASIC processors, controlling functions ranging from anti-lock brakes to airbag deployment. Many of these automotive ASICs are massively produced and are thus low-priced parts. In fact, many ASICs that implement specific functions appealing to a wide market have evolved into COTS products called application-specific standard products (ASSPs).

Unfortunately, most of the high performance embedded systems described in this book may never be massively produced. Therefore, the designer should scrutinize the application to determine whether its size, weight, and power (SWAP) specifications can be met with FPGAs or GPPs. For example, a careful hardware and algorithm co-design (see Chapter 8) may lower the system requirements into the realms of FPGAs and GPPs. However, a high performance system developer will sooner or later face the challenge of developing a system with requirements that can only be met with the use of ASICs.

The objective of this chapter is to provide an overview of ASIC technology. Readers who are interested in learning ASIC design skills in depth can continue their study with textbooks in this area (Vai 2000; Weste and Harris 2004; Wolf 2002). ASIC development involves highly specialized expertise in a broad range of disciplines, so contracting an external design house to provide the service is often a valid, or even preferable, alternative to maintaining a full-service design team. System planners/developers can use this chapter to acquire a high-level understanding of ASIC technology, which is often critical in the effective procurement of ASIC development services.

This chapter begins with an overview of the complementary metal oxide semiconductor (CMOS) technology. The ASIC design methodology will then be explained. Two different approaches to ASIC design will be described: full-custom and synthesis. The chapter will conclude with a case study of two high performance ASICs designed at MIT Lincoln Laboratory.

9.2   INTEGRATED CIRCUIT TECHNOLOGY EVOLUTION

An integrated circuit (IC) is a tiny semiconductor chip on which a complex of electronic components (e.g., transistors) and their interconnections are fabricated with a set of pattern-defining masks. The concept of integrating multiple electronic devices and their interconnections entirely on a small semiconductor chip was proposed shortly after the transistor was invented in 1948. ICs began to become commercially available in the early 1960s. IC design and fabrication technologies have been developed at a phenomenal rate since then.

There are several ways to chronicle the evolution of IC technology. In 1965, Gordon Moore (a cofounder of Intel) made an observation and prediction on the development of semiconductor technology (Moore 1965; Schaller 1997). His prediction is that the effectiveness of semiconductor technology, as roughly measured by the maximum number of transistors on a chip, approximately doubles every 18 months. This forecast, now commonly known as "Moore's Law," has been accurate until around 2000. After 2000, the exponential growth rate described by Moore's Law has continued, even though it has somewhat slowed down to two to three times every three years. If Moore's Law were applicable to the airline industry, a flight from New York to Paris in 1978 that cost $900 and took seven hours would now cost about $0.01 and take less than one second.

The growth of IC technology is also evident in the process technology, called a technology node, which has been scaled down from around 10 μm (1 μm = 10^{-6} m) in the 1970s to 65 nm (1 nm = 10^{-9} m) in 2007. The International Technology Roadmap for Semiconductors (ITRS) has provided the industry with guidance on semiconductor improvement trends associated with various technology nodes (ITRS 2007). Figure 9-1 shows the evolution of CMOS technology nodes. In general, a smaller technology node allows smaller feature sizes to be fabricated, thus producing smaller and faster transistors.

Until recently, DRAM (dynamic random access memory) devices have been the driver of technology scaling. The smallest half-pitch of contacted metal lines allowed in a DRAM fabrication process is used to define the technology node (see Figure 9-2), which is treated as a single, simple indicator of overall industry progress in IC feature scaling. However, the 2005 ITRS has recognized that DRAM is no longer the sole driver of technology scaling and has stopped using the technology node as a pace indicator. Instead, it now independently measures the technology

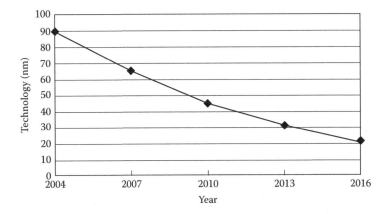

FIGURE 9-1 Evolution of CMOS technology nodes.

paces of DRAM, MPU (microprocessor)/ASIC, and flash memory products. Tables 9-1 and 9-2 summarize the trends of DRAM and MPU/ASIC devices, respectively.*

With each new advance along the Moore's Law curve, there have been new predictions as to when either fundamental physics or practical manufacturability concerns will bring the exponential scaling trend to an end. At present, scaling to the 45 nm and 32 nm technology nodes seems to be limited more by economic and engineering issues than by physics. However, it is readily apparent that there are diminishing returns in circuit performance as the technology becomes more deeply scaled. In addition, today's leading-edge transistors are so small that the gate dielectrics are only a few atomic layers thick, and there are so few dopant atoms in the metal oxide semiconductor (MOS) channel that the devices are subject to wide statistical variation. Significant changes to both transistor designs and circuit architectures will eventually be required to allow for the continued advancement of semiconductor technology.

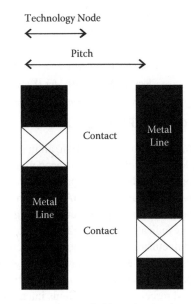

FIGURE 9-2 ITRS definition of technology node.

Hundreds of millions of transistors are routinely integrated into a single chip. The creation of such a complicated chip involves a large number of activities such as logic design, modeling, simulation, testing, and fabrication, each of which can in turn be divided into many tasks. A broad spectrum of knowledge ranging from solid-state physics to logic circuit design is required to successfully carry out all these steps. The rapid advancement of manufacturing capability has allowed for a higher density of logic to be placed within a given chip area. However, without the methodologies specifically developed for the design of ASICs, the potential of advanced manufacturing techniques would not have been realized. We will discuss ASIC design methodology later in this chapter.

---

* Dr. James Anderson, MIT Lincoln Laboratory, contributed these two tables, which were derived using public domain information from the ITRS 2006 update.

TABLE 9-1
DRAM Trend

Year of Production	2007	2010	2013	2016
DRAM 1/2 pitch (nm)	65	45	32	22
Chip size at production (mm^2)	110	93	93	93
Gbits/cm^2 at production	1.94	4.62	9.23	18.46
Gbits/chip	2.15	4.29	8.59	17.18

TABLE 9-2
MPU/ASIC Trend

Year of Production	2007	2010	2013	2016
MPU/ASIC 1/2 pitch (nm)	68	45	32	22
Transistors (million)/chip	1,106	2,212	4,424	8,848
Max. watts @ volts	189@1.1	198@1.0	198@0.9	198@0.8
Clock frequency (MHz)	9,285	15,079	22,980	39,683
Clock frequency (MHz) for 189 W power	9,285	14,394	21,935	37,879

9.3 CMOS TECHNOLOGY

Silicon, germanium, and carbon, all in column 14 of the periodic table, are elemental semiconductors. The relative ease of devising an economically manufacturable technology has led to the almost universal adoption of silicon as the primary semiconductor for the electronics industry. Compound materials such as gallium arsenide (GaAs), silicon germanium (SiGe), and silicon carbide (SiC) are also used, especially in high-frequency analog designs and, in the case of SiC, for high-temperature components. All signs indicate that, in the foreseeable future, silicon will continue to be the preferred material of IC manufacturing, with compound semiconductors providing specific benefits to their particular niche markets. Beyond Moore's Law there might be a new paradigm of carbon-based electronics (either in the form of nanotubes or graphite substrates), but currently those technologies are more relevant to advanced research than to systems integrators (Georgia Institute of Technology 2006).

Two types of silicon-based device, the bipolar junction transistor (BJT) and the metal oxide semiconductor field-effect transistor (MOSFET), have been used to construct logic circuits. The complementary MOSFET (CMOS) technology, which employs both n-channel (nMOS) and p-channel (pMOS) transistors to form logic circuits, has significant advantages over the bipolar technology and dominates digital logic ICs.

A MOSFET occupies a smaller area than a BJT, leading to a higher circuit density. The very high input impedance allows a MOSFET to be modeled as a switch. Furthermore, high fanouts can be readily achieved by providing appropriate drivers. These features simplify the design and analysis of logic circuits. CMOS circuits provide large logic swings and thus excellent noise margins. Also, CMOS circuits operate with low power, enabling a high integration density.

On the other hand, while the CMOS technology has many preferable features, logic circuits based on BJTs, such as transistor-transistor logic (TTL) and emitter-coupled logic (ECL), have the advantages of high speed and large current capability. Both BJTs and MOSFETs may coexist in the same circuit. This technology, called the BiCMOS technology, is considerably more expensive than is regular CMOS technology, yet it offers the high speed of BJTs and the low power advantage of CMOS technology in building mixed-signal (i.e., analog and digital) circuitry. To learn more about BiCMOS and mixed-signal technologies, see Sedra and Smith (1998).

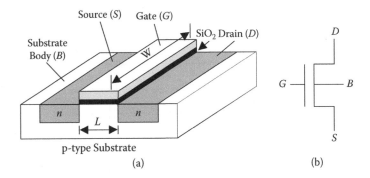

FIGURE 9-3 Physical structure (a) and circuit symbol (b) of an nMOS transistor.

9.3.1 MOSFET

There are two types of semiconductors, n-type and p-type. The n-type semiconductor, which has an abundance of electron "carriers," is used as the substrate of pMOS transistors. On the other hand, the p-type semiconductor, being rich in holes (positive charge carriers created by the absence of valence band electrons), provides the substrate for nMOS transistors. The CMOS technology, which dominates digital logic ICs, operates with low power by employing both nMOS transistors and pMOS transistors.

Figure 9-3 shows the physical structure of an nMOS transistor, along with its circuit symbol. There are four nodes in this device structure: a source (S), a drain (D), a gate (G), and a body substrate (B). Two heavily doped n regions are created in the substrate to form the source and the drain. Since the structure of a MOSFET is symmetrical, the source and drain are interchangeable. The doped polysilicon gate sits on top of a channel area between the source and drain with a thin layer of insulator (SiO_2) sandwiched in between.

For proper operation, the p-type substrate must be connected to the lowest potential in the circuit so that the p-n junctions formed between the source/drain and the substrate are reverse-biased. If a p-n junction is forward-biased, the CMOS circuit may be induced into a latch-up condition, which causes a large abnormal current to flow through.

In a highly simplified view for digital circuits, an nMOS transistor operates in one of two states. Each nMOS transistor has a threshold voltage V_{th} (e.g., 0.7 V), which is a physical property determined by the design and technology. An nMOS transistor is turned off by a gate-to-source voltage (V_{gs}) that is below its threshold voltage V_{th}. Alternatively, the nMOS transistor turns on and provides a resistive conducting path (on the order of $10^3 \Omega$) when $V_{gs} > V_{th}$. It is this simple switch operation that renders the ease of designing with MOSFETs and thus the popularity of the technology.

The pMOS transistor is a complementary device to the nMOS transistor. It is similar to its nMOS counterpart except that it is built by doping p-type dopants into an n-type substrate. In order to avoid latching up, the substrate must be connected to the highest potential in the circuit to avoid p-n junctions from being turned on. A pMOS transistor has a negative threshold voltage (e.g., −0.7 V). The switching conditions of a pMOS transistor are complementary to an nMOS transistor.

CMOS technology requires both n-channel and p-channel MOSFETs to coexist on the same substrate. A portion of the substrate must thus be converted to accommodate the transistors of the opposite type. The n-well CMOS technology uses a p-type wafer as the substrate in which n-channel transistors can be formed directly. P-channel transistors are formed in an n-well, which is created by converting a portion of the p-type wafer from being hole (p-type carrier) rich into electron (n-type carrier) rich. Other possibilities include the p-well CMOS technology, which is rarely available, the twin-well CMOS technology, and the silicon-on-insulator (SOI) technology (Marshall and Natarajan 2002). Twin-well and SOI CMOS technologies offer significant advantages in noise isolation and latch-up immunity. In addition, SOI technology offers significant speed and power

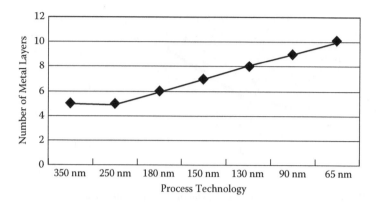

FIGURE 9-4   Number of metal layers in various CMOS technologies.

advantages. However, these technologies are considerably more expensive, both in terms of NRE and fabrication costs.

Besides feature sizes, another important parameter of a CMOS technology is the number of available routing layers for component interconnections. Routing layers insulated from each other by dielectric allow a wire to cross over another line without making undesired contacts. The silicon dioxide (SiO_2), commonly used in older technologies as insulation, is being replaced with materials with lower dielectric permittivity, thus reducing the parasitic capacitance of the interconnections and resulting in faster, lower-power circuits. Connecting paths called vias can be formed between two routing layers to make interlayer connections as required. A typical CMOS process has five or more layers available for the routing purpose. In many modern processes, one or more of the upper metal layers is designed to allow for thicker wires. This design allows for high currents to be routed in power buses and clock trees, and also allows for implementation of transmission lines and radio frequency (RF) passives (e.g., capacitors). Figure 9-4 shows the representative number of routing layers available at each technology node.

9.4   CMOS LOGIC STRUCTURES

Logic functions can be implemented using various types of CMOS structures. Most notable is the distinction between static logic and dynamic logic. A dynamic logic circuit is usually smaller in size since it produces its output by storing charge in a capacitor. The output of a dynamic logic circuit decays with time unless it is refreshed periodically, either by reinforcing the stored charge or overwriting with new values. In contrast, a static logic circuit holds its output indefinitely. The static logic can be implemented in a variety of structures. The complementary logic and pass-gate logic are described later in this chapter.

9.4.1   STATIC LOGIC

The general structure of a CMOS complementary logic circuit contains two complementary transistor networks. We will use the inverter circuit shown in Figure 9-5 to explain this structure. Figure 9-5 also illustrates the relationship between a CMOS inverter and its physical implementation. The pMOS transistor connects the output (Z) to V_{DD} (i.e., logic 1) while the nMOS transistor connects Z to V_{SS} (i.e., logic 0). When the input a is logic 1, the nMOS transistor turns on and the output is pulled down to 0. In contrast, if the input is logic 0, the pMOS transistor turns on and the output is pulled up to 1. Since only one half of the circuit is conducting when the input is a stable 1 or 0, ideally there is no current flow between V_{DD} and V_{SS}. This implies that the inverter should not dissipate power when the input is stable. During the input transition from 1 to 0 or vice versa, charging and discharging of the output node happens and power dissipation occurs. This is an important

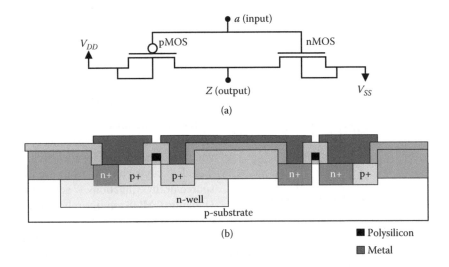

FIGURE 9-5 CMOS inverter: (a) schematic diagram; (b) cross-section structure.

feature of low-power CMOS logic circuits. The major power consumption comes from circuit switching.

A CMOS logic structure contains complementary pull-up and pull-down transistor circuits controlled by the same set of inputs. Due to the complementary relationship between the pull-up and pull-down circuits, an input combination that turns on the pull-down turns off the pull-up, and vice versa. This logic structure allows complex functions (e.g., a binary adder) to be implemented as composite gates. The design details of complementary logic circuits are beyond the scope of this book. Interested readers are referred to the references by Vai (2000), Weste and Harris (2004), and Wolf (2002).

In a complementary logic circuit, the source of its output is always a constant logic signal: V_{DD} (1) for the pull-up network and V_{SS} (0) for the pull-down network. Pass-gate logic, on the other hand, allows input signals themselves (and their complements) to be passed on to the circuit output. This behavior usually simplifies the circuit.

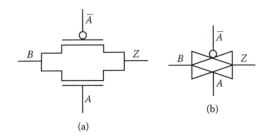

FIGURE 9-6 Pass-gate logic for a 2-to-1 multiplexer.

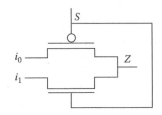

FIGURE 9-7 Transmission-gate (t-gate): (a) structure and (b) symbol.

A simple pass-gate logic example is the 2-to-1 multiplexer shown in Figure 9-6. The control signal S selectively passes input signals (i_0 and i_1) to its output Z. Pass-gate logic circuits come with a caveat. A pMOS transistor cannot produce a full strength 0 due to its threshold voltage requirements. The drain voltage cannot go below ($V_{ss} + |V_{th}|$), which is referred to as a weak 0. Similarly, an nMOS transistor can only produce a weak 1 ($V_{DD} - |V_{th}|$). Therefore, a pass-gate circuit does not always produce a full logic swing. In practice, a subsequent complementary logic stage following the pass-gate logic stage can be used to restore the strength of the output signal.

The transistors in a pass-gate logic circuit can be replaced with transmission gates to achieve full-scale logic levels. A transmission gate (t-gate) is built by connecting a pMOS transistor and an nMOS transistor in parallel. The structure and the symbol of the t-gate are shown in Figure 9-7.

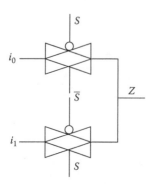

FIGURE 9-8 T-gate logic implementation of a 2-to-1 multiplexer.

Figure 9-8 shows the result of converting the 2-to-1 multiplexer of Figure 9-6 into a t-gate logic circuit. Note that an additional inverter (not shown) is needed to generate control signal $\bar{S}$.

9.4.2 DYNAMIC CMOS LOGIC

CMOS circuit nodes are commonly modeled as capacitors. A node capacitance, the value of which is determined by the circuit geometry, includes contributions from transistor gates, source/drain junctions, and interconnects. Dynamic CMOS circuits rely on these capacitor nodes to store signals. A capacitor node holds its charge indefinitely when there is no leakage current. In practice, all CMOS circuits have leakage currents so the capacitors will eventually be discharged. The storage nodes in a dynamic circuit thus need to be periodically refreshed. Refreshing is typically done in synchronization with a clock signal. Since the storage nodes must be refreshed before their values deteriorate, the clock should be running at or above a minimum frequency, which is on the order of tens of MHz. Most circuits operate at much higher speeds so no specific refreshing arrangement will be needed.

Figure 9-9 illustrates a dynamic latch as an example. The data are stored at the input node capacitance of the inverter. The benefit of dynamic circuits is exemplified in the comparison of this circuit with its static counterpart, also shown in Figure 9-9.

In summary, static logic is designed to perform logic-based operations, and dynamic logic is created to perform charge-based operations. Compared to dynamic logic, static logic needs more devices to implement and thus consumes larger area and power. However, the functionality of a dynamic circuit is limited by its most leaky node. This fact makes the leakage statistics of a given technology particularly relevant, and also limits the applicability of these circuits in environments that tend to increase leakage, such as in radiation and higher-temperature applications. Therefore, the use of dynamic logic in space-borne applications is limited.

9.5 INTEGRATED CIRCUIT FABRICATION

The construction of a high-production, state-of-the-art IC fabrication facility costs billions of dollars. Only a few big companies (e.g., Intel, IBM, etc.) perform an end-to-end chip-production process, which includes chip design, manufacture, test, and packaging. Many other companies are fabless (fabrication less) and do not have in-house manufacturing facilities. Although they design and test the chips, they rely on third-party silicon foundries for actual chip fabrication. A silicon

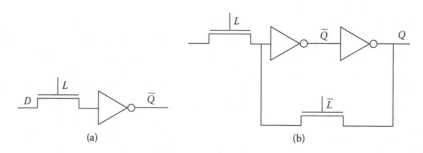

FIGURE 9-9 Comparison between (a) dynamic and (b) static latches.

foundry is a semiconductor manufacturer that makes chips for external customers; IBM, AMI, TSMC, UMC, etc., are some of the major foundries.

MOSIS is a low-cost prototyping and small-volume production service established in 1981 for commercial firms, government agencies, and research and educational institutions around the world (see the MOSIS website at http://mosis.org for more information). Most of these clients do not have volume large enough to form an economically meaningful relationship with a silicon foundry.

MOSIS provides multiproject wafer (MPW) runs on a wide variety of semiconductor processes offered by different foundries. The cost of fabricating prototype quantities is kept low by aggregating multiple designs onto one mask set so overhead costs associated with mask making, wafer fabrication, and assembly can be shared. As few as 40 die can be ordered from MOSIS within the regularly scheduled multiproject runs. Dedicated runs, which can be scheduled to start at any time, are also available for larger quantities. In Europe, the EU Europractice IC Service offers an MPW service comparable to MOSIS (see the Europractice IC Service website at http://www.europractice-ic.com for more information). For sensitive designs related to United States government applications, the Department of Defense Trusted Programs Office has also recently established a contract with the Honeywell Kansas City Plant to provide secure access to MPW prototyping (Honeywell 2007).

The circuit layout is often sent, or taped out, to a silicon foundry for fabrication in the format of GDSII (Graphic Data Stream) (Wikipedia, GDSII 2007). The newer Open Artwork System Interchange Standard (OASIS) format has been developed to allow for a more compressed and easy-to-process representation of large circuit designs that contain 100 million transistors or more (Wikipedia, OASIS 2007). The layout data are first verified by the foundry for design rule compliance. A post-processing procedure is then used to process the layout data. In layout post-processing, the actual data to be used in photomask generation are produced. New design layers are added and data on existing layers are modified. For example, resolution enhancement techniques (RET) may be used to obtain suitable lithographic performance. Fill patterns and holes may also be added to interconnect levels to provide sufficient pattern uniformity for chemical-mechanical polishing (CMP) and etch.

A photomask set contains one or more masks for each lithographic layer in a given technology. A typical mask consists of a chrome pattern on a glass plate, with the chrome pattern printed at 4× or 5× magnification with respect to the desired IC pattern. This is imaged using a reduction lens onto a silicon wafer substrate that has been coated with a photosensitive material called photoresist. The exposure pattern is stepped across the wafer to form an array of circuit layouts. Modern wafers are typically 200 mm to 300 mm in diameter, and the maximum stepped pattern size, i.e., the maximum IC die size, is typically on the order of 25 mm by 25 mm.

The creation of an IC on a silicon wafer involves two major types of operations: doping impurities into selected wafer regions to locally change electrical properties of the silicon and depositing and patterning materials on the wafer surface. Doping is typically performed by lithographically opening windows in the photoresist and implanting with a selected impurity. Doping is used for the definition of MOS source/drain regions. The threshold voltage and other transistor properties can also be tuned by doping.

Another important process is to build a layer of patterned material on top of the wafer. This process is used to build transistor gates (doped polysilicon) and routing metal lines (aluminum or copper). The patterning process may be additive or subtractive.

A typical CMOS process requires the use of 20 or more masks, depending on the number of polysilicon and metal layers needed. The physical layout designer creates these masks by drawing a layout which is a composite view of the masks stacked together. It is often convenient for a layout designer to think and work directly with the layout as viewed from the top of a circuit. The layout design is thus two-dimensional from the viewpoint of an IC designer. In fact, designers do not control the depth dimension of an IC.

Figure 9-10 shows the layout of a NAND gate. One of the four transistors in this NAND gate is called out in the layout for the following explanation. From the viewpoint of IC design, a MOS

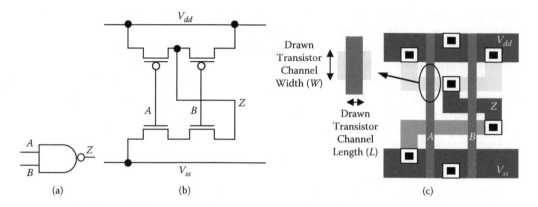

FIGURE 9-10 A CMOS NAND gate: (a) symbol; (b) circuit schematic; and (c) layout.

transistor is characterized by its channel length (L) and width (W) defined by the dimensions of its gate (or channel). Since the switching speed of a transistor is inversely proportional to its channel length, it is usually desirable to keep the channel length L as short as the design rules would allow.

In addition to specifying the shortest channel length, design rules, which are created and supplied by silicon foundries, also place constraints on feature widths, separations, densities, etc. They capture the physical limitations of a fabrication process to ensure that a design that conforms to the design rules can be successfully fabricated. Therefore, design rules release IC designers from the details of fabrication and device physics so they can concentrate on the design instead. Note that the successful fabrication of an IC does not automatically imply that it meets the design criteria. It simply means that the transistors and their connections specified in the layout are operational.

In their 1980 classic text, Mead and Conway (1980) developed a set of simplified design rules now known as the scalable γ-based design rules, which are valid for a range of fabrication technologies. Different fabrication technologies apparently require different design rules. The scalable γ-based design rules are possible because they are created to be sufficiently conservative for a range of fabrication technologies. This performance limitation is so critical in deep-submicron (≤ 0.18 μm) technologies that scalable γ-based design rules are no longer effective.

9.6 PERFORMANCE METRICS

An IC is evaluated by its speed, power dissipation, and size, not necessarily in this order. Since a function often has more than one way of implementation, these criteria can be used, either individually or in combination, to select a structure that suits a specific application. The layout determines the chip area. Before layout, the number of transistors can be used to estimate the size. The speed and power can be estimated by a simplified RC (resistor and capacitor) switch model using an estimate of circuit parasitic resistance and capacitance values. Parameters should be acquired for a selected fabrication process and used to estimate parasitic resistance and capacitance values.

9.6.1 SPEED

Parasitic capacitors, resistors, and inductors are incidentally formed when any circuit is implemented. The first source of parasitic resistance and capacitance can be found in the MOSFETs themselves. A transistor in its off state has a source-drain resistance that is high enough to be considered as an open circuit in general applications. A turned on transistor is modeled as a resistance charging or discharging a capacitance.

Figure 9-11 shows a very simple RC model to estimate delay times. The transistor that is involved with charging/discharging the load C_L is represented by a channel resistance (R_n or R_p). The main

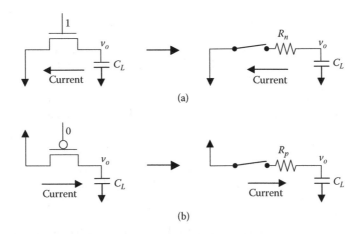

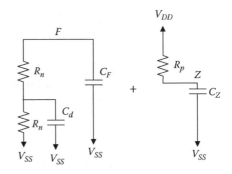

FIGURE 9-11   RC timing models: (a) nMOS transistor; (b) pMOS transistor.

sources of capacitance C_L in a CMOS circuit are transistors (gate, source, and drain areas), wiring, and overlapping structures. While not shown in Figure 9-11, the effect of interconnecting wires, especially when they are long, can be modeled by their parasitic capacitance and resistance and incorporated into the RC model.

The voltage v_o has a transient behavior of the form $v_o = V_{DD}\,(e^{-t/RC})$ [v_o: 1 → 0; Figure 9-11(a)] or $v_o = V_{DD}\,(1 - e^{-t/RC})$ [v_o: 0 → 1; Figure 9-11(b)], where V_{DD} is the positive rail voltage (i.e., power supply voltage). In either case, the voltage v_o crosses $0.5V_{DD}$ at $t = 0.7RC$, where R is R_n or R_p and C is C_L. The value $0.7RC$ can thus be used as a first-order estimation of signal delay time.

The use of this RC timing model is illustrated in the following example to estimate the worst-case low-to-high propagation delay of the circuit shown in Figure 9-12. The RC equivalent circuit for calculating the low-to-high propagation delay at Z, $t_{PLH}(Z)$, is shown in Figure 9-13. R_p and R_n are the pMOS and nMOS effective channel resistance, respectively. C_F and C_Z are the total capacitance at nodes F and Z, respectively. Capacitance C_d is the parasitic capacitance between the two pull-down transistors in the first stage of the circuit.

From the RC timing model,

$$t_{PHL}(F) = 0.7(R_n C_d + 2R_n C_F)\text{, so}$$

$$t_{PLH}(Z) = 0.7((R_n C_d + 2R_n C_F) + R_p C_Z).$$

The first two terms in $t_{PLH}(Z)$ constitute the delay from the input to node F [i.e., $t_{PHL}(F)$] and

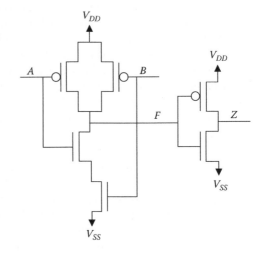

FIGURE 9-12   Propagation delay estimation example.

FIGURE 9-13   RC model for determining the low-to-high delay time at node Z.

the third term is the delay between nodes F and Z. Note that the distributed RC network equation has been applied to calculate $t_{PLH}(Z)$.

In most cases, the logic function implemented by a CMOS logic circuit depends solely on the connectivity of its transistors rather than on the sizes of its transistors. In fact, a circuit consisting of only minimum-size transistors will function correctly. However, the performance of such a minimum-size design may not meet the design requirements such as speed. A considerable design effort is often spent on sizing the transistors in an established circuit topology to meet design specifications.

9.6.2 Power Dissipation

Power dissipation is an important IC performance evaluation criterion. For battery-powered systems, it is desirable to minimize both active and standby power dissipation. Even when the application can afford a high power budget, it is necessary to deal with the heat generated by the millions of switching devices.

CMOS complementary logic circuits have historically been known for their low power dissipation, particularly when compared to bipolar transistor or NMOS-only logic. The primary reason is that when the CMOS logic gate is in a steady state, regardless of whether the output is a 1 or a 0, only half of the circuit is conducting so there is no closed path between V_{DD} and V_{SS}.

When a CMOS logic gate is in transition, both the pull-up and pull-down networks are turned on so a current flows between V_{DD} and V_{SS}. This is called the short-circuit current, which contributes to the dynamic power dissipation in the CMOS circuit. A well-designed circuit operating with well-behaved signals of reasonably fast rise time and fall time would go through the transition quickly. The short-circuit power dissipation is thus traditionally less significant when compared to the dynamic power caused by the current that flows through the transistors to charge or discharge a load capacitance C_L.

CMOS circuits dissipate power by charging and discharging the various load capacitances whenever they are switched. The average dynamic power dissipation due to charging and discharging capacitance C_L is $P_D = \alpha\, f\, C_L V_{DD}^2$, where f is the clock frequency and α is the activity factor accounting for the fraction of logic nodes that actually change their values in a clock cycle.

The traditional assumption that a CMOS circuit has little or no static power no longer holds in deep submicron ($\leq$180 nm) technologies. As devices are scaled to improve switching speed and supply voltages are scaled to improve active power, the transistor threshold voltages are also made lower. The off-state leakage current of deep submicron transistors would increase approximately 10 times for every 100 mV reduction in threshold voltage. So, if the threshold voltage of a high performance device is reduced from, for example, 500 mV to 200 mV for low gate voltage operations, then there is clearly a considerable off-state leakage current. Therefore, low-power designs often use low threshold voltages only in the critical paths where performance is particularly essential.

Furthermore, the quick transition assumption also breaks down as wires on chip become narrower and more resistive. CMOS gates at the end of long resistive wires may see slow input transitions. Long transitions cause both the pull-up and pull-down networks to partially conduct, and current flows directly from V_{dd} to V_{ss}. This effect can be mitigated by avoiding weakly driven, long, skinny wires. Using higher threshold voltages on the receiving end of signals with slow rise and fall times can also reduce the amount of time in which both the nMOS and pMOS paths are on.

Low-power circuits can also be achieved through design decisions. For example, a circuit may be divided into different power domains using different supply voltages. In addition, power conservation can be achieved by selectively turning off unused circuit portions, dynamically adjusting the power supply voltage, and varying the clock frequency.

9.7 DESIGN METHODOLOGY

The ASIC design methodology developed by Mead and Conway and others in the late 1970s released IC designers from the details of semiconductor manufacturing, so they could concentrate their efforts on coping with the circuit functionality. This design methodology, illustrated in Figure 9-14 as an exchange of information between an ASIC designer and a silicon foundry, is largely responsible for the success of the ASIC industry in the last three decades. The third participant of the ASIC industry is the electronic design automation (EDA) tool vendor who develops software tools to support design activities. EDA vendors often optimize their tools for specific fabrication processes.

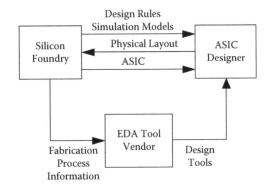

FIGURE 9-14 Relationship between a silicon foundry, an ASIC designer, and an EDA tool vendor.

The most important design principle emphasized in the ASIC design methodology is to divide and conquer. Instead of dealing with an entire ASIC circuit altogether at the same time, the designer partitions it into smaller and thus more manageable parts. These parts may further be broken down into even smaller building blocks. The partitioning of a system into increasingly smaller subsystems so that they can be handled efficiently is called the hierarchical design methodology. EDA tools have been developed to automate the steps in a hierarchical design.

9.7.1 FULL-CUSTOM PHYSICAL DESIGN

The circuit design must ultimately be realized in silicon through the generation of mask data. A set of mask layouts for a building block or even a complete IC can be manually implemented. This approach, called a full-custom design, creates a layout of geometrical entities indicating the transistor dimensions, locations, and their connections.

Computer-aided-design (CAD) tools have been developed to facilitate the design of full-custom ICs. For example, a design rule checker can be used to verify that a layout conforms to the design rules, and a routing tool can be employed to perform the wiring of the transistors.

The advantage of a full-custom design is that it allows the designer to fully control the circuit layout so that it can be optimized. However, these benefits only come at the cost of a very high design complexity, and thus a very high design cost. The full-custom design approach is thus usually reserved for small circuits such as the library cells to be described below, and the performance-critical part of a larger circuit. In some cases when a circuit such as a microprocessor is to be mass-produced, it may be worth the many man-months necessary to lay out a chip with a full-custom approach to achieve optimized results. Full-custom design examples will be shown at the end of this chapter.

9.7.2 SYNTHESIS PROCESS

Many ASIC designs leverage the physical design of logic building blocks available in the form of a library consisting of standard cells and intellectual property (IP) cores. The ASIC design process then is focused on optimizing the placement and the interconnection of these building blocks to meet the design specifications.

With millions of transistors involved, it is extremely difficult to manually lay out the entire chip. In order to mitigate the complexity of designing at the physical level, ASIC design usually begins by specifying a high-level, behavioral representation of the circuit. This involves a description of how the circuit should communicate with the outside world. Typical issues at this representation level include the number of input-output (I/O) terminals and their relations. A well-defined behavioral description of a circuit has a major benefit. It allows the designer to optimize a design by choosing a circuit from a set of structurally different, yet functionally identical, ones that conform to the

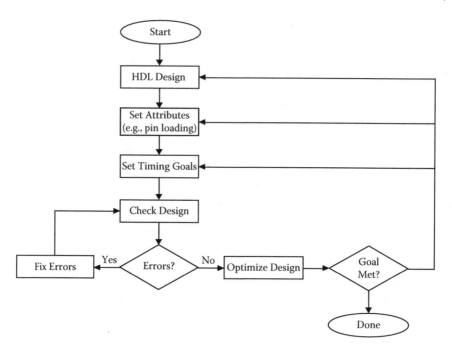

FIGURE 9-15   Synthesis steps.

desired behavioral representation. It is common to use hardware description languages (HDLs) to describe the design.

Two mainstream HDLs are commonly used: Verilog and VHDL. These two languages provide similar capabilities, but require a different formalism. VHDL syntax is based on the Ada programming language and has historically been favored for defense applications. Verilog is based on the C language and has become popular for consumer applications. It is not unusual to mix Verilog and VHDL in a design process to get the benefits of the properties of each language. For example, the design architecture can be done in VHDL to take advantage of its system description capability while the testing infrastructure can be created using Verilog to take advantage of its C-like language features.

More recently, other "higher-level" design languages (e.g., SystemC, online at http://www.systemc.org) have allowed for more powerful design approaches at the behavioral level, thus allowing for effective conceptualization of larger systems. One major benefit of SystemC is that it allows for more hardware/software co-design. The SystemVerilog extension of the Verilog HDL allows for a co-simulation of Verilog and SystemC blocks. Verilog analog mixed-signal (AMS) extensions have also been developed to support the high-level abstraction of analog and mixed-signal circuits in a system.

The behavioral description often combines what will eventually be a combination of both hardware and software functionality into a single representation. Once a behavioral representation of a system has been coded, one must ensure that it is complete, correct, and realizable. This is done through a combination of simulation, emulation, and formal analysis.

Designing at the behavioral level requires a synthesis process, which is analogous to software compilation. The synthesis steps are shown in Figure 9-15. The synthesis process converts an initial HDL design into a structural representation that closely corresponds to the eventual physical layout. This process uses a library of basic logic circuits, called standard cells, and other IP cores that have already been designed and characterized for the target fabrication technology. A layout is then created by arranging and interconnecting them using computer-aided placement and routing tools. This is not unlike the design of a printed circuit board system using components from standard logic families.

Standard-cell libraries are commercially available for specific fabrication technologies. Sources of cell libraries are silicon foundries and independent design houses. A library usually contains

V_{SS} V_{DD} **Standard Cell** V_{SS} V_{DD}

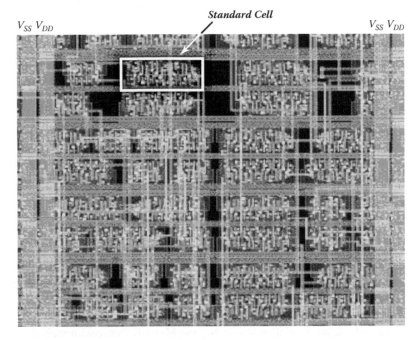

FIGURE 9-16 Example standard-cell layout.

basic logic gates (e.g., inverter, NAND, NOR, etc.), which can be assembled by the designer to form desired functions, and predesigned, ready-to-use functional blocks (e.g., full-adder, register, etc.). Standard cells optimized for different performance metrics (e.g., power or speed) are available. Examples of more complex IPs include memory cores, processor cores, embedded FPGA structures, specialty I/O structures, etc.

Standard cells in a library are designed to have an identical height, which is called a standard height. In contrast, the widths of standard cells are determined by their functions. A more complicated standard cell needs a larger area so its width is longer than that of a less complicated one. The identical heights of standard-cells allow them to be conveniently arranged in a row when a physical layout is created. The arrangement of cells and the distribution of power in a standard-cell layout are shown in Figure 9-16. The power rails (V_{DD} and V_{SS}) run vertically and connect the power lines of the standard-cell rows to an external power supply.

Besides basic standard cells, more sophisticated modules (processors, memory, etc.) are often available in the form of IP cores, which are optimized, verified, and documented to allow efficient reuses. Note that in addition to the cost of acquiring the IP cores themselves, considerable effort must be dedicated to their validation in the targeted design and the construction of an acceptable interface. Chapter 11 provides a detailed description of incorporating IPs into a design project.

The netlist modules produced by the synthesis process have to be arranged on the available chip area and all the associated signals connected in a manner that successfully realizes the timing specifications for the design. The task of arranging the modules on the layout, called the placement process, attempts to determine the best location for each module. The criteria for judging a placement result include the overall area of the circuit and the estimated interconnection lengths (which determine propagation delays). A routing process then takes the placement result and automatically completes the interconnections.

9.7.3 Physical Verification

Regardless of whether a layout is created manually or automatically, its correctness must be verified. A layout-versus-schematic (LVS) procedure extracts a circuit netlist from the physical layout, which is verified against the original netlist. LVS can be performed with different levels of abstraction. For example, a cell-level LVS considers standard cells as building blocks and extracts a netlist of standard cells by analyzing their physical placement and routing. A mask-level LVS produces from a physical layout a transistor netlist. The user can also select the level of parasitic (resistance, capacitance, and inductance) extraction: no parasitic, lumped elements, or distributed elements. A more detailed extraction gives a more realistic representation of the circuit, but its simulation will be more time-consuming.

9.7.4 Simulation

A chip design must be simulated many times at different levels of abstraction to verify its functionality and to predict its performance (i.e., speed and power). As a circuit can be represented at different levels, circuit simulation can also be done at multiple levels ranging from transistor level to behavioral level. Simulation can also be used to determine other effects such as the voltage drop on the power rails.

For small circuits, transistor-level simulation can be used; the basic elements in such a simulation are transistors, resistors, capacitors, and inductors. This level of simulation gives the most accurate result at the cost of long simulation times. Alternatively, transistors can be modeled as switches with propagation delays in a switch-level simulation. This level significantly reduces the simulation time and frequently yields acceptable results.

However, transistor-level simulation is an indispensable tool to more accurately determine circuit behaviors. Among various transistor-level simulation programs, those based on SPICE (Simulation Program with Integrated Circuit Emphasis), developed at the University of California–Berkeley, are by far the most popular (Roberts and Sedra 1997). SPICE has been developed into a number of free or low-cost and commercial products. Approaches have been developed to speed up transistor-level simulations. One common approach is to represent transistor behaviors with lookup tables, which are calibrated to accurately represent the physical models.

If there is no need to determine the internal condition of a building block, it can be modeled at the behavioral level. Mixed-level simulation, in which different parts of a system are represented differently, is commonly used to analyze a complex system. The critical parts of the system can be modeled at the circuit level and/or the switch level while the rest of the circuit can be modeled at the behavioral level. For circuits in which precharacterized IP cores are used, a static timing analysis, in which the delay is calculated by adding up all the delay segments along a signal path, may be performed.

The performance simulation depends on accurate device (e.g., transistor) models and an accurate parasitic extraction. Furthermore, the simulation must take into consideration the effects of statistical process verification, such as gate length variation due to the limitation of lithographic resolution and threshold voltage variation due to dopant statistics. The post-layout processing such as automatically generated fill patterns (to meet density design rules) should be considered as it gives rise to additional parasitics that do not exist in the initial design.

Another important role of simulation is to verify that the circuit will function correctly despite the statistical variability of various fabrication process parameters. Every aspect of the fabrication technology is subject to its own set of systematic and random variations. Statistical models for these effects should be used to ensure that timing and other constraints can be met across all likely process, voltage, and temperature variations. Local effects, such as supply rail voltage drop and local self-heating, often are important aspects of a design that must be modeled using physical layout data.

9.7.5 Design for Manufacturability

Regardless of the approach by which a design is created, its physical implementation must comply with the targeted technology design rules. In addition to traditional width and spacing rules, deep-submicron technologies also require the layout to comply with various design-for-manufacturability (DFM) rules such as pattern density and feature placement restrictions. In fact, these DFM constraints should be considered early in the technology selection phase as they may limit the realizable structures in the final physical design. Some important DFM rules include minimum-density rules, antenna rules, and electrical rules.

The minimum-density rules arise from the use of chemical-mechanical polishing (CMP) to achieve planarity. Effective CMP requires that the variations in feature density on the polysilicon (for transistor gates and short distance routing) layer and metal layers (for all routing) be restricted. On these layers, open areas must be filled with dummy patterns to comply with the minimum-density rules.

Floating interconnects (metal or polysilicon) can act as antennas, attracting ions and thus picking up charge during the fabrication process. The accumulated charge can damage the thin gate oxide and cause performance and reliability problems. The antenna rules, also known as process-induced damage rules, check ratios of amounts of material in two layers from the same node. They are used to limit the damage of the thin gate oxide during the manufacturing process because of charge accumulation. Electrical rules, which check the placement of substrate contacts and guard bands, are specified to protect against latch-up conditions.

9.8 PACKAGES

Packaging of an ASIC component must be considered early in the design process to ensure that power and signal connections between the IC and the outside world are sufficient. Figure 9-17 illustrates that a chip consists of the circuit design itself and a ring of bonding pads, called a pad frame. The signal and power connections of the circuit are connected to the bonding pads. The size and pitch constraints of bonding pads typically produce center-to-center pitches on the order of 100 µm. This relatively large size is necessary to accommodate the operating tolerances of automatic bonding equipment. Smaller pad center-to-center pitches can be accommodated at increased cost. A signal buffer and an electrostatic discharge (ESD) protection circuit are provided for each connection between a bonding pad and the circuit. Bonding wires are used to connect the bonding pads to the pins of a package after the chip is mounted inside.

The development of packages over the last three decades was aimed at two categories: those that contain one chip, namely single-chip modules (SCM), and those that can support more than one chip, called multichip modules (MCM). MCM can support up to and in excess of 100 chips. Packages are made of either plastic or ceramic. The package can be connected to the printed circuit board with pins or with pads/balls.

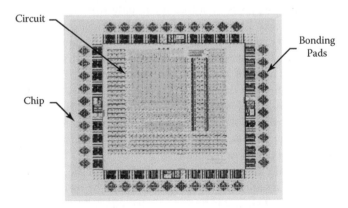

FIGURE 9-17   Chip layout including a circuit and a ring of bonding pads (pad frame). The chip was designed by Dr. William S. Song, MIT Lincoln Laboratory.

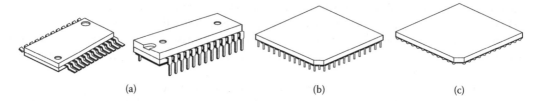

FIGURE 9-18   Example packages: (a) dual in-line; (b) pin-grid array; and (c) ball-grid array .

Figure 9-18 shows several common package examples. In order to fully utilize the processing power of a chip, enough I/O pins must be provided. Two types of traditional dual in-line packages (surface-mounted and pin-through-hole mounted), shown in Figure 9-18(a), have a severe limitation on the number of pins available. Modern chips often require other packages that can provide more pins.

Figure 9-18(b) shows a pin-grid array (PGA) package, which arranges pins at the bottom of the package and can provide 400 or more pins. Figure 9-18(c) shows a more advanced package called a ball-grid array (BGA), which can provide 1000 or more pins. This package has a similar array to the PGA, but the pins have been replaced with solder bumps (balls). Besides the number of pins, other important issues, such as the capability for heat dissipation, must be considered. Manufacturers' specifications should be carefully evaluated before a package is selected.

9.9   TESTING

The testing of a chip is an operation in which the chip under test is exercised with carefully selected test patterns (stimuli). The responses of the chip to these test patterns are captured and analyzed to determine if it works correctly. A faulty chip is one that does not behave correctly. The incorrect operation of a chip may be caused by design errors, fabrication errors, and physical failures, which are referred to as faults.

In some cases, the tester is only interested in whether the chip under test behaves correctly. For example, chips that have been fully debugged and put in production normally require only a pass-or-fail test. The chips that fail the test are simply discarded. This type of testing is referred to as fault detection.

In order to certify a prototype chip for production, the test must be more extensive in order to exercise the circuit as much as possible. The test of a prototype also requires a more thorough test procedure called fault location. If incorrect behaviors are detected, the causes of the errors must be identified and corrected.

An important problem in testing is test generation, which is the selection of test patterns. A combinational circuit with n inputs is fault-free if and only if it responds to all 2^n input patterns correctly. Testing a chip by exercising it with all its possible input patterns is called an *exhaustive test*. This test scheme has an exponential time complexity so it is impractical except for very small circuits. For example, 4.3×10^9 test patterns are needed to exhaustively test a 32-input combinational circuit. Assume that a piece of automatic test equipment (ATE) can feed the circuit with test patterns and analyze its response at the rate of 10^9 patterns per second (1 GHz). The test will take only 4.3 seconds to complete, which is long but may be acceptable. However, the time required for an exhaustive test quickly grows as the number of inputs increases. A 64-input combinational circuit needs 1.8×10^{19} test patterns to be exhaustively tested. The same piece of test equipment would need 570 years to go over all these test patterns.

The testing of sequential circuits is even more difficult than the testing of combinational circuits. Since the response of a sequential circuit is determined by its operating history, a sequence of test patterns rather than a single test pattern would be required to detect the presence of a fault. There are also other problems in the testing of a sequential circuit, such as the problem of bringing the circuit into a known state and the problem of timing verification.

The first challenge in testing is thus to determine the smallest set of test patterns that allows a chip to be fully tested. For chips that behave incorrectly, the second challenge is to diagnose, or locate, the cause of the bad response. This operation is difficult because many faults in a chip are equivalent so they are indistinguishable by output inspection.

Fortunately, a truly exhaustive test is rarely needed. In addition, it is often sufficient to determine that a functional block, instead of an individual signal line or transistor, is the cause of an error. We begin with a discussion of popular fault models that allow practical test procedures to be developed.

9.9.1 Fault Models

As noted above, except for very small circuits, it is impractical to pursue an exhaustive test. Instead, a test should consist of a set of test patterns that can be applied in a reasonable amount of time. This test should provide the user with the confidence that the chip under test is very likely to be fault free if it passes the test. An important issue in the development of a test procedure is thus to evaluate the effectiveness of a test. The quality of a test can be judged by an index called *fault coverage*. Fault coverage is defined as the ratio between the number of faults a test detects and the total number of possible faults. This is usually determined by means of a simulated test experiment. This experiment, which is called a *fault simulation*, uses a software model of the chip to determine its response to the test when faults are present. A fault is detected by a test pattern if the circuit response is different from the expected fault-free response.

Fault models are created to facilitate the generation of test patterns. A fault model represents a subset of the faults that may occur in the chip under test. Several fault models have been developed for representing faults in CMOS circuits. These models can be divided into logic fault models, delay fault models, and current-based fault models.

The most widely used logic fault models are the stuck-at fault, stuck-open fault, and bridging fault models. We will explain these models below. Delay fault models incorporate the concept of timing into fault models. Examples of delay fault models are the transition delay and path delay fault models. The current-based fault models were developed by recognizing the very low leakage current of a CMOS circuit. Many defects, such as opens, shorts, and bridging, result in a significantly larger current flow in the circuit.

The stuck-at fault model assumes that a design error or a fabrication defect will cause a signal line to act as if it were shorted to V_{SS} or V_{DD}. If a line is shorted to V_{SS}, it is a constant 0 and is named a stuck-at-0 (s-a-0) fault. On the other hand, if a line is shorted to V_{DD}, it is a constant 1 and is called a stuck-at-1 (s-a-1) fault. The stuck-at fault model is most effective if it is used at the inputs and outputs of a logic unit such as a logic gate, a full adder, etc. The application of this fault model in a test is to force a signal line to 1 for the s-a-0 fault and to 0 for the s-a-1 fault. The response of the circuit is then analyzed.

The stuck-open (s-op) fault model attempts to model the behaviors of a circuit with transistors that are permanently turned off. The result of having transistors that would not be turned on is unique to CMOS circuits. A stuck-open fault changes a CMOS combinational circuit into a sequential circuit. A two-step test is thus required. The first step brings the signal line being tested to an initial value. The second step then carries out the test in a way similar to the testing of stuck-at faults.

A bridging fault model represents the accidental connection of two or more signal lines in a circuit. The most common consequence of a bridging fault is that the shorted signal lines form wired logic so the original logic function is changed. It is also possible that the circuit may become unstable if there is an unwanted feedback path in the circuit. Bridging faults can be tested by applying opposite values to the signal lines being tested.

9.9.2 Test Generation for Stuck-at Faults

Test generation deals with the selection of input combinations that can be used to verify the correct operation of a chip. Many automatic test pattern generation (ATPG) algorithms are based on the "single-fault assumption." This assumption assumes that at most one fault exists at any time so the test generation complexity can be significantly reduced.

Consider a circuit with a function $F(X) = F(x_1, \ldots, x_n)$, where X is an input vector representing n inputs $x_1, \ldots, x_n$. Suppose we would like to find a test pattern to detect a single stuck-at fault occurring at an internal circuit node k (i.e., $k \notin X$). The first observation is that node k must be set to 0 in order to detect k: s-a-1 or 1 to detect k: s-a-0. The second observation is that a test pattern X_k qualified to detect the specific stuck-at fault must satisfy the following condition. When X_k is applied to the circuit, the fault-free response $F(X_k)$ must be different from the incorrect output $F'(X_k)$ caused by the stuck-at fault at k. This is the basic principle of test generation.

Normally it is impossible to directly inject a value at an internal node of a chip. It is thus necessary to find an input combination X_k that can set k to the desired value. If we can set the value of a node of a chip, either directly in the case of an input node, or indirectly in the case of an internal node, the node is said to be controllable. It is also impractical to physically probe the internal nodes of a chip for their values. In order to observe an internal node, some path must be chosen to propagate the effect of a fault to the chip output. The test pattern X_k must be chosen to sensitize a path from the node under test to an observable output. If the value of a node can be determined, either directly in the case of an output, or indirectly in the case of an internal node, it is said to be observable.

9.9.3 DESIGN FOR TESTABILITY

An ASIC naturally has limited controllability and observability. One principle on which all IC designers agree is that a design must be made testable by providing adequate controllability and observability. These properties must be well planned for in the design phase of the chip and not as an afterthought. This practice is referred to as design for testability (DFT).

The test of a sequential circuit can be significantly simplified if its state is controllable and observable. If we make the registers storing the state values control points, the circuit controllability is improved. On the other hand, if we make the registers observation points, the circuit observability is increased.

This is usually done by modifying existing registers so that they double as test points. In a test mode, the registers can be reconfigured to form a scan register (i.e., a shift register). This allows test patterns to be scanned in and responses to be scanned out. A single long scan register may cause a long test time since it takes time to scan values in and out. In this case, multiple scan registers can be formed so that different parts of the circuits can be tested concurrently. Even though a scan-based approach is normally applied by using the same register cells that are used to implement the desired logical function, additional registers can also be added solely for the purpose of DFT.

IEEE has developed a standard (IEEE Std1149.1) for specifying how circuitry may be built into an integrated circuit to provide testability. The circuitry provides a standard interface through which communication of instructions and test data is done. This is called the IEEE Standard Test Access Port and Boundary-Scan Architecture.

Another problem of a sequential circuit testing is that the circuit must be brought into a known state. If the initialization (i.e., reset) of a circuit fails, it is very difficult to test the circuit. Therefore, an easy and foolproof way to initialize a sequential circuit is a necessary condition for testability. The scan-based test-point DFT approach also allows registers to be initialized by scanning in a value.

A number of other DFT techniques are also possible. These include the inclusion of switches to disconnect feedback paths and the partitioning of a large combination circuit into small circuits. Remember that the cost of testing a circuit goes up exponentially with its number of inputs. For example, partitioning a circuit with 100 inputs into 2 circuits, each of which has 50 inputs, can reduce the size of its test pattern space from 2^{100} to 2^{51} (2×2^{50}).

Most DFT techniques usually require additional hardware to be included to the design. This modification affects the performance of the chip. For example, the area, power, number of pins, and delay time are increased by the implementation of a scan-based design. A more subtle point is that

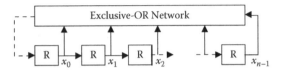

FIGURE 9-19 Linear feedback shift register.

DFT increases the chip area and logic complexity, which may reduce the yield. A careful balance between the amount of testability and its penalty on performance must be applied.

9.9.4 BUILT-IN SELF-TEST

Built-in self-test (BIST) is the concept that a chip can be provided with the capability to test itself. There are several ways to accomplish this objective. One way is that the chip tests itself during normal operation. In other words, there is no need to place the chip under test into a special test mode. We call this the on-line BIST. We can further divide on-line BIST into concurrent on-line BIST and nonconcurrent on-line BIST. Concurrent on-line BIST performs the test simultaneously with normal functional operation. This is usually accomplished with coding techniques (e.g., parity check). Nonconcurrent BIST performs the test when the chip is idle.

Off-line BIST tests the chip when it is placed in a test mode. An on-chip pattern generator and a response analyzer can be incorporated into the chip to eliminate the need for external test equipment. A few components that are used to perform off-line BIST are discussed below.

Test patterns developed for a chip can be stored on chip for BIST purposes. However, the storage of a large set of test patterns increases the chip area significantly and is impractical. A pseudorandom test is carried out instead. In a pseudorandom test, pseudorandom numbers are applied to the circuit under test as test patterns and the responses compared to expected values. A pseudorandom sequence is a sequence of numbers that is characteristically very similar to a random number, but the numbers are generated mathematically and are deterministic. The expected responses of the chip to these patterns can be predetermined and stored on chip. Later, we discuss the structure of a linear feedback shift register (LFSR), which can be used to generate a sequence of pseudorandom numbers.

The storage of the chip's correct responses to pseudorandom numbers also has to be avoided for the same reason as that for avoiding the storage of test patterns. An approach called signature analysis was developed for this purpose. A component called a signature register can be used to compress all responses into a single vector (signature) so that the comparison can be done easily. Signature registers are also based on LFSRs.

Figure 9-19 shows the general structure of a linear feedback shift register. All register cells (R) are synchronized by a common clock (not shown). The exclusive-or network in the feedback path performs modulo-2 addition of the values (x_0 to x_{n-1}) in the register cells. The value at each stage is a function of the initial state of the LFSR and of the feedback path input. As an example, the LFSR shown in Figure 9-20 produces a sequence of seven nonzero binary patterns.

Providing the LFSR in Figure 9-20 with an external input to the exclusive-or feedback network creates a three-bit signature analyzer, which is shown in Figure 9-21. It is shown that two sequences, one correct and one incorrect, coming from a chip under test, produce different signatures after they are clocked through the linear feedback shift register.

The signature can thus be used to indicate the presence of a fault. Instead of storing and comparing a long sequence of data, a signature

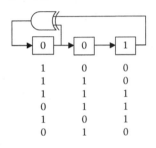

0	0	1
1	0	0
1	1	0
1	1	1
0	1	1
1	0	1
0	1	0

FIGURE 9-20 LFSR set up as a pseudorandom pattern generator.

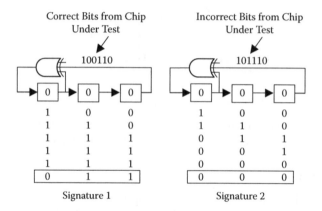

FIGURE 9-21   Signature analyzer example.

is all that is needed to carry out a built-in self-test. However, there is a caveat in this approach. More than one sequence can produce the same signature as the correct sequence. Techniques have been developed to determine the length of the signature as well as the exclusive-or network to improve the confidence level of a correct signature (Bardel, McAnney, and Savir 1987).

9.10   CASE STUDY

This chapter is now concluded by a case study of high performance ASIC development. The chosen ASIC example is the high performance, low-power subband channelizer chip set developed circa 2000 at MIT Lincoln Laboratory for wideband adaptive radar and communications applications. This chip set consists of a polyphase filter (PPF) chip and a fast Fourier transform (FFT) chip.

In order to meet the high computational throughput requirement with low power consumption, a VLSI (very-large-scale integration) bit-level systolic array technology was used (see Chapter 12). By putting tens of thousands of simple 1-bit signal processors on a single die, a very high computational throughput was achieved. In addition, the architecture is highly pipelined and has a highly regular structure that is very well suited to a full-custom implementation. In addition, because the computational structure is mostly based on a small number of simple 1-bit processors, more time could be spent in optimizing the speed, area, and power consumption of these cells. Therefore, very high computational throughput and low power consumption were achieved simultaneously.

In order to achieve low power consumption, low-power dynamic logic circuits were used whenever possible. The supply voltage was also reduced to 1.0 V from 2.5 V. In order to deliver the high speed with low power supply voltage, the layout was hand optimized. Transistor sizes were minimized whenever possible to reduce the power consumption, and special attention was paid to the minimization of interconnection parasitic capacitances.

Each PPF chip contains two banks of polyphase filters; each consists of 128 12-tap polyphase filters, as shown in Figure 9-22. Each bank receives input from two different analog-to-digital converters (ADCs) to process two different channels. The polyphase processor chip was designed so that it can be connected seamlessly to the companion FFT filter chip to perform a short-time Fourier transformation. The outputs of the two filter banks go to the real and imaginary parts of the FFT input, respectively. By processing two real number data streams with one complex FFT, this implementation saves hardware by enabling one FFT chip to process two different ADC channels. The PPF chip was fabricated using a 0.25 micron CMOS process. The PPF die shown in Figure 9-23(a) has 394 pins and approximately six million transistors.

The FFT chip performs a 128-point FFT on the PPF chip output, which is shown in Figure 9-24. This chip consists of seven radix-2 FFT butterfly stages. Any of the butterfly stages can be bypassed to implement smaller FFTs, including 2-, 4-, 8-, 16-, 32-, and 64-point FFTs. Each butterfly stage also

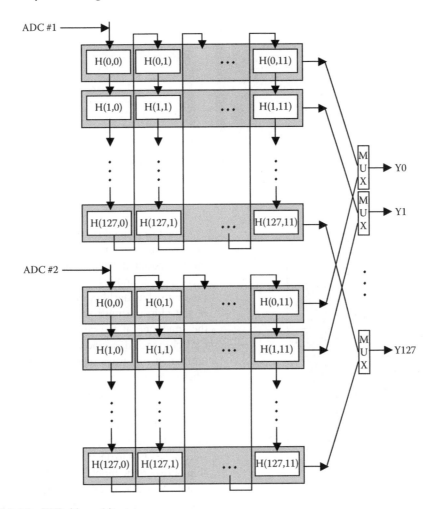

FIGURE 9-22 PPF chip architecture.

has an option to scale the output by half (i.e., divide by two) so that block floating-point computation can be performed with predetermined scaling factors. The FFT chip has 128 input ports and 128 output ports. Two's complement numbers (16 bits) are used for the inputs, outputs, and twiddle factors. The FFT chip was fabricated along with the PPF chip using the same 0.25 micron CMOS process. The FFT die shown in Figure 9-23(b) has approximately 3.5 million transistors and 400 pins.

The chip set was tested with input signals equivalent to 800 million samples per second (MSPS) ADC output. At 480 MSPS, the chip set performs 54 billion operations per second (GOPS) and consumes approximately 1.3 W with a 1-volt power supply. That is equivalent to the power efficiency of 41 GOPS/W.

The design objective of obtaining the highest performance achievable with the fabrication process has limited the use of additional circuitry to enhance the testability of the chips. However, high degrees of controllability and observability are made possible by the functionality of the chips.

As mentioned, each stage of the 128-point FFT chip can be individually bypassed. This bypass mode provides the controllability and observability required to test each individual stage. For example, test vectors can be applied to the inputs of stage 3 by bypassing stages 1 and 2. Also, the response of stage 3 to the test vectors can be observed at the chip output by setting stages 4, 5, 6, and 7 to bypass mode. The PPF chip does not have a bypass mode. However, the same effect is obtained by setting certain taps to zero. Again, a high degree of controllability and observability is achieved.

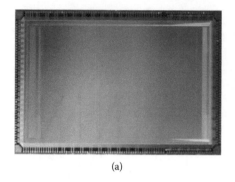

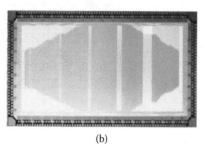

(a)   (b)

FIGURE 9-23   (a) PPF die; (b) FFT die.

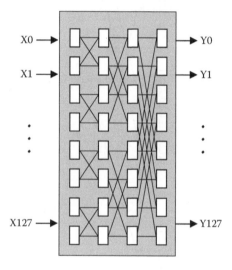

FIGURE 9-24   FFT chip architecture.

The test procedure of these chips is divided into two phases. The first phase is involved with the evaluation and debugging of prototypes. The second phase applies a pseudo-exhaustive test to the chips. During the prefabrication simulation of the chips, stimuli were developed to verify the functions of the chips. The stimuli for the FFT chip include a sine wave input, a noise-contaminated sine wave input, and a random input. These inputs were used along with real FFT coefficients and random coefficients to provide a set of functional tests.

In addition, in order to facilitate the calibration of the test setup, a number of other tests were developed to determine the timing relationship between the input/output data and the clock signal. A "walking-one" test, in which one and only one output pin produces a logic 1 at a time and this logic 1 "walks" from one output pin to the next, was developed to check the bypass mode of the chip. This allowed the routing between the butterfly stages, the test setup, and the packaging (i.e., package and bonding) to be verified. Most importantly, this simple test allowed the timing relationship between the input/output data and the clock signal to be determined, thus allowing for proper adjustment of the test setup timing.

In the case of the PPF chip, a "walking-one" test was also created by setting only selected coefficients to nonzero values. Choosing appropriate inputs and coefficients allowed the chip to operate in a mode similar to the bypass mode of the FFT chip. The benefits of this test were also similar. The timing relationship between the input/output and the clocks, the routing inside the chip, the packaging, and the test setup were checked in this test. Functional tests included the use of real coefficients and random coefficients, along with noise-contaminated sine wave input and random input.

A set of pseudo-exhaustive tests has been developed for the production testing of both chips. Due to the large number of inputs, a true exhaustive test of 2^n possible input vectors is impractical. Instead, a pseudo-exhaustive test was developed to exhaustively test each partition in the circuit. This scheme can test almost any permanent faults with significantly less than 2^n test vectors.

The main computations of these two chips are multiplications and additions. Both multiplication and addition are based on 1-bit full adder cells. The pseudo-exhaustive testing technique thus partitions the circuit under test into interconnecting 1-bit full adder cells.

It is well known that a number of full adder cells forming a ripple-carry adder are C-testable, i.e., they can be pseudo-exhaustively tested with a number of tests that do not depend on the size of the additions. In the case of addition, each full-adder cell has three inputs and the test vectors are designed to ensure that each cell will receive all eight input combinations (000 – 111) during the testing procedure.

In the case of multiplication, the full-adder cell is augmented with an AND function that computes the single-bit product. The number of inputs in each cell is thus increased to four. Again, the test ensures that each cell receives all 16 combinations (0000 – 1111) to perform pseudo-exhaustive testing.

A challenge in the testing of these chips is that the observability of the chips was reduced by the way in which the $2n$-bit products were rounded off into n-bits. Special considerations were applied to ensure that any fault detected by a test vector would cause the chip output to be altered and thus observed. A fault coverage analysis was performed to show that the chips were appropriately tested.

No special test vectors were designed for verifying the operations of the registers. However, the fact that the above tests depend on the correct operations of the registers allowed them to be automatically tested along with the multiplication and addition units.

9.11 SUMMARY

This chapter has covered all the important aspects of ASIC development. Beginning with a background on the CMOS technology, it has explained the logic circuit structures and their design methodologies. Any single subject discussed in this chapter deserves its own book for a thorough description. It is definitely not the intention of this chapter to condense an ASIC design textbook into a single chapter, even though the reader might decide to use it as one. Interested readers are encouraged to consult other books and technical papers, some of which are cited at the end of this chapter, to further develop their knowledge of this fascinating technology that has revolutionized the lives and culture on this planet.

REFERENCES

Bardel, P.H., W.H. McAnney, and J. Savir. 1987. *Built-In Test for VLSI: Pseudorandom Techniques.* New York: John Wiley & Sons.

Georgia Institute of Technology. 2006. Carbon-based electronics: researchers develop foundation for circuitry and devices based on graphite. Available online at http://gtresearchnews.gatech.edu/newsrelease/graphene.htm. Accessed 22 May 2007.

Honeywell. Kansas City Plant. Trusted Access Programs. Available online at http://www.honeywell.com/sites/kcp/trusted_access.htm. Accessed 13 August 2007.

International Technology Roadmap for Semiconductors (ITRS). 2007. Available online at http://www.itrs.net.

Marshall, A. and S. Natarajan. 2002. *SOI Design: Analog, Memory, and Digital Techniques.* Norwell, Mass.: Kluwer Academic Publishers.

Mead, C.A. and L.A. Conway. 1980. *Introduction to VLSI Systems.* Boston: Addison-Wesley.

Moore, G. 1965. Cramming more components onto integrated circuits. *Electronics Magazine* 38(8): 114–117.

Roberts, G.W. and A.S. Sedra. 1997. *SPICE,* 2nd edition. Oxford, U.K.: Oxford University Press.

Schaller, R.R. 1997. "Moore's Law: past, present, and future." *IEEE Spectrum* 34(6): 53–59.

Sedra, A.S. and K.C. Smith. 1998. *Microelectronic Circuits,* 4th edition. Oxford, U.K.: Oxford University Press.

Vai, M.M. 2000. *VLSI Design.* Boca Raton: CRC Press LLC.

Weste, N.H.E. and D. Harris. 2004. *CMOS VLSI Design: A Circuits and Systems Perspective,* 3rd edition. Boston: Addison-Wesley.

Wikipedia. GDS II stream format. 2007. Available online at http://en.wikipedia.org/wiki/GDSII. Accessed 22 May 2007.

Wikipedia. Open Artwork System Interchange Standard (OASIS™). 2007. Available online at http://en.wikipedia.org/wiki/Open_Artwork_System_Interchange_Standard. Accessed 22 May 2007.

Wolf, W. 2002. *Modern VLSI Design: System-on-Chip Design,* 3rd edition. Upper Saddle River, N.J.: Prentice Hall.

10 Field Programmable Gate Arrays

Miriam Leeser, Northeastern University

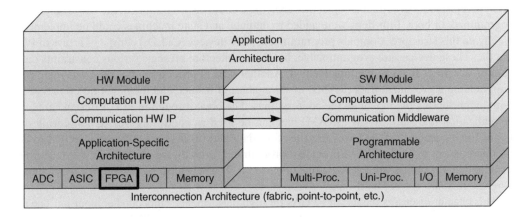

This chapter discusses the use of field programmable gate arrays (FPGAs) for high performance embedded computing. An overview of the basic hardware structures in an FPGA is provided. Available commercial tools for programming an FPGA are then discussed. The chapter concludes with a case study demonstrating the use of FPGAs in radar signal processing.

10.1 INTRODUCTION

An application-specific integrated circuit (ASIC) is an integrated circuit customized for a particular use, and frequently is part of an embedded system. ASICs are designed using computer-aided design (CAD) tools and then fabricated at a foundry. A field programmable gate array (FPGA) can be viewed as a platform for implementing ASIC designs that does not require fabrication. FPGAs can be customized "in the field," hence the name. While the design flow for ASICs and that for FPGAs are similar, the underlying computational structures are very different. Designing an ASIC involves implementing a design with transistors. An FPGA provides structures that can be "programmed" to implement many of the same functions as on a digital ASIC. The transistors on an FPGA have already been designed to implement these structures. FPGAs are based on memory technology that can be written to reconfigure the device in order to implement different designs. A pattern of bits, called a bitstream, is downloaded to the memory structures on the device to implement a specific design.

FPGAs have been around since the mid-1980s. Since the mid-1990s, they have increasingly been applied to high performance embedded systems. There are many reasons for FPGAs' growing popularity. As the number of transistors that can be integrated onto a device has grown, FPGAs have been able to implement denser designs and thus higher performance applications. At the same time, the cost of ASICs has risen dramatically. Most of the cost of manufacturing an ASIC is in the generation of a mask set and in the fabrication (see Chapter 9). Of course, an FPGA is also an ASIC and requires mask sets and fabrication. However, since the cost of an FPGA is amortized over many designs, FPGAs can provide high performance at a fraction of the cost of a state-of-the-art ASIC design. This reusability is due to the main architectural distinction of FPGAs: an FPGA can be configured to implement different designs at different times. While this reconfigurability introduces increased overhead, a good rule of thumb is that an FPGA implemented in the latest logic family has the potential to provide the same level of performance as an ASIC implemented in the technology of one previous generation. Thus, FPGAs can provide high performance while being cost-effective and reprogrammable.

Historically, programmable logic families were built on the programmable logic array (PLA) model, in which combinational logic is implemented directly in sum-of-products (SOP) form. For different types of programmable logic, the AND logic could be programmed, the OR logic could be programmed, or both. Flip-flops were added to outputs, and state machines could be implemented directly on this structure. Complex programmable logic devices (CPLDs) consist of arrays of PLAs on a die. These devices are programmable, but the architecture presents serious constraints on the types of structures that can be implemented. Another type of programmable logic was realized by using random access memories (RAMs) to store tables of results, and then looking up the desired result by presenting the inputs on the address lines. The breakthrough in FPGA design came with the realization that static RAM (SRAM) bits can be used as control as well as for storing data. Before FPGA designs, SRAM was used to store data and results only. By using SRAM to control interconnect as well, FPGAs provide a programmable hardware platform that is much more versatile than the programmable logic that preceded it. The new term *reconfigurable* was coined to designate this new type of hardware.

This chapter discusses the use of FPGAs for high performance embedded computing. The next section provides an overview of the basic hardware structures in an FPGA that make the device reconfigurable. Then the discussion moves to the architecture of a modern FPGA device, which is called a programmable system-on-a-chip since it integrates embedded components along with the reconfigurable hardware. Available commercial tools for programming an FPGA are discussed next. Tools are increasingly important for the productivity of designers as well as for the efficiency of the resulting designs. A case study from radar processing is presented, and the chapter concludes with a discussion of future challenges to implementing high performance designs with reconfigurable hardware.

10.2   FPGA STRUCTURES

An application implemented on an FPGA is designed by writing a program in a hardware description language (HDL) and compiling it to produce a bit stream that can be downloaded to an FPGA. This design process resembles software development more than hardware development. The major difference is that the underlying structures being programmed implement hardware. This section introduces FPGA technology and explains the underlying structures used to implement digital hardware. Section 10.5 presents the tools used to program these structures. There are several companies that design and manufacture FPGAs, for example, Altera [http://www.altera.com], Lattice [http://www.latticesemi.com], Actel [http://www.actel.com], and Xilinx [http://www.xilinx.com]. The architecture of the FPGA structures from the Xilinx Corporation is used as an example in this chapter. Other companies' architectures are similar.

10.2.1   BASIC STRUCTURES FOUND IN FPGAS

Let's consider the architecture of Xilinx FPGAs. The objective is to show the relationship of the structures present in these chips to the logic that is implemented on them. We start with the basic structures and discuss more advanced features in Section 10.3. The structures described in this

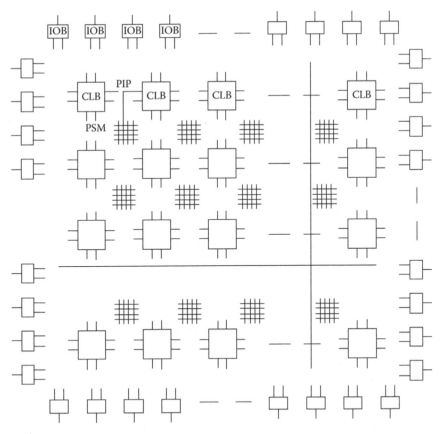

FIGURE 10-1 Overview of the Xilinx FPGA. I/O blocks (IOBs) are connected to pads on the chip, which are connected to the chip-carrier pins. Several different types of interconnect are shown, including programmable interconnect points (PIPs), programmable switch matrices (PSMs), and long line interconnect.

chapter have all been simplified for ease of understanding. Specific details of actual structures, which are considerably more detailed, are available from the manufacturer.

A Xilinx FPGA chip is made up of three basic building blocks:

- CLB: The configurable logic blocks (CLBs) are where the computation of the user's circuit takes place.
- IOB: The input/output blocks (IOBs) connect I/O pins to the circuitry on the chip.
- Interconnect: Interconnect is essential for wiring between CLBs and from IOBs to CLBs.

The Xilinx chip is organized with its CLBs in the middle, its IOBs on the periphery, and lots of different types of interconnect. Interconnect is essential to support the ability of the chips to implement different designs and to ensure that the resources on the FPGA can be utilized efficiently. An overview of a Xilinx chip is presented in Figure 10-1. Each CLB is programmable and can implement combinational logic or sequential logic or both. Data enter or exit the chip through the IOBs. The interconnect can be programmed so that the desired connections are made. Distributed configuration memory which stores the "program" of the FPGA (not shown in the figure) controls the functionality of the CLBs and IOBs, as well as the wiring connections. Implementation of the CLBs, interconnect, and IOBs are described in more detail below.

The CLBs, which implement the logic in an FPGA, are distributed across the chip. A CLB is made up of slices. The underlying structure for implementing combinational logic in a slice is the lookup table (LUT), which is an arrangement of memory cells. The truth table of any function is

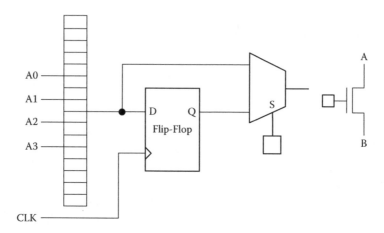

FIGURE 10-2 On the left, a simplified CLB logic slice containing one 4-input lookup table (LUT) and optional DFF. The 16 one-bit memory locations on the left implement the LUT. A0 through A3 are the inputs to the LUT. One additional bit of memory is used to configure the MUX so the output comes either directly from the LUT or from the DFF. On the right is a programmable interconnect point (PIP). LUTs, PIPs, and MUXes are three of the components that make FPGA hardware programmable.

downloaded to the LUT. The correct results are computed by simply "looking them up" in the table. Changing the contents of the memory cells changes the functionality of the hardware.

The basic Xilinx logic slice contains a 4-input LUT for realizing combinational logic. The result of this combinational function may be used directly or may be stored in a D flip-flop. The implementation of a logic slice, with 4-input LUT and optional flip-flop on the LUT output, is shown on the left in Figure 10-2. Note that the multiplexer (MUX) can be configured to output the combinational result of the LUT or the result of the LUT after it has been stored in the flip-flop by setting the memory bit attached to the MUX's select line. The logic shown is configured by downloading 17 bits of memory: the 16 bits in the lookup table and the one select bit for the MUX. By using multiple copies of this simple structure, any combinational or sequential logic circuit can be implemented.

A slice in the Xilinx Virtex family CLB is considerably more complicated; however, the basic architecture and the way it is programmed are the same. Extra logic and routing are provided to speed up the carry chain for an adder since adders are such common digital components and often are on the critical path. In addition, extra routing and MUXes allow the flip-flops to be used independently from the LUTs as well as in conjunction with them. Features that support using the LUT as a RAM, a read only memory (ROM), or a shift register are also provided.

Once the CLBs have been configured to implement combinational and sequential logic components, they need to be connected to implement larger circuits. This requires programmable interconnect, so that an FPGA can be programmed with different connections depending on the circuit being implemented. The key is the programmable interconnect point (PIP) shown on the right in Figure 10-2. This simple device is a pass transistor with its gate connected to a memory bit. If that memory bit contains a one, the two ends of the transistor are logically connected; if the memory bit contains a zero, no connection is made. By appropriately loading these memory bits, different wiring connections can be realized. Note that there is considerably more delay across the PIP than across a simple metal wire. Flexibility versus performance is the trade-off when one is using programmable interconnect.

The example FPGA architecture has CLBs arranged in a matrix over the surface of a chip, with routing channels for wiring between the CLBs. Programmable switch matrices (PSMs) are implemented at the intersection between a row and column of routing. These switch matrices support multiple connections, including connecting a signal on a row to a column (shown in bold in

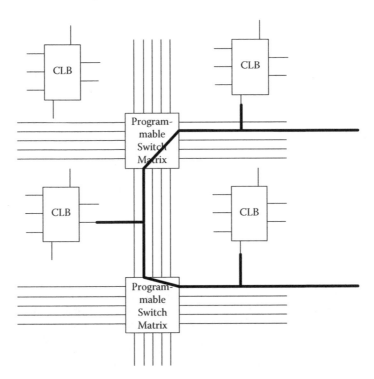

FIGURE 10-3 Programmable interconnect, including two programmable switch matrices (PSMs) for connecting the output of one CLB to the input of two other CLBs.

Figure 10-3), signals passing through on a row, and signals passing through on a column. Figure 10-3 shows a signal output from one CLB connecting to the inputs of two others. This signal passes through two PSMs and three PIPs, one for each CLB connection.

While the programmable interconnect makes the FPGA versatile, each active device in the interconnection fabric slows the signal being routed. For this reason, early FPGA devices, in which all the interconnect went through PIPs and PSMs, implemented designs that ran considerably slower than did their ASIC counterparts. More recent FPGA architectures have recognized the fact that high-speed interconnect is essential to high performance designs. In addition to PIPs and PSMs, many other types of interconnect have been added. Many architectures have nearest-neighbor connections in which wires connect from one CLB to its neighbors without going through a PIP. Lines that skip PSMs have been added. For example, double lines go through every other PSM in a row or a column, quad lines go through every fourth PSM, etc. Long lines have been added to support signals that span the chip. Special channels for fast carry chains are available. Finally, global lines that transmit clock and reset signals are provided to ensure these signals are propagated with little delay. All of these types of interconnect are provided to support both versatility and performance.

Finally, we need a way to get signals into and out of the chip. This is done with IOBs that can be configured as input blocks, output blocks, or bidirectional I/Os that switch between input and output under the control of a signal in a circuit. The output enable (OE) signal enables the IOB as an output. If OE is high, the output buffer drives its signal out to the I/O pad. IF OE is low, the output function is disabled, and the IOB does not interfere with reading the input from the pad. The OE signal can be produced from a CLB, thus allowing the IOB to sometimes be enabled as an output and sometimes not. No input enable is required. A pin is always in input mode if its output is not enabled. Additionally, IOBs contain D-type flip-flops for latching the input and output signals. The latches can be bypassed by appropriately programming multiplexers. A simplified version of the IOB is shown in Figure 10-4. An actual IOB in a commercial FPGA contains additional circuitry to

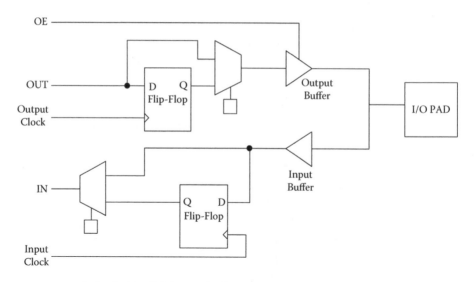

FIGURE 10-4 Simplified version of the IOB. IOBs can be configured to input or output signals to the FPGA. OE enables the output buffer, so an output signal is driven on the I/O pad. If OE is low, the IOB functions as an input block. Buffers and additional circuitry (not shown) deal with electrical signals from the I/O pad.

properly deal with such electrical issues as voltage and current levels, ringing, and glitches that are important when interfacing the chip to signals on a circuit board.

CLBs, IOBs, and interconnect form the basic architecture for implementing many different designs on a single FPGA device. The configuration memory locations, distributed across the chip, need to be loaded to implement the appropriate design. For Xilinx and Altera FPGAs, these memory bits are SRAM and are loaded on power-up. Special I/O pins that are not configurable by the user are provided to download the configuration bits that define the design to the FPGA. Other manufacturers' devices use different underlying technologies from SRAM (such as anti-fuse and flash memory) to provide programmability and reconfigurability.

10.3   MODERN FPGA ARCHITECTURES

The previous section described the basic building blocks of an FPGA architecture. Modern FPGA architectures, as exemplified by the Xilinx Virtex II and Virtex 4 and 5 families [http://www.xilinx.com] as well as the Altera Stratix family [http://www.altera.com], expand upon this basic architecture. The current generation of products features large embedded blocks that cannot be efficiently implemented in CLB logic. These include memories, multipliers, digital signal processing (DSP) blocks, and even processor cores. This section describes some of these advances in FPGA products. The architectures provided by the Altera and Xilinx corporations are compared for each feature. Future directions include more and larger embedded blocks, as well as more coarsely grained FPGA architectures.

10.3.1   EMBEDDED BLOCKS

Certain commonly used functions, such as small memories or arithmetic units, are difficult or inefficient to program using the available CLBs on an FPGA. For this reason, modern FPGA architectures now feature embedded logic blocks that perform these common functions with more efficiency in terms of both speed and space used.

Small RAMs are used in many hardware designs. While some manufacturers' CLBs have the ability to be configured as small RAMs, putting several CLBs together to form a RAM of useful size, along with the associated control logic, can quickly absorb an undesirable number of CLBs.

Embedded memories provide an easy-to-use interface and enough memory to be useful in common applications. Because they are implemented directly in hardware and not mapped to CLBs, a larger amount of on-chip memory can be made available. These memories are faster to access than off-chip memories. The Xilinx Virtex II Pro family provides 18K-bit embedded memories, called block RAMs, that can be configured with different word widths varying from 1 bit up to 36 bits, and may be single-ported or dual-ported.

The Altera Stratix family provides a hierarchy of memory sizes, including 512 bit blocks, 4K bit blocks, and 512K bit blocks of SRAM. Modern devices from both manufacturers are capable of providing over 1 MB of memory on chip.

Digital signal processing (DSP) applications are often good targets for implementation on FPGAs; because of this, FPGA manufacturers provide embedded blocks designed to be useful for implementing DSP functions. Multipliers are the most common example, though other dedicated arithmetic units are available as well. Multipliers take up a large area if implemented using CLBs, and embedded multipliers are much more efficient in both area and speed. Some embedded DSP blocks also feature logic that is designed for streaming data applications, further increasing the efficiency of DSP applications. In its Virtex 4 SX series, Xilinx provides embedded DSP blocks with an 18×18 bit multiplier and an adder/subtractor. The block is designed to be easily configured as a multiply accumulator. The DSP block is designed close to the embedded memory to facilitate the rapid transfer of data. Altera has similar DSP blocks available on its Stratix family of devices. Each DSP block can support a variety of multiplier bit sizes (9×9, 18×18, 36×36) and operation modes (multiplication, complex multiplication, multiply-accumulation, and multiply-addition).

Most FPGA designs are a combination of software and hardware, and involve FPGA fabric communicating with a microprocessor. Embedded processor cores provide the ability to integrate this functionality on a single chip. The key challenge is to efficiently interface the embedded processor with the reconfigurable fabric. Xilinx's Virtex II Pro and Virtex 4 FX families of devices feature one or two embedded PowerPC cores on an FPGA. Intellectual property (IP) for the processor local bus (PLB), as well as on-chip memory (OCM) connections, is provided. The Virtex 4 family also provides support for additional instructions that are configurable by the designer and implemented directly using CLBs with the auxiliary processing unit (APU) interface. Similar to the other embedded blocks previously discussed, the PowerPC takes up space on the FPGA chip even if one is not using it, but provides a high clock rate to improve computational efficiency. The major advantage of integrating the processor on the chip is to reduce the latency of communication between processor and FPGA logic.

An alternative to using up chip real estate with an embedded core is to build a processor directly on the FPGA fabric. This processor has the advantage of being customizable to the specific application being run on it, as well as being easier to interface to than an embedded core. In addition, a designer can choose any number of cores to instantiate. The main disadvantage is that the clock speed of these "soft" processors is slower than that of the embedded processors. Even this can be an advantage, however, since the clock speed can match the speed of the rest of the fabric. Both Xilinx and Altera offer IP soft processing cores customized for their particular hardware architectures. A designer can customize these soft cores even more by choosing the instructions to support. Altera has chosen to support a soft processor only, while Xilinx offers both hard- and soft-core processing options. Altera's soft core is called the Nios II; Xilinx's soft core is the Microblaze. There are three versions of the Nios core currently available from Altera; one is optimized for performance, another is optimized for low cost, and the third blends performance and cost as a goal. In addition to the Microblaze, Xilinx offers the Picoblaze 8-bit microcontroller.

10.3.2 Future Directions

The increased number of transistors that can be integrated onto one die, thanks to Moore's Law, point toward more and larger embedded blocks in the future. In addition to larger and more complex memories, multipliers, and processors, we can expect to see new architectures emerging. Researchers are investigating architectures that process wider bit widths than the single-bit-width-oriented architecture of current FPGAs. In these architectures, word widths can vary from a few bits to a 16- or 32-bit word. Tiled architectures are another area of active research. Tiles may consist of word-oriented processing, simple central processing units (CPUs), or a mix of reconfigurable and processor-based architectures. FPGA architectures of the future are likely to resemble multicore architectures.

As feature sizes on silicon die shrink, the number of defects on a chip continues to rise. The ability to reconfigure a chip after it has been manufactured will take on increasing importance in order to improve the yield rate at nanoscale geometries. Thus, reconfigurable hardware is likely to become part of many architectures that are much more static today.

10.4   COMMERCIAL FPGA BOARDS AND SYSTEMS

Designs containing FPGAs can be implemented on custom boards or on commercially available accelerator boards that are provided by a number of different manufacturers. A board usually integrates one or more FPGA chips, memories of different types and sizes, and support for interfaces for a host processor. Expansion modules for additional memory or other hardware such as analog-to-digital converters are also available. Many vendors also provide tools for programming their boards. Tools for programming FPGAs are discussed in the next section.

Example commercial manufacturers of FPGA boards include Annapolis Microsystems [http://www.annapmicro.com], Mercury Computer Systems [http://www.mc.com], TekMicro Quixilica [http://www.tekmicro.com/products/productcategory.cfm?id=2&gid=5], and Nallatech [http://www.nallatech.com]. Annapolis features boards that communicate with a host PC over PCI or VME connections in a master/slave configuration. Annapolis boards contain one or more FPGAs, both SRAM and DRAM, and can be interfaced, through expansion slots, to other hardware, including analog-to-digital converters (ADCs), low-voltage differential signaling (LVDS) interconnect, etc. The Quixilica family of FPGA-based products is provided with multiple ADCs and DACs to support sensor processing applications. Their boards communicate with a VXS backplane and support multichannel fiber supporting protocols such as Serial FPDP. Nallatech also features boards that interface with PCs using PCI or VME busses. Nallatech systems are based on a modular design concept in which the designer chooses the number of FPGAs, amount and type of memory, and other expansion cards to include in a system. Mercury Computer Systems has an FPGA board that closely integrates FPGAs with PowerPCs in a system with high-speed interconnect between all the processing elements. The Mercury system interfaces with a host PC via a VME backplane.

Recent trends focus on complete systems featuring traditional CPUs and reconfigurable processors arranged in a variety of cluster geometries. Nallatech and Mercury Computer Systems have products that can be viewed in this light. The Cray XD1 [http://www.cray.com], SGI RASC product [http://www.sgi.com/products/rasc/], and the MAP processor from SRC [http://www.srccomp.com/MAPstations.htm] are all entrants into this new market featuring entire systems provided in the form of heterogeneous clustered processors. These systems show great promise for the acceleration of applications.

For all these configurations, tools are essential to allow programmers to make efficient use of the available processing power. All of the manufacturers mentioned provide some level of software and tools to support their hardware systems. The next section discusses tools available for programming systems with FPGAs. The focus is on general-purpose solutions, as well as tools from specific manufacturers aimed at supporting their proprietary systems.

10.5   LANGUAGES AND TOOLS FOR PROGRAMMING FPGAS

FPGA designers do not program the underlying hardware directly; rather, they write code much as a software programmer does. Synthesis tools translate that code into bitstreams that can be downloaded to the reconfigurable hardware. The languages differ in the level of abstraction used to program the hardware. The most commonly used languages are intended specifically for hardware designs. Recent trends are to use high-level languages to specify the behavior of a hardware design and more sophisticated synthesis tools to translate the specifications to hardware. The challenge here is to get efficient hardware designs. Another approach is to use libraries of predesigned components that have been optimized for a particular family of devices. These components are usually parameterized so that different versions can be used in a wide range of designs. This section presents some of the available design entry methods for FPGA design and discusses the tools that support them. Many of these languages and tools support design flows for ASICs as well as for FPGA designs.

10.5.1 HARDWARE DESCRIPTION LANGUAGES

The most common method for specifying an FPGA design is to use an HDL. There are two dominant choices in this field, VHDL and Verilog. Both languages have the power of international standards and working groups behind them. Choosing between the two languages is similar to choosing between high-level languages for programming a PC; VHDL and Verilog can both ultimately provide the same functionality to the designer.

VHDL was developed in the 1980s by the Department of Defense as a way of documenting the behavior of complex ASICs. It is essentially a subset of the ADA programming language, with extensions necessary to describe hardware constructs. Programs that could simulate the described behavior soon followed, and the IEEE standard (1076-1987) was published shortly thereafter. Verilog was originally a proprietary language, developed in 1985 with the intention of providing a C-like language for modeling hardware. After changing ownership several times, Verilog eventually became an open IEEE standard (1364-1995) similar to VHDL. These standards have since undergone several updates. Verilog and VHDL are the most popular choices for hardware designers today, although many researchers are working on higher-level languages that target FPGA hardware; these are discussed in the next section.

While both VHDL and Verilog support different levels of abstraction, most FPGA design specifications are done at the register transfer level (RTL). Hardware description languages require the developer to keep in mind the underlying structures that are particular to hardware design. While their syntax is at least reminiscent of high-level software languages, the specification of a circuit in an HDL is different from writing a software program. Software programs have a sequential execution model in which correctness is defined as the execution of each instruction or function in the order that it is written. Decision points are explicit and common, but the movement of data is implicit and is left to the underlying hardware. Memory accesses are inferred, and processors provide implicit support for interfaces to memory. By contrast, hardware designs consist of blocks of circuitry that all run concurrently. Decision points are usually avoided. When needed, special control constructs are used with the goal of keeping overhead to a minimum. Data movement is written explicitly into the model, in the form of wires and ports. Memory and memory accesses must be explicitly declared and handled.

Two tools are necessary to support the development of hardware programs in an HDL. A simulation package (such as ModelTech's ModelSim package [http://www.model.com/]) interprets the HDL and allows testing on desktop PCs with relatively short compilation times. Synthesis packages (such as Synplicity's Synplify [http://www.synplicity.com/products/fpga_solutions.html], Xilinx's ISE [http://www.xilinx.com/ise/logic_design_prod/foundation.htm], or Altera's QuartusII [http://www.altera.com/products/software/products/quartus2/qts-index.html]) perform the longer process of translating the HDL-specified design into a bitstream.

10.5.2 HIGH-LEVEL LANGUAGES

More recently, there has been a movement to adapt high-level software languages such as C directly to describe hardware. The goal of these languages is to make hardware design resemble programming more and to leave dealing with special hardware structures to the accompanying tools. While much research continues to be conducted in this field, several solutions have been developed that combine the familiarity of high-level languages with certain features to guide the hardware mapping process.

SystemC [http://www.systemc.org] is a set of library routines and macros implemented in C++ that allow a hardware designer to specify and simulate hardware processes using a C++ syntax. The benefits of this approach include the ability to use object-oriented coding techniques in development and the use of a standard C++ compiler to produce simulatable executables. Similar to HDLs, SystemC models are specified as a series of modules that connect through ports. In addition, SystemC supports more flexibility in terms of the number of usable data types and the dynamic allocation of memory. A formal specification of SystemC (IEEE1666-2005) was recently accepted, and further developments are in progress. Synthesis tools that allow the translation of SystemC designs into the Electronic Design Interchange Format [EDIF, an industry-standard netlist format (Kahn and Goldman 1992)] are currently available, though the technology is still relatively new.

Handel-C [http://www.celoxica.com/technology/c_design/handel-c.asp] is an extended subset of ANSI C that allows hardware designers to specify their design with a C syntax (not C++). It does not allow many standard features of C, such as dynamic memory allocation, string and math functions, or floating-point data types. However, it can be synthesized directly to EDIF netlists for implementation on FPGAs. It also supports explicitly described parallelism, macros, communication channels, and RAM and ROM data types. Handel-C is the basis of the design tools available from Celoxica [http://www.celoxica.com].

Accelchip [http://www.accelchip.com] is aimed at designers who develop DSP algorithms in MATLAB [http://www.mathworks.com]. Accelchip generates synthesizable code blocks at the register transfer level for common MATLAB DSP functions; these can be output in VHDL or Verilog. Accelchip also automatically converts data types from floating-point to fixed-point before implementation. Accelchip uses a combination of synthesis techniques common to the language-based approaches, as well as library-based solutions (described in the next section). Accelchip has recently been acquired by the Xilinx Corporation and now interfaces to Xilinx System Generator [http://www.xilinx.com/ise/optional_prod/system_generator.htm]. IP blocks developed by Accelchip for more complex functions are available in the Accelware library.

10.5.3 LIBRARY-BASED SOLUTIONS

Many algorithms that are implemented on FPGAs are similar to each other. They often contain computational blocks that are also used in other algorithms. This can lead to scenarios in which a developer writes HDL code for a function that has been implemented many times before. FPGA manufacturers now provide libraries of commonly used cores that have already been tuned for performance and/or speed for their particular chips. Developers can choose the blocks they need from these libraries without needing to "reinvent the wheel." This option can lead to reduced design cycles as well as increased performance of the finished design. The core libraries provided are often parameterizable by number of inputs, bit widths of inputs and outputs, level of pipelining, and other options.

Both Xilinx [http://www.xilinx.com/ipcenter/index.htm] and Altera [http://www.altera.com/products/ip/ipm-index.html] provide libraries of parameterized functions implemented efficiently on their chips. Commonly implemented blocks, such as arithmetic units or specialized memories, are available from the library as parameterized macros. The output is an HDL module that implements the function for simulation and a data file that contains the optimized bitstream implementation of that function. Developers can then take these files and include them in their HDL design and synthesis processes. Some of these cores are available free, while more complex cores require the designer to pay for their use. The functions implemented include IP in the categories of DSP functions (filters, FFT, etc.), communications (support for SONET, Bluetooth, 802.11, etc.), and embedded processors. These cores allow the FPGA designer to more simply include highly optimized function blocks in their design.

Xilinx System Generator and Altera DSP Builder [http://www.altera.com/products/software/products/dsp/dsp-builder.html] take the library approach one step further. Developers can create entire applications using a graphical interface. Both Xilinx System Generator and Altera DSP Builder use the MathWorks Simulink environment [http://www.mathworks.com/products/simulink/] for this graphical environment. Simulink provides an interactive graphical environment based on a customizable set of block libraries (block set). Designs can be specified in a hierarchical manner. Both Xilinx System Generator and Altera DSP Builder use the Simulink environment and provide a block set of parameterizable cores similar to the Xilinx CoreGen or Altera MegaCore libraries.

A significant advantage provided by Simulink is a simulation environment that allows designers to thoroughly test their designs before synthesizing a bitstream. The System Generator or DSP Builder block sets provide the mechanism for translating the design defined in Simulink to the chosen FPGA implementation. The output is a design that can be passed into synthesis tools, thus insulating the designer from writing any HDL code.

Annapolis Microsystems' CoreFire [http://www.annapmicro.com/corefire.html] is similar in philosophy to Xilinx System Generator, but particularly targets Annapolis boards. Developers can create entire applications in the CoreFire package using a graphical interface, simply by choosing several computation blocks and indicating how they are connected together. No HDL need be written, as CoreFire outputs information which can be read directly by the synthesis tools. Debug modules can be included in the design from the beginning. Testing and simulation support is built into the tool.

CoreFire users give up the control of writing their design in an HDL in exchange for a marked decrease in development time. For applications that can be broken down into the library elements provided, this can be a good route for rapid algorithm development. High performance is still possible due to the optimized nature of the precompiled kernels. By targeting FPGA boards, CoreFire can incorporate knowledge of memory interfaces and thus potentially improves on the performance of Xilinx System Generator.

This section has mentioned just a few of the many languages and tools available for programming FPGAs. Many companies and researchers are working in this area, and the solutions available for programming FPGAs and FPGA-based systems are changing rapidly.

10.6   CASE STUDY: RADAR PROCESSING ON AN FPGA

This section provides an example of the process of mapping an algorithm onto reconfigurable hardware. Highlighted are the most important design trade-offs: those which most affect the performance of the finished implementation. The case study used is from radar processing. Modern radar systems collect huge amounts of data in a very short time, but processing of the data can take significantly longer than the collection time. This section describes a project to accelerate the processing of radar data by using a supercomputing cluster of Linux PCs with FPGA accelerator boards at each node. The goal is to exploit the coarse-grained parallelism of the cluster and the fine-grained parallelism provided by the FPGAs to accelerate radar processing.

10.6.1   PROJECT DESCRIPTION

In general, the purpose of radar imaging is to create a high-resolution image of a target area. This is accomplished by bombarding the target area with one or more electromagnetic pulses. The radiation that is received by the radar sensor in response to these pulses can be processed and turned into the desired image. This case study is based on a synthetic aperture radar (SAR) system, in which multiple radar observations from different angles are used to generate much higher resolution than would be possible with a single observation (Soumekh 1999).

Our processing algorithm takes as its input a list of projections that are received by the radar sensor. Each projection is a response to a single outgoing radar pulse, so there is one projection for every radar pulse. Through standard signal processing methods, each response is converted into a time-indexed array of digital values.

The computation involved in turning these projections back into an image that makes sense to the human eye consists of a few simple steps. For each projection, we must determine which pixels in the reconstructed image were affected by the original radar pulse. There is a correlation between the time index of the received projection and the distance from the radar sensor to a pixel in the reconstructed image. Once this correlation has been determined, the correct sample in the projection can be used to determine that projection's contribution to the pixel. The correlation can be precomputed and stored in a table for lookup at runtime. Coherently summing every projection's contribution provides the final value of that pixel.

The goal of the project was to implement this algorithm (known as backprojection) on the aforementioned supercomputing cluster. A successful implementation will provide the best possible speedup over single-node, software-only solutions. The next few sections describe the process of finding parallelism, managing I/O, and extracting as much performance from this algorithm as possible.

10.6.2 PARALLELISM: FINE-GRAINED VERSUS COARSE-GRAINED

The backprojection algorithm is an excellent choice for implementation on reconfigurable computing resources because it exhibits several axes along which operations can occur in parallel. The value of any pixel in the final image is completely independent of the value of any other pixel; that is, there is no data dependency between any two pixels. This means that the summation of multiple pixels can occur in parallel. Running multiple accumulators at a time is an example of exploiting fine-grained parallelism to achieve speedup.

Also exploited is a standard hardware design technique known as pipelining, which increases the clock rate and throughput at the expense of a small amount of latency. In this case, the process of correlating a pixel to a time index of the projection will occur in the first cycle, the lookup in the projection array in the second, and the accumulation in the last. These three operations are overlapped on different projections. The clock rate is set to the clock rate of the memories that provide input data; we cannot run any faster than this as the memories are the critical path of this design.

Because we are targeting a parallel processing platform, we can also divide the problem across multiple computation nodes. This is known as coarse-grained parallelism and is an additional source of speedup in our implementation. There are two ways to break up the problem: divide either the data or the processing. In this case, we will divide the data. The hardware implemented on each node is identical to every other node, but each node works on a different portion of the problem. Dividing the processing can be effective on certain classes of algorithms, but typically involves significantly more communication between nodes. This often leads to reduced performance due to the limitations of internode communication in such systems.

10.6.3 DATA ORGANIZATION

The input data to our chosen portion of the radar processing algorithm are a series of radar pulses. The goal is to generate an image as the output. Since we have decided to divide the data across nodes, we have two options: divide the input data (projections) or the output data (target image). If we divide the input data, then each node will compute one radar pulse's contribution to the final image. This will require an additional step at the end to collect all of the individual images and combine them. If, on the other hand, we divide the output image into small pieces, then each node operates on a portion of each input pulse. Since all of the pixels are independent, combining the images at the end of the process simply involves placing them next to each other and stitching together a final image. No additional computation is necessary.

The size of the subimage that each node will process is limited by the size of the available memories on each FPGA accelerator board. A "ping-pong" implementation, in which two copies of the image are kept on each board, was used. On each pass through the pipeline, several radar pulses are processed and the results accumulated to generate an intermediate result. On a single pass, the intermediate image is read from one memory, new values are added to it, and the next intermediate image computed so far is written to the second memory. In the next step, the two memories switch roles, so now we read from the second memory and write to the first memory. The control logic to allow this process is relatively small and keeps the data flowing efficiently to the processing pipeline.

Each step of the process involves reading a portion of an input radar pulse. Because we are able to implement multiple pipelines, we must read a portion of several input radar pulses at each computation step. These data must be constantly refreshed before each step, meaning that data must be sent from the host to the accelerator board. It is important that this data transfer be as efficient and small as possible because I/O transfer rates are often a bottleneck. Thus, the host is given the responsibility of filtering out the input data so that only the portion of the pulse needed by a particular node is sent to it. The portion of the pulses necessary for each node is small enough that the input data can fit in small on-chip RAMs, thus improving the access time to the input data and allowing one to feed a large number of accumulator pipelines since many of these on-chip RAMs (on the order of 100) can be used in a single design.

10.6.4 EXPERIMENTAL RESULTS

Now that we have determined how the hardware will work and how the data will be divided, the software to run on the PCs and control the operation of the system can be written. A master/slave control system is used to alleviate file system I/O bottlenecks. The master program reads the input data from the file system and divides them into the portions that will be needed by each slave program. Message passing via Myrinet [http://www.myri.com/myrinet/overview] is used to transfer the data to each slave. Slaves then initialize their FPGA boards and begin a loop through which a portion of the input data is transferred to the FPGA and then processed. When all of the input data have been processed, the result image is sent from the FPGA to the slave PC. The master PC collects all the partial images from the slave PCs, puts them together, and writes an output data file to the file system.

To gauge performance, we compared the runtime of a single-node version of the program that ran completely in software to the runtime of a 32-node version that also used the FPGAs. Overall we were able to achieve over 200× speedup over the software-only solution (Cordes et al. 2006).

10.7 CHALLENGES TO HIGH PERFORMANCE WITH FPGA ARCHITECTURES

FPGAs are often used as algorithm accelerators—that is, co-processors that can perform part of the computation of an algorithm more efficiently than a standard microprocessor can. High performance designs demand efficient implementation from the designer, but the performance can also be impacted significantly by features of the product being used.

10.7.1 DATA: MOVEMENT AND ORGANIZATION

One feature whose performance can dominate overall system performance is the bus interface, which connects the controlling microprocessor to the FPGA. Modern bus speeds frequently do not keep up with the amount of processing work that can be done on an FPGA of the same generation. There is much current work on high-speed interfaces such as InfiniBand [http://www.infinibandta.org/home], PCI-Express and PCI-X [http://www.pcisig.com/specifications], RapidIO [http://www.rapidio.org/home], and HyperTransport [http://www.hypertransport.org]. Still, it is important that the amount of data communicated between the FPGA and the controller be minimized as much as possible. It is also advantageous to provide concurrent execution and transfer, such that the data processing begins while later input data are still being transferred to the board. An excellent example of this principle is streaming data applications, in which data are passed in through one I/O port on the FPGA board, processed, and sent out again through another I/O port. FPGAs generally provide very high performance on these sorts of applications, provided that enough input data are available to keep the computation fabric busy at all times.

For nonstreaming applications, the arrangement of input data in on-board and on-chip memories can also make a large difference in application performance. Arrays of constants (such as filter coefficients) that must be read again and again perform better when they are kept in memories with very low latency. On-chip memories with single-cycle access times (such as Xilinx BlockRAMs) are useful in this case. Conversely, keeping such parameters in memories that require multiple clock cycles to access can severely reduce the performance of an algorithm.

10.7.2 Design Trade-offs

Very high performance FPGA implementations often feature multiple identical pipelines, performing the same computation on multiple pieces of data. This type of implementation is a good way to increase the performance of an FPGA design but places even more emphasis on the available memory bandwidth. Most modern FPGA accelerator boards provide multiple memory ports for exactly this reason, but care must be taken to arrange the input data in an efficient fashion to take advantage of them.

FPGA designs are rarely space-bound; that is, there is often more logic that could be included in the design but some other factor (such as limited memory bandwidth) prevents it from being used. In this case, there are often implementation techniques that can be used to increase performance at the cost of space. Obviously, these techniques are most useful when there is space available that would otherwise be wasted. An example of this technique is control-bound algorithms, in which a decision between two computations must be made depending on a piece of input data. It may be advantageous to perform both possible computations and choose between the results, rather than making the choice before performing a single computation.

10.8 SUMMARY

This chapter has introduced field programmable gate arrays and described the structures in SRAM-based FPGA technology that provide reprogramming capability. A few available boards and systems containing FPGAs, as well as pointers to some programming tools available to designers, have been described. Introduced was a case study: radar processing on a high performance cluster with FPGA nodes used to form the radar image. In this case study, the FPGA-based solution demonstrated a 200× improvement over a software-only solution. New hardware and software solutions for FPGAs are appearing all the time in this rapidly changing field, and the potential for runtime speedup in a wide array of scientific, medical, and defense-related applications is enormous.

This chapter just brushed the surface of the reconfigurable computing field. Several papers and books are available about the emergence of FPGA architectures, including Brown and Rose (1996), Bostock (1989), and Trimberger (1994). More recent surveys of systems (Hauck 1998; Tessier and Burleson 2001) and tools (Compton and Hauck 2002) are also available.

ACKNOWLEDGMENTS

The author would like to thank Benjamin Cordes and Michael Vai for their contributions to this chapter. The radar case study was implemented by Benjamin Cordes and Albert Conti, with help from Prof. Eric Miller and Dr. Richard Linderman.

REFERENCES

Bostock, G. 1989. Review of programmable logic. *Journal of Microprogramming and Microsystems* 13(1): 3–15.

Brown, S. and J. Rose. 1996. FPGA and CPLD architectures: a tutorial. *IEEE Design and Test of Computers* 13(2): 42–57.

Compton, C. and S. Hauck. 2002. Reconfigurable computing: a survey of systems and software. *ACM Computing Surveys* 34(2): 171–210.

Cordes, B., M. Leeser, E. Miller, and R. Linderman. 2006. Improving the performance of parallel backprojection on a reconfigurable supercomputer. *Proceedings of the Tenth Annual High Performance Embedded Computing Workshop.* Available online at http://www.ll.mit.edu/HPEC/agendas/proc06/agenda.html.

Hauck, S. 1998. The roles of FPGAs in reprogrammable systems. *Proceedings of the IEEE* 86(4): 615–638.

Kahn, H.J. and R.F. Goldman. 1992. The electronic design interchange format EDIF: present and future. *Proceedings of the 29th ACM/IEEE Conference on Design Automation* 666–671.

Soumekh, M. 1999. *Synthetic Aperture Radar Signal Processing with MATLAB Algorithms.* New York: Wiley-Interscience.

Tessier, R. and W. Burleson. 2001. Reconfigurable computing for digital signal processing: a survey. *VLSI Signal Processing* 28(1): 7–27.

Trimberger, S.M. 1994. *Field-Programmable Gate Array Technology.* Boston: Kluwer Academic Publishers.

11 Intellectual Property-Based Design

Wayne Wolf, Georgia Institute of Technology

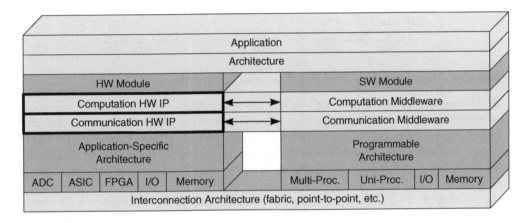

This chapter surveys various types of intellectual property (IP) components and their design methodologies. The chapter closes with a consideration of standards-based and IP-based design.

11.1 INTRODUCTION

As embedded computing systems become more complex, they can no longer be created by a single designer or a team. Instead, hardware and software components are acquired as intellectual property (IP)—designs that are used as-is or modified for system designs. Intellectual property, when properly used, reduces design time, increases compatibility with official and *de facto* standards, and minimizes design risk.

Embedded computing systems can take many forms (Wolf 2000). Systems-on-chips (SoCs) (Jerraya and Wolf 2004) are widely used in cell phones, automotive and consumer electronics, etc. Board-level designs may be used in industrial and military applications. Increasingly, board-level designs are being moved to large field programmable gate arrays (FPGAs) (see Chapter 10). Standard hardware platforms, such as personal computers (PCs) or personal digital assistants (PDAs), may also house software that is used for embedded applications. All these types of embedded computing systems can make use of intellectual property.

Figure 11-1 shows one reason why IP-based design has become popular for integrated circuit design. The graph shows data presented by Sematech in the mid-1990s that compared the growth in semiconductor manufacturing capability (i.e., Moore's Law) versus designer productivity. While the size of manufacturable chips grows exponentially, designer productivity grows at a much slower rate. IP-based design helps designers reuse components, thus increasing productivity. Software productivity has also been inadequate for several decades (see Section IV of this book).

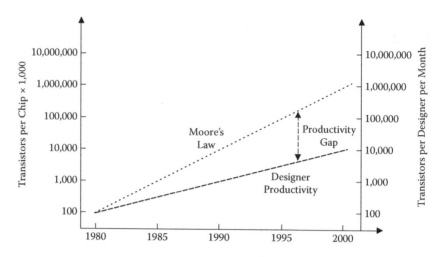

FIGURE 11-1   Hardware manufacturing capability versus design productivity.

IP components are used everywhere. To give just one example, over one billion instantiations of ARM processors have been manufactured [http://www.arm.com]. ARM processors are delivered as IP and incorporated into other designs. IP-based designs can also be very complex. ST Microelectronics estimates that home entertainment devices that support all the major audio standards (MP3, Dolby digital, etc.) must include about one million lines of software, most or all of which will be acquired as IP from third parties.

In this chapter, the term IP is used to mean a component design, either hardware or software, that is acquired for use in a larger system design. The IP may come from any of several sources: an internal source that does not require a cash transfer; the open market; the public domain; or through open-source agreements. Lawyers consider items such as patents and copyrights, as well as component designs, to be intellectual property. Although a designer may need to purchase a license to a patent to use a particular piece of IP, the main interest here is in the actual design artifact.

The next section surveys the various types of intellectual property components used in embedded computing system design. A survey of the types of sources that may supply IP is then presented, followed by a brief review of the range of possible licensing terms. Various categories of components, such as central processing units (CPUs) and operating systems (OSs), are discussed in more detail. The design methodologies for IP-based systems are discussed next. The chapter closes with a consideration of standards-based and IP-based design.

11.2   CLASSES OF INTELLECTUAL PROPERTY

As with software testing methodologies, intellectual property can be classified as *black box* or *clear box* (sometimes known as *white box*). A clear box design is given as source to the designer; it may be read and understood and perhaps modified. A black box IP module, in contrast, is provided in some nonreadable form, such as binary code or optimized logic.

Hardware IP comes in two varieties. Hard IP is a complete physical design that has been placed and routed. Soft IP, in contrast, may be logic or other forms that have not yet been physically implemented. Hard IP is generally smaller, faster, and more energy efficient thanks to its careful physical design. However, moving a piece of hard IP to a new fabrication technology requires considerable manual effort. Soft IP is less efficient but can be ported to a new fabrication process much more quickly.

Many types of hardware modules can be provided as intellectual property. Memory modules must be carefully crafted and are generally not designed from scratch. Input/output (I/O) devices are prime candidates for reuse. Busses must be compatible with existing devices, so relatively few standards become widely used, making busses well suited to packaging as IP. Central processing units also implement standard instruction sets that can be packed as IP. As it will become clear, new technologies allow custom CPUs to be configured and delivered as intellectual property.

Microprocessor manufacturers often provide design data for evaluation boards, in many cases free of charge. The schematics and layout of the evaluation board design are provided as starting points for the design of new boards for specific products.

Software IP is also widely used in the design of embedded systems. The sale of microprocessors and digital signal processors (DSPs) is in part predicated on a flow of associated software IP for that processor.

Basic input/output system (BIOS) code, broadly speaking, is one type of software IP that is often supplied with microprocessors. The code to boot the processor, operate timers, etc., is straightforward but dependent upon the details of the processor.

Algorithmic libraries are another important category of software IP; these libraries are in some cases purchased and in other cases supplied free of charge by microprocessor manufacturers. Digital signal processing (DSP) is one area in which key functions are generally implemented in standard libraries (see Chapter 17). These functions are well defined and often have one best implementation on a given processor. Once an efficient library version of a standard algorithm is available, there is often little to gain by rewriting it.

Standards-based libraries are similar to algorithmic libraries in that they are well defined and need to be efficiently implemented. The network stack is an important example of a standards-based library.

Operating systems are perhaps the most prevalent form of software intellectual property. A wide range of embedded operating systems, ranging from minimal to full-featured, is available for many different processors and platforms.

Middleware is a growing category of IP in embedded systems. As embedded computing devices implement more complex applications, designs increasingly rely on middleware that provides common functions. Middleware may be used for purposes such as security, license management, and data management.

Applications themselves, particularly applications based upon standards, are increasingly delivered as intellectual property. Audio and video compression standards are examples of standards-based applications (or, at least, the codecs are major components of multimedia applications).

11.3 SOURCES OF INTELLECTUAL PROPERTY

Intellectual property may come from several types of sources. The terms under which IP is acquired vary, but each type of institution generally has its own style of transaction.

Some intellectual property is created and sold by companies that specialize in intellectual property. On the hardware side, ARM [http://www.arm.com] and MIPS [http://www.mips.com] are examples of IP houses that sell processor designs. On the software side, Green Hills [http://www.ghs.com] and Wind River [http://www.windriver.com] are examples of companies that sell operating systems and other software IP.

Semiconductor houses also provide many types of IP. They may, for example, license processors or other parts of chips to customers who want to build custom chips. IBM Microelectronics' use of PowerPC is a prime example of a large IP block provided by a semiconductor house [http://www-306.ibm.com/chips/techlib/techlib.nsf/productfamilies/PowerPC]. Chip suppliers also supply software for their chips, sometimes free of charge and other times at an additional cost.

A great deal of hardware and software IP is available on the World Wide Web as shareware. Open-Cores is a home for the designs of CPUs, I/O devices, boards, and other hardware units [http://www.opencores]. eCos is an example of an open-source real-time operating system (RTOS) [http://ecos.sourceware.org]. These IP designs are often distributed under the Gnu General Public License.

The VSI Alliance (VSIA) was formed by electronic design automation (EDA) companies to promote standards related to IP-based design [http://www.vsia.org]. VSIA develops standards for the interfaces exported by IP components for system-on-chip design. The Open Core Protocol International Partnership (OCP-IP) provides standards for sockets for system-on-chip design [http://www.ocpip.org]. These sockets allow IP modules to be interconnected in a system-on-chip design.

11.4   LICENSES FOR INTELLECTUAL PROPERTY

Unless a piece of IP is in the public domain, the user of that IP must arrange to license the design. A wide variety of licensing options exist in the marketplace.

A license fee paid to a commercial IP house may be structured in several ways. A simple license would require a one-time payment that would allow the licensee to make an unlimited number of artifacts using the design. This arrangement provides very-well-understood costs that do not grow with product volume. Alternatively, the licensor may ask for a royalty per unit sold. The licensor may also ask for an up-front license fee in addition to a royalty.

Open-source IP is often distributed under the Gnu General Public License (GPL), which was originally crafted for the Gnu software effort. Such IP is not in the public domain—the GPL imposes some restrictions on what the user can and cannot do with the IP.

11.5   CPU CORES

CPUs are a major category of intellectual property. The type of CPU used influences every aspect of software design. The choice of a CPU architecture and model is one of the most important in the design of an embedded computing system.

ARM is the dominant embedded CPU architecture today. Thanks to the fact that ARM processors are used in a large majority of the world's cell phones, about 500 million ARM processors ship each year. ARM has a family of code-compatible processors at varying points in the performance/size space. Some ARMs are delivered as hard IP while others are provided as soft IP.

The ARM family members span a wide range of features and performance based on a common instruction set. Some of the features, such as memory management units (MMUs), are fairly standard. The Thumb extension to the instruction set provides 16-bit encodings of instructions that can be used to reduce the memory image size of programs. Jazelle is an extension to accelerate Java. SecurCore provides features for cryptography and security.

In the next paragraphs the members of the ARM family [http://www.arm.com/products/CPUs/families.html] are reviewed.

The ARM7 is the smallest member of the ARM family. The largest of the ARM7 cores has a cache and an MMU. However, most of the ARM7 models have neither a cache nor an MMU. ARM7 is primarily designed for integer operations. It can run up to 130 Dhrystone MIPS in a 0.13 micron process. One of the ARM7 cores provides Thumb and Jazelle.

ARM9 is designed around a five-stage pipeline that provides up to 300 Dhrystone MIPS in a 0.13 micron process. All the models of ARM9 provide caches and MMUs. The ARM9E variant is designed for digital signal processing and real-time applications. Not all ARM9E models provide caches; those that do, allow caches of varying sizes to be attached to the processor. Similarly, not all ARM9E processors provide MMUs.

ARM10E uses a six-stage instruction pipeline. It runs at up to 430 Dhrystone MIPS in a 0.13 micron process. All variants provide an MMU. Some variants allow the cache size to be varied at design time while one variant uses a fixed-size cache. An optional co-processor implements floating-point operations.

ARM11 is a broad family of high performance microprocessors, all of which feature variable-sized caches and MMUs. One model provides TrustZone security. All models provide floating-point units. The ARM11 MPCore is a multiprocessor built around the ARM11 core. It can support up to four processors in a shared-memory system; the shared memory can be configured to allow different types of access to various parts of the memory space from the constituent processors.

The MIPS architecture has been widely used in both general-purpose and embedded systems [http://www.mips.com/content/Products/Cores/32-BitCores]. MIPS processors are widely used in video games and other graphics devices. MIPS provides both 32-bit and 64-bit cores.

Most versions of MIPS32 use five-stage pipelines, but some high-end processors use eight-stage pipelines. All the MIPS32 cores are synthesizable; some also come in hard IP versions. MIPS provides caches that can be configured in several different ways at runtime; the MIPS4K can be configured with no cache. All MIPS32 models provide memory management units. Floating-point co-processors are available on some models.

PowerPC has been used as an IP core as well as a finished product. For example, the IBM/Sony/Toshiba Cell processor includes a PowerPC core. PowerPC was jointly developed by IBM and Motorola; now each company produces its own versions separately.

A variety of CPU cores are also available as shareware or other forms of open-source IP at sites such as OpenCores.

Configurable processors are synthesized based on user requirements. (The term *reconfigurable* is applied to FPGAs whose personality can be changed in the field. Configurable processors, in contrast, are configured at design time.) A configurable processor can be thought of as a framework on which various CPU components can be added. The basic structure of the processor does not change, but a wide variety of options can be added: cache configuration, bus configuration, debugging options, and new instructions. As a result, the size of a configurable processor can vary widely depending on how it is configured.

Configurable processors are generated by tools that accept a set of configuration parameters from the designer. The tool generates a hardware description language model for the processor. The configuration tool generally does not optimize the logic of the CPU, but the configuration tool must be complex to ensure that all the options can be implemented and combined properly. Configuration tools should also provide compilers and debuggers if the configured processor is to be useful.

Tensilica provides the Xtensa family of configurable processors [http://www.tensilica.com]. Xtensa can be configured in many ways: cache configuration, memory management, floating-point capability, bus interfaces, multiprocessing interfaces, I/O devices, and instructions. The TIE language is used to describe new instructions. The Xtensa configuration tool runs on Xtensa servers and delivers a synthesizable Verilog model of the processor, a simulator, and a compilation tool suite.

ARC provides 600 and 700 series cores, with the 700 series aimed at higher performance applications [http://www.arc.com]. Some models of the 600 series do not provide caches to make them more predictable for real-time code. Mid-level 600 and 700 series models provide caches. High-end 700 series CPUs provide memory management units as well. The ARC XY is designed for DSP applications and supports multiple memory-bank operation. A floating-point unit is also available.

ARC cores are configured with the ARChitect Processor Configurator. It allows the designer to select many parameters: register file size, number of interrupts, endianness, cache size, cache configuration, closely coupled memory size, new instructions, DSP XY additions, peripherals, bus interfaces, and debug features.

ASIP Meister was developed by a consortium of universities and laboratories in Japan [http://www.eda-meister.org/asip-meister/]. It provides a tool suite to configure custom application-specific instruction processors (ASIPs) and their associated software development tools.

11.6   BUSSES

Busses are chosen in part based on the CPU, but the bus also influences the choice of I/O devices and memories that must connect to the bus.

The AMBA bus standard [http://www.arm.com/products/solutions/AMBAHomePage.html] was developed by ARM but has been opened for general use. The AMBA standard includes two types of busses. The AMBA High-Speed Bus (AHB) is used for memory and high-speed devices.

This bus supports pipelining, burst transfers, split transactions, and multiple bus masters as ways to improve bus performance. The AMBA Peripherals Bus (APB) is used for lower-speed devices.

CoreConnect was developed for PowerPC [http://www-03.ibm.com/chips/products/coreconnect/]. The Processor Local Bus (PLB) is the high-speed bus for memory transactions; the On-Chip Peripheral Bus (OPB) is used for I/O devices; and a device control register bus (DCR) is used to convey configuration and status.

Sonics supplies the SiliconBackplane III for on-chip interconnect [http://www.sonicsinc.com]. The network is designed to be configured to connect a set of agents. The backplane provides both data and control communication so that processing elements can be decoupled from each other.

11.7   I/O DEVICES

Many standard I/O devices—timers, general-purpose I/O (GPIO) blocks, display drivers—are used in embedded systems. I/O devices are well suited to acquisition as IP. A major factor in the choice of an I/O device is the bus interface it supports. Because the AMBA bus standard is open, many devices have been designed to connect to AMBA busses. If a different bus is used, a bus adapter can be designed to connect an I/O device with a different bus interface.

As with CPUs, many devices are available as shareware. OpenCores is a good source for information on shareware I/O devices [http://www.opencores.org/].

11.8   MEMORIES

Memories are an important category of IP for system-on-chip design. Many SoCs include a large amount of memory. The architecture of the memory system is often the most important determinant of the system's real-time performance, average performance, and power consumption. The architecture of the memory system is constrained by the types of memory blocks that are available.

Memory IP is often supplied by or in concert with the semiconductor manufacturer since memory circuits must often be carefully tuned to take advantage of process characteristics. Memory IP is often supplied in the form of a generator that creates a layout based upon a number of parameters, such as memory size and aspect ratio.

11.9   OPERATING SYSTEMS

Operating systems must be carefully designed and ported to hardware platforms. Though some designers still insist on creating their own operating systems or schedulers, embedded systems increasingly rely on operating systems acquired as IP. These operating systems are often referred to as real-time operating systems (RTOSs) because they are designed to meet real-time deadlines. Perhaps the biggest challenge in developing RTOSs is porting them to the large number of hardware platforms that are used in embedded systems.

A great many commercial RTOSs are available. QNX from QNX, OS-9 from Microware, VxWorks from Wind River, and Integrity from Green Hills Software are just a few examples of commercial RTOSs.

Many open-source operating systems have been developed. FreeRTOS [http://www.freertos.org] is a very small footprint operating system that runs on a variety of processors and boards. eCos [http://ecos.sourceware.org/] is another RTOS that has been ported to a large number of platforms.

Linux is used in many embedded computing systems [http://www.linux.org/]. One advantage of Linux is that it does not require royalty payments. Many designers also like to be able to examine and possibly modify the source code. Several companies, including Red Hat [http://www.redhat.com] and MontaVista [http://www.mvista.com], provide Linux ports and services for embedded platforms.

11.10   SOFTWARE LIBRARIES AND MIDDLEWARE

Software libraries are often supplied with CPUs. Some libraries may be available from third parties. Software libraries perform standard functions that need to be efficiently implemented. For example, signal processing functions are often provided as IP components.

Middleware has long been used in servers and desktop units to provide higher-level services for a variety of programs. Middleware is general enough to be useful to many programs but is more abstract than basic operating system functions like scheduling and file systems. As embedded systems become more complex, designers increasingly use middleware to structure embedded applications. Because embedded systems need to be efficient, embedded middleware stacks tend to be shorter than general-purpose stacks. An emerging stack for embedded computing includes the following features:

- At the lowest level, drivers and a board support package abstract hardware details.
- The operating system provides scheduling, file access, power management services, etc.
- A communication layer provides communication both within the processor and across the network.
- One or more programming application programming interfaces (APIs) provide services for a category of applications, such as multimedia.
- Applications are built on top of the API stacks.

11.11   IP-BASED DESIGN METHODOLOGIES

IP-based design requires somewhat different methodologies than are used when designing with smaller, more uniformly defined components. Three major elements of an IP-based design methodology can be identified. First, the designer must be able to search and identify the appropriate pieces of IP to be used. Second, a set of selected components must be tested for compatibility and, if necessary, augmented or modified. Third, the IP components must be integrated with each other and with custom-designed components. These basic steps are common to both hardware and software IP.

Search tools for IP libraries have been proposed several times, particularly for software engineering. IP-specific search tools may make use of ontologies created specifically for the category of IP being searched; they may also try to extract information automatically from modules. However, many designers still use text-search or Web-search tools to find appropriate IP. In many cases, once some basic technology choices have been made, the universe of candidate IP components is small enough that sophisticated search tools offer little advantage.

IP components often need to be used in environments that do not exactly match their original interfaces. Wrappers are often used in both hardware and software design to match the interface of an IP component to the surrounding system. A wrapper performs data transformations and protocol state operations to match the two sides of the interface.

Bergamaschi et al. (2001) describe an IP-based methodology for system-on-chip designs. Their methodology builds on the IBM CoreConnect bus. Their Coral tool uses virtual components to describe a class of real components—a class of PowerPCs may be described by a single virtual PowerPC component, for example. The designer describes the system using virtual components. Those virtual components are instantiated into real components during system realization. Coral synthesizes glue logic; it also checks the compatibility of components at the interface boundaries. They use wrappers to integrate third-party components into designs.

Cesario and Jerraya (2004) developed a design flow called ROSES to implement a component abstraction-based design flow for multiprocessor systems-on-chips. The ROSES methodology maps a virtual architecture onto a target platform:

- Hardware wrappers interface hardware IP cores and CPUs to the communication network. The communication network itself is an IP component.
- Software tasks are interfaced to the processor using an embedded middleware stack.
- Communication between tasks is mapped onto the hardware and software wrapper functions, depending on the allocation of tasks to hardware and software.

A generic channel adapter structure generates an interface between processors and the communications infrastructure. Each channel adapter is a storage block with an interface on each side, one to the communications infrastructure and another to the processor.

Memories deserve special attention because the memory system architecture is often a key part of the hardware architecture. The adapter interfaces to the memory blocks on one side and the communications infrastructure on the other side. An arbiter may be required to control access to the memory system.

On the software side, an operating system library includes OS components that can be selected to create a small, configured operating system. A code selector looks up dependencies between services; a code expander generates final C and assembly code for the OS depending on the required modules.

IP integration creates problems for simulation as well. De Mello et al. (2005) describe a distributed co-simulation environment that handles heterogeneous IP components. Martin and Chang (2003) discuss methodologies for IP-based design of systems-on-chips.

11.12 STANDARDS-BASED DESIGN

Application standards for multimedia, communications, etc. provide an important venue for IP-based design. These standards create large markets for products. Those markets may have tight market windows that require IP-based hardware design to speed products to the market. These products also make use of software IP to develop their systems.

The impetus to view applications as intellectual property comes from the now common practice of standards committees to build a reference implementation as part of the definition of the standard. Where the standards document was once the principal standard and any software developed during the standardization process was considered secondary, today the reference implementation is commonly considered to be a key part of the specification. Reference implementations conform to the standard, but they generally do not provide enhanced or optimized versions of the subsystems of the standard. Most standards allow for variations in how certain steps are designed and implemented so that vendors can differentiate those products; reference implementations do not address such enhancements. The reference implementation may or may not be available through open-source agreements. For example, several open-source versions of MPEG-2 video codecs are available, but the MPEG-4 standards committee did not release an open-source implementation of that standard.

A reference implementation is a specialized form of clear-box IP. On the one hand, the design team has the full source of the implementation and can change it at will. On the other hand, there is some incentive to make as few changes as necessary, since changes incur the possibility of introducing bugs. Using a reference implementation as software IP requires careful analysis of the source code for performance and energy consumption. Reference implementations are generally not designed with performance or energy in mind, and significant changes to the software structure—for example, reducing or eliminating dynamic memory allocation—may be necessary to meet the system's nonfunctional requirements. The implementation must also be carefully verified. Any algorithmic changes to the standard must be verified as units. The entire system must also be checked for global conformance to the standard.

11.13 SUMMARY

Modern hardware and software design rely on intellectual property designed by others. IP-based design introduces some new problems because the implementation may not be available; even when the implementation is available, the user may not have enough time to fully understand its details. Standards can provide some useful abstractions about IP, but design methodologies must be adapted to handle both the available abstractions and some unavailable data.

REFERENCES

Bergamaschi, R., S. Bhattacharya, R. Wagner, C. Fellenz, M. Muhlada, F. White, W.R. Lee, and J.-M. Daveau. 2001. Automating the design of SOCs using cores. *IEEE Design and Test of Computers* 18(5): 32–45.

Cesario, W.O. and A.A. Jerraya. 2004. Component-based design for multiprocessor systems-on-chips. Chapter 13 in *Multiprocessor Systems-on-Chips*, A.A. Jerraya and W. Wolf, eds. San Francisco: Morgan Kaufman.

de Mello, B.A., U.R.F. Souza, J.K. Sperb, and F.R. Wagner. 2005. Tangram: virtual integration of IP components in a distributed cosimulation environment. *IEEE Design and Test of Computers* 22(5): 462–471.

Jerraya, A.A. and W. Wolf, eds. 2004. *Multiprocessor Systems-on-Chips*. San Francisco: Morgan Kaufman.

Martin, G. and H. Chang, eds. 2003. *Winning the SoC Revolution: Experiences in Real Design*. Norwell, Mass.: Kluwer.

Wolf, W. 2000. *Computers as Components: Principles of Embedded Computing System Design*. San Francisco: Morgan Kaufman.

12 Systolic Array Processors

M. Michael Vai, Huy T. Nguyen, Preston A. Jackson,
and William S. Song, MIT Lincoln Laboratory

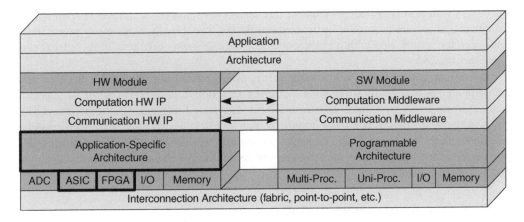

This chapter discusses the design and application of systolic arrays. A systematic approach for the design and analysis of systolic arrays is explained, and a number of high performance processor design examples are provided.

12.1 INTRODUCTION

Modern sensor systems, such as the canonical phased-array radar introduced in Chapter 3, have extremely demanding processing requirements. In order to meet the required throughput, embedded system designers have been exploring both temporal and spatial parallelisms in signal processing algorithms to boost performance. A good example of massively parallel, application-specific processors is the systolic array first proposed for signal processing applications by H. T. Kung and Charles Leiserson (1978). The objective of this chapter is to demonstrate the exploration of massive parallelism in applications. The development of application-specific processors is illustrated by means of explaining a systematic approach to design systolic arrays.

A systolic array consists of an arrangement of processing elements (PEs), optimally designed and interconnected to explore parallel processing and pipelining in the desired signal processing task. The operation of a systolic array is analogous to the blood-pumping operation of a heart. Under the control of a clock signal, each processing element receives its input data from one or more "upstream" neighbors, processes them, and presents the result to "downstream" processing elements to be used at the next clock cycle.

Performance is gained by having data flow synchronously across a systolic array between neighbors, usually with multiple data streams flowing in different directions. With its multiple processing elements and data streams, the design of a high performance systolic array is understandably difficult. Indeed, even the task of studying the operations of a given systolic array is often challenging.

Nevertheless, it is useful to note that with respect to physical implementation, there are several favorable characteristics in systolic arrays.

Individual processors in a "general-purpose" parallel processor are typically modeled after a data-path and control architecture (e.g., a microprocessor architecture). In contrast, the processing elements in a systolic array are highly specialized for the application. Their functionalities are often set at the level of basic arithmetic operations, such as multiplications and additions. Furthermore, a systolic array typically needs only a few different types of processing elements. Therefore, these processing elements can be optimized (e.g., by a full-custom design approach, see Chapter 9) for high performance. The interconnections and data flow between processing elements often follow a simple and regular pattern. In fact, many signal processing applications can be mapped into systolic arrays with primarily nearest-neighbor interconnections, thereby significantly simplifying the placement and routing procedure in the physical design flow. Finally, since every processing element operates in lock step with a clock signal in a systolic array, very few, if any, global control signals are needed.

The design and application of systolic arrays are discussed in this chapter. We first present the development of a beamformer to demonstrate an intuitive design process of an application-specific processor. A systematic approach for the design and analysis of systolic arrays is then explained using the design of a finite impulse response (FIR) filter. We then introduce a number of high performance processor design examples, including a real-time fast Fourier transform (FFT), a high performance QR decomposition for adaptive beamforming, and a very-large-scale integration (VLSI) bit-level systolic array FIR filter.

12.2 BEAMFORMING PROCESSOR DESIGN

This section begins with a description of the design of an application-specific beamformer using intuition. Beamforming is the combining of signals from a set of sensors to simulate one sensor with desirable directional properties. The radiation pattern of the sensor array can be altered electronically without its physical movement. For transmission, beamforming may be used to direct energy toward a desired receiver. When used for reception, beamforming may be used to steer the sensor array toward the direction of a signal source. In a more advanced form called *adaptive beamforming*, the receiving pattern can be adaptively adjusted according to the condition in order to enhance a desired signal and suppress jamming and interference.

It is beyond this chapter's scope to fully explain the theory of beamforming operation. A simplified definition of beamforming is given below to set the stage of the discussion. Readers who are interested in learning more are referred to Van Veen and Buckley (1988).

In Figure 12-1, a beamformer operates on the outputs of a phased sensor array with the objective of electronically forming four beams. In order to form a beam, the sensor signals are multiplied by a set of complex weights (where the number of weights equals the number of sensors) before they are combined (summed). Steering the direction of a beam, therefore, only involves changing the weight set.

Mathematically, the beamforming of n sensors is the process of computing the inner product of weight vector $W = [w_1\ w_2\ ...\ w_n]$ and sensor outputs $X = [x_1\ x_2\ ...\ x_n]$, indicated by the following equation:

$$\text{Beam}(t) = \sum_{i=1}^{n} w_i x_i(t) \ ,$$

where $t = 0$ to m is the index of the beam output sequence.

Note that complex number arithmetic is typically involved in the beamforming equation. The computational burden of the above equation is thus n complex multiplications and n complex addi-

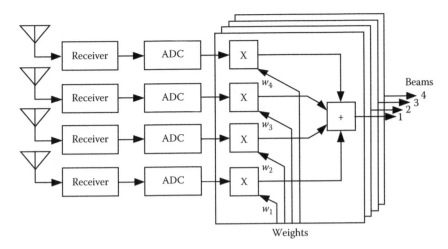

FIGURE 12-1 Beamforming operations.

tions, where n is the number of sensors participating in the beamforming. A complex multiplication requires four real multiplications and two real additions. A complex addition is formed by two real additions. For the purpose of developing a sense of the computational requirement of a beamforming operation, assume a data rate of 1 GSPS (giga-samples per second) and $n = 4$; thus, the computation of a complex beam would require 32 GOPS (giga-operations per second).

Typically, multiple beams pointing at different directions are formed. For the purpose of illustration, assume that the beamforming is performed to form four individual beams with $n = 4$ sensors. The forming of four beams would require a computation throughput of 128 GOPS, a number that justifies the use of an application-specific processor in a SWAP (size, weight, and power) constrained platform such as a UAV (unmanned aerial vehicle). Chapter 8 has a detailed description of selecting the implementation technology for demanding front-end processors.

Apparently, each beam should be formed with different weights independently of the other beams. A straightforward implementation of a beamformer is, therefore, to provide four beamformer (BF) modules to form four concurrent beams in parallel. Such a configuration is shown in Figure 12-2.

A quick inspection of the above configuration reveals a major challenge, which is to deliver signals from the sensors to the BF modules. It is not difficult to imagine, when more than a few sen-

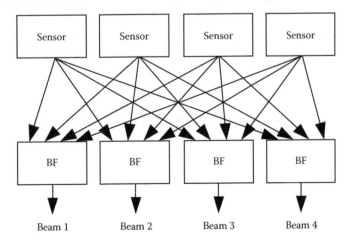

FIGURE 12-2 Parallel beamformer implementation.

sors participate in the beamforming, the routing of signals between the sensors and the BF modules would be extremely complicated, both mechanically and electrically.

We now begin to discuss the development of an application-specific processor for a real-time streaming beamforming operation. The beamforming operation of Figure 12-1 is described below as a form of matrix-vector multiplication. In this equation, four beams y_{it} (beam number i = 1, 2, 3, and 4; time index t = 0, 1, 2, …) are created by applying four weight sets w_{ij} (beam number i = 1, 2, 3, and 4; sensor number j = 1, 2, 3, and 4) to the signals of four sensor channels x_{jt} (sensor number j = 1, 2, 3, and 4; time index t = 0, 1, 2, …). Note that the columns in the multiplier matrix $[x_{jt}]$ and product matrix $[y_{it}]$ correspond to time samples of the sensor outputs and beamforming outputs, respectively.

$$\begin{bmatrix} y_{10} & y_{11} & y_{12} & \cdots \\ y_{20} & y_{21} & y_{22} & \cdots \\ y_{30} & y_{31} & y_{32} & \cdots \\ y_{40} & y_{41} & y_{42} & \cdots \end{bmatrix} = \begin{bmatrix} w_{11} & w_{12} & w_{13} & w_{14} \\ w_{21} & w_{22} & w_{23} & w_{24} \\ w_{31} & w_{32} & w_{33} & w_{34} \\ w_{41} & w_{42} & w_{43} & w_{44} \end{bmatrix} \times \begin{bmatrix} x_{10} & x_{11} & x_{12} & \cdots \\ x_{20} & x_{21} & x_{22} & \cdots \\ x_{30} & x_{31} & x_{32} & \cdots \\ x_{40} & x_{41} & x_{42} & \cdots \end{bmatrix}$$

The above matrix multiplication operation can be described as the following pseudocode:

```
for (t=0; ; t++){
        for (i=0; i<4; i++){
                y[i][t]=0;
                for (j=0; j<4; i++){
                        y[i][t]=y[i][t]+w[i][j]*x[j][t];
                }
        }
}
```

The objective is to develop this algorithm into an efficient processor for beamforming. Please note that the outer loop represents an infinite loop to process the continuous streaming in signals. On the other hand, also note that the beam outputs can be independently computed for each time index t. Therefore, an efficient way to calculate the beam outputs at time t should be developed to process the data continuously for streaming data. These beamforming operations are represented by the inner loops of the above listing.

A little observation reveals that the inner loop computation can be implemented as a pipeline operation, in which y_t (i.e., $y[t]$) is initialized to be 0 and is streamlined through a pipeline. On its way, it collects and accumulates the $w_i \times x_{it}$ (i.e., $w[i]*x[i][t]$) terms and comes out of the pipeline as the final results. This suggests the parallel pipeline structure shown in Figure 12-3.

Figure 12-3 shows how to use four pipelines to compute one output for each of four beams. Since the sensor outputs (x_{it}'s) are used in different pipeline stages, apparently buffers are needed to hold the samples until they are used in the pipelined computation. It is desirable to modify this architecture so that it can keep up with the input data rate. In other words, the architecture will be able to accept one data point every clock cycle. Furthermore, the design should avoid the use of a complicated memory/buffer access scheme.

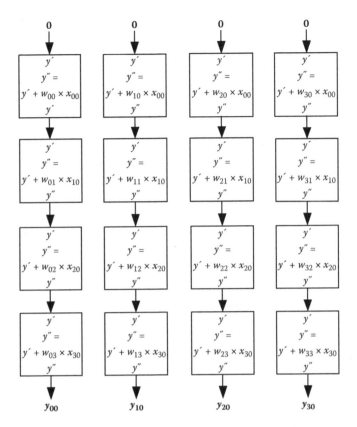

FIGURE 12-3 Parallel pipeline architecture for beamforming.

Figure 12-4 shows this modification. Note that this is essentially the same four pipelines with a few additional registers (i.e., the D-blocks) for timing. New interconnections are also added in the horizontal direction.

The array requires only one type of processing element and delay elements (i.e., registers). Note that the processing elements do not have a delay element in the horizontal direction, even though it is likely to need a buffer to recondition the signal in a practical design. Each processing element has a storage for a weight. Its function is explained in the call-out diagram in Figure 12-4.

12.3 SYSTOLIC ARRAY DESIGN APPROACH

While the intuitive development of an application-specific beamformer has been demonstrated, in many cases it is desirable to explore the parallelism in a given algorithm in a systematic approach. The algorithm can then be converted into a massively parallel architecture. The number of processing elements in the array, the functionality of processing elements, their interconnection pattern, and the data flow pattern can be systematically determined.

Such an approach, which develops a systolic array by transforming the parallelism in an algorithm into an array of processing elements, is now explained. A mathematical procedure can be used to facilitate this transformation (Kung 1988). In the following explanation, the systolic array design flow is demonstrated by going through, step by step, the design procedure of a systolic FIR filter.

Given a time-invariant input signal $x(n)$, the operation of a pth-order FIR filter is defined as the convolution of $x(n)$ and the filter coefficients, $h(k)$, as follows:

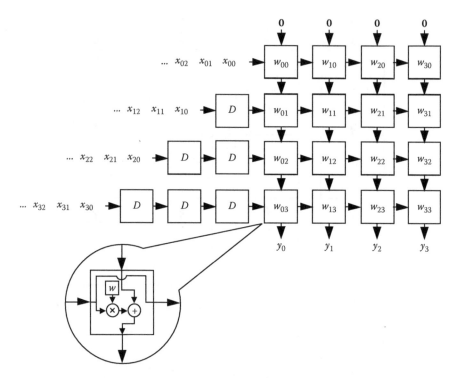

FIGURE 12-4 Modified parallel pipeline architecture for beamforming.

$$y(n) = \sum_{k=0}^{P-1} h(k)x(n-k).$$

The above equation can be expressed into a sequence of computation steps to produce $y(n)$. Without loss of generality, the pseudocode listed below computes the first m output of this FIR filter (i.e., $y(0)$ to $y(m-1)$).

```
for n=0 to m-1
        y(n)=0;
        for k=0 to (p-1)
                y(n)=y(n)+h(k)*x(n-k);
        end k;
end n;
```

Data dependency is evident in the above listing. Consider the statement y(n)=y(n)+h(k)*x(n-k), the variable y(n) on the receiving side of the = sign is available only after the operation on the feeding side is complete. Therefore, there is an implicit temporal distinction between the receiving y(n) and the feeding y(n), which represent the values of y(n) before and after the operation, respectively. In order to clearly express the relationship represented by these variables for the exploration of parallelism, a time dimension is introduced into the statement for explicit temporal dependency. In the current example, a time index k is introduced into variable y(n) and the statement under discussion becomes y(n,k+1)=y(n,k)+h(k)*n(n-k). This results in the pseudocode listed below.

Without loss of generality, the discussion continues by assuming that the filter has four taps (i.e., $p = 4$). This assumption has been reflected in the following listing. The filter output is produced as y(n,4).

```
for n=0 to m-1
        y(n,0)=0;
        for k=0 to 3
                y(n,k+1)=y(n,k)+h(k)*x(n-k);
        end k;
end n;
```

The above pseudocode is an example of a single-assignment representation, in which no variable is assigned with values more than once. The single-assignment representation of a signal processing algorithm is the first step in developing a systolic array. Operations of an algorithm that have no dependence on each other can be executed concurrently in a parallel architecture. Based on the single-assignment code, the dependence between data is then explored for maximum parallelism using a dependence graph.

A dependence graph is one that shows the dependence of computations occurring in an algorithm and is a tool for exploring parallelism in an algorithm. The number of dimensions in a dependence graph is determined by the nested loop in the algorithm. The filter example has a two-dimensional dependence graph since the nested loop has two levels. The dependence graph axes are labeled as n (number of y's) and k (number of taps), respectively, following the indices used in the nested loop. The operation y(n,k+1)=y(n,k)+h(k)*x(n-k) is then represented as a node, which is shown in Figure 12-5.

These computation nodes are placed in the dependence graph for each pair of n, k, as shown in Figure 12-6. The single-assignment property guarantees that no loops or cycles exist in the dependence graph of a single assignment code. A few extra nodes needed to deal with the initial filter operations (grayed out in Figure 12-6) have been added for graph regularity. The correctness of the first three y outputs is maintained by setting $x(-1)$, $x(-2)$, and $x(-3)$ to 0. The benefits of these nodes will become clear when the projection from a dependence graph to an array of processing elements is explained.

The projection process begins by selecting a permissible schedule for the computation nodes in a dependence graph. A schedule is permissible if data dependence is retained. In other words, if node p depends on node q, then node q cannot be placed on a time step that occurs earlier than node p. A permissible schedule of the FIR filter dependence graph is shown in Figure 12-6 by a set of parallel and equally spaced lines. Note that a dependence graph may have multiple permissible schedules, each of which may lead into a unique systolic array implementation.

With the selected permissible schedule in Figure 12-6, each computation node only needs the "shared" signals [$h(k)$ and $x(n)$] to be available at its execution time. This allows the conversion of "global" interconnections into "local" ones, resulting in a simpler physical implementation. Figure 12-7 shows the dependence graph modified to contain only local communication paths.

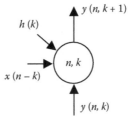

FIGURE 12-5 FIR filter computation node.

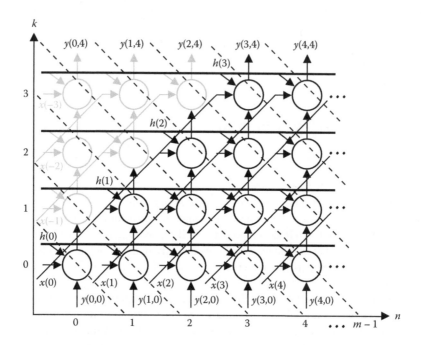

FIGURE 12-6   FIR filter dependence graph.

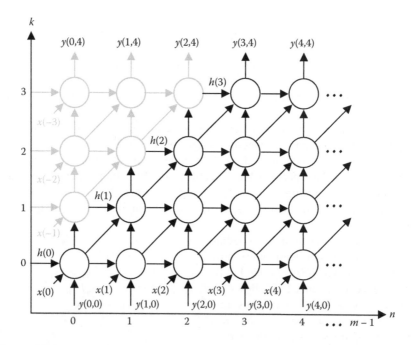

FIGURE 12-7   FIR filter dependence graph with only local communication paths.

The FIR filter description pseudocode is rewritten below to reflect the changes made to its dependence graph.

```
for n=0 to m-1
        y(n,0)=0;
        for k=0 to 3
                if (k==0)
                        x(n,k)=x(n);
                else
                        x(n,k)=x(n-1,k-1);
                end if;
                if (n==0)
                        h(n,k)=h(k);
                else
                        h(n,k)=h(n-1,k);
                end if;
                y(n,k+1)=y(n,k)+h(n,k)*x(n,k);
        end k;
end n;
```

The dependence graph suggests an array architecture to implement the FIR filter. The nodes can be implemented as PEs connected according to the dependence graph. However, in this design, the PEs will not have a high utilization rate.

The data dependence graph is now projected into a signal flow graph, which is a more concise representation of the algorithm to be implemented. The projection is demonstrated using the current example. If the dependence graph contains an array of operations, the following questions need to be answered when a systolic array is designed:

- Which PE in the systolic array should an operation (i.e., a node in a dependence graph) be mapped to?
- What is the correct scheduling of the operations (i.e., the arcs in a dependence graph) in the PEs of the systolic array?

When there is more than one way of projecting a dependence graph into an array processor, the quality of the projection can be evaluated by considering the throughput, the PE utilization rate, and the complexity of PE interconnections.

The projection requires selection of a permissible schedule, which has been explained and shown in Figure 12-6. The permissible schedule is represented in Figure 12-8 by a set of hyperplanes specified by a vector S that points in the direction of time flow.

In the dependence graph of the FIR filter, only

$$S = \begin{bmatrix} 1 \\ 1 \end{bmatrix},$$

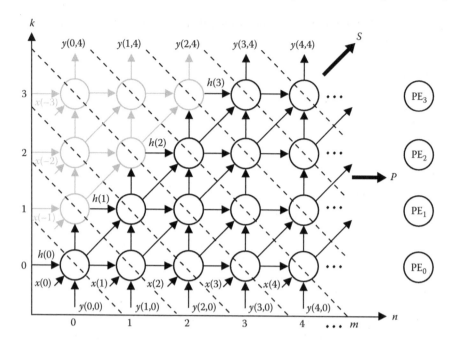

FIGURE 12-8 Relationship between hyperplanes and projection.

a direction from the lower left corner of the graph toward the upper right corner, is a permissible schedule. Note that unit vectors are used to indicate directions. In general, there may be more than one permissible set of hyperplanes. In order to mathematically check the validity of this schedule, compute and verify that $S^T \times e \geq 0$, in which e is the direction of any dependence arc in the dependence graph. Satisfying this condition implies that no dependence arcs are going against the flowing direction of the hyperplanes. Ideally, it is desirable to have $S^T \times e > 0$, since the equal-to-zero condition indicates that the parallel operations have to be implemented sequentially. In the above figure, there are three types of dependence arcs:

$$e_1 = \begin{bmatrix} 1 \\ 0 \end{bmatrix}, \ e_2 = \begin{bmatrix} 0 \\ 1 \end{bmatrix}, \text{ and } e_3 = \begin{bmatrix} 1 \\ 1 \end{bmatrix}.$$

$$S^T e_1 = \begin{bmatrix} 1 & 1 \end{bmatrix} \times \begin{bmatrix} 1 \\ 0 \end{bmatrix} = 1 > 0, \ S^T e_2 = \begin{bmatrix} 1 & 1 \end{bmatrix} \times \begin{bmatrix} 0 \\ 1 \end{bmatrix} = 1 > 0, \text{ and } S^T e_3 = \begin{bmatrix} 1 & 1 \end{bmatrix} \begin{bmatrix} 1 \\ 1 \end{bmatrix} = 2 > 0.$$

Based on this chosen scheduling of operations, now a projection vector P can be chosen. It is apparent that the projection vector must not be parallel to the hyperplanes since all the parallel operations on the same hyperplane will then be mapped to the same PE and become sequential. Mathematically, this condition can be expressed as $S^T \times P > 0$, which means that the projection vector must not be orthogonal to the hyperplane normal vector.

Choose

$$P = \begin{bmatrix} 1 \\ 0 \end{bmatrix}$$

to be the projection vector in Figure 12-8, the validity of which can be checked by computing

$$S^T \times P = \begin{bmatrix} 1 & 1 \end{bmatrix} \begin{bmatrix} 1 \\ 0 \end{bmatrix} = 1 > 0 .$$

The result of projection, as shown in Figure 12-8, is the four PEs on the right-hand side of the figure. While the mapping can be carried out manually for a simple data dependence graph, a systematic approach is desirable for more complex operations to avoid the possibility of making errors. The mapping of node activities according to a selected projection vector P can be determined by

$$c' = H^T \times c ,$$

where c is the location of a node in the data dependence graph, c' is the location of a PE to which the node is to be mapped, and H^T is an $n \times (n-1)$ node mapping matrix (n is the number of dimensions in the data dependence graph) that is orthogonal to the projection vector P (i.e., $H^T \times P = 0$).

In the FIR example, choose

$$H = \begin{bmatrix} 0 \\ 1 \end{bmatrix}, \; H^T \times P = \begin{bmatrix} 0 & 1 \end{bmatrix} \begin{bmatrix} 1 \\ 0 \end{bmatrix} = 0.$$

Each node in the data dependence graph can be identified as

$$\begin{bmatrix} n \\ k \end{bmatrix} .$$

The mapping of the nodes into the array processor created according to P is found to be

$$c' = \begin{bmatrix} 0 & 1 \end{bmatrix} \times \begin{bmatrix} n \\ k \end{bmatrix} = k .$$

For example, all the activities in a row, e.g., nodes (0,0) , (1,0) , (2,0) ..., are mapped into PE_0.

The arcs in a data dependence graph describe the dependencies between operations. They are used to determine the interconnection pattern between PEs in a systolic array. In addition to the interconnections between PEs, this step also determines the delays (i.e., registers) that must be provided on an interconnection. The need of delays on interconnections is obvious if the possibility of having a dependence between PEs that operate in parallel is considered. A pipeline structure is an example of such a situation. In summary, two decisions are made in this mapping step:

1. The interconnection e' in the systolic array corresponding to a dependence e
2. The number of delays, $D(e')$, required in the interconnection

Both of these can be found as follows:

$$\begin{bmatrix} D(e') \\ e' \end{bmatrix} = \begin{bmatrix} S^T \\ H^T \end{bmatrix} \times e .$$

Applying this operation to the arcs,

$$\begin{bmatrix} D(e') \\ e' \end{bmatrix} = \begin{bmatrix} 1 & 1 \\ 0 & 1 \end{bmatrix} \times e_1 = \begin{bmatrix} 1 & 1 \\ 0 & 1 \end{bmatrix} \times \begin{bmatrix} 1 \\ 0 \end{bmatrix} = \begin{bmatrix} 1 \\ 0 \end{bmatrix}.$$

The result of direction 0 indicates that it is going to stay in the same PE to form a feedback loop. One delay should be provided. It is apparent that this is a register:

$$\begin{bmatrix} D(e') \\ e' \end{bmatrix} = \begin{bmatrix} 1 & 1 \\ 0 & 1 \end{bmatrix} \times e_2 = \begin{bmatrix} 1 & 1 \\ 0 & 1 \end{bmatrix} \times \begin{bmatrix} 0 \\ 1 \end{bmatrix} = \begin{bmatrix} 1 \\ 1 \end{bmatrix}.$$

The result shows that the signal is mapped into a positive direction in the systolic array. One delay is needed.

$$\begin{bmatrix} D(e') \\ e' \end{bmatrix} = \begin{bmatrix} 1 & 1 \\ 0 & 1 \end{bmatrix} \times e_3 = \begin{bmatrix} 1 & 1 \\ 0 & 1 \end{bmatrix} \times \begin{bmatrix} 1 \\ 1 \end{bmatrix} = \begin{bmatrix} 2 \\ 1 \end{bmatrix}.$$

Again, this is a connection in the positive direction. Two delays are needed. The reader should be able to verify these mappings by studying the mapping itself.

The last step is to determine when and where to apply inputs and to collect outputs. The equation is

$$\begin{bmatrix} t(c') \\ c' \end{bmatrix} = \begin{bmatrix} S^T \\ H^T \end{bmatrix} \times c \ ,$$

where c is the node location in the dependence graph, c' is the PE location, and $t(c')$ is the time to apply or collect data. Note that the $t(c')$ is expressed in a relative time unit.

$$\text{y:}\ \begin{bmatrix} t(c') \\ c' \end{bmatrix} = \begin{bmatrix} 1 & 1 \\ 0 & 1 \end{bmatrix} \times \begin{bmatrix} n \\ 0 \end{bmatrix} = \begin{bmatrix} n \\ 0 \end{bmatrix}$$

$$\text{h:}\ \begin{bmatrix} t(c') \\ c' \end{bmatrix} = \begin{bmatrix} 1 & 1 \\ 0 & 1 \end{bmatrix} \times \begin{bmatrix} 0 \\ k \end{bmatrix} = \begin{bmatrix} k \\ k \end{bmatrix}$$

$$\text{x:}\ \begin{bmatrix} t(c') \\ c' \end{bmatrix} = \begin{bmatrix} 1 & 1 \\ 0 & 1 \end{bmatrix} \times \begin{bmatrix} n \\ 0 \end{bmatrix} = \begin{bmatrix} n \\ 0 \end{bmatrix}$$

The output is

$$y(n,4):\begin{bmatrix} t(c') \\ c' \end{bmatrix} \cdot = \begin{bmatrix} 1 & 1 \\ 0 & 1 \end{bmatrix} \times \begin{bmatrix} n \\ 4 \end{bmatrix} = \begin{bmatrix} n+4 \\ 4 \end{bmatrix}.$$

This systolic array FIR filter is shown in Figure 12-9. Several snapshots of the data distribution inside the systolic array are shown in Figure 12-10, which also shows the data movements and computations.

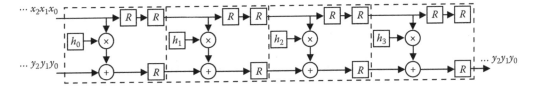

FIGURE 12-9 Systolic array FIR filter.

12.4   DESIGN EXAMPLES

Three case studies—a QR decomposition processor, a real-time FFT, and a bit-level systolic array design methodology—are presented in this section. The QR decomposition processor is an example of implementing a well-known systolic array design in a field programmable gate array (FPGA) with limited resources. The second design example is also a case of FPGA implementation. The real-time streaming FFT is designed for the correlation processor described in Chapter 8. The challenge in this design is the use of an FPGA, which has a maximum clock speed of ~150 MHz, to process analog-to-digital conversion (ADC) data in the rate of 1.2 GSPS. The designer developed an innovative FFT architecture consisting of multiple pipelines. With this architecture, long FFTs (up to ~32K points) can be completely implemented in FPGAs without requiring external memory. The last case study presents a bit-level systolic array design methodology created to support the development of front-end radar signal processors.

12.4.1   QR DECOMPOSITION PROCESSOR

In addition to aligning the phases of multiple sensor signals so that they add up coherently, adaptive beamforming places a null in the desired beam pattern to cancel noise and interference coming from a specific heading direction. QR decomposition is a critical step in computing the adaptive beamforming weights. This first design case illustrates a high performance FPGA implementation of QR decomposition for adaptive beamforming radar applications, in which both processing throughput and latency are important.

A small set of interference and noise samples is first collected to form an $n \times m$ sample matrix X, where m is the number of sensors and n is the number of samples. A covariance matrix C is formed as

$$C = X^H \bullet X , \tag{12.1}$$

where X is the sample matrix itself and X^H is the conjugate transpose of X. The following linear system of equations shows the relationship between a steering vector V supplied by the user to point the beam at a desired direction, the weights W adapted to cancel the interference, and the covariance matrix C,

$$V = C \bullet W . \tag{12.2}$$

Solving Equation (12.2) for the values of W requires the computation of the inverse of matrix C to perform

$$W = C^{-1} \bullet V . \tag{12.3}$$

The following matrix manipulation is commonly used to avoid the high complexity of doing matrix inversion in a hardware implementation. First QR decomposition is used to transform the sample matrix into $X = Q \bullet R$, where Q is an orthonormal matrix (i.e., $Q^H \bullet Q$ = identity matrix) and R is an upper triangular matrix (i.e., all elements below the diagonal are zeros). Substitute X into Equation (12.2)

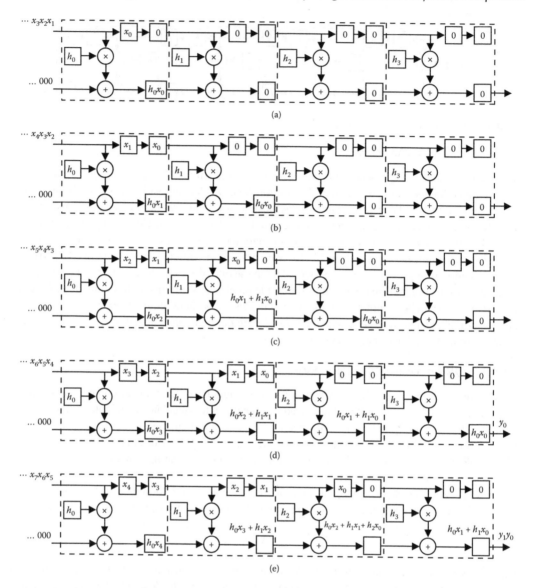

FIGURE 12-10　Snapshots of data flow in a systolic array FIR filter: (a) $t = 0$, (b) $t = 1$, (c) $t = 2$, (d) $t = 3$, and (e) $t = 4$.

$$V = (X^H \bullet X) \bullet W = (R^H \bullet Q^H \bullet Q \bullet R) \bullet W = R^H \bullet R \bullet W. \tag{12.4}$$

The adaptive weights W can be determined with Equation (12.4) by solving two linear systems of equations:

$$V = R^H \bullet Z \quad \text{and} \quad Z = R \bullet W.$$

Since both R and R^H are upper triangular matrices, the equations can be solved with back substitutions, which are commonly referred to as a double substitution. The adaptive weights W are then applied to the sensor data for beamforming and interference cancellation.

A number of algorithms have been developed to perform QR decomposition (Horn and Johnson 1985). The Householder algorithm, due to its computational efficiency, is often the choice of

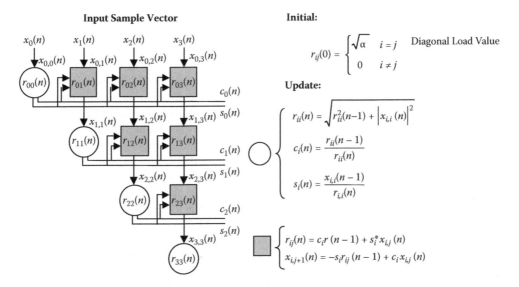

FIGURE 12-11 McWhirter algorithm.

software implementation. However, its hardware implementation is inefficient since the variables have to be stored and accessed in a shared storage area. The computations must have their data distributed locally to optimize the benefits of a hardware parallel processor. The McWhirter array, which implements the Givens rotation to perform QR decomposition, uses only near-neighbor data communication and is thus suitable for systolic array implementation (McWhirter, Walke, and Kadlec 1995). The computations and signal flow of the McWhirter algorithm are illustrated in Figure 12-11.

The array shown in Figure 12-11 consists of m (m is the number of sensors, which is four in this example) boundary nodes (circles), and $m(m-1)/2$ internal nodes (squares). The computations implemented in these nodes are also provided in Figure 12-11. Each boundary node updates its $r_{ii}(n)$ value, which is initialized to $\sqrt{\alpha}$, a loading factor. At each step, the new value $r_{ii}(n)$ and the previous value $r_{ii}(n-1)$ are used to generate the Givens rotation parameters $c_i(n)$ and $s_i(n)$ for the internal nodes in the same row. The computed parameters c_i and s_i are passed on the internal nodes to compute values $r_{ij}(n)$ and $x_{i+1,j}(n)$, the latter of which are delivered to the next row below. The array is typically fed with five m samples for an adequate representation of the noise and interference. The final r_{ij} values form the upper triangular matrix R in Equation (12.4), which is now ready for the computation of adaptive weights W by means of a double substitution.

Figure 12-12 shows a systolic implementation of the McWhirter array, which inserts registers (i.e., delays) to convert the signal paths from a boundary node to the internal nodes into nearest-neighbor communications. The correct computation is preserved by "retiming" the input pattern, which is shown in Figure 12-12 as a skewed sample matrix X. The use of only nearest-neighbor communications allows the array size to be scaled without impacting the communication speed, which is a very desirable property in hardware design.

The McWhirter array in Figure 12-12 has a systolic triangular signal flow graph. However, an FPGA may not have sufficient resource to directly implement the full systolic array. A straightforward solution to this issue is to implement only the first row and reuse it to perform the computation in the other rows as the array schedule is stepped through. This approach results in many nodes idling when used on the lower rows of the array, thus reducing the overall efficiency to about 50%. A mapping to a linear array was proposed to fold the array and allow the idle processors to operate on a second dataset while the first dataset is still in process (Walke 2002). Interleaving the processing of two datasets doubles the hardware utilization efficiency to full capacity. Figure 12-13(b) shows the scheduling of a three-node linear array so that it performs the computations of a 5×5 trian-

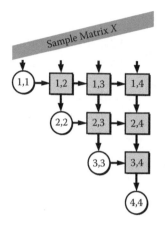

FIGURE 12-12 Systolic implementation of the McWhirter array.

gular array. The utilization rate of the nodes is doubled from that of the five-node linear array in Figure 12-13(a).

Another benefit of the McWhirter array is that it can be modified to perform, in addition to the QR decomposition, all other adaptive beamforming steps (i.e., double substitution and beamforming). Figure 12-14 shows a modified array, which operates in two modes. The first mode is the original QR decomposition. The second mode is one in which the same array can be used to perform back substitution with an upper triangular matrix stored in the array and a vector fed in from the top. Also, a new column of processing nodes (hexagons) is provided for the computation of a dot multiplication.

A complete adaptive beamforming has to perform $K = Y \bullet W = Y \bullet R^{-1} \bullet (R^{H})^{-1} \bullet V$, where in addition to R and V, which have been defined, K is the beam result, and Y is the sensor data to be used for beamforming.

A sample matrix X is fed into the array for QR decomposition. The result R is produced in the array. The array is then switched, by a mode control signal, into performing a back substitution to solve for vector $Z = (R^{H})^{-1} \bullet V$, the result of which is stored in the rightmost column of the array for a later dot product computation. The sensor data to be used for beamforming is fed into the array to solve for $U = Y \bullet R^{-1}$, the result of which is sent to the rightmost column to be dot multiplied with Z. The dot product is the adaptive beam K.

The scheduling of array operations with a five-node linear array is illustrated in Figure 12-14(b). An innovative approach to pipeline the operations in the linear array for lower latency and better throughput is described in Nguyen et al. (2005). The pipelined version is estimated to be 16 times faster than the non-pipelined design.

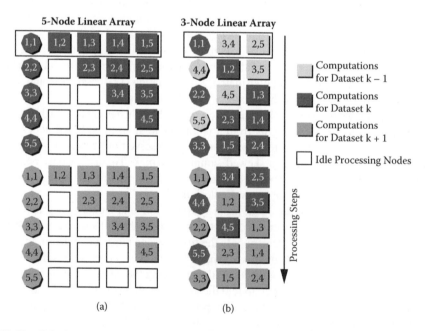

FIGURE 12-13 Scheduling of two linear arrays: (a) five-node array and (b) three-node array.

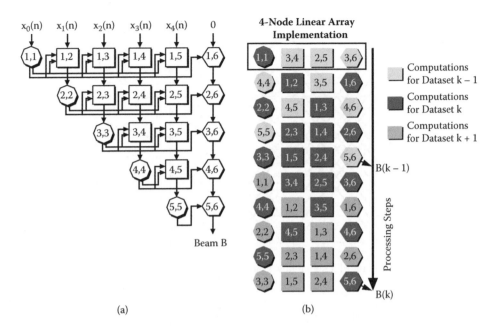

FIGURE 12-14 Adaptive beamforming with a McWhirter array: (a) signal flow graph; (b) scheduling of a five-node linear array.

12.4.2 REAL-TIME FFT PROCESSOR

MIT Lincoln Laboratory has developed a systolic FFT architecture for FPGA and application-specific integrated circuit (ASIC) implementation (Jackson et al. 2004). This architecture has been designed to process the output of an ADC in real time, even when the processor operates at a clock rate (e.g., 150 MHz) that is lower than the ADC sample rate (e.g., 1.2 GSPS). This architecture allows an 8192-point real-time FFT operating at 1.2 GSPS to completely fit on a single FPGA device.

An FFT operation converts a time-domain signal into frequency domain, in which filtering and correlation can be computed more efficiently. Figure 12-15 shows the data flow graph of a 16-point FFT, which can be used to develop an FFT array processor. In general, an n-point FFT consists of $\log_2(n)$ pipeline stages ($n = 16$ in Figure 12-15). The basic operation in each stage is a butterfly computation, which performs multiplications and additions with a pair of data points and produces results for the next stage. In Figure 12-15, for example, x_0 and x_8 form a pair of data points, go through a butterfly computation, and produce y_0 and y_8. The efficient operation of such an array FFT architecture requires that all n data points ($x_0 - x_{15}$) are available simultaneously.

In a streaming operation, a single data point is produced by the ADC every clock cycle. A memory buffer can be used to collect the n data points and feed them to an array FFT processor constructed according to Figure 12-15. The processor thus operates at a rate n times slower than the ADC data rate.

An optimal 16-point FFT architecture for a streaming operation is shown in Figure 12-16. The architecture is pipelined to accept one data point every clock cycle. It has three types of building blocks: a switch, a FIFO (first-in, first-out memory), and a butterfly unit. The operating principle of this pipelined FFT architecture is as follows. The FIFOs are sized so that the data can be steered by the switches, which operate in one of two modes (straight and cross) to pair up at the right place in the right time.

Figure 12-17 uses a few snapshots to illustrate the operations of the pipelined FFT architecture. The FIFOs are labeled with x_i, y_i, z_i, and w_i, respectively, according to the signal flow graph in Figure 12-15. In the first-stage butterfly unit, the data path that x_0 and x_8 go through is identified in Figure 12-17(a) (before the butterfly computation) and Figure 12-17(b) (after the butterfly computa-

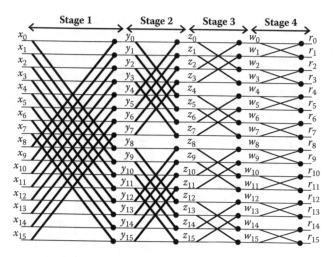

FIGURE 12-15   Sixteen-point FFT signal flow graph.

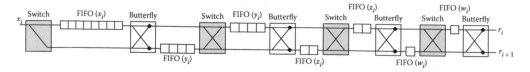

FIGURE 12-16   Pipelined FFT architecture for streaming operation.

tion). Similarly, Figures 12-17(c) and 12-17(d) show the switch positions when y_0 and y_4 are entering and exiting the second butterfly unit, respectively.

The processor must operate at the ADC data rate. The maximum clock rate of the increasingly popular FPGAs is currently around 300–400 MHz, which is a significant mismatch from wideband ADCs operating at above 1 GSPS. Figure 12-18 shows a solution developed to address this issue.

In Figure 12-18, the pipeline architecture of Figure 12-17 is parallelized so that the processor can operate at a lower speed. The Lincoln Laboratory FFT processor was originally developed to process the output of an ADC sampling at 1.2 GSPS. The real number ADC data is first converted into a complex number stream with the data rate of 600 MSPS (conversion not shown in Figure 12-18), which is still considered to be too fast for an FPGA implementation.

Four parallel pipelines are provided to reduce the speed to 150 MHz (600 MHz / 4 = 150 MHz). A high-speed (600 MHz) demultiplexer (DMUX) distributes the incoming data (x_i) into four parallel streams, each of which has a data rate of 150 MSPS. Therefore, the FPGA-implemented processor only needs to operate at 150 MHz to keep up with the ADC outputs.

The operating principle of the parallel pipelined architecture remains the same. It has to coordinate the data flow so that the right pairs of data points meet in the right place at the right time to perform the butterfly computations. A snapshot of the data distribution in the architecture is captured in Figure 12-18, which shows the processing of a data frame (0–15) and the first four data points in the next frame (16–19).

The architecture is very suitable for hardware implementation. First, the entire circuit consists of only a few types of building blocks. Second, except for the last two stages, the building blocks are connected with local interconnections that significantly simplify placement and routing. The architecture can be readily scaled to perform longer FFTs by adding more stages. Furthermore, more parallel pipelines can be used to further slow down the clock rate. This property allows an exploration of the design space that spans from a single pipeline to a structure with $N/2$ pipelines. Notice that the number of stages that require non-local interconnection, while being a function of the number of parallel pipelines, is independent of the FFT length.

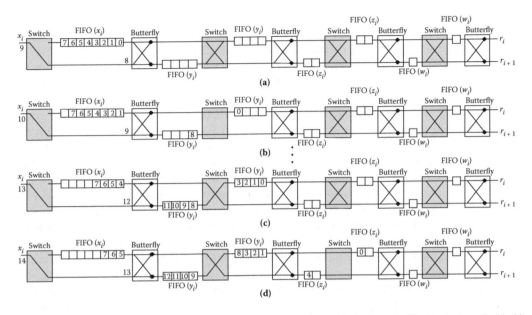

FIGURE 12-17 Snapshots of pipelined FFT processor operations: (a) clock cycle 10; (b) clock cycle 11; (c) clock cycle 14; and (d) clock cycle 15.

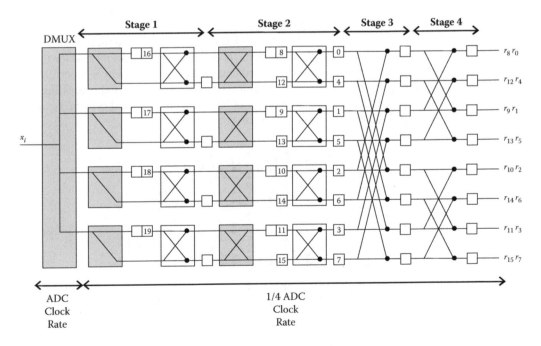

FIGURE 12-18 Parallel pipelined 16-point FFT architecture.

12.4.3 Bit-Level Systolic Array Methodology

This section describes a bit-level systolic array methodology developed for massively parallel signal processors at Lincoln Laboratory. This methodology is based on the fully efficient bit-level systolic array architecture (Wang, Wei, and Chen 1988), which is applicable to the development of filters, inner product computations, QR decomposition, polyphase FIR filters, nonlinear filters, etc. Lincoln Laboratory has applied this methodology to design VLSI signal processors that deliver very high computational throughput (e.g., thousands of GOPS) with very low power (e.g., a few watts) and very small form factors (Song 1994).

Figure 12-19 shows an example architecture of a FIR filter created with the bit-level systolic array methodology, the details of which can be found in Wang, Wei, and Chen (1988). This filter architecture demonstrates the benefits of this methodology, which are direct results of systolic array properties. Among these benefits, the most significant one is that the nonrecurring expenses (NRE) of full-custom systolic arrays are significantly lower than the NRE of general circuits.

The systolic array FIR filter is built using a few types of simple one-bit processors. A very high computational throughput can be achieved by putting tens of thousands of simple one-bit signal processors on a single die. In addition, the architecture is very well suited for VLSI implementation since it has a highly regular structure and utilizes primarily nearest-neighbor interconnections. As there are only a small number of simple one-bit processors to be designed, the designer can use custom design techniques (see Chapter 9) to optimize speed, area, and power consumption. Therefore, very high computational throughput and low power consumption can be achieved simultaneously.

Another important benefit of a systolic array architecture is its scalability. For example, the filter architecture, as shown in Figure 12-19, has only four 4-bit taps. However, the number of taps and number of bits can be readily changed by varying the number of rows and columns in the array. This scalability enables the automatic generation of systolic arrays with different design parameters (e.g., word size, number of taps, etc.). The regularity of systolic arrays also enables an accurate estimate of speed, area, and power consumption before the arrays' physical design is completed. This capability is invaluable in design space exploration.

A library of custom cells has been developed at MIT Lincoln Laboratory to support the bit-level systolic array methodology. Figure 12-20 summarizes the result of a study performed for this bit-level systolic array methodology. The study concluded that the design complexity and performance requirements in sensor applications are driven by a small number of kernels and processing functions. The fact that these kernels share a common set of bit-level computational cells makes it possible to perform their detailed design, optimization, and modeling.

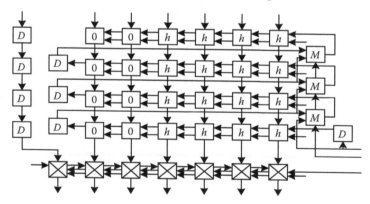

FIGURE 12-19   Bit-level systolic array FIR filter.

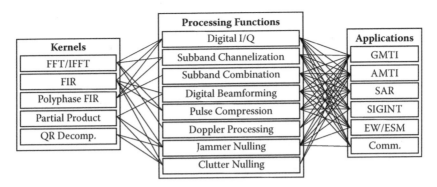

FIGURE 12-20   Kernels and processing functions for sensor applications.

12.5   SUMMARY

The development in VLSI technology has revolutionized the design of front-end processors in sensor applications. This chapter discussed the design methodology of high performance, low-cost, application-specific array processors, which often use systolic architectures to explore the parallelism of an application to gain performance. In contrast to general-purpose parallel computers, which have a number of relatively powerful processors, systolic array processors typically employ the extensive parallel processing and pipelining of low-level processing elements to sustain high throughput. Two important properties justify the use of a full-custom design process to design systolic array processors: (1) only a few different types of processing elements, which are usually small and simple, are required and (2) the simple and regular interconnections between processing elements, most of which are nearest-neighbor interconnections, simplify the placement and routing process.

REFERENCES

Horn, R.A. and C.R. Johnson. 1985. *Matrix Analysis.* Cambridge, U.K.: Cambridge University Press.

Jackson, P., C. Chan, C. Rader, J. Scalera, and M. Vai. 2004. A systolic FFT architecture for real time FPGA Systems. *Proceedings of the Eighth Annual High Performance Embedded Computing Workshop.* MIT Lincoln Laboratory, Lexington, Mass. Available online at http://www.ll.mit.edu/HPEC/agenda04.htm.

Kung, H.T. and C.E. Leiserson. 1978. Systolic arrays (for VLSI). *Sparse Matrix Proceedings 1978* 256–282. Philadelphia: Society for Industrial and Applied Mathematics (SIAM).

Kung, S.Y. 1988. *VLSI Array Processors.* Englewood Cliffs, N.J.: Prentice Hall.

McWhirter, J.G., R.L. Walke, and J. Kadlec. 1995. Normalised Givens rotations for recursive least squares processing. *IEEE Signal Processing Society Workshop on VLSI Signal Processing VIII* 323–332.

Nguyen, H., J. Haupt, M. Eskowitz, B. Bekirov, J. Scalera, T. Anderson, M. Vai, and K. Teitelbaum. 2005. High-performance FPGA-based QR decomposition. *Proceedings of the Ninth Annual High Performance Embedded Computing Workshop.* MIT Lincoln Laboratory, Lexington, Mass. Available online at http://www.ll.mit.edu/HPEC/agendas/proc05/agenda.html.

Song, W.S. 1994. VLSI bit-level systolic array for radar front-end signal processing, *Conference Record of the 28th Asilomar Conference on Signals, Systems and Computers* 2: 1407–1411.

Van Veen, B.D. and K.M. Buckley. 1988. Beamforming: a versatile approach to spatial filtering. *IEEE ASSP Magazine* 5(2): 4–24.

Walke, R. 2002. Adaptive beamforming using QR in FPGA. *Proceedings of the Sixth Annual High Performance Embedded Computing Workshop.* MIT Lincoln Laboratory, Lexington, Mass. Available online at http://www.ll.mit.edu/HPEC/agendas/agenda02.html.

Wang, C.-L., C.-H. Wei, and S.-H. Chen. 1988. Efficient bit-level systolic array implementation of FIR and IIR digital filters. *IEEE Journal on Selected Areas in Communications* 6(3): 484–493.

Section IV

Programmable High Performance Embedded Computing Systems

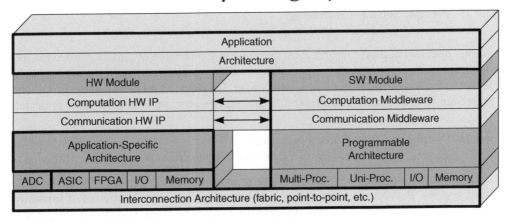

Chapter 13 Computing Devices
Kenneth Teitelbaum, MIT Lincoln Laboratory

This chapter presents embedded computing devices, present and future, with a focus on the attributes that differentiate embedded computing from general-purpose computing. Common metrics and methods used to compare devices are discussed. The fast Fourier transform is used as an illustrative example. Software programmable processors and field programmable gate arrays are surveyed and evaluated.

Chapter 14 Interconnection Fabrics
Kenneth Teitelbaum, MIT Lincoln Laboratory

This chapter discusses technologies used to interconnect embedded processor computing devices. The anatomy of a typical interconnection fabric and some simple topologies are covered. Network bisection bandwidth, its relation to total exchange problem, and an illustrative fast Fourier transform example are covered. Networks that can be constructed from multiport switches are described. The VXS standard is presented as an emerging high performance interconnect standard.

Chapter 15 Performance Metrics and Software Architecture
Jeremy Kepner, Theresa Meuse, and Glenn E. Schrader, MIT Lincoln Laboratory

This chapter presents HPEC software architectures and evaluation metrics. A canonical HPEC application (synthetic aperture radar) is used to illustrate basic concepts. Different types of parallelism are reviewed, and performance analysis techniques are discussed. A typical programmable multicomputer is presented, and the performance trade-offs of different parallel mappings on this computer are explored using key system performance metrics. The chapter concludes with a discussion of the impact of different software implementations approaches.

Chapter 16 Programming Languages
James M. Lebak, The MathWorks

This chapter examines programming languages for high performance embedded computing. First, principles of programming embedded systems are discussed, followed by a review of the evolution of programming languages. Specific languages used in HPEC systems are described. The chapter concludes with a comparison of the features and popularity of the various languages.

Chapter 17 Portable Software Technology
James M. Lebak, The MathWorks

This chapter discusses software technologies that support the creation of portable embedded software applications. First, the concept of portability is explored, and the state of the art in portable middleware technology is surveyed. Then, middleware that supports portable parallel and distributed programming is discussed, and advanced techniques for program optimization are presented.

Chapter 18 Parallel and Distributed Processing
Albert I. Reuther and Hahn G. Kim, MIT Lincoln Laboratory

This chapter discusses parallel and distributed programming technologies for high performance embedded systems. Parallel programming models are reviewed, and a description of supporting technologies follows. Illustrative benchmark applications are presented. Distributed computing is distinguished from parallel computing, and distributed computing models are reviewed, followed by a description of supporting technologies and illustrative application examples.

Chapter 19 Automatic Code Parallelization and Optimization
Nadya T. Bliss, MIT Lincoln Laboratory

This chapter presents a high-level overview of automated technologies for taking an embedded program, parallelizing it, and mapping it to a parallel processor. The motivation and challenges of code parallelization and optimization are discussed. Instruction-level parallelism is contrasted to explicit parallelism. A taxonomy of automatic code optimization approaches is introduced. Three sample projects, each in a different area of the taxonomy, are highlighted.

13 Computing Devices

Kenneth Teitelbaum, MIT Lincoln Laboratory

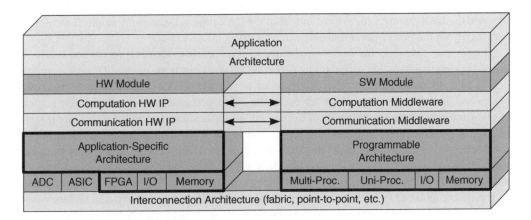

This chapter presents embedded computing devices, present and future, with a focus on the attributes that differentiate embedded computing from general-purpose computing. Common metrics and methods used to compare devices are discussed. The fast Fourier transform is used as an illustrative example. Software programmable processors and field programmable gate arrays are surveyed and evaluated.

13.1 INTRODUCTION

This chapter continues the discussion of the anatomy of a programmable high performance embedded computer, focusing on computing devices—the computational engines that drive these systems.

One of the earliest examples of a programmable embedded computing device is the Whirlwind computer—the first real-time embedded digital computer built by MIT Lincoln Laboratory ca. 1952, as part of the SAGE (Semiautomatic Ground Environment) air defense system (Everett 2001). It tracked radar data from multiple sites, displayed an integrated air picture, and calculated intercept courses for fighter aircraft. Whirlwind had a 16-bit word length, was capable of executing about 50,000 operations per second, and had 16 Kb of magnetic core memory. It comprised approximately 10,000 vacuum tubes, filling several large rooms, and had a mean time between failures of several hours.

Since Whirlwind, with the development of the transistor, integrated circuits, and the microprocessor, technology has brought embedded computing devices to a new era. Fueled by more than 40 years of exponential growth in the capacity of microelectronic devices, modern microprocessors can execute tens of billions of floating-point arithmetic operations per second and can store millions of bytes of information—all on a single chip less than half the size of a deck of playing cards. As a result, embedded computing devices have become pervasive. Today, we find computers embedded in cell phones, automobile engines, automated teller machines, entertainment systems, home appliances, medical devices and imaging systems, avionics, navigation, and guidance systems, as well as radar and sonar sensor systems.

Embedded computing devices have some unique constraints that they inherit from the systems in which they are embedded. The programs executing on these devices must meet real-time constraints. They must interface to the system in which they are embedded, with high-speed input/output (I/O) often a pressing concern. These systems may also be severely constrained in form factor and available power. A signal processor for an unmanned aerial vehicle (UAV)-borne radar, for example, must be able to input streaming sensor data, process the incoming data in real time to a set of target reports, and downlink the data to the ground. The radar signal processor (along with the radar and the UAV avionics suite) must fit within the UAV payload space and weight constraints and operate only on the power generated on board the UAV.

This chapter discusses the anatomy of embedded computing devices, present and future, with a focus on the attributes that differentiate embedded computing from general-purpose computing. Section 13.2 discusses common metrics in order to provide a framework for the comparison of devices and describes a methodology for evaluating microprocessor performance in the context of the real-time computing requirements of the embedded application. A simple fast Fourier transform (FFT) example is used for illustration. Section 13.3 looks at some current computing devices that have been used in embedded applications, focusing on their constituent components (e.g., vector processors and on-chip memory caches) and their ability to deliver computational performance to embedded applications. Finally, Section 13.4 focuses on the future of embedded computing devices and the emerging trend to exploit the ever-growing number of transistors that can be fabricated on a chip to integrate multiple computing devices on a single chip—possibly leading one day to the availability of complete multicomputer systems on a chip.

13.2   COMMON METRICS

As a prelude to any discussion of commercial off-the-shelf (COTS) computing devices for embedded applications, it is necessary to consider the standards by which the suitability of these devices to their intended application will be assessed. To foster a quantitative assessment, a set of metrics is required. The metric most commonly used is the computation rate, and for signal processing applications it is typically expressed as the number of arithmetic operations to be executed per unit time (rather than the number of computer instructions per second). Typical units are MOPS (millions of operations per second) or GOPS (billions of operations per second). When these arithmetic operations are to be executed using floating-point arithmetic, the metrics are expressed as MFLOPS (millions of floating-point operations per second) or GFLOPS (billions of floating-point operations per second).

13.2.1   ASSESSING THE REQUIRED COMPUTATION RATE

Implementation of any embedded application typically begins with a determination of the required computation rate by analysis of the algorithm(s) to be executed, a process sometimes referred to as "counting FLOPS." As an example of the process, consider the problem of executing an FFT on a continuous stream of complex data. An N-point (N a power of two) radix-2 FFT consists of $(N/2)\log_2 N$ butterflies, each of which requires four real multiplies and six real additions/subtractions—a total of 10 real arithmetic operations per butterfly. The N-point FFT thus requires a total of $5N\log_2 N$ real arithmetic operations. If the data have been sampled at a rate F_s (samples per second), it takes time $T = N/F_s$ (seconds) to collect N points to process. To keep up with the input data stream, it is necessary to execute one N-point FFT every T seconds. The resulting computation rate required to process the incoming data is simply the number of operations to be executed divided by the time available for computation: $5F_s\log_2 N$ (real arithmetic operations per second). The computation rate in this simple example is, of course, dependent on the sampling rate and the size of the transform—the faster the sampling rate, or the larger the transform size, the greater the required computation rate. For a sampling rate of 100 MSamples/s, for example, and a 1024 (1K) point FFT, the resulting required computation rate would be 5 GOPS. The presence of multiple parallel channels of data would push the required computation rate proportionally higher.

More complex algorithms are typically attacked by breaking the algorithm into constituent components, or computational kernels, and estimating the computation rates for the individual kernels, multiplying by the number of times each kernel needs to be executed, and summing over all of the various kernels. Spreadsheets are valuable tools for this process. A more detailed treatment of estimating the workloads for commonly used signal processing kernels can be found in Arakawa (2003).

13.2.2 QUANTIFYING THE PERFORMANCE OF COTS COMPUTING DEVICES

Once the required computation rate for real-time implementation of a particular algorithm has been assessed, the next step is to consider the capability of the computing device(s) that will execute the algorithm. There are several relevant metrics to consider.

Peak Computation Rate: In order to get an upper bound on achievable performance, the peak computation rate can be calculated by multiplying the number of arithmetic operations that the processor can execute each clock cycle by the maximum clock rate at which the processor can operate. For example, if a hypothetical processor could execute four floating-point arithmetic operations per clock cycle and ran at a maximum clock rate of 1 GHz, its peak computation rate would be 4 GFLOPS. This is the maximum theoretical performance that the processor could achieve if it were able to keep all of its arithmetic units fully occupied on every clock cycle—a practical impossibility. This is the number most often specified on manufacturer data sheets.

Sustained Computation Rate: The actual performance that can be achieved is, of course, dependent on algorithm, processor architecture, and the compiler used, and must be established through benchmarking. Taking the number of arithmetic operations to be executed by the benchmark and dividing by the benchmark execution time provides a useful measure of the sustained computation rate. Note that only arithmetic operations have been considered here. Overhead, such as data movement, array index calculation, looping, and branching, has not been included in either the calculation of the sustained computation rate or the required computation rate. As a result, the sustained computation rate is often significantly less than the peak computation rate, but provides an accurate assessment of achievable performance by a particular processor on a particular algorithm.

As an example, consider a hypothetical benchmark that computed 10,000 1K-point FFTs in 0.5 s. The number of arithmetic operations to be executed can be calculated using the formula in the preceding section and is equal to $512*10^6$. Dividing by 0.5 s yields a sustained computation rate of approximately 1 GFLOPS. Benchmarks are often provided by manufacturers for common kernels like the FFT, but the user will typically have to benchmark less common kernels to get an accurate assessment of algorithm performance on a particular machine.

Achievable Efficiency: The ratio of the sustained computation rate to the peak computation rate is often referred to as the achievable efficiency. In the previous example, with the FFT benchmark sustaining 1 GFLOPS on a 4 GFLOPS (peak) processor, the achievable efficiency would be 25%.

Power Efficiency: In an embedded application, in which power and space are at a premium, the power efficiency—obtained by dividing the computation rate by the power consumed—is often a useful metric. It not only permits estimation of the amount of power that will be consumed by an embedded operation, but also has a strong impact on the physical size of the embedded system, given typical cooling constraints.

Communication-to-Computation Ratio: Real-time embedded signal processing systems are often concerned with processing streaming data from an external source, and their ability to keep up with the I/O stream is a function of both the processor and the algorithm's balance between computation and I/O. Consider the simple radix-2 FFT example above. An N-point FFT (N a power of 2) requires N input samples and N output samples to perform $5N\log_2 N$ real arithmetic operations.

If we assume single-precision floating-point arithmetic, each complex sample will require 8 bytes to represent. Expressed as a simple ratio of communication-to-computation, the FFT will require $16/(5\log_2 N)$ bytes of I/O for every arithmetic operation executed. For a 1K-point FFT this is 0.32 bytes/operation, and it decreases slowly ($O\{\log_2 N\}$) with increasing N. If the computing hardware cannot supply data at this rate, the processor will stall waiting for data and the achieved efficiency will suffer.

13.3   CURRENT COTS COMPUTING DEVICES IN EMBEDDED SYSTEMS

COTS processors for embedded computing can be grouped into two general categories: general-purpose microprocessors that owe their legacy to desktop and workstation computing, and the group of digital signal processor (DSP) chips that are essentially special-purpose microprocessors that have been optimized for low-power real-time signal processing. The first group of microprocessors can be further subdivided into those microprocessors that owe their legacy to the Intel architecture (produced by both Intel and AMD) and the IBM personal computer; those processors that owe their legacy to the IBM/Motorola PowerPC architecture and the Apple MacIntosh computers; and other architectures such as the MIPS processor and Sun UltraSPARC. The DSPs can be further divided on the basis of whether they support both integer and floating-point or integer-only arithmetic. Of the floating-point DSPs, Analog Devices' SHARC (Super Harvard ARchitecture Computer) family and Texas Instruments' TMS320C67xx family are widely used. A detailed comparison of COTS processors for embedded computing is included in Table 13-1.

TABLE 13-1

COTS Microprocessor Comparison

	Freescale PowerPC 7447A (low power)	IBM PowerPC 970FX (G5)	Intel Pentium M	Intel Pentium 4	Analog Devices' Tiger SHARC TS-201	Texas Instruments TMS 320c6727
Word Length (bits)	32	64	IA-32	EM64T	32	
Process (nm)	90	90	90	90		130
Clock (GHz)	1.2	2.2	2	3.4	0.6	0.3
FLOPS per Clock (single precision)	8	8	4	4	6	6
Peak GFLOPS	9.6	17.6	8	13.6	3.6	1.8
FLOPS per Clock (double precision)	2	4	2	2	n/a	
Peak GFLOPS	2.4	8.8	4	6.8	n/a	
Core Voltage	1.1	1.2	1.26–1.356	1.2–1.3375	1.2	1.4
Power (W)	9.3	56	27	86	3.1	1.6
Power Efficiency (MFLOPS/W)	1032	314	296	132	1145	1146
L1 Instruction Cache (KB)	32	64	12	12	n/a	4
L1 Data Cache (KB)	32	32	16	16	n/a	4
L2 Data Cache (KB)	512	512	2048	2048	n/a	256
On-Chip DRAM (KB)	n/a	n/a	n/a	n/a	3072	n/a
Bus Speed (MHz)	166	1350	533	800		100
Maximum Data Rate (off chip) (Mb/s)	1328	10,800	4264	6400	1000	400
Bytes/OP	0.14	0.63	0.53	0.42	0.28	0.22

Note: Data for this table were obtained from documentation from the respective manufacturers; for more information see the manufacturers' websites.

13.3.1 GENERAL-PURPOSE MICROPROCESSORS

Modern microprocessors are typically superscalar designs, capable of executing multiple instructions simultaneously, in a heavily pipelined fashion. Multiple functional units are employed, including integer arithmetic logic units (ALUs), floating-point units, and vector processing units, with parts of different instructions executing simultaneously on each. For embedded signal processing, it is typically the vector processing unit that executes the bulk of the real-time algorithm, while the integer units are occupied with auxiliary calculations such as array indexing.

13.3.1.1 Word Length

Since the first 4-bit microprocessors, microprocessor word lengths have increased with decreasing feature sizes. Longer word lengths bring two principal benefits: improved arithmetic precision for numerical calculations and larger addressable memory spaces. During the 1980s, when 16-bit Intel processors were the workhorses of IBM PCs, MS-DOS was limited to accessing 640K of memory. While, initially, this seemed like more memory than could practically be used, eventually it became a severe limitation, handicapping the development of software applications for the IBM PC. Development of 32-bit Intel processors (the IA-32 architecture) increased the amount of directly addressable memory to 4 GB. The newest Intel processors are 64-bit machines. The Xeon and some Pentium processors employ Intel's EM64T technology, which is a set of 64-bit extensions to the IA-32 architecture originally developed by AMD for the AMD64 processor. The EM64T machines are capable of addressing up to 1 TB of physical memory. Intel's Itanium and Itanium-2 processors employ a fundamentally new 64-bit architecture (IA-64) and are capable of addressing up to 1 petabyte (1024 TB) of memory. In the PowerPC family, the newer G5 (fifth generation) processor, the IBM970 FX, is capable of addressing 4 TB of physical memory.

13.3.1.2 Vector Processing Units

Most general-purpose microprocessors offer some type of vector processing unit. The PowerPC vector processing unit is called Altivec by Freescale (a subsidiary of Motorola spun off to provide embedded processing to the automotive, networking, and wireless communication industries), VMX (Vector/SIMD Multimedia eXtension) by IBM, and the Velocity Engine by Apple. The Altivec is a single-instruction multiple-data (SIMD) stream vector processor that operates on 128-bit vectors that can be treated as a vector of four single-precision floating-point values, eight 16-bit integer values, or sixteen 8-bit integer values. Vector operations generally operate in parallel on all elements of a vector (i.e., a vector add of 16-bit integers performs eight parallel additions on each of the eight 16-bit elements of two 128-bit vectors, writing the result to a third 128-bit vector) although some instructions can operate within a single vector (such as summation of all of the elements of a vector, useful in computing vector dot products). Altivec is capable of performing simultaneous vector multiply and add, and can, therefore, compute a maximum of eight floating-point operations each clock cycle, sixteen 16-bit integer operations each clock, or thirty-two 8-bit integer operations each clock.

Intel and AMD have a similar capability with Intel's Streaming SIMD Extensions (SSE) unit. The SSE2 instructions have added the capability to support double-precision arithmetic on vectors of two 64-bit floating-point numbers, and the SSE3 instructions provided support for intravector arithmetic such as summation of the elements of a vector. Unlike in the Altivec, however, multiply and add cannot be executed simultaneously, and so the SSE vector processor can execute a maximum of four floating-point operations per cycle, two double-precision floating point operations per cycle, eight 16-bit integer operations per cycle, and sixteen 8-bit integer operations per cycle.

13.3.1.3 Power Consumption versus Performance

The power consumption, P, of a complementary metal oxide semiconductor (CMOS) device is generally given as $P = CfV^2$, where C is the gate capacitance, f is the clock frequency, and V is the supply voltage. The clock frequency is generally determined by some fixed number of gate delays, and, since the gate delay is inversely proportional to the supply voltage, the clock frequency will generally be directly proportional to the supply voltage. The result is a cubic dependence of power on supply voltage; lowering the supply voltage (and correspondingly the clock frequency) is a common approach to producing lower-power (but also slower) devices for power-sensitive applications. The low-power version of the PowerPC 7447A and the low-voltage Xeon are examples using this practice. Lowering the core voltage of the 7447A slightly from 1.3 V to 1.1 V reduces the power consumption by approximately half, from 19 W to 9.3 W. It also reduces the clock frequency from 1.5 GHz to 1.2 GHz, resulting in a decrease in the peak computation rate from 12 GFLOPS to 9.6 GFLOPS. Although the peak computation rate has decreased, the power efficiency (the peak computation rate normalized by power consumption) has actually increased from 632 MFLOPS/W to 1032 MFLOPS/W.

Many newer devices designed for mobile computing applications dynamically exploit this relationship between power, supply voltage, and clock frequency. In these devices, the processor can, under software control, change supply voltage and clock speed from a range of predetermined values to optimize power utilization. Intel's SpeedStep technology (Intel 2004) is an example of this, and the Pentium M, for instance, can vary core voltage and clock frequency, resulting in variable power consumption over a range of 6 W–24.5 W (Intel 2007). A Pentium M-based notebook computer, for example, might sense when it was operating on battery versus AC and lower its core voltage, and correspondingly its clock frequency, effectively sacrificing some performance for extended battery life. For embedded applications that are constrained to run off battery power, active power management techniques of the type discussed here have potential application. When the workload is essentially fixed and there are hard real-time constraints (e.g., processing streaming sensor data), one could imagine optimizing battery life by turning the power down to the point at which performance was just adequate to meet real-time requirements. At the opposite end of the spectrum, when the workload is variable (e.g., post-detection processing and tracking of sensor contacts), the processor could adjust its power/performance operating point based on demand, thus conserving battery life in the process.

13.3.1.4 Memory Hierarchy

Operands for the vector processor (and other on-chip processing units) and results computed by the vector processor are typically stored in memory. In order to achieve high efficiency, it is necessary to keep the processing units busy, in turn requiring that memory data access rates must be high and memory latency must be low. Since this is difficult to achieve for large amounts of memory, most modern microprocessors employ a memory hierarchy that consists of modestly sized blocks of high-bandwidth, low-latency on-chip memory coupled with larger blocks of lower-bandwidth, longer-latency off-chip memory. Often (but not always), on-chip memory is a cached version of off-chip memory, with a copy of the most recently accessed memory locations being stored on chip. When an operand is required that has been previously accessed and is still in cache (a cache hit), it can be fetched quickly. If the operand is not in cache (a cache miss), it must be fetched from off-chip memory—a much slower process. When computed results are written to memory, they are typically written to the cache and replace a memory location previously cached in the event of a cache miss. Multilevel caches are usually employed with smaller L1 (level 1) caches for both instructions and data, and larger L2 (level 2) data caches typically on chip. Some microprocessors have L3 caches as well, either on or off chip.

Caching schemes work best when there is substantial locality of reference (i.e., memory locations are accessed repeatedly in a short time interval). The efficiency achievable by embedded processing algorithms can be exquisitely sensitive to both the amount of cache memory and the fraction of accesses that come from cache versus external memory (the cache hit rate).

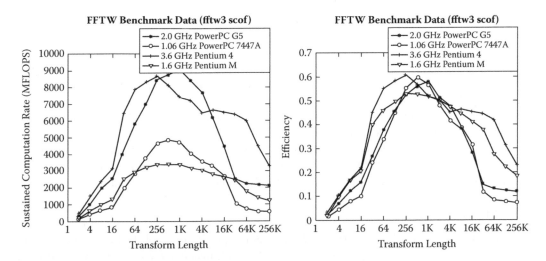

FIGURE 13-1   Single-precision complex FFT benchmark data (FFTW).

13.3.1.5   Some Benchmark Results

An examination of published FFT benchmark data can illustrate some of the points discussed in this section. The "Fastest Fourier Transform in the West" (FFTW) benchmark is a free software benchmark that efficiently implements the discrete Fourier transform (DFT), taking advantage of the vector units (e.g., Altivec, SSE/SSE2) on some processors. The benchmark software as well as the benchmark results for a variety of processors can be found on the web (http://www.fft.org), and a more detailed description of the benchmark and results can be found in Frigo and Johnson (2005). A selected subset of the benchmark results is included in Figure 13-1 (single-precision complex FFT) and Figure 13-2 (double-precision complex FFFT). The results are plotted both as sustained MFLOPS and normalized by the peak computation rate to get efficiency as a function of the transform length for four different processors (PowerPC G5, PowerPC 7447A, Pentium 4, Pentium M).

The principal differences in raw MFLOPS are directly attributable to differences in clock frequency and the number of arithmetic operations per clock. When normalized to calculate efficiency, the (single-precision) results for all four processors are virtually identical. They exhibit bell-shaped curves with a relatively narrow range of optimum efficiency, typically around 50%–60% for single-precision, at approximately 1K point transforms. Efficiency for small FFTs is relatively poor because

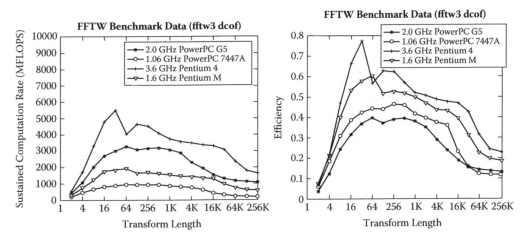

FIGURE 13-2   Double-precision complex FFT benchmark data (FFTW).

of the overhead of the subroutine call compared to the small number of arithmetic operations that need to be executed. Efficiency for large FFTs is poor because of cache issues. The PowerPC processors here have L2 cache sizes of 512 KB, which would store 32K complex samples. Since an out-of-place N-point transform would typically keep two N-point vectors in memory at once, the largest FFT that could remain completely in cache would be 16K points. Beyond this transform length, the benchmark data shows expectedly poor efficiency. The Intel processors have larger caches, 2 MB, that would hold the data for a 64K point transform. As a result, the central region of efficient operation in the benchmark data is a little wider for the Intel processors.

The double-precision data show somewhat greater variability between processors, possibly because the SSE2 vector unit in the Intel processors can operate on double-precision data while the Altivec cannot and the PowerPC processors must rely on their floating-point units.

13.3.1.6　Input/Output

Access to external (off-chip) memory and I/O devices is typically accomplished via the use of a chip set that interfaces to the microprocessor's bus, and it is ultimately the bus clock speed and width (bytes) that determine the maximum data transfer rate in and out of the microprocessor. Since the bus clock speeds are typically related to the processor clock frequency, in theory, faster processors should be able to move data in and out of the processors more quickly. It is the balance between I/O and processing here that is of greatest interest. The communication-to-computation ratios for some COTS microprocessors are tabulated in Table 13-1, ranging from about 0.14 for the 7447A up to about 0.63 for the 970FX. Compare these data to the previously presented simple FFT example in which we needed to supply about 0.34 bytes of I/O for every arithmetic operation executed in order to stream data in and out of a 1K point FFT.

13.3.2　Digital Signal Processors

Compared to general-purpose microprocessors, the class of DSP chips are optimized for power-sensitive embedded signal processing applications. DSP clock rates are slower than those of their general-purpose cousins, resulting in significantly lower power (a few watts or less compared to tens of watts) at the cost of lower peak computation rates. The approach to management of on-chip memory on DSP chips is often different from that for general-purpose microprocessors. Data may not be cached, and it is left to the programmer to explicitly orchestrate the movement of data from off-chip sources (external memory, I/O) and synchronize data movement and program execution. On-chip direct memory access (DMA) controllers are typically provided to assist in this purpose. For example, while the DSP chip was executing an FFT out of on-chip memory, the DMA controller would move data from the last FFT from on-chip memory off the chip (either to an output device or to external memory) and move data for the next FFT into on-chip memory. Compared to cache-based schemes, this approach can be very efficient, but requires considerable effort from the programmer.

13.4　FUTURE TRENDS

In 1965, Gordon Moore observed that the number of transistors per integrated circuit was growing exponentially, roughly doubling every year (Moore 1965). The rate of growth has slowed, but we have maintained exponential growth over more than four decades. It is interesting to note that Moore's now famous observation, which has come to be known as Moore's Law, was made at a time when fewer than 100 functions could be integrated on a single chip. The first 4-bit microprocessor, Intel's 4004, was still five years in the future. Modern 64-bit microprocessors can have in excess of 100,000,000 transistors on a single chip—an increase of approximately six orders of magnitude.

Over the same time period, the number of arithmetic operations executed by programmable microprocessors on each clock cycle has increased by less than one order of magnitude—a com-

TABLE 13-2

International Technology Roadmap for Semiconductors Projections (2005)

	2007	2010	2013	2016	2019
Feature size (nm)	65	45	32	22	16
Millions of Transistors per Chip	1,106	2,212	4,424	8,848	17,696
Clock Frequency (MHz)	9,285	15,079	22,980	39,683	62,443
Power Supply Voltage (V)	1.1	1	0.9	0.8	0.7
Power (W)	189	198	198	198	198

paratively glacial pace. The consequence of this architectural shortfall is that the peak computation rate of microprocessors, the number of arithmetic operations executed per second, has increased at a rate primarily determined by the increase in clock frequency achievable with the continuing reduction in feature size, with only occasional improvements due to architectural innovation (e.g., short-vector extensions). The bounty of Moore's Law, the ever-increasing supply of transistors, has gone primarily into on-chip memory caches to address the memory latency issues attendant with the increasing clock frequencies. As we look toward the future, we must find a way to harvest this bounty, to scale the number of operations executed per clock period with the number of transistors and maintain exponential growth of the peak computation rate of embedded computing devices.

13.4.1 TECHNOLOGY PROJECTIONS AND EXTRAPOLATING CURRENT ARCHITECTURES

Currently, the job of analyzing trends in the semiconductor industry and forecasting future growth of the capacity and performance of integrated circuits has been taken up by a multinational group of semiconductor experts who annually publish a 15-year look ahead for the semiconductor industry known as the International Technology Roadmap for Semiconductors, or simply ITRS. This road map considers developments in lithography, interconnects, packaging, testing, and other relevant technology areas to predict the evolution of die size, number of transistors per chip, clock frequency, core voltages, power consumption, and other semiconductor attributes. A few of the ITRS 2005 projections for attributes that directly relate to microprocessor performance prediction are shown in Table 13-2 for the 2007–2012 timeframe [ITRS http://www.itrs.net/home.html]. These projections assume a new generation of lithography every three years with a reduction in feature size corresponding to a doubling of the number of transistors with every generation—about one-third the rate originally predicted by Moore. The road map also calls for a continuing increase in clock frequency and a continuing decrease in core voltage.

These projections are applied to extrapolate the microprocessor performance depicted in Figure 13-3. Perhaps the figure of merit of greatest interest here is not computation rate, but computation

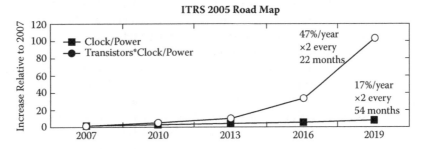

FIGURE 13-3 Extrapolating microprocessor performance.

rate normalized by power consumption—the power efficiency. To a large extent, power efficiency determines the computational resources that can be brought to bear on embedded computing problems, which are often constrained by form factor (which is limited by thermal dissipation) and available power. If we extrapolate current architectures, with the number of arithmetic operations per clock cycle held constant, the power efficiency will trend as the clock frequency for constant power. The first of the two curves in Figure 13-3 shows this trend, and it is slowly increasing at a rate of approximately 17% per year, doubling roughly every 54 months. If we assume architectural innovation, letting the number of arithmetic operations per clock cycle increase proportionally to the number of transistors available on the chip, the power efficiency will trend as the number of transistors times the clock frequency for constant power, seen in the second curve in Figure 13-3. Based on the ITRS projections, this power efficiency is increasing at a rate of 47% per year, doubling every 22 months—clearly a much more rapid pace. By the year 2019, the difference between these two curves, the potential reward for architectural innovation, is more than an order of magnitude improvement in power efficiency.

13.4.2　Advanced Architectures and the Exploitation of Moore's Law

The real question, then, is how best to scale arithmetic operations per clock with the number of transistors per device. Emerging architectures exhibit several approaches. Mainstream microprocessor vendors are now offering multiple complete processor cores on a single chip. IBM has taken a different approach with its Cell processor, which is based on a single core plus multiple VMX-like vector processors. ClearSpeed's multithreaded array processor (MTAP) is essentially a SIMD array of smaller processing elements (PEs). The Defense Advanced Research Projects Agency (DARPA) Polymorphic Computing Architectures (PCA) program is exploring tile-based architectures that can "morph" between stream-based and multithreaded paradigms based on application requirements (Vahey et al. 2006). Other approaches to enhancing embedded computing capabilities employ the use of numeric co-processors, and there has been some work trying to exploit graphics processing units (GPUs) as co-processors. In addition, some embedded processing architectures are now including field programmable gate arrays (FPGAs) in the same fabric with general-purpose microprocessors.

13.4.2.1　Multiple-Core Processors

Dual-core processors—essentially two complete microprocessors on a single chip—have become very popular, with all of the major microprocessor manufacturers now offering dual-core versions of their popular CPUs. Intel has dual-core versions of its Pentium and Xeon processors, including Pentium-D, the ultralow-voltage dual-core Xeon ULV, which is aimed at embedded applications, as well as their new Core2 Duo processor. AMD now has a dual-core Opteron as well. Freescale is offering a dual-core PowerPC, the MPC8641D, which consists of dual e600 PowerPC cores, PowerQUIC system controller, and PCI Express/serial RapidIO interfaces on one chip. The IBM PowerPC 970MP is a dual-core version of the 970FX. Quad-core processors are on the horizon as well. Intel has announced two new quad-core processors, the quad-core Xeon 5300 aimed at the server market and the quad-core Core2 Extreme processor aimed at the desktop market. Intel's quad-core processors consist of a pair of dual-core dice integrated onto a single substrate.

If we extrapolate this trend forward, it is not hard to imagine complete multicomputers on a chip, but there are issues of scalability which must be addressed first. The dual-core processors are typically interconnected on chip and share an external bus for access to I/O and memory. The Intel dual-core processors even share L2 cache (the quad-core processors share two L2 caches, one on each die, between four processors). In order to maintain the balance between computation and communication, the bus bandwidth must increase proportionally with the number of on-chip cores. Historically, however, bus bandwidths have increased much more slowly than has the number of transistors. From a programming perspective, the burden of distributing embedded applications between processors falls

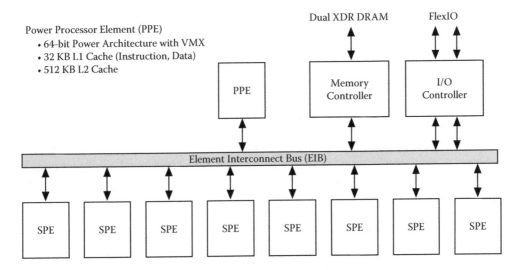

Power Processor Element (PPE)
- 64-bit Power Architecture with VMX
- 32 KB L1 Cache (Instruction, Data)
- 512 KB L2 Cache

Synergistic Processor Element (SPE)
- 128-bit SIMD (8 FLOPS/clock cycle)
- 256 KB Local Store

FIGURE 13-4   IBM Cell BE processor.

to the programmer. As the number of on-chip cores increases, pressure for the development of tools for automatic compilation to these increasingly parallel architectures will intensify.

13.4.2.2   The IBM Cell Broadband Engine

Cell BE, the processing engine of the Playstation 3 set-top game box, is the product of a joint venture between IBM, Sony, and Toshiba (Ruff 2005). Unlike the dual-core chips discussed in the previous section, which are targeted at general-purpose processing, the Cell processor is designed for high performance graphics acceleration. This is reflected in its architecture, shown in Figure 13-4. The Cell processor consists of a PowerPC core, referred to as the power processor element (PPE), with eight synergistic processor elements (SPEs), which are essentially VMX-like vector processing engines. The SPEs are interconnected via the element interconnect bus (EIB), which facilitates the exchange of data between the SPEs and the PPE, external memory, and I/O.

Like VMX, the SPE is capable of executing up to eight single-precision floating-point operations per clock. Eight SPEs operating in parallel bring the total to 64 operations per clock. At a 3 GHz clock rate, the Cell processor has a peak computation rate of 192 GFLOPS, easily an order of magnitude better than the rate for single-core microprocessors.

How well Cell's architecture fares at other compute-intensive but nongraphical embedded applications remains to be seen, but initial results are encouraging. Mercury Computer Systems has begun to offer Cell-based products and has benchmarked the cell on large FFTs with impressive results (Cico et al. 2006). Mercury Computer Systems has also investigated the performance of space-time adaptive processing (STAP), a radar signal processing algorithm, on Cell, estimating that a sustained computation rate of around 90 GFLOPS might be feasible for STAP (Cico, Greene, and Cooper 2005). MIT Lincoln Laboratory has also investigated Cell performance for common signal processing kernel functions with encouraging results (Sacco et al. 2006).

13.4.2.3   SIMD Processor Arrays

Rather than employing multiple small SIMD processors like the Cell BE, ClearSpeed uses an approach that consists of a single large SIMD array. The CSX 600 MTAP architecture consists of

96 poly-execution (PE) cores arranged in a SIMD fashion (ClearSpeed 2007; ClearSpeed 2006). The PE cores communicate with each other via a nearest-neighbor interconnection referred to as a "swazzle path." The device is implemented in 130 nm CMOS and runs at a 250 MHz clock. Each PE core can perform a single flop on each clock cycle and runs at a theoretical peak of 24 GFLOPS. Each device consumes approximately 10 W of power, yielding very high power efficiency, on the order of 2400 MFLOPS/W, more than twice that of even the most capable general-purpose microprocessors found in Table 13-1. ClearSpeed also offers an accelerator board containing two devices for a total of almost 50 GFLOPS while consuming on the order of 25 W. A parallel C compiler is available, several math libraries (e.g., BLAS, LAPACK, FFTW) are supported, and a VSIPL Core Lite library is under development (Reddaway et al. 2004). Initial benchmark results for radar pulse compression have achieved efficiencies of about 23% on a 64 PE test chip consuming less than 2 W (Cameron et al. 2003).

13.4.2.4 DARPA Polymorphic Computing Architectures

DARPA's Polymorphic Computing Architectures program is focused on developing malleable computer microarchitectures that can be adapted, or "morphed," in real time to optimize performance by changing the computer architecture to match evolving computational requirements during the course of a mission. The PCA program is developing several tiled-architecture processors, including the RAW chip (MIT), TRIPS microprocessor (University of Texas at Austin), and the MONARCH processor (Raytheon/IBM).

The MONARCH processor (Vahey et al. 2006), for example, is implemented in an IBM 90 nm process and comprises an array of 12 arithmetic clusters, 31 memory clusters, and 6 RISC (reduced instruction set computer) processors. These microarchitecture elements can be arranged to support processing in two fundamentally different modes: a streaming mode in which data flows through the arithmetic/memory clusters in a static, predetermined configuration, or a multithreaded mode as a group of multiple RISC processors each with 256-bit SIMD processing units. The processor can "morph" between the two configurations in real time. Each arithmetic cluster can execute eight multiplies and eight add/sub per clock, for a total of 192 arithmetic operations per clock, which at a 333 MHz clock rate works out to 64 GFLOPS peak. Raytheon is currently estimating 3–6 GFLOPS/W.

13.4.2.5 Graphical Processing Units as Numerical Co-processors

Graphical processing units are special-purpose application-specific integrated circuits whose function is to accelerate the graphics pipeline (the process of converting three-dimensional objects into a two-dimensional raster scan), a particularly compute-intensive aspect of computer video games. Increasingly, the shading process—the part of the graphics pipeline in which pixel color is determined from texture information and lighting—is being implemented in GPUs by processors with some limited programmability (Fernando et al. 2004). Standardized application programming interfaces (APIs) for these programmable processors are emerging, the most notable of these being OpenGL and DirectX. Some researchers have begun to experiment with using GPUs to accelerate nongraphics functions, exploiting the programmable nature of the pixel/vertex shaders on these devices. In August 2004, ACM held a workshop on this topic, the Workshop on General-Purpose Processing on Graphics Processors, in Los Angeles. Two of the central issues, of course, are these questions: How much faster, if at all, are GPUs at general-purpose computing compared to general-purpose CPUs, and is it worth the overhead of breaking up the problem and moving data in and out of the GPU in an attempt to accelerate the problem? While the answers are likely to be highly algorithm dependent, some initial results suggest that for applications like matrix multiply, which are similar to the graphics operations for which the GPUs are designed, the GPU is about 3× faster than are typical CPUs for large problem sizes (Thompson, Hahn, and Oskin 2002). The next frontier

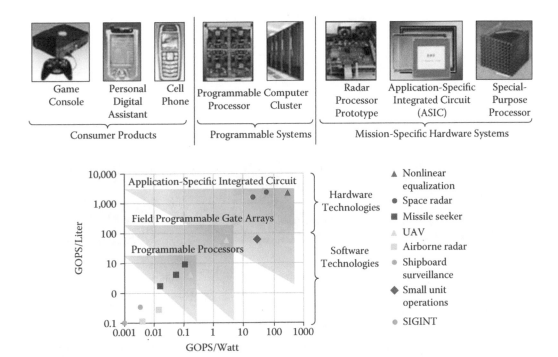

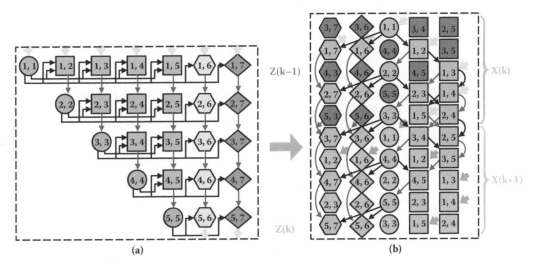

COLOR FIGURE 1-3 Embedded processing spectrum.

COLOR FIGURE 5-24 FPGA mapping for a systolic Givens QR decomposition: (a) the logical data flow and operations in the systolic array; and (b) the folded array, in which the execution of the green datagram is overlapped with the brown from the previous time frame, and orange from the next time frame.

Image Processing Pipeline

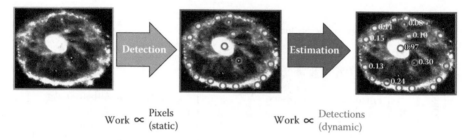

Work ∝ Pixels (static)   Work ∝ Detections (dynamic)

Static Parallel Implementation

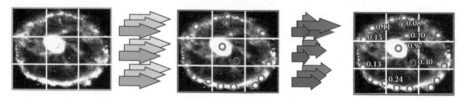

Load: balanced   Load: unbalanced

• Static parallelism implementations lead to unbalanced loads

COLOR FIGURE 5-26 Parallelism in an image processing algorithm. The algorithm first detects celestial objects in the image and then performs parameter-estimation tasks (location, size, luminance, spectral content) on the detections. The algorithm-to-architecture mapping first exploits the data parallelism in the image to perform detection. If the detections are not distributed evenly amongst the processors, a load imbalance will occur in the estimation phase.

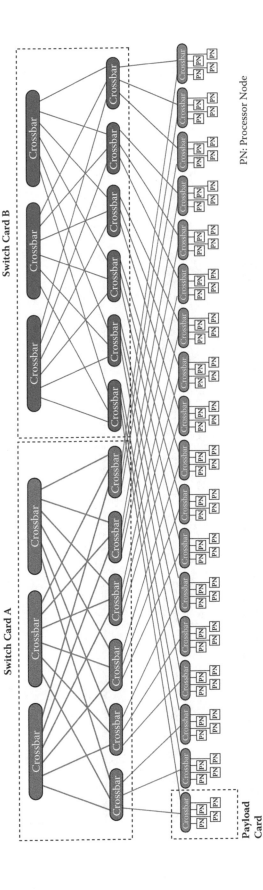

COLOR FIGURE 14-15(a)   Example three-level VXS interconnect.

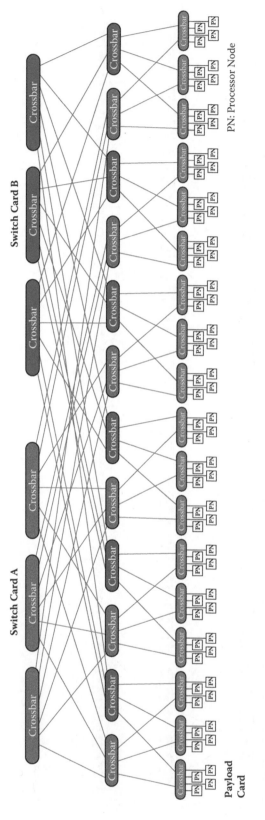

COLOR FIGURE 14-15(b) Same interconnect redrawn as least-common-ancestor network.

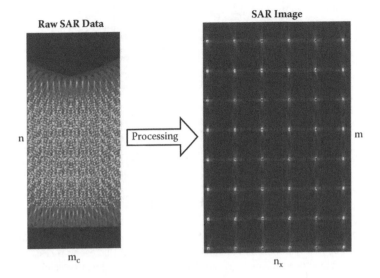

COLOR FIGURE 15-2 Unprocessed (left) and processed (right) SAR data. The area that reflects a single pulse is large and an image of this raw data is very blurry (left). A SAR system provides multiple looks at the same area of the ground from multiple viewing angles. Combining these different viewing angles together produces a much sharper image (right). (From Bader et al., Designing scalable synthetic compact applications for benchmarking high productivity computing systems, *CTWatch Quarterly* 2(4B), 2006. With permission.)

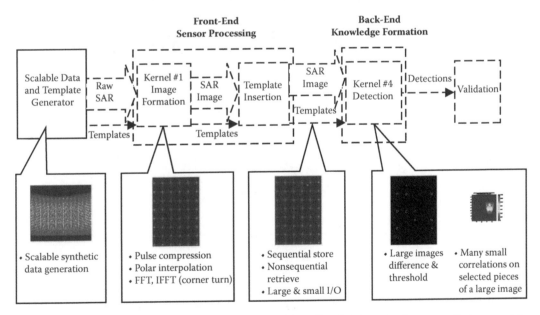

COLOR FIGURE 15-4 Compute Only mode block diagram. Simulates a streaming sensor that moves data directly from front-end processing to back-end processing. (From Bader et al., Designing scalable synthetic compact applications for benchmarking high productivity computing systems. *CTWatch Quarterly* 2(4B), 2006. With permission.)

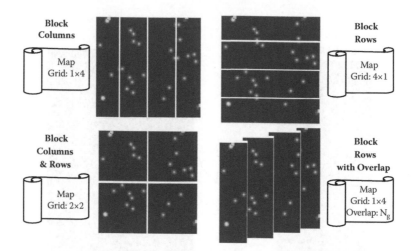

COLOR FIGURE 15-12 Global array mappings. Different parallel mappings of a two-dimensional array. Arrays can be broken up in any dimension. A block mapping means that each processor holds a contiguous piece of the array. Overlap allows the boundaries of an array to be stored on two neighboring processors.

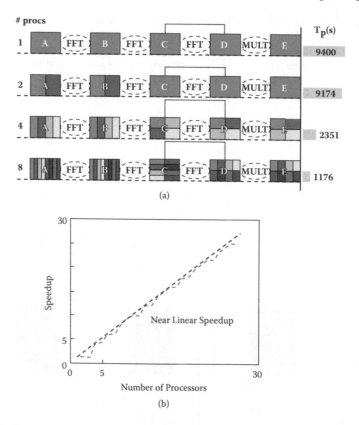

COLOR FIGURE 19-9 pMapper mapping and speedup results. These results were obtained for a low-latency architecture that would be consistent with a real-time embedded processor. Note how pMapper chooses mapping to balance communication and computation. At two processors, only arrays A and B are distributed as there is no benefit to distributing the other arrays. At four processors, C is distributed to benefit the matrix multiple operation and not the FFT operation. (From Travinin, N. et al., pMapper: automatic mapping of parallel Matlab programs, *Proceedings of the IEEE Department of Defense High Performance Computing Modernization Program Users Group Conference*, pp. 254–261. © 2005 IEEE.)

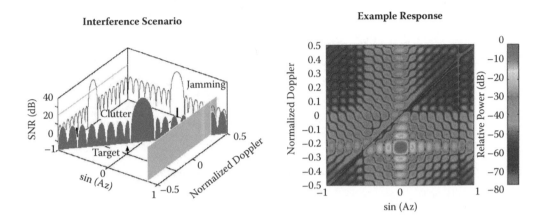

COLOR FIGURE 20-5 Space-time adaptive processing to suppress airborne clutter. (Ward 1994, reprinted with permission of MIT Lincoln Laboratory.)

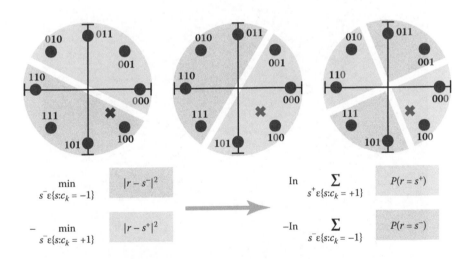

COLOR FIGURE 22-4 8-ary PSK maximum-likelihood demodulation.

COLOR FIGURE 23-8   Feature point cloud with outlier points flagged in red.

COLOR FIGURE 23-11   Highway interchange; tracked features are shown in red, moving targets are shown in green.

TABLE 13-3

FPGA Characteristics

	Virtex XCV1000	Virtex XCV3200E	Virtex IIXC2V8000	Virtex II ProXC2VP100	Virtex 4XC4VFX140 (XC4VSX55)	Virtex 5XC5VLX330T
Linewidth (nm)	220	180	150	130	90	65
Logic Slices	12,288	32,488	46,592	44,096	63,168 (24,576)	51,840
Block RAM Bits	131,072	851,968	3,024,000	7,992 Kb	9,936 Kb (384 Kb)	11,664 Kb
Dedicated Computational Blocks	n/a	n/a	168 18-bit multipliers	444 18-bit multipliers	192 (512) Xtreme DSP slices: 18-bit multiplier, accumulator, adder	192 DSP 48E slices: 25 × 18 bit multiplier, accumulator, adder
Clock Speed (MHz)	n/a	n/a	210	300	500	550
Peak GOPS	n/a	n/a	35	133	288 (768)	316

Note: Data for this table were obtained from Xilinx documentation; more information can be found at the Xilinx website.

is the development of a mathematical function library for GPUs that has a standard API. Campbell (2006) has made some progress here using the emerging VSIPL++ standard as a framework.

13.4.2.6 FPGA-Based Co-processors

Increasingly, FPGAs are finding their way into embedded processing systems. While often they are used for interfacing purposes, FPGAs can also play a valuable role off-loading certain calculations from general-purpose processors. Sometimes an FPGA used in the form of a preprocessor might execute operations on high-bandwidth data streams in order to reduce the data rate into a general-purpose programmable processor, or the FPGA might take the form of a numeric co-processor, in which a general-purpose processor might off-load computationally intensive calculations to the FPGA.

Developed initially for implementing digital logic functions, FPGAs have become extremely capable computing engines thanks to Moore's Law scaling of device parameters. Xilinx Virtex FPGAs, for example, are SRAM-based devices consisting of a sea of programmable logic cells called slices made up of 4-input look-up tables (LUTs) and registers arranged in grid fashion and interconnected by a programmable interconnect. Newer Virtex FPGAs also include memory blocks and DSP blocks, which include dedicated multipliers and adders, as part of the basic fabric. Serial I/O blocks and in some cases (Virtex-II, Virtex-II Pro) PowerPC cores are also included. Table 13-3 shows how the capability of the Virtex FPGA has evolved over time with decreasing feature sizes. In the newer devices with dedicated multiplier and DSP blocks, hundreds of arithmetic operations can be executed on each clock cycle, yielding hundreds of GOPS per device. Floating-point arithmetic can also be implemented by using multiple blocks per floating-point operation, with additional logic slices used for rounding and normalization. It is, of course, up to the FPGA designer how to implement processing functions using the basic building blocks provided, and here, despite recent advances in design tools, the process is still much more like hardware design than software programming and can be quite labor intensive. On the plus side, it is also possible to carefully optimize specific designs, resulting in very high performance, highly efficient designs that use a significantly greater fraction of the available resources than is possible with a general-purpose programmable processor.

13.5 SUMMARY

This chapter has focused on computing devices for high performance embedded computing (HPEC) applications as part of a broader discussion of the anatomy of a programmable HPEC system. These embedded computing devices have several special constraints, including the requirement to operate in real time, the need to interface (I/O) with the system in which they are embedded, form-factor constraints (they must fit within the envelope of these embedded systems), and constraints on power utilization and/or thermal dissipation.

This chapter has presented an approach for assessing real-time computing requirements by counting the number of arithmetic operations per second to be executed by the real-time algorithm and comparing this number to the sustained computation rate of the processor (the peak computation rate discounted by the achievable efficiency—a function of both the processor and the algorithm it is executing). The chapter discussed some common metrics for evaluating processor performance and looked in some detail at the anatomy of a typical COTS microprocessor. Also discussed were the SIMD vector processing engines that support the bulk of the computational workload for these processors, how the memory hierarchy can have a significant impact on achievable performance, and the importance of balancing I/O and computation in order to efficiently utilize the computational resources of the processor. For power-constrained systems, it was shown that lower-voltage, slower processors can exhibit significantly improved power efficiency (MFLOPS/W), making them attractive choices for embedded applications.

Looking toward the future, we have seen how the number of transistors per microprocessor chip evolves over time according to an exponential growth curve widely known as Moore's Law. We have also seen that continuing the historical paradigm of increasing the microprocessor clock frequency and increasing cache sizes will produce only modest future gains in microprocessor performance, and that a new paradigm, which applies these transistors to increasing the number of arithmetic operations per clock cycle, is needed in order to continue aggressive scaling of computation rate from generation to generation. Several emerging approaches have been discussed, ranging from putting multiple complete microprocessor cores on a single chip or increasing the number of SIMD/vector units per chip, all the way to tiled or SIMD processor array architectures employing nearly 100 processors per chip. These new architectures will require advances in the development of software tools for efficient, automatic parallelization of software in order to become truly effective and widely accepted.

REFERENCES

Arakawa, M. 2003. *Computational Workloads for Commonly Used Signal Processing Kernels*. MIT Lincoln Laboratory Project Report SPR-9. 28 May 2003; reissued 30 November 2006.

Cameron, K., M. Koch, S. McIntosh-Smith, R. Pancoast, J. Racosky, S. Reddaway, P. Rogina, and D. Stuttard. 2003. An ultra-high performance architecture for embedded defense signal and image processing applications. *Proceedings of the Seventh Annual High Performance Embedded Computing Workshop*. MIT Lincoln Laboratory, Lexington, Mass. Available online at http://www.ll.mit.edu/HPEC/agenda03.htm.

Campbell, D. 2006. VSIPL++ acceleration using commodity graphics processors. *Proceedings of the Tenth Annual High Performance Embedded Computing Workshop*. MIT Lincoln Laboratory, Lexington, Mass. Available online at http://www.ll.mit.edu/HPEC/agendas/proc06/agenda.html.

Cico, L., J. Greene, and R. Cooper. 2005. Performance estimates of a STAP benchmark on the IBM Cell processor. *Proceedings of the Ninth Annual High Performance Embedded Computing Workshop*. MIT Lincoln Laboratory, Lexington, Mass. Available online at http://www.ll.mit.edu/HPEC/agendas/proc05/agenda.html.

Cico, L., R. Cooper, J. Greene, and M. Pepe. 2006. Performance benchmarks and programmability of the IBM/Sony/Toshiba cell broadband engine processor. *Proceedings of the Tenth Annual High Performance Embedded Computing Workshop*. MIT Lincoln Laboratory, Lexington, Mass. Available online at http://www.ll.mit.edu/HPEC/agendas/proc06/agenda.html.

ClearSpeed Technology. 2007. CSX processor architecture, white paper. Available online at http://www.clearspeed.com/docs/resources/ClearSpeed_Architecture_Whitepaper_Feb07v2.pdf.

ClearSpeed Technology. 2006. CSX600 datasheet. Available online at http://www.clearspeed.com/docs/resources/CSX600_Product_Brief.pdf.

Everett, R.R. 2001. *Building the SAGE System—The Origins of Lincoln Laboratory.* MIT Lincoln Laboratory Heritage Lecture Series.

Fernando, R., M. Harris, M. Wloka, and C. Zeller. 2004. Programming graphics hardware. The Eurographics Association, Aire-la-ville, Switzerland.

Frigo, M. and S.G. Johnson. 2005. The design and implementation of FFTW3. *Proceedings of the IEEE* 93(2): 216–231.

Intel Corporation. 2004. Enhanced Intel SpeedStep technology for the Intel Pentium M processor. Intel white paper.

Intel Corporation. 2007. Intel Pentium 4 processors for embedded computing—overview. Available online at http://www.intel.com/design/intarch/pentium4/pentium4.htm.

International Technology Roadmap for Semiconductors, 2005 edition. Executive summary. Available online at http://www.itrs.net/home.html.

Moore, G. 1965. Cramming more components onto integrated circuits. *Electronics* 38(8).

Reddaway, S., B. Atwater, P. Bruno, D. Latimer, R. Pancoast, P. Rogina, and L. Trevito. 2004. Hardware benchmark results for an ultra-high performance architecture for embedded defense signal and image processing applications. *Proceedings of the Eighth Annual High Performance Embedded Computing Workshop.* MIT Lincoln Laboratory, Lexington, Mass. Available online at http://www.ll.mit.edu/HPEC/agenda04.htm.

Ruff, J.F. 2005. Cell broadband engine architecture and processor. IBM Systems and Technology Group, Austin, Tex. Formerly available online at the IBM website.

Sacco, S.M., G. Schrader, J. Kepner, and M. Marzilli. 2006. Exploring the Cell with HPEC Challenge benchmarks. *Proceedings of the Tenth Annual High Performance Embedded Computing Workshop.* MIT Lincoln Laboratory, Lexington, Mass. Available online at http://www.ll.mit.edu/HPEC/agendas/proc06/agenda.html.

Thompson, C.J., S. Hahn, and M. Oskin. 2002. Using modern graphics architectures for general-purpose computing: a framework and analysis. *Proceedings of the 35th Annual IEEE/ACM International Symposium on Microarchitecture* 306–317.

Vahey, M., J. Granacki, L. Lewins, D. Davidoff, J. Draper, G. Groves, C. Steele, M. Kramer, J. LaCoss, K. Prager, J. Kulp, and C. Channell. 2006. MONARCH: a first generation polymorphic computing processor. *Proceedings of the Tenth Annual High Performance Embedded Computing Workshop.* MIT Lincoln Laboratory, Lexington, Mass. Available online at http://www.ll.mit.edu/HPEC/agendas/proc06/agenda.html.

14 Interconnection Fabrics*

Kenneth Teitelbaum, MIT Lincoln Laboratory

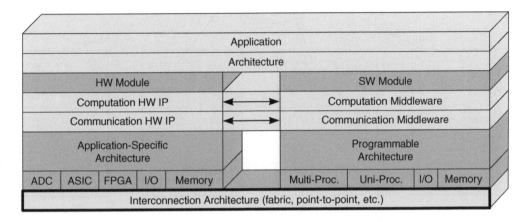

This chapter discusses technologies used to interconnect embedded processor computing devices. The anatomy of a typical interconnection fabric and some simple topologies are covered. Network bisection bandwidth, its relation to total exchange problem, and an illustrative fast Fourier transform example are covered. Networks that can be constructed from multiport switches are described. The VXS standard is presented as an emerging high performance interconnect standard.

14.1 INTRODUCTION

Continuing the discussion of the anatomy of a programmable high-performance embedded computer, this chapter focuses on the interconnection fabric used for communications between processors in a multiprocessor context. The fundamental challenge is providing an interprocessor communication fabric that is scalable to a large number of processing nodes in such a way that the achievable communication bandwidths grow along with the communication traffic attendant with multiple processing nodes.

Commercial off-the-shelf (COTS) hardware in this area has evolved along two principal lines: (1) box-to-box interconnects for connecting servers within a cluster using gigabit Ethernet (with 10 GbE on the horizon), InfiniBand, Myrinet, or Quadrics over either copper or fiber; and (2) board-to-board and intraboard connections within a chassis (to support applications with higher computational density requirements) using RapidIO, InfiniBand, or PCI Express. In either case, the underlying technology is fundamentally similar, based on switched-serial interconnects, and the fundamental differences are in packaging. Without loss of generality, this chapter focuses on the latter application (intrachassis), drawing heavily on the emerging VME Switched Serial (VXS) standard as an example.

---

* Portions of this chapter and some figures and tables are based on Teitelbaum, K., Crossbar Tree Networks for Embedded Signal Processing Applications, *Proceedings of the Fifth International Conference on Massively Parallel Processing*, pages 201–202, © 1998 IEEE. With permission.

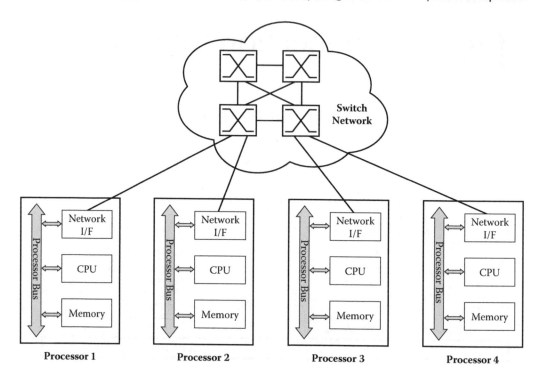

FIGURE 14-1   Typical interconnection fabric.

The remainder of this section provides the background and motivation for the ensuing discussion, beginning with a description of the anatomy of a typical interconnection fabric and some simple topologies. The importance of the network bisection bandwidth as a metric and its relation to the well-studied total exchange problem is described, and a simple two-dimensional fast Fourier transform (FFT) example (which requires a total exchange between processing steps) is considered, illustrating the effect of network parameters on scalability. The second section elaborates on network topologies that can be constructed from multiport switches, which are the basis of most of the COTS hardware available today. The third section discusses the emerging VXS standard, beginning with a description of switched-serial communications technology and concluding with a discussion regarding the topologies supported by the VXS standard.

14.1.1   ANATOMY OF A TYPICAL INTERCONNECTION FABRIC

The block diagram of typical switched-serial interconnection fabric is shown in Figure 14-1.

Communication between microprocessors in high performance embedded computing systems is typically point-to-point (i.e., processor-to-processor) along a path controlled by some number of switches. The bandwidth across any single point-to-point link is a function of the technology employed. Multiple links may be active concurrently, and the aggregate bandwidth across all of these possible paths is one often-used (although possibly misleading) metric of network capacity.

Each processor or computer connected to the fabric has a network interface, which may be a network card in a computer, an interface chip on a multiprocessor board, or even an interface core directly on the microprocessor chip itself. On the processor side, the network interface is typically connected to the processor and memory via the processor bus. On the network side, the network interface is typically connected directly to the network switches, either by copper or fiber-optic cable for box-to-box type connections, or via copper traces on chassis backplanes or printed circuit boards for board-to-board and intraboard connections. The network interface must retrieve data to be sent from processor memory and transmit that data serially as a stream of packets, according

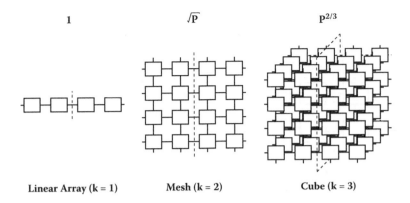

FIGURE 14-2   Bisection width of some simple networks.

to the protocol of the network. Similarly, it must reassemble received data from the network and deposit the information in the processor's memory.

The switches that compose the network must route data through the network based on routing information contained in the data packets. The switches examine the incoming packets, extract the routing information, and route the packets to the appropriate output port. The number and degree (number of ports) of the switches and their interconnection patterns constitute the network topology, which, to a very large extent, determines the scalability of the network.

14.1.2   NETWORK TOPOLOGY AND BISECTION BANDWIDTH

If we consider a network of processors divided into two equal halves, then the bisection bandwidth of the network is the bandwidth at which data may be simultaneously communicated between the two halves. This is given by the product of the bisection width (the number of communication links that pass through the network bisection) and the bandwidth of each link. Bisection bandwidth has units of the number of data items per unit time and is typically expressed in MB/s (millions of bytes per second) or GB/s (billions of bytes per second). For a network to be *scalable*, the number of links in the network bisection must increase as the number of processors in the network increases. This is a highly desirable property.

As an example, several simple networks are illustrated in Figure 14-2. These networks belong to a family referred to as a k-dimensional grid, or sometimes a k-dimensional cube or hypercube. A linear array has one dimension, a mesh has two dimensions, and a cube has three dimensions. There is a one-to-one correspondence between processors and network switches, which are typically of low degree. Parallel processors have been built based on each of these topologies. The network bisector of each network is illustrated, and it is seen to have dimension k-1; thus, the bisector of a cube is a plane and the bisector of the mesh is a line. As a result, it is possible to write a simple expression for the bisection bandwidth of the k-dimensional grid as a function of the parameter k. Note that the k-dimensional grid is scalable for k > 1. The linear array is not scalable because the number of links in the bisection is always one, regardless of the number of processors in the network.

14.1.3   TOTAL EXCHANGE

Many processing algorithms are mapped onto parallel processors in a data parallel fashion, such that each processor performs the same operation but on a different set of data. This simplifies program development and minimizes interprocessor communication requirements. The difficulty with this approach lies in the observation that not all algorithms are parallel in the same dimensions, and this fact necessitates remapping the data between algorithms. For example, consider an algorithm that operates on the columns of a matrix first and the rows of the matrix next. If the data are

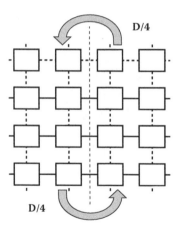

FIGURE 14-3　Corner-turning on a 4 × 4 mesh.

distributed column-wise onto the parallel processor, a remapping of data will be required in order to operate on the matrix rows. This remapping, essentially a matrix transpose, will require each processor to communicate with every other processor, remapping the entire matrix. The implementation of matrix transposes on parallel processors is a well-studied problem in the computer science literature, where it is often referred to as a total exchange. The time it takes a processor to perform a total-exchange operation depends on the bisection bandwidth of the processor, as illustrated in Figure 14-3, using a mesh network as an example.

In this example, data are distributed on a 4 × 4 mesh processor, with each processor having 1/16th of the total data volume D. In order to perform the total exchange, each processor must send 1/16th of its local data (1/256th of the total data volume) to each of the other 15 processors. One-half of each processor's data will have to be transmitted across the network bisection. In total, one half of the total data volume ($D/2$) must be transmitted through the network bisection, $D/4$ in each direction. The message size for each transfer will be D/n^2, where n is the number of processing nodes in the network. As the network required to solve a particular problem in a given time becomes larger, the size of the data-exchange messages becomes smaller rapidly, and the fixed overhead causes the sustained bandwidth across each communication link to decrease. At some point, the decrease in link bandwidth more than offsets the increase in the number of links in the network bisection, and the bisection bandwidth of the network decreases with increasing network size. This effectively limits the size of the processing network that may be productively employed.

14.1.4　PARALLEL TWO-DIMENSIONAL FAST FOURIER TRANSFORM—A SIMPLE EXAMPLE

Let us consider the parallel implementation of a two-dimensional FFT of size N, running on a parallel processor with P nodes. For simplicity, we will consider P an integer power of two. The dataset for this example is an $N \times N$ complex matrix X, and the two-dimensional FFT consists of an N-point one-dimensional FFT applied to each column of X, followed by a one-dimensional FFT applied to each row of X. The initial distribution of data across the processor is column-wise, with each processor getting N/P columns of X. Each processor must compute N/P one-dimensional N-point FFTs and then exchange data with each of the other processors so that each processor has N/P of the transformed rows—a total exchange. Following this, each processor computes N/P one-dimensional FFTs on the row data. The total compute time will be the time required for a single node to perform $2N/P$ N-point one-dimensional FFTs plus the time required to exchange the data. To estimate the compute time, we assume a workload of $10(N/2)\log_2 N$ real arithmetic operations per FFT for a workload per processor of $10N^2\log_2 N/P$ operations per processor. Based on the discussion in Chapter 19, a peak computational rate of 8 GFLOPS (billions of floating-point operations per second) and an efficiency of 12.5% are assumed for the processing hardware, resulting in a sustained computational rate of 1 GFLOPS. Assuming 8 bytes per element of X, the total volume of data to be communicated is $8N^2$ bytes, which is to be transmitted as P^2 messages of size $8N^2/P^2$. Half of this message traffic must pass through the network bisection. Assuming a k-dimensional grid as the network topology, the bisection width will be $P^{(k-1)/k}$. For reasons that will be discussed in Section 14.3.1 of this chapter, we will model the link performance with two parameters: a peak data rate, r, of 800 MB/s and an assumed overhead (hardware and software), o, of one microsecond. The time required to transmit a b byte message is given by $b/r + o$.

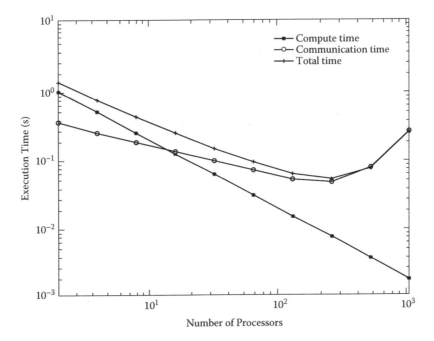

FIGURE 14-4   Total execution time for a parallel 4K × 4K point two-dimensional FFT.

The computation time, the communication time, and the total execution time are plotted in Figure 14-4 for a 4096 × 4096 point two-dimensional FFT on a two-dimensional mesh connected processor. As the number of processors, P, increases, the computation time decreases monotonically. Because the number of communication links in the network bisection increases with the number of processors, the communication time decreases initially, but then begins to increase as the decreasing message size begins to adversely impact the link bandwidth. As a result, the total execution time has a clear minimum, and increasing the number of processors beyond this point (about 256 processors in this example) produces no useful benefit.

Dividing the single-processor execution time by the total execution time shown in Figure 14-4 yields the effective speedup and is shown in Figure 14-5 along with the ideal speedup, which represents a linear increase in speedup with unity slope. The effective speedup is shown for three different network topologies: linear array (one-dimensional), mesh (two-dimensional), and cube (three-dimensional). These curves each exhibit a peak that corresponds to the minima in the execution times. The slope of the speedup curves in the region where the execution time is decreasing and the maximum speedup are both strong functions of the scalability of the network bisection bandwidth with increasing P. The linear array, which is not scalable, has essentially a flat speedup curve, implying that adding additional processors does nothing to accelerate processing—clearly not desirable. The mesh and cube both scale well, with the cube offering some improvement over the mesh, although at an increased cost in network hardware.

14.2   CROSSBAR TREE NETWORKS

In this section, we will consider the interconnection of multiple processing nodes via a dynamically switched network of multiport crossbar switches. A p-port crossbar switch can connect up to p processing nodes, supporting $p/2$ simultaneous messages. Since the complexity of a crossbar switch grows as p^2, it is impractical to build large machines with many processing nodes interconnected by a single crossbar switch. As a consequence, large machines are constructed using networks of smaller crossbar switches. One of the simplest networks that can be imagined is the binary tree illustrated in Figure 14-6.

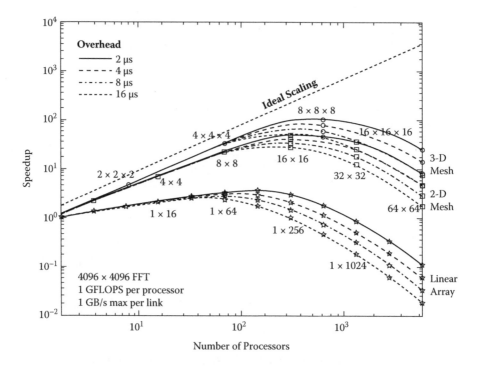

FIGURE 14-5   Effective speedup for 4K × 4K two-dimensional FFT example.

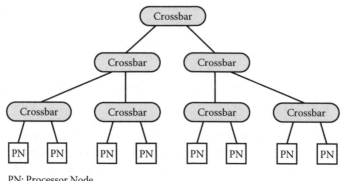

PN: Processor Node

FIGURE 14-6   Binary tree network.

The switches in the tree are arranged in multiple levels with each switch having connections going up to its "parent" node, and down to its "children." In a binary tree, each switch has a single parent and two children. The uppermost level of the tree is called the root and has no parent. The children of the switches at the lowest level of the tree are the processing nodes. The principal drawback to binary tree switching networks is that the root typically experiences the greatest message traffic. The bisection bandwidth of the network is equal to the bisection bandwidth of the root, which is a single link, regardless of the number of processors. As a result, binary trees are not scalable. As a means to avoid the communication bottleneck at the root of a binary tree, Leiserson (1985) proposed a network he called a *fat tree*, which is illustrated in Figure 14-7. Like the binary tree, the fat tree is constructed using binary switches, each switch having one parent and two children. In the fat tree, however, additional parallel communication paths are provided as the root is approached. If the number of parallel paths connecting to the parent of a given switch is chosen to be equal to the number of processing nodes that have that switch as an ancestor, then the network will be perfectly scalable and will have a bisection bandwidth

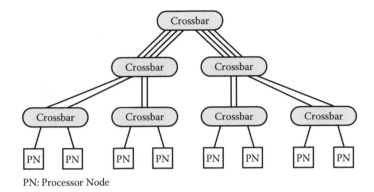

PN: Processor Node

FIGURE 14-7   Binary fat-tree network.

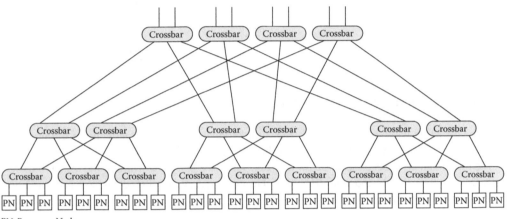

PN: Processor Node

FIGURE 14-8   Generalized crossbar tree network.

that increases linearly with the number of processors. The drawback, of course, is that the switches become more complex and difficult to build as they approach the root. The CM-5, built by Thinking Machines, Inc., was an example of a massively parallel processor built around a fat-tree architecture.

The crossbar tree networks in more modern parallel processors are extensions of the basic fat-tree concept, as illustrated in Figure 14-7. These networks belong to the family of least-common-ancestor networks described by Scherson and Chien (1993). The growth of very-large-scale integration technology has resulted in the availability of crossbar switches of many ports, facilitating the construction of non-binary tree networks. In addition, multiple switches in parallel are used to provide increased bandwidth. Nodes closer to the root have more switches paralleled to handle the increased traffic. We will characterize these networks by three parameters: u, the number of "up" links connecting a switch to its parent nodes; d, the number of "down" links connecting a switch to its children; and q, the number of levels in the network. The tree illustrated in Figure 14-8 has switches with two parents ($u = 2$) and three children ($d = 3$) and has three levels ($q = 3$). In this tree, the number of parent and child links for each switch is constant. In some cases, the allocation of switch ports (parent versus child) may vary between levels of the tree. In this case, we will subscript the parameters (e.g., d_i, u_i). Note that the binary tree is generated by $d = 2$, $u = 1$.

14.2.1   NETWORK FORMULAS

The basic properties of this family of crossbar trees can be calculated from a few simple formulas that are listed in Table 14-1. Two sets of formulas have been derived: one for the general case in

TABLE 14-1
Formulas for Crossbar Tree Networks

	Uniform	General
Processors	d^q	$\displaystyle\prod_{i=1}^{q} d_i$
Bisection Width (links)	$\dfrac{du^{q-1}}{2}$	$\dfrac{d_q}{2}\displaystyle\prod_{i=1}^{q-1} u_i$
I/O Links	u^q	$\displaystyle\prod_{i=1}^{q} u_i$
Switches	$\dfrac{d^q - u^q}{d - u}$	$\displaystyle\sum_{k=1}^{q} U_{q-k-1} D_k$ $U_j = \dfrac{\displaystyle\prod_{i=1}^{j} u_i}{u_j}$ $D_j = \dfrac{\displaystyle\prod_{i=1}^{j} d_i}{d_j}$

which d and u can vary from level to level and one for the uniform case in which d and u are the same for each level of the tree.

The bisection width of the crossbar tree is the product of the bisection width of a single switch ($d/2$) and the number of parallel switches at the root level of the tree, which, in turn, is the product of the individual u_is for each level below the root. Thus, the bisection width of the tree will increase only linearly with d, but exponentially with u. The cost associated with increasing u is the additional switches and interconnects required.

14.2.2   SCALABILITY OF NETWORK BISECTION WIDTH

Assigning integer values to u, d, and q will result in families of networks with varying numbers of processing nodes and bisection widths, as illustrated in Figure 14-9. Three clusters of curves are shown. One cluster (dotted line) is for networks with 4 children per switch ($d = 4$), one cluster (solid line) is for networks with 16 children per switch ($d = 16$). Within each cluster of curves, the number of parents per switch (u) is varied. The relationship between bisection bandwidth and number of processing nodes is also shown for the k-dimensional grid (dashed line) for comparison. These curves, plotted in log-log coordinates, are straight lines of varying slope. The greater the slope the more rapidly bisection bandwidth increases with increasing processor size, and the more scalable the network. Note that the slope of the line increases as u is increased.

In order to better understand the behavior of these curves, let us consider the case in which $u = d^k$, that is, when the number of parent and child nodes are related through the parameter k, $k <= 1$. Substituting into the formula for bisection bandwidth from Table 14-1, and denoting the bisection width as β, the number of processors P yields

$$\beta = \frac{d^{1-k}}{2} P^k$$

$$\log \beta = k \log P + \log\left(\frac{d^{1-k}}{2}\right).$$

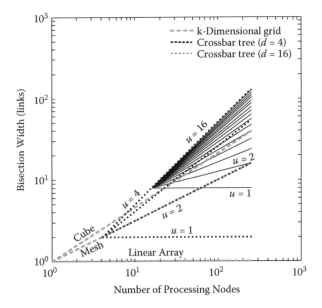

FIGURE 14-9   Scalability of crossbar tree networks.

TABLE 14-2
Special Cases of Crossbar Tree Networks

$k = 0$	$k = 1/2$	$k = 2/3$	$k = 1$
$\beta = \dfrac{d}{2}$	$\beta = \dfrac{\sqrt{d}}{2}P^{1/2}$	$\beta = \dfrac{\sqrt[3]{d}}{2}P^{2/3}$	$\beta = \dfrac{P}{2}$
$d = 2$ scales like linear array (1D)	$d = 4$ scales like mesh (2D)	$d = 8$ scales like cube (3D)	(scales linearly)

This is the equation of a straight line with slope k in log-log coordinates. The slope, then, of the lines in Figure 14-9 is just: $k = \log_d(u)$, and the most scalable networks are those with the greatest k.

By choosing k and d appropriately, crossbar tree networks can be constructed which exhibit scalability identical to the k-dimensional grid. This is illustrated in Table 14-2, which lists the equivalent networks for the linear array, mesh, and cube.

14.2.3   UNITS OF REPLICATION

For embedded multipurpose processors that are VME-based, the processing nodes and crossbar switches are packaged into boards, and multiple boards are packaged into a chassis. These are the basic units of replication (UOR) that are packaged together to construct systems. Typically, we want to place as many processing nodes on a board as possible, but we are restricted in terms of the number of connector pins and, therefore, the number of I/O links available for expansion of the crossbar network.

As an example, consider packaging 16 processing nodes onto a single 9U VME card, and packaging 16 cards in a chassis. This would provide a total of 256 processing nodes. At the same time, we assume standard VME connectors J1, J2, and J3, each of which has 96 pins for a total of 288 pins. If we desire to maintain VME compatibility, we will have approximately 160 pins free with which to connect our crossbar tree. Since we want to maximize our link bandwidth, we will assume

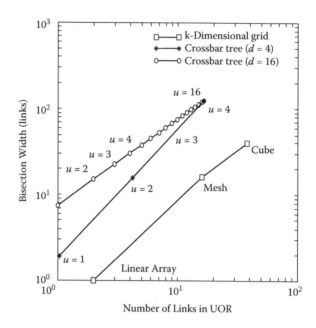

FIGURE 14-10   Sensitivity of various network topologies to I/O constraints.

that our crossbar links are fully parallel and require slightly greater than 32 bits each (allowing for handshaking). We will, therefore, be able to bring at most four links off board and ask which topology will yield the greatest bisection bandwidth given that constraint.

In Figure 14-10, several candidate topologies have been explored. For each topology, the bisection width (links) of a network of 256 processors is calculated and plotted as a function of the number of links that need to be brought off each 16-processor board to complete the network. The general trend evident in Figure 14-10 is that when the number of links in the unit of replication is constrained, crossbar tree networks exhibit superior scalability and superior bisection bandwidth when compared to the k-dimensional grid. In fact, for the parameters given, the only k-dimensional grid that would have four or fewer links coming off each board is the linear array, which is not scalable at all. It is, however, possible to construct a crossbar tree network ($d = 4$, $u = 2$) that has exactly four links coming off each board and that scales equivalently to a mesh (two-dimensional). This may help to explain why crossbar architectures are prevalent in the world of VME-based embedded processors.

14.2.4   PRUNING CROSSBAR TREE NETWORKS

In some commercial implementations, crossbar tree networks may be *pruned* to remove unwanted branches, as illustrated in Figure 14-11. Pruning *down*links reduces the number of processors and may be helpful in constructing processors with an odd number of processing nodes (as in a chassis with only a few boards). Pruning *up*links can reduce the number of I/O links required in a unit of replication (board or chassis).

Another example, shown in Figure 14-12, clearly illustrates the impact of pruning crossbar trees on the scalability of bisection bandwidth. In the left-hand plot, two curves are shown: the solid curve represents a fully populated array and the solid starred curve illustrates pruning of downlinks to achieve an intermediate number of processors. The downlinks are pruned from the root of the tree. For these parameters, the slope of the solid curve is 2/3, and this network scales equivalently to a three-dimensional grid (cube). The solid curve is evaluated only for P (the number of processors) equal to d^n where n is an integer. At these values of P, the solid curve and the solid starred curve

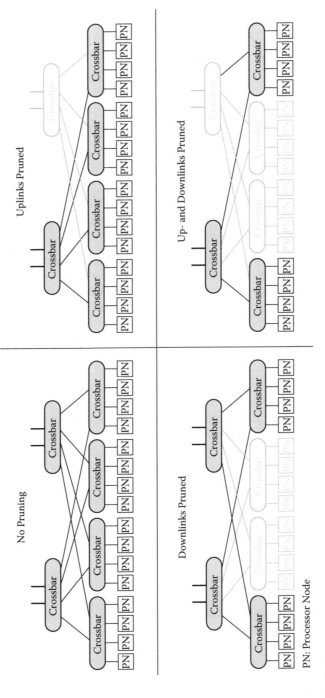

PN: Processor Node

FIGURE 14-11 Pruning crossbar tree networks.

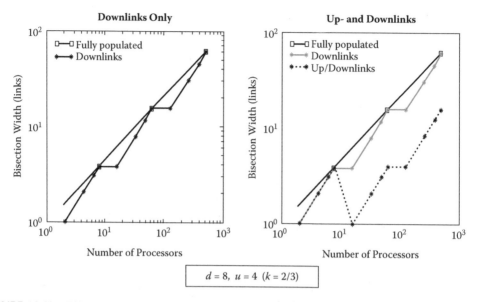

FIGURE 14-12   Effect of pruning on scalability of bisection bandwidth.

coincide. In between these values, the solid starred curve exhibits a staircase type of behavior. Note that as the number of processors is doubled from d^n to $2d^n$, the bisection bandwidth of the network remains constant. Continuing to increase the number of processors from $2d^n$ to d^{n+1} results in a linear improvement in bisection bandwidth. The exact behavior of the staircase depends on the values of u and d. For some configurations, it is possible for the bisection bandwidth to decrease while the number of processors is doubled from d^n to $2d^n$. It is interesting to note that the bisection bandwidth of the (three-dimensional) cube has the same type of behavior for values of P, which are not cubes of integers.

The effect of pruning uplinks is illustrated in the right-hand plot of Figure 14-12. Here we have pruned uplinks at the lowest level of the tree, restricting u_1 to be one instead of four. The effect is noted at the dip in the dotted curve where the size of the processor is doubled (from 8 to 16 processing nodes), but the bisection bandwidth is reduced by a factor of four. The bisection bandwidth continues to scale from this point onward, but is always a factor of four lower than had we not pruned the uplinks (solid and solid starred curves). This type of behavior, obviously, is highly undesirable.

In order for the bisection bandwidth of a network to be a *nondecreasing* function of the number of processors, the following relationship must hold true:

$$u_q - 1 \geq \frac{d_q - 1}{d_q}.$$

The ratio on the right-hand side of this inequality is simply the ratio by which the number of children has been pruned in going from a tree of height $q - 1$ to a tree of height q. If the tree is not pruned, and $d_q = d_{q-1}$, this inequality will be satisfied. When the tree is pruned, the inequality will be satisfied as long as each switch in the level below the root has a number of parents greater than this ratio.

The scenario in which this becomes an issue most frequently is in going from a single board to two boards or from a single chassis to two chassis. In this case, d_q, the number of children of the root level, is effectively two, and the ratio d_{q-1}/d_q will be greater than one in all cases except for the binary tree. If we also prune the parent links at the board/chassis level (the level below the root) to minimize the number of connector pins or the amount of interchassis cabling, it becomes very difficult to maintain the ratio required for nondecreasing bisection bandwidth.

14.3 VXS: A COMMERCIAL EXAMPLE

As a concrete example, we consider the emerging VXS standard, which represents the latest attempt at increasing interprocessor communication bandwidth on the VME bus by leveraging recent advances in switched-serial interconnection fabrics. VXS is designed to be "fabric agnostic," which means that it will support multiple switched-serial fabrics, including serial RapidIO, InfiniBand, and PCI Express. The VXS standard defines two circuit card types that plug into a common backplane: a payload card and a switch card. A typical payload card might have multiple processors, their associated network interfaces, and a switch to interconnect them. The payload cards have a new P0 connector that brings multiple serial links off the card for intercard communication. Direct card-to-card connections can be made on the backplane, or the individual cards can be connected to a switch card that facilitates the construction of larger networks. Multiple topologies are possible. Section 14.3.1 discusses the fundamentals of the switched-serial links, and Section 14.3.2 discusses the supported topologies.

14.3.1 LINK ESSENTIALS

The link and the factors that determine link bandwidth are considered first. A typical link in an interconnection fabric is depicted in Figure 14-13, which illustrates the path of a message exchanged between application processes residing on two different computers.

The message data to be sent, as well as the message source and destination, are typically read by the network interface over the processor bus. The network interface then breaks the message data up into a series of packets, each of which contains a header, payload, and trailer. The header contains housekeeping and routing information, the payload contains a block of data from the message, and the trailer might contain a cyclic redundancy check (CRC) to be used by the recipient for error-detection purposes.

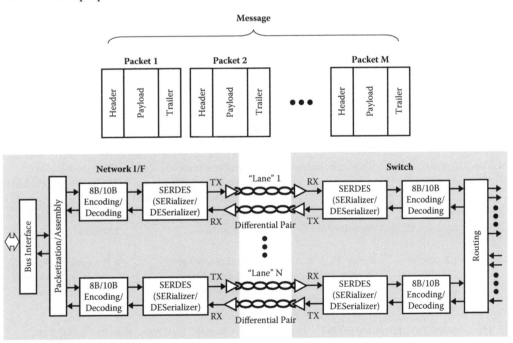

FIGURE 14-13   A typical interconnection fabric link.

Packet data are typically sent serially, often employing multiple low-voltage differential signaling (LVDS) pairs referred to as *lanes*. Typically applied is 8B/10B encoding, which maps each byte of data to be sent into a 10-bit symbol prior to transmission.

The transmitted packet is received at the switch, where it is routed based on the destination address contained in the packet header. Typically, the switch contains a routing table (established as part of the system configuration process) that identifies the switch output port corresponding to the destination address. This deterministic routing approach is simple and results in in-order delivery of message packets at the destination, but cannot adaptively route around busy nodes. Once routed to the appropriate output port, the packet is reserialized and transmitted.

At the destination node, the received serial data are parallelized, the 10-bit symbols decoded, and the symbols from multiple lanes combined. The CRC code is checked, and if it matches the CRC of the packet received, an acknowledge packet is sent to the sender (a not-acknowledged packet is sent if the CRC does not match and the sender will retransmit the packet if it receives a not-acknowledge or fails to receive an acknowledge within a certain period of time). The received packets are then reassembled into a complete message and transferred to memory via the processor bus.

The achievable link bandwidth is a function of several factors:

1. The signaling rate of the serial data
2. The efficiency of the 8B/10B encoding (80%)
3. The number of parallel differential pairs (lanes)
4. The protocol efficiency (the number of payload data bits divided by the total number of bits in the packet including header and trailer): close to 100% for large payloads, can be significant for small payloads (very short messages)

Typical link bandwidths for some COTS fabrics are shown in Table 14-3. Serial RapidIO is a board-to-board and intraboard fabric used by Mercury Computer Systems on their VXS and compact PCI systems. It is also supported by Freescale, with RapidIO network interfaces embedded in some PowerPCs. PCI Express was developed for use within PCs to provide higher bandwidth access between the CPU and bus peripherals such as the graphics processing unit (GPU), hard drives, USB

TABLE 14-3
Typical COTS Interconnection Fabrics

	Serial RapidIO*	PCI Express*	InfiniBand§
Signaling Rate	1.25, 2.5, 3.125 GHz	2.5 GHz	2.5 GHz
Lanes	1X, 4X	2X, 4X, 8X, 12X, 16X, 32X	1X, 4X, 12X
Encoding	8B/10B	8B/10B	8B/10B
Payload/Overhead	256 byte maximum/16 byte	variable size packets 128–4096 bytes, 4096 KB max payload, 22 byte overhead	0–4096 byte payload 82 byte overhead (RDMA)
Protocol Efficiency	94% max	99.5% max	98% max
Effective Data Rate (4X, 2.5 GHz)	940 MB/s	995 MB/s	980 MB/s
Switch Chip Examples	8 ports, 4X Mercury, Tundra	48 channels/9 ports (configurable) PLX Technology	8 12X ports, 96 channels-configurable (Mellanox) 8 4X ports (32 channels) Red Switch

* From RapidIO, PCI Express, and Gigabit Ethernet Comparison, RapidIO Trade Association, Rev. 3, May 2005.
§ From InfiniBand Architecture Specification, Vol. 1, Rel. 1.2.

devices, etc. It is supported on VXS systems, compact PCI, and ATCA (Advanced Telecom Computing Architecture), a standard similar to VXS supported within the telecommunications industry. InfiniBand is a box-to-box fabric aimed at cluster computing and processor-to-storage interconnects. It is also supported on both VXS and ATCA.

The time required to transmit a message across the network can usually be approximated as $b/r + o$, where b is the total message size in bits, r is the data rate in bits per second, and o is the overhead. Sources of overhead include software processes running on the source and destination processors, as well as processing in the network interfaces and switches. The sustainable data rate is, of course, the size of the message divided by the data transfer time and can be expressed as $r/(o + or/b)$. For small messages, the sustainable data rate tends to zero, and for large messages, the sustainable data rate asymptotically approaches the maximum data rate. For $b = b_{1/2} = or$, the sustainable data rate is exactly half of the maximum. Measurements on network hardware reveal that InfiniBand, for example, with a theoretical peak data rate of 980 MB/s will asymptotically approach about 830 MB/s (Liu et al. 2004). The sustainable data rate drops to about half of this for messages around 1 KB in length, suggesting an overhead of about 1 microsecond.

14.3.2 VXS-Supported Topologies

The P0 connector on each payload card can carry two 4X serial links, and some simple networks can be constructed directly without a switch card. For example, up to three cards can be completely connected (each card connected to the other two), up to four cards can be connected in a mesh, and larger networks can be interconnected in a ring, but the bisection bandwidth of such a network will not scale with an increasing number of processor cards. To create larger, scalable networks, one or more switch cards are required and are typically connected in a star or dual-star configuration. A common backplane configuration for a 20-slot VME chassis supports 18 payload cards and two switch cards in a dual-star configuration (Figure 14-14). Each payload slot has one 4X serial link connected to each of the switch cards. Each switch card interconnects 18 payload cards. Since current switch technology supports up to about eight ports per switch, multiple switches per switch card are typically employed, resulting in a multilevel interconnect essentially similar to the least-com-

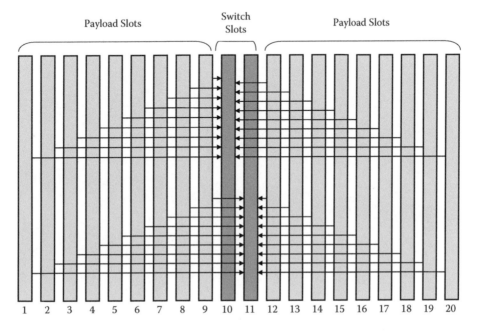

FIGURE 14-14 VXS dual-star backplane configuration.

mon-ancestor networks previously discussed. A hypothetical example is shown in Figure 14-15(a) and consists of 18 payload cards, each with four processors (a total of 72 processors), and two switch cards. A three-level interconnect consisting of six-port switches is sufficient to connect all of the processors. The first-level interconnect is on the payload cards; the second- and third-level interconnects are on the switch cards. The same network can be redrawn as in Figure 14-15(b), and from this representation, it is clear that this is a least-common-ancestor network with $\mathbf{d} = [6\ 4\ 3]^T$, $\mathbf{u} = [0\ 3\ 2]^T$.

14.4   SUMMARY

This chapter has discussed interconnection fabrics from the perspective of a programmable high performance embedded computer. The chapter built upon the discussion of computing devices in Chapter 13, exploring the fundamental technologies for interconnecting computing devices and exchanging data between them. We have seen the importance of bisection bandwidth on efficient parallel program execution and the importance of scaling bisection bandwidth with increasing machine size. To a great extent, bisection bandwidth is a function of network topology, and we have studied least-common-ancestor networks as representative of networks found in COTS technology. We have seen that typical COTS interconnection fabrics consist of compute nodes and switches interconnected by serial data links and have looked at RapidIO, InfiniBand, and PCI Express as examples of current technology. We have also looked at VXS as a COTS example for packaging high performance embedded computing systems in a fabric-agnostic way.

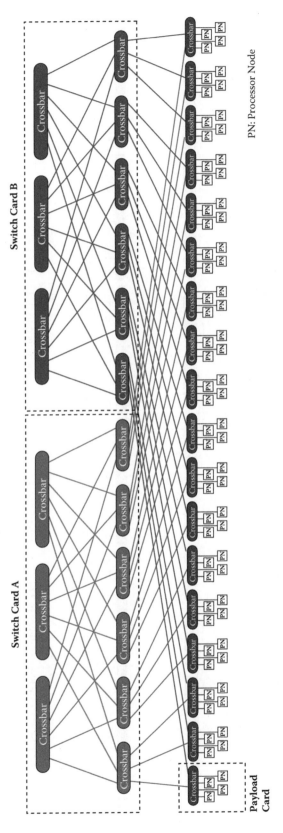

FIGURE 14-15(a)   (Color figure follows p. 278.) Example three-level VXS interconnect.

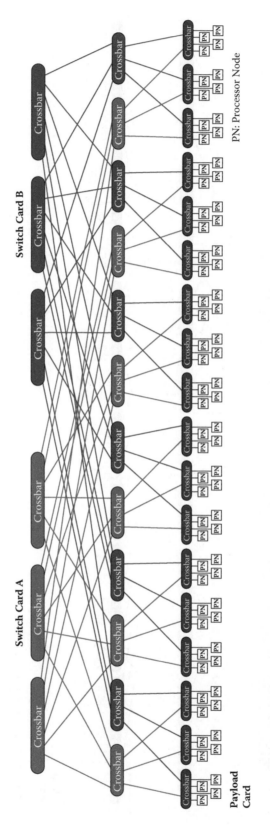

FIGURE 14-15(b)   (Color figure follows p. 278.) Same interconnect redrawn as least-common-ancestor network.

REFERENCES

Leiserson, C.E. 1985. Fat trees: networks for hardware-efficient supercomputing. *IEEE Transactions on Computers* C-34(10): 892–901.

Liu, J., B. Chandrasekaran, W. Yu, J. Wu, D. Buntinas, S. Kini, D.K. Panda, and P. Wyckoff. 2004. Microbenchmark performance comparison of high-speed cluster interconnects. *IEEE Micro* 24(1): 42–51.

Scherson, I.D. and C.-K. Chien. 1993. Least common ancestor networks. *Proceedings of the Seventh International Parallel Processing Symposium* 507–513.

REFERENCES

15 Performance Metrics and Software Architecture

Jeremy Kepner, Theresa Meuse, and
Glenn E. Schrader, MIT Lincoln Laboratory

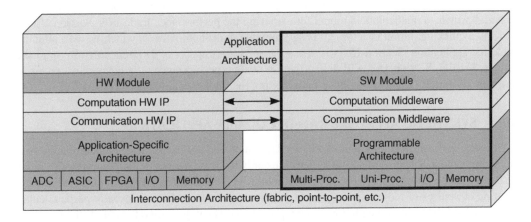

This chapter presents HPEC software architectures and evaluation metrics. A canonical HPEC application (synthetic aperture radar) is used to illustrate basic concepts. Different types of parallelism are reviewed, and performance analysis techniques are discussed. A typical programmable multicomputer is presented, and the performance trade-offs of different parallel mappings on this computer are explored using key system performance metrics. The chapter concludes with a discussion of the impact of different software implementation approaches.

15.1 INTRODUCTION

High performance embedded computing (HPEC) systems are amongst the most challenging systems in the world to build. The primary sources of these difficulties are the large number of constraints on an HPEC implementation:

Performance: latency and throughput
Efficiency: processing, bandwidth, and memory
Form Factor: size, weight, and power
Software Cost: code size and portability

This chapter will primarily focus on the performance metrics and software architectures for implementing HPEC systems that minimize software cost while meeting as many of the other requirements as possible. In particular, we will focus on the various software architectures that can be used to exploit parallel computing to achieve high performance. In this context, the dominating factors in the HPEC software architecture are

Type of parallelism: data, task, pipeline, and/or round-robin
Parallel programming model: message passing, threaded and/or global arrays
Programming environment: languages and/or libraries

The approaches for dealing with these issues are best illustrated in the context of a concrete example.

Section 15.2 gives an overview of a canonical HPEC application (synthetic aperture radar or SAR) taken from the HPEC Challenge benchmark suite [http://www.ll.mit.edu/hpecchallenge]. [Note: For a more detailed description, see Appendix A.] The rest of the chapter is organized as follows. Section 15.3 will describe the different types of parallelism that can be applied to the application and provide a mathematical model for exploring the performance trade-offs. Section 15.4 will provide a quick definition of a typical programmable multi-computer on which we will attempt to build the application. Section 15.5 discusses the software impacts of different software implementations. Finally, Section 15.6 will define the key system performance, efficiency, form factor, and software cost metrics that we will use for assessing the implementation.

15.2   SYNTHETIC APERTURE RADAR EXAMPLE APPLICATION

SAR is one of the most common modes in a radar system and one of the most computationally stressing to implement. The goal of a SAR system is usually to create images of the ground from a moving airborne radar platform. The basic physics of a SAR system (Soumekh 1999) begins with the radar sending out pulses of radio waves aimed at a region on the ground that is usually perpendicular to the direction of motion of the platform (see Figure 15-1). The pulses are reflected off the ground and detected by the radar. Typically, the area of the ground that reflects a single pulse is quite large and an image made from this raw unprocessed data is very blurry (see Figure 15-2). The key concept of a SAR system is that it moves between each pulse, allowing multiple looks at the same area of the ground from *different* viewing angles. Combining these different viewing angles together produces a much sharper image (see Figure 15-2). The resulting image is as sharp as one taken from a much larger radar with a "synthetic" aperture the length of the distance travelled by the platform.

There are many variations on the mathematical algorithms used to transform the raw SAR data into a sharpened image. In this chapter, we will focus on the variation referred to as "spotlight" SAR. Furthermore, we will look at a simplified version of this algorithm that focuses on the most computationally intensive steps of SAR processing that are common to nearly all SAR algorithms. Our example is taken from the "Sensor Processing and IO Benchmark" from the HPEC Challenge benchmark suite (see http://www.ll.mit.edu/hpecchallenge).

The overall block diagram for this benchmark is shown in Figure 15-3. At the highest level it consists of three stages:

SDG: Scalable Data Generator. Creates raw SAR inputs and writes them to files to be read in by Stage 1.
Stage 1: Front-End Sensor Processing. Reads in raw SAR inputs, turns them into SAR images, and writes them out to files.
Stage 2: Back-End Knowledge Formation. Reads in pairs of SAR images, compares them, and then detects and identifies the difference.

Although the details of the above processing stages vary dramatically from radar to radar, the core computational details are very similar: input from a sensor, followed by processing to form an image, followed by additional processing to find objects of interest in the image.

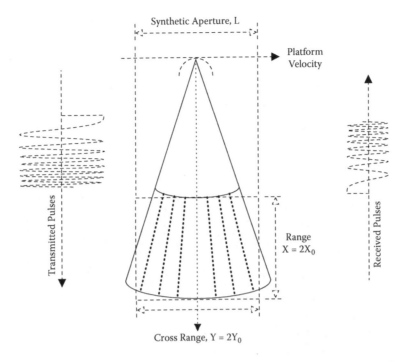

FIGURE 15-1 Basic geometry of a SAR system. A SAR system sends out pulses of radio waves aimed at a region on the ground that is usually perpendicular to the direction of motion of the platform. The pulses are reflected off the ground and detected by the radar. The direction parallel to the transmission of the pulses is referred to as the "range" or "down range" direction. The direction perpendicular to the transmission of the pulses is referred to as the "cross range" direction. (From Bader et al., Designing scalable synthetic compact applications for benchmarking high productivity computing systems, *CTWatch Quarterly* 2(4B), 2006. With permission.)

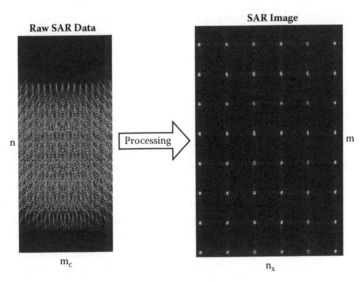

FIGURE 15-2 (**Color figure follows p. 278.**) Unprocessed (left) and processed (right) SAR data. The area that reflects a single pulse is large and an image of this raw data is very blurry (left). A SAR system provides multiple looks at the same area of the ground from multiple viewing angles. Combining these different viewing angles together produces a much sharper image (right). (From Bader et al., Designing scalable synthetic compact applications for benchmarking high productivity computing systems, *CTWatch Quarterly* 2(4B), 2006. With permission.)

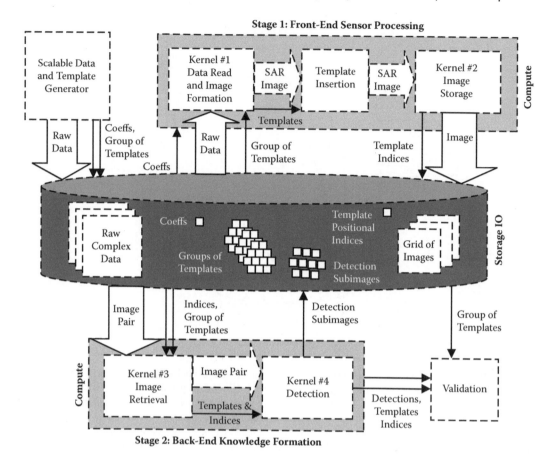

FIGURE 15-3 System mode block diagram. SAR system mode consists of Stage 1 front-end sensor processing and Stage 2 back-end knowledge formation. In addition, there is significant IO to the storage system. (From Bader et al., Designing scalable synthetic compact applications for benchmarking high productivity computing systems, *CTWatch Quarterly* 2(4B), 2006. With permission.)

15.2.1 OPERATING MODES

This particular SAR benchmark has two operating modes (Compute Only and System). The "Compute Only Mode" represents the processing performed directly from a continuously streaming sensor (Figure 15-4). In this mode, the SDG is meant to simulate a sensor data buffer that is filled with a new frame of data at regular intervals, T_{input}. In addition, the SAR image created in Stage 1 is sent directly to Stage 2. In this mode, the primary architectural challenge is providing enough computing power and network bandwidth to keep up with the input data rate.

In "System Mode" the SDG represents an archival storage system that is queried for raw SAR data (Figure 15-3). Likewise, Stage 1 stores the SAR images back to this archival system and Stage 2 retrieves pairs of images from this storage system. Thus, in addition to the processing and bandwidth challenges, the performance of the storage system must also be managed. Increasingly, such storage systems are the key bottleneck in sensor processing systems. Currently, the modeling and understanding of parallel storage systems are highly dependent on the details of the hardware. To support the analysis of such hardware, the SAR benchmark has an "IO Only Mode" that allows for benchmarking and profiling. The theoretical modeling and analysis of parallel file systems are beyond the scope of this chapter, which will primarily focus on the "Compute Only Mode."

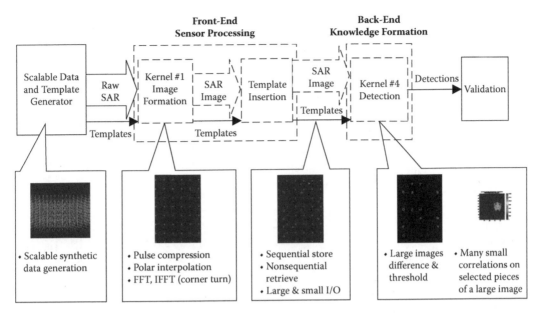

FIGURE 15-4 (Color figure follows p. 278.) Compute Only Mode block diagram. Simulates a streaming sensor that moves data directly from front-end processing to back-end processing. (From Bader et al., Designing scalable synthetic compact applications for benchmarking high productivity computing systems. *CTWatch Quarterly* 2(4B), 2006. With permission.)

15.2.2 COMPUTATIONAL WORKLOAD

The precise algorithmic details of this particular SAR processing chain are given in Appendix A. For the purposes of mapping the algorithm onto a parallel computer, the only relevant pieces are listed below:

Core Data Structures: The array(s) at each stage that will consume the largest amount of memory and upon which most of the computations will be performed.

Computational Complexity: The total operations performed in the stage and how they depend upon the algorithmic parameters.

Degrees of Parallelism: The parallelism inherent in the stage and how it relates to the core data structures.

Data IO: The amount of data that needs to be moved into and out of the computational stage.

These pieces of the HPEC Challenge SAR benchmark are summarized in Figure 15-5 and Table 15-1. In Stage 1, the data are transformed in a series of steps from a $n \times m_c$ single-precision complex valued array to a $m \times n_x$ single-precision real valued array. The "degrees of parallelism" described at each stage refers to the amount of parallelism within each stage. This is sometimes referred to as "fine grain" parallelism. There is also pipeline or task parallelism, which exploits the fact that each step in the pipeline can be performed in parallel, with each step processing a frame of data. Finally, there is also coarse grain parallelism, which happens because separate SAR images can be processed independently. This is equivalent to setting up multiple pipelines.

In Stage 1, the processing is along either the rows or the columns, which defines how much parallelism can be exploited. In addition, when the direction of parallelism switches from rows to columns or columns to rows, then this means that a "corner turn" of the matrix must be performed. On a typical parallel computer a corner turn requires every processor to talk to every other processor. These corner turns often are natural boundaries along which to create different stages in a

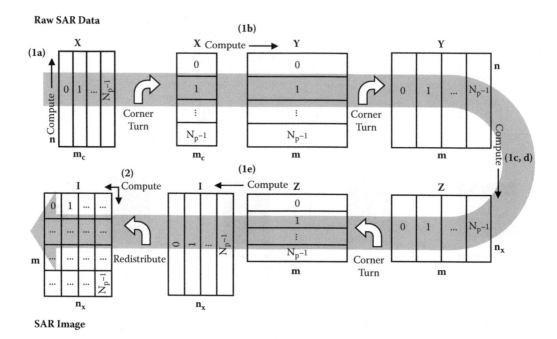

FIGURE 15-5   Algorithm flow. In Stage 1, the data are transformed in series of steps from a $n \times m_c$ single-precision complex valued array to a $n_x \times m$ single-precision real valued array. At each step the processing is along either the rows or the columns.

TABLE 15-1

Algorithm Complexity and Parallelism

Stage	Data Array	Compute Complexity	Degrees of Parallelism	Input Samples	Output Samples	Operations/Sample
SDG	$X : C_{32bit}^{n \times m_c}$	NA	NA	NA	nm_c	NA
1a	$X : C_{32bit}^{n \times m_c}$	$5nm_c \lg(n)$	m_c	nm_c	nm_c	$5\lg(n)$
1b	$Y : C_{32bit}^{n \times m}$	$5nm_c \lg(m_c)$ $+10nm\lg(m)$	n	nm_c	nm	$5\lg(m_c)$ $+10\lg(m)$
1c	$Z : C_{32bit}^{n_x \times m}$	$40nmn_K$	m	nm	$n_x m$	$40n_K$
1d	$Z : C_{32bit}^{n_x \times m}$	$5n_x m\lg(m)$	m	$n_x m$	$n_x m$	$5\lg(m)$
1e	$I : R_{32bit}^{m \times n_x}$	$5n_x m\lg(n_x)$	n_x	$n_x m$	$n_x m$	$5\lg(n_x)$
2	$I : R_{32bit}^{m \times n_x}$ $T : R_{32bit}^{n_{let} \times n_{rot} \times n_{font}^2}$	$\frac{1}{8}n_{let}n_{rot}n_{font}^4 n_{target}$	n_{target}	$n_x m$	$n_{font}^2 n_{target}$	$\frac{n_{let}n_{rot}n_{font}^4}{8n_{target}}$

Note: Computational complexity, degrees of parallelism, IO, and operations per sample for each stage of the SAR algorithm.

parallel pipeline. For example, it makes sense to combine Stages 1c and 1d (now referred to as 1cd) since they do not involve changing the direction of parallelism (see Table 15-1). Thus, in Stage 1 there are four steps (1a, 1b, 1cd, and 1e), which require three corner turns. This is typical of most SAR systems.

In Stage 2, pairs of images are compared to find the locations of new "targets" (denoted by n_{target}). In the case of the SAR benchmark, these targets are just $n_{font} \times n_{font}$ images of rotated capital letters that have been randomly inserted into the SAR image. The region of interest (ROI) around each target is then correlated with each possible letter and rotation to determine its identity, its rotation, and its location in the SAR image. The parallelism in this stage is determined by n_{target} and can be along the rows or columns or both, as long as enough overlapping edge data are kept on each processor to do the correlations. These edge pixels are sometimes referred to as "overlap," "boundary," "halo," or "guard" cells.

The input bandwidth is a key parameter in describing the overall performance requirements of the system. The input bandwidth BW^i_{input} (in samples/second) for each processing stage i is given by

$$BW^1_{input} = nm_c/T_{input} ,$$
$$BW^2_{input} = n_x m/T_{input} .$$

$$(15.1)$$

A simple approach for estimating the overall required processing rate is to multiply the input bandwidth by the number of operations per sample required. Looking at Table 15-1, if we assume

$$n \approx n_x \approx 8000 , \quad m_c \approx m \approx 4000 , \quad n_{target} \approx \frac{n_x m}{8 n^2_{font}} ,$$

then the operations (or work) done per data input sample W^i_{sample} can be approximated by

$$W^1_{sample} \approx 10 \lg(n) + 20 \lg(m_c) + 40 \approx 400 ,$$
$$W^2_{sample} \approx \frac{1}{8} n_{let} n_{rot} n^2_{font} \approx 1000 .$$

$$(15.2)$$

Thus the rate of performance goal R^i_{goal} is approximately

$$R^1_{goal} \approx W^1_{sample} BW^1_{input} \approx 25 \times 10^9 / T_{input} ,$$
$$R^2_{goal} \approx W^2_{sample} BW^2_{input} \approx 16 \times 10^9 / T_{input} .$$

$$(15.3)$$

T_{input} varies from system to system, but can easily be much less than a second, resulting in large compute performance goals. Satisfying these performance goals often requires a parallel computing system.

The file IO requirements in "System Mode" or "IO Only Mode" are just as challenging as the computation. In this case the goal is to read and write the files as quickly as possible. During Stage 1, a file system must read large input files and write large image files. Simultaneously, during Stage 2, the image files are selected at random and read. After Stage 2 detection is performed, many very small "thumbnail" images around the targets are written. This diversity of file sizes and the need for simultaneous read and write are very stressing and often require a parallel file system.

Processor Sizing

Often the first step in the development of a system is to produce a rough estimate of how many processors will be needed. This step often occurs during development, perhaps early on in the system design phase. Frequently processor sizing estimates are used to determine the type of processing technology to use (programmable or non-programmable) and approximately how much a solution will cost to implement. A basic processor sizing estimate consists of

Computational Complexity Analysis. Estimates the total amount of work that needs to be done by the application (e.g., see Equation 15.2).

Throughput Analysis. Converts the total work done into a rate at which this work must be performed (e.g., see Equation 15.3).

Processor Performance. Estimates of what the performance of the algorithm will be on a single processor.

After the above analysis has been completed, the nominal processor size (in terms of number of processors) is estimated by dividing the required throughput by the processing rate

$$ N_P^{est} \approx \sum_i \frac{R_{goal}^i}{\varepsilon_{comp}^i R_{peak}} = \frac{R_{goal}^1}{\varepsilon_{comp}^1 R_{peak}} + \frac{R_{goal}^2}{\varepsilon_{comp}^2 R_{peak}}, \tag{15.4} $$

where ε_{comp}^i is the efficiency at which a processor with peak performance R_{peak} (in operations/second) can perform the work in stage i. Note: This procedure only gives a rough approximation and should always be used as such.

15.3 DEGREES OF PARALLELISM

The parallel opportunities at each stage of the calculation discussed in the previous section show that there are many different ways to exploit parallelism in this application. These different types of parallelism can be roughly grouped into three categories (see Figure 15-6):

Coarse Grained. This is the highest level of parallelism and exploits the fact that each raw SAR input can be processed independently of the others. This form of parallelism usually requires the least amount of communication between processors, but requires the most memory and has the highest latency.

Task/Pipeline. This decomposes different stages of the processing into a pipeline. The output of one stage is fed into the input of another so that at any given time each stage is working on a different SAR dataset.

Data Parallelism. This is the finest grain parallelism and decomposes each SAR image into smaller pieces.

For the specific problem at hand, we will look at exploiting all of these different types of parallelism. However, our analysis will only extend to using these to one level of depth. In other words, the most complex case we will consider is multiple coarse-grained pipelines, where each pipeline has multiple data parallel steps. We will not examine the very common, but even more complex case, where individual steps within a pipeline are further broken up into more coarse grain sub-pipelines with a further number of additional data parallel sub-steps.

We will parameterize parallel implementation as follows:

$N_{coarse} \equiv$ number of different problems or pipelines,
$N_{stage} \equiv$ number of stages in the pipeline,
$N_{stage}^i \equiv$ the number of processors used at stage i in the pipeline.

These parameters must satisfy the constraint that the sum of all the processors used at every stage in every pipeline is equal to the total number of processors N_P:

$$N_P = N_{coarse} \sum_{i=1}^{N_{stage}} N_{stage}^i .$$

There are several important special cases of the above description. First is the pure coarse grain parallel case, where each processor is given an entire SAR dataset to process and neither fine grain parallelism nor pipeline parallelism is exploited:

$$N_{coarse} = N_P, \; N_{stage} = 1, \; N_{stage}^1 = 1 .$$

Next is the pure pipeline parallel case. This case is limited in that the total number of processors cannot be more than the number of stages in the pipeline:

$$N_{coarse} = 1, \; N_{stage} = N_P, \; N_{stage}^i = 1 .$$

Finally, there is the pure data parallel case where only fine grain parallelism is exploited:

$$N_{coarse} = 1, \; N_{stage} = 1, \; N_{stage}^1 = N_P .$$

15.3.1 Parallel Performance Metrics (no communication)

In the absence of any communication and IO cost, all of these parallel approaches will take the same time to process a frame of data (where we assume that there are N_{frame} instances),

$$T_{comp}(N_P) = \frac{W^{tot}}{\varepsilon_{comp} R_{peak} N_P} \propto \frac{W^{tot}}{N_P} ,$$

where W^{tot} is the total computational operations or work required to process a frame of data at each stage (see Table 15-1)

$$W^{tot} \equiv \sum_{i=1}^{N_{stage}} W^i = W^{1a} + W^{1b} + W^{1cd} + W^{1e} + W^2 .$$

The computational speedup will be linear in the number of processors:

$$S_{comp}(N_P) \equiv \frac{T_{comp}(1)}{T_{comp}(N_P)} = N_P .$$

Likewise, in the zero communication case, the latency (i.e., the time to get the first answer) is the number of stages times the longest time it takes to process any one stage,

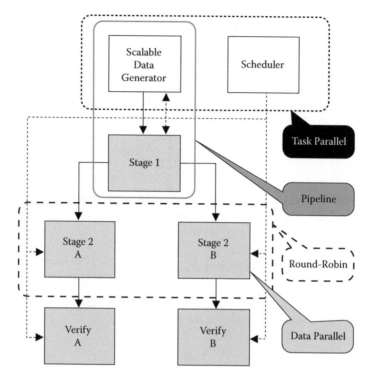

FIGURE 15-6 Types of parallelism. Within the SAR benchmark there are many different types of parallelism that can be exploited: task (coarse grain), pipeline, round-robin (coarse grain), and data (fine grain).

$$T_{latency}(N_P) = N_{stage} \max_i (T^i_{latency}(N^i_{stage})),$$

where

$$T_{latency}(N^i_{stage}) = \frac{W^i_{stage}}{\varepsilon_{comp} R_{peak} N^i_{stage}}.$$

This is because no one stage can progress until the previous stage has also finished. The latency speedup in this case is

$$S_{latency}(N_P) \equiv \frac{T_{latency}(1)}{T_{latency}(N_P)}.$$

The total memory required for the different approaches is

$$M(N_P) = N_{coarse} \sum_{i=1}^{N_{stage}} \frac{M^i_{stage}}{N^i_{stage}}.$$

Perhaps more importantly is the max amount of memory required by any particular processor,

$$M_{cpu-max}(N_P) = \max_i (M^i_{stage}/N^i_{stage}).$$

15.3.2 Parallel Performance Metrics (with communication)

Of course, assuming no communication and storage costs is a bad assumption and essentially the entire goal of designing a parallel implementation is to minimize these costs. Nevertheless, the above models are useful in that they provide upper bounds on performance. In addition, the speedup and memory footprint definitions are unaltered when communication is included.

In the case of fine grain parallelism, communication is required between steps where the directionality of the parallelism changes, which in turn requires that a corner turn is performed to reorganize the data to exploit the parallelism. The time to communicate data between steps i and $i+1$ is given by

$$T_{comm}^{i \to i+1}(N_{stage}^i \to N_{stage}^{i+1}) = \frac{D^{i \to i+1}}{BW_{eff}(N_{stage}^i \to N_{stage}^{i+1})},$$

where $D^{i \to i+1}$ is the amount of data (in bytes) moved between steps i and $i+1$ and $BW_{eff}(N_{stage}^i \to N_{stage}^{i+1})$ is the effective bandwidth (in bytes/second) between the processors, which is given by

$$BW_{eff}(N_{stage}^i \to N_{stage}^{i+1}) = \varepsilon_{comm}^{i \to i+1} BW_{peak}(N_{stage}^i \to N_{stage}^{i+1}).$$

$BW_{peak}(N_{stage}^i \to N_{stage}^{i+1})$ is the peak bandwidth from the processors in stage i to the processors in stage $i+1$. A more detailed model of this performance requires a model of the processor interconnect, which is given in the next section.

The implication of the communication on various cases is as follows. For the pure pipeline case (assuming that computation and communication can be overlapped), we have

$$T_{tot}(N_P) = \max_i(\max(T_{latency}^i(1), T_{comm}^{i \to i+1}(1 \to 1)))/N_{coarse},$$

$$T_{latency}(N_P) = 2 N_{stage} \max_i(\max(T_{latency}^i(1), T_{comm}^{i \to i+1}(1 \to 1))).$$

If communication and computation cannot be overlapped, then the inner max(,) function is replaced by addition. In this case, we see that the principal benefit of pipeline parallelism is that it is determined by the max (instead of the sum) of the stage times. The price for improved performance is increased latency, which is proportional to the number of computation and communication stages. There is a nominal impact on the memory footprint.

For the pure data parallel case, we have

$$T_{tot}(N_P) = \frac{T_{comp}(N_P) + T_{comm}(N_P)}{N_{coarse}},$$

$$T_{latency}(N_P) = T_{comp}(N_P) + T_{comm}(N_P),$$

where

$$T_{comm}(N_P) = \sum_i T_{comm}^{i \to i+1}(N_P \to N_P).$$

In general, in a pure data parallel approach, it is hard to overlap computation and communication. The principal benefit of data parallelism is that the compute time is reduced as long as it is not offset by the required communication. This approach also reduces the latency in a similar fashion and provides a linear reduction in the memory footprint.

For the combined pipeline parallel case, we have

$$T_{tot}(N_P) = \max_i(\max(T_{comp}(N_{stage}^i), T_{comm}^{i \to i+1}(N_{stage}^i \to N_{stage}^{i+1})))/N_{coarse},$$

$$T_{latency}(N_P) = 2N_{stage} \max_i(\max(T_{comp}(N_{stage}^i), T_{comm}^{i \to i+1}(N_{stage}^i \to N_{stage}^{i+1}))).$$

This approach provides an overall reduction in computation time and allows for overlapping computation and communication, and a reduction in the memory footprint. As we shall see later, this is often the approach of choice for parallel signal processing systems.

15.3.3 Amdahl's Law

Perhaps the most important concept in designing parallel implementations is managing overhead. Assume the total amount of work that needs to be done can be broken up into a part that can be done in parallel and a part that can only be done in serial (i.e., on one processor):

$$W_{tot} = W_{||} + W_{|}.$$

The execution time then scales as follows:

$$T_{comp}(N_P) \propto W_{||}/N_P + W_{|},$$

which translates into a speedup of

$$S_{comp}(N_P) = \frac{W_{tot}}{W_{||}/N_P + W_{|}}.$$

If we normalize with respect to W_{tot}, this translates to

$$S_{comp}(N_P) = \frac{1}{w_{||}/N_P + w_{|}},$$

where $w_{||} = W_{||} / W_{tot}$ and $w_{|} = W_{|} / W_{tot}$. In the case when N_P is very large, the maximum speedup achievable is

$$S_{max} = S_{comp}(N_P \to \infty) = w_{|}^{-1}.$$

For example, if $w_1 = 0.1$, then the maximum speedup is 10 (see Figure 15-7). This fundamental result is referred to as Amdahl's Law and the value $w_{|}$ is often referred to as the "Amdahl fraction" of the application. Amdahl's Law highlights the need to make every aspect of a code parallel. It also applies to other overheads (e.g., communication) that cause no useful work to be done. Finally, Amdahl's Law is also a useful tool for making trade-offs. Half the maximum speedup is achieved at $N_P = w_{|}^{-1}$:

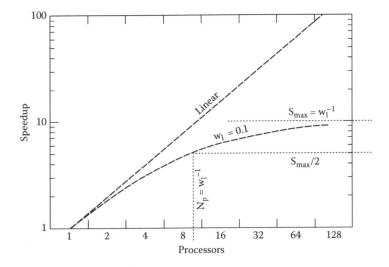

FIGURE 15-7 Amdahl's Law. Plot showing speedup of a program with an Amdahl fraction of $w_{\parallel} = 0.1$. Dotted lines show the max speedup $S_{max} = w_{\parallel}^{-1} = 10$ and the half speedup point $N_P = w_{\parallel}^{-1}$.

$$S_{comp}(N_P = w_{\parallel}^{-1}) = w_{\parallel}^{-1} / (w_{\parallel} + 1) \approx \frac{1}{2} S_{max} .$$

If $w_{\parallel} = 0.1$, then using more than 10 processors will be of marginal utility. Likewise, if an application needs a speedup of 100 to meet its performance goals, then $w_{\parallel} < 0.01$.

15.4 STANDARD PROGRAMMABLE MULTI-COMPUTER

Embedded computers are tightly coupled to their applications and are physically connected to their input sensors. A canonical embedded system is shown in Figure 15-8. It consists of control processor, interface, and network, and a signal processor, interface, and network. Typically, the control processor, interface, and network are selected for programmability, reliability, and flexibility. In contrast, the signal processor, interface, and network are often driven by performance consider-

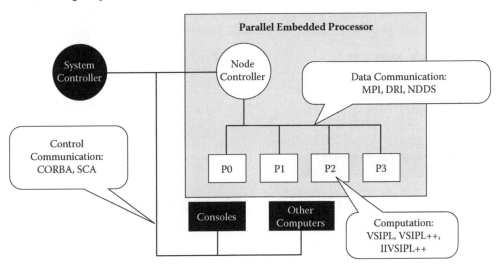

FIGURE 15-8 Canonical embedded system architecture.

ations. If just one processor cannot process the input data stream, then the signal processor must incorporate multiple processors. Sometimes this is referred to as an embedded multi-computer. If a programmable signal processor is selected (as opposed to one implemented in hardware), then parallel computing is the only approach for increasing performance to meet the timing requirements.

The canonical parallel computer architecture consists of a number of nodes connected by a network. Each node consists of a processor and memory. The memory on the node may be further decomposed into various levels of cache, main memory, and disk storage (see Figure 15-9). There are many variations on this theme. A node may have multiple processors sharing various levels of cache, memory, and disk storage. In addition, there may be external global memory or disk storage that is visible to all nodes. An important implication of this canonical architecture is the "memory hierarchy" (Figure 15-9). The memory hierarchy concept refers to the fact that the memory "closer" to the processor is much faster (the bandwidth is higher and latency is lower). However, the price for this performance is capacity. The canonical ranking of the hierarchy is typically as follows:

- Processor registers
- Cache L1, L2, ...
- Local main memory
- Remote main memory
- Local disk
- Remote disk

Typically, each level in the hierarchy exchanges a 10× performance increase for a 10× reduction in capacity. The "steepness" of the memory hierarchy refers to the magnitude of the performance reduction as one descends down the memory hierarchy. If these performance reductions are large, we say the system has a very steep memory hierarchy.

One benchmark specifically designed to probe the memory hierarchy is the HPC Challenge Benchmark suite (Luszczek, Dongarra, and Kepner 2006). HPC Challenge benchmarks have been chosen to cover a range of memory access patterns and stress different parts of the memory hierarchy. Top 500 performance is mostly dominated by local matrix multiply operations. STREAM requires no communication, is dominated by local vector operations, and stresses local processor to memory bandwidth. The fast Fourier transform (FFT) is also dominated by all-to-all communications, but for very large messages. Random Access is dominated by all-to-all communications of very small messages.

The concept of the memory hierarchy is probably the most important idea to keep in mind when developing a high performance implementation. More specifically, a high performance imple-

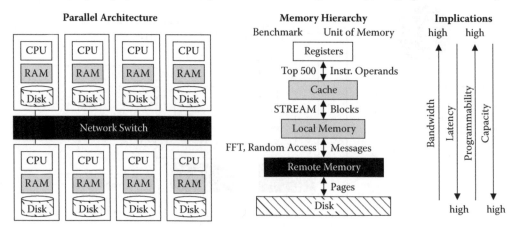

FIGURE 15-9   Canonical parallel architecture and memory hierarchy.

mentation of a sensor processing system is one that best mitigates the performance impacts of the memory hierarchy. This requires programmers to have a clear picture of the system in their minds so that they can understand the precise performance implications. The basic rule of thumb is to construct an implementation that minimizes both the total volume and number of data movements up and down the hierarchy.

15.4.1 Network Model

The communication network that connects the processors in a parallel system is often the most critical piece in designing a parallel implementation of an application. The starting point of modeling a communication network is point-to-point performance. This can be effectively measured by timing how long it takes two processors to send messages of different sizes to each other. The result is a standard curve (Figure 15-10), which can mostly be characterized by a single function $T(m)$, which is the time it takes to send a message of size m. The function $T_{comm}(m)$ is typically well described by the following simple two-parameter formula:

$$T_{comm}(m) = \text{Latency} + \frac{m}{\text{Bandwidth}} \ . \tag{15.5}$$

Bandwidth, which is typically measured in bits or bytes per second, is the maximum rate at which data can flow over a network. Bandwidth is typically measured by timing how long it takes to send a large message between two processors and dividing the message size by the total time:

$$\text{Bandwidth} = \lim_{m \to \infty} \frac{T(m)}{m} \ . \tag{15.6}$$

Latency is how long it takes for a single bit to be sent from one processor to the next. It is usually measured by timing a very short message and is typically computed from the formula

$$\text{Latency} = \lim_{m \to 0} T(m) \ . \tag{15.7}$$

These two parameters are important because some programs require sending a few very large messages (and are limited by bandwidth), and some programs require sending lots of very small messages (and are limited by latency).

Figure 15-10 shows the latency and instantaneous bandwidth for a typical cluster network as a function of message size. The important features of this curve, which are typical of most networks, are the leveling off in latency at small messages and the leveling off in bandwidth that occurs at large messages. While knowing the bandwidth and latency of a system is helpful, what it is even more helpful is comparing these parameters with respect to the computing power of the processor. More specifically, dividing the processor speed (typically measured in floating-point operations per second—FLOPS) by the bandwidth (measured in 8 byte floats per second) yields the number of FLOPS that can be performed in the time it takes to send a message of a given size (see Figure 15-10). This "inverse bandwidth" number provides a good guide to the minimum number of operations that need be performed on a value to amortize the cost of communicating that value.

In this case, for small messages nearly 10,000 operations need to be performed on each 8 byte element. At large messages, only 100 operations need to be performed. These data clearly illustrate a key parallel programming design paradigm: it is better to send fewer large messages than many smaller messages.

What is particularly nice about describing the network relative to the processor performance is that these numbers are relatively constant for most of the systems of interest. For example, these

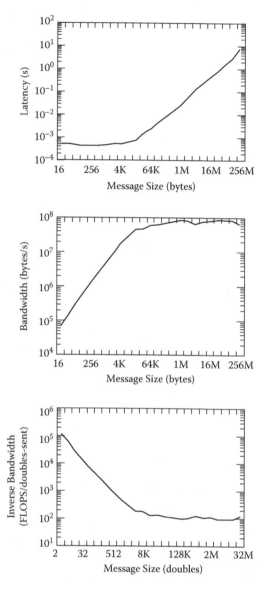

FIGURE 15-10 Network performance. Latency, bandwidth, and "inverse bandwidth" as a function of message size for a typical cluster network. Inverse bandwidth is shown in terms of the number of FLOPS/doubles-sent.

values (10^5 for small messages and 10^2 for large messages) are typical of many systems and do not vary much over time. The absolute processor and network speeds can change dramatically over time, but these ratios remain relatively constant. For example, while processor and network speeds have improved by almost 10^3 over the last several decades, the relative performance has changed by only a factor of 10.

These parameters suggest that if a parallel program uses large messages, it should be performing >100 operations on each number before communicating the number over the network. Likewise, if a program sends small messages, it should be performing >10,000 operations on each number before sending. Doing less than these amounts will tend to result in a parallel program that does not perform very well because most of the time will be spent sending messages instead of performing computations.

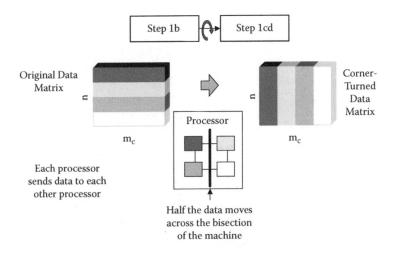

FIGURE 15-11   Corner turn.

Using the simple point-to-point model, we can predict the time for a more complex operation such as a "corner turn" (see Figure 15-11). This requires all-to-all communication where a set of processors P_1 sends a messages of size m to each of a set of processors P_2:

$$T_{CornerTurn} = \frac{P_1 P_2 (\text{Latency} + m/\text{Bandwidth})}{Q},$$

where B is the bytes per message and Q is the number of simultaneous parallel paths from processors in P_1 to processors in P_2. Total amount of data moved in this operation is $mP_1 P_2$.

15.5   PARALLEL PROGRAMMING MODELS AND THEIR IMPACT

The parallel programming model describes how the software is going to implement the signal processing chain on a parallel computer. A good software model allows the different types of parallelism to be exploited: data parallelism, task parallelism, pipeline parallelism, and round-robin. In addition, a good implementation of the parallel programming model allows the type of parallelism to be exploited to change as the system is built and the performance requirements evolve.

There are a number of different parallel programming models that are commonly used. We will discuss three in particular: threaded, messaging, and global arrays, which are also called partitioned global address spaces (PGAS).

The threaded model is the simplest parallel programming model. It is used when a problem can be broken up into a set of relatively independent tasks that the workers (threads) can process without explicitly communicating with each other. The central constraint of the threaded model is that each thread only communicates by writing into a shared-memory address space that can seen by all other threads. This constraint is very powerful and is enormously simplifying. Furthermore, it has proved very robust and the vast majority of parallel programs written for shared-memory systems using a small number of processors (i.e., workstations) use this approach. Examples of this technology include OpenMP [http://www.openmp.org], POSIX Threads or pthreads [http://www.pasc.org/plato/], and Cilk [http://supertech.csail.mit.edu/cilk/].

The message passing model is in many respects the opposite of the threaded model. The message passing model requires that any processor be able to send and receive messages from any other processor. The infrastructure of the message passing model is fairly simple. This infrastructure is most typically instantiated in the parallel computing community via the Message Passing Interface

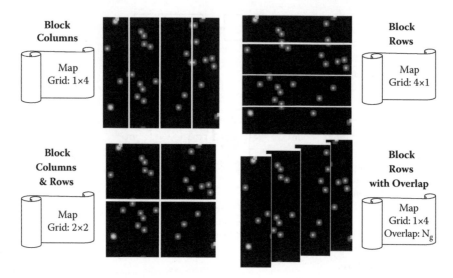

FIGURE 15-12 (**Color figure follows p. 278.**) Global array mappings. Different parallel mappings of a two-dimensional array. Arrays can be broken up in any dimension. A block mapping means that each processor holds a contiguous piece of the array. Overlap allows the boundaries of an array to be stored on two neighboring processors.

(MPI) standard [http://wwww.mpi-forum.org]. The message passing model requires that each processor have a unique identifier and must know how many other processors are working together on a problem (in MPI terminology these are referred to as the processor "rank" and the "size" of the MPI world). Any parallel program can be implemented using the message passing model. The primary drawback of this model is that the programmer must manage every individual message in the system, which can often require a great deal of additional code and can be extremely difficult to debug. Nevertheless, there are certain parallel programs that can only be implemented with a message passing model.

The PGAS model is a compromise between the two models. Global arrays impose additional constraints on the program, which allow complex programs to be written relatively simply. In many respects it is the most natural parallel programming model for signal processing because it is implemented using arrays, which are the core data type of signal processing algorithms. Briefly, the global arrays model creates distributed arrays in which each processor stores or owns a piece of the whole array. Additional information is stored in the array so that every processor knows which parts of the array the other processors have. How the arrays are broken up among the processors is specified by a Map (Lebak et al. 2005). For example, Figure 15-12 shows a matrix broken up by rows, columns, rows and columns, and columns with some overlap. The different mappings are useful concepts to have even if the global array model is not being used. The concept of breaking up arrays in different ways is one of the key ideas in parallel computing. Computations on global arrays are usually performed using the "owner computes" rule, which means that each processor is responsible for doing a computation on the data it is storing locally. Maps can become quite complex and express virtually arbitrary distributions.

In the remainder of this section, we will focus on PGAS approaches.

15.5.1 HIGH-LEVEL PROGRAMMING ENVIRONMENT WITH GLOBAL ARRAYS

The pure PGAS model presents an entirely global view of a distributed array. Specifically, once created with an appropriate map object, distributed arrays are treated the same as non-distributed ones. When using this programming model, the user never accesses the local part of the array and all operations (such as matrix multiplies, FFTs, convolutions, etc.) are performed on the global structure.

The benefits of pure global arrays are ease of programming and the highest level of abstraction. The drawbacks include the need to implement parallel versions of every single function that may exist in a serial software library. In addition, these functions need to be supported for all possible data distributions. The implementation overhead of a full global arrays library can be quite large.

Fragmented PGAS maintains a high level of abstraction but allows access to local parts of the arrays. Specifically, a global array is created in the same manner as in pure PGAS; however, the operations can be performed on just the local part of the array. Later, the global structure can be updated with locally computed results. This allows greater flexibility. Additionally, this approach does not require function coverage or implementation of parallel versions of all existing serial functions. Furthermore, fragmented PGAS programs often achieve better performance by eliminating the library overhead on local computations.

The first step in writing a parallel program is to start with a functionally correct serial program. The conversion from serial to parallel requires users to add new constructs to their code. In general, PGAS implementations tend to adopt a separation-of-concerns approach to this process which seeks to make functional programming and mapping a program to a parallel architecture orthogonal. A serial program is made parallel by adding maps to arrays. Maps only contain information about how an array is broken up onto multiple processors, and the addition of a map should not change the functional correctness of a program. An example map for the pMatlab [http://www.ll.mit.edu/pMatlab] PGAS library is shown in Figure 15-12. A pMatlab map (see Figure 15-13) is composed of a grid specifying how each dimension is partitioned, a distribution that selects either a block, cyclic, or block-cyclic partitioning, and a list of processors that defines which processors actually hold the data.

The concept of using maps to describe array distributions has a long history. The ideas for pMatlab maps are principally drawn from the High Performance Fortran (HPF) community (Loveman 1993; Zosel 1993), MIT Lincoln Laboratory Space-Time Adaptive Processing Library (STAPL) (DeLuca et al. 1997), and Parallel Vector Library (PVL) (Lebak et al. 2005). A map for a numerical array defines how and where the array is distributed (Figure 15-12). PVL also supports task parallelism with explicit maps for modules of computation. pMatlab and VSIPL++ explicitly only support data parallelism; however, implicit task parallelism can be implemented through careful mapping of data arrays.

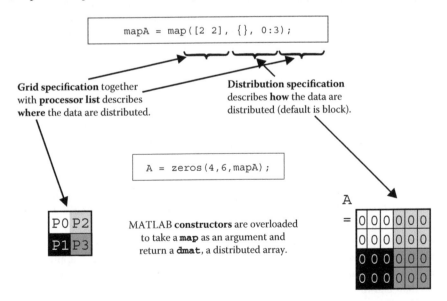

FIGURE 15-13 Anatomy of a map. A map for a numerical array is an assignment of blocks of data to processing elements. It consists of a grid specification (in this case a 2 × 2 arrangement), a distribution (in this case {} implies that the default block distribution should be used), and a processor list (in this case the array is mapped to processors 0, 1, 2, and 3).

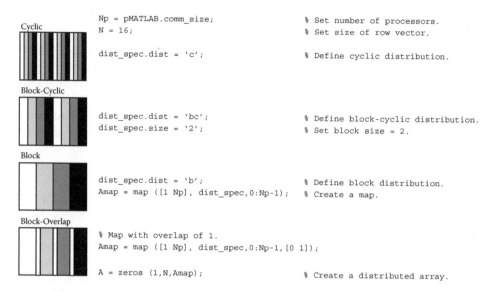

```
                    Np = pMATLAB.comm_size;              % Set number of processors.
Cyclic              N = 16;                              % Set size of row vector.

                    dist_spec.dist = 'c';                % Define cyclic distribution.


Block-Cyclic

                    dist_spec.dist = 'bc';               % Define block-cyclic distribution.
                    dist_spec.size = '2';                % Set block size = 2.

Block

                    dist_spec.dist = 'b';                % Define block distribution.
                    Amap = map ([1 Np], dist_spec,0:Np-1);  % Create a map.

Block-Overlap

                    % Map with overlap of 1.
                    Amap = map ([1 Np], dist_spec,0:Np-1,[0 1]);

                    A = zeros (1,N,Amap);                % Create a distributed array.
```

FIGURE 15-14 Block cyclic distributions. Block distribution divides the object evenly among available processors. Cyclic distribution places a single element on each available processor and then repeats. Block-cyclic distribution places the specified number of elements on each available processor and then repeats.

For illustrative purposes, we now describe the pMatlab map. The PVL, VSIPL++, as well as many other PGAS implementations, use a similar construct. The pMatlab map construct is defined by three components: (1) grid description, (2) distribution description, and (3) processor list. The grid description together with the processor list describes where the data object is distributed, while the distribution describes how the object is distributed (see Figure 15-13). pMatlab supports any combination of block-cyclic distributions up to four dimensions. The API defining these distributions is shown in Figure 15-14.

Block distribution is the default distribution, which can be specified explicitly or by simply passing an empty distribution specification to the map constructor. Cyclic and block-cyclic distributions require the user to provide more information. Distributions can be defined for each dimension and each dimension could potentially have a different distribution scheme. Additionally, if only a single distribution is specified and the grid indicates that more than one dimension is distributed, that distribution is applied to each dimension.

Some applications, particularly image processing, require data overlap, or replicating rows or columns of data on neighboring processors. This capability is also supported through the map interface. If overlap is necessary, it is specified as an additional fourth argument. In Figure 15-14, the fourth argument indicates that there is 0 overlap between rows and 1 column overlap between columns. Overlap can be defined for any dimension and does not have to be the same across dimensions.

While maps introduce a new construct and potentially reduce the ease of programming, they have significant advantages over both message passing approaches and predefined limited distribution approaches. Specifically, pMatlab maps are scalable, allow optimal distributions for different algorithms, and support pipelining.

Maps are scalable in both the size of the data and the number of processors. Maps allow the user to separate the task of mapping the application from the task of writing the application. Different sets of maps do not require changes to be made to the application code. Specifically, the distribution of the data and the number of processors can be changed without making any changes to the algorithm. Separating mapping of the program from the functional programming is an important design approach in pMatlab.

Maps make it easy to specify different distributions to support different algorithms. Optimal or suggested distributions exist for many specific computations. For example, matrix multiply operations are most efficient on processor grids that are transposes of each other. Column and row-wise FFT operations produce linear speedup if the dimension along which the array is broken up matches the dimension on which the FFT is performed.

Maps also allow the user to set up pipelines in the computation, thus supporting implicit task parallelism. For example, pipelining is a common approach to hiding the latency of the all-to-all communication required in parallel FFT. The following pMatlab code fragment elegantly shows a two-dimensional pipelined FFT run on eight processors:

```
Ymap1b = map([4 1],{},[0:3]);  % Row map on ranks 0,1,2,3
Ymap1c= map([1 4],{},[4:7]);   % Col map on ranks 4,5,6,7
Y1b = complex(zeros(n,m));      % Create Y for step 1b
Y1c = complex(zeros(n,m));      % Create Y for step 1c
...                             % Fill Y with data
Y1b = fft(Y1b,{},1);            % FFT rows (ranks:0,1,2,3)
Y1c(:,:) = Y1b;                 % Cornerturn
Y1c = fft(Y1c,{}21);            % FFT cols (ranks:4,5,6,7)
```

The above fragment shows how a small change in the maps can be used to set up a pipeline where the first half of the processors perform the first part of the FFT and the second half perform the second part. When a processor encounters such a map, it first checks if it has any data to operate on. If the processor does not have any data, it proceeds to the next line. In the case of the FFT with the above mappings, the first half of the processors (ranks 0 to 3) will simply perform the row FFT, send data to the second set of processors and skip the column FFT, and proceed to process the next set of data. Likewise, the second set of processors (ranks 4 to 7) will skip the row FFT, receive data from the first set of processors, and perform the column FFT.

15.6 SYSTEM METRICS

At this point, we have described a canonical application and a canonical parallel signal processor. In addition, we have nominally parameterized how the application might be mapped onto a parallel architecture. The actual selection of a parallel mapping is decided by the constraints of the system. This section presents a more formal description of some of the the most common system metrics: performance, efficiency, form factor, and software cost.

15.6.1 PERFORMANCE

Performance is the primary driver for using a parallel computing system and refers to the time it takes to process one dataset in a series of datasets. In our application it refers to time to process a SAR image through the entire chain. Performance is usually decomposed into latency and throughput.

Latency refers to the time it takes to get the first image through the chain. Latency is fundamentally constrained by how quickly the consumer of the data needs the information. Some typical latencies for different systems are

- microseconds: Real-time control/targeting
- milliseconds: Operator in the loop
- seconds: Surveillance systems supporting operations
- minutes: Monitoring systems
- hours: Archival systems

For the SAR application, we will pick a latency target $(T_{latency}^{goal})$ and we may choose to express it in terms T_{input}. For this example, let us set an arbitrary latency goal of

$$T_{latency}^{goal}/T_{input} \approx 10 .$$

Throughput is the rate at which the images can be processed. Fundamentally, the data must be processed at the same rate it is coming into the system; otherwise it will "pile up" at some stage in the processing. A key parameter here is the required throughput relative to what can be done on one processor. For this example, let us set an arbitrary throughput goal of

$$S_{comp}^{goal} = T_{comp}^{goal}/T_{comp}(1) \approx 100 .$$

15.6.2 FORM FACTOR

One of the unique properties of embedded systems is the form factor constraints imposed by the application. These include the following:

Size. The physical volume of the entire signal processor including its chassis, cables, cooling and power supplies. The linear dimensions (height, width, and depth) and the total volume are constrained by the limitations of the platform.

Weight. The total weight of the signal processor system.

Power. Total power consumed by the signal processor and its cooling system. In addition, the voltage, its quality, and how often it is interrupted are also constraints.

Heat. The total heat the signal processor can produce that can be absorbed by the cooling system of the platform.

Vibration. Continuous vibration as well as sudden shocks may require additional isolation of the system.

Ambient Air. For an air-cooled system the ambient temperature, pressure, humidity, and purity of the air are also constraints.

IO Channels. The number and speed of the data channels coming into and out of the system.

The form factor constraints are very dramatically based on the type of platform: vehicle (car/truck, parked/driving/off-road), ship (small boat/aircraft carrier), aircraft [small unmanned air vehicle (UAV) to Jumbo jet]. Typically, the baseline for these form factor constraints is what can be found in an ideal environmentally controlled machine room. For example, if the overall compute goal requires at least 100 processing nodes, then in an ideal setting, these 100 processors will require a certain form factor. If these 100 processors were then put on a truck, it might have the following implications:

Size. 30% smaller volume with specific nonstandard dimensions $\Rightarrow$ high-density nodes with custom chassis $\Rightarrow$ increased cost.

Weight. Minimal difference.

Power. Requires nonstandard voltage converter and uninterruptible power supply $\Rightarrow$ greater cost and increased size, weight, and power.

Heat. Minimal difference.

Vibration. Must operate on road driving conditions with sudden stops and starts $\Rightarrow$ vibration isolators and ruggedized disk drives $\Rightarrow$ greater cost and increased size, weight, and power.

Compute (2)	Compute (2)	Compute (2)	Compute (2)	Compute (2)
Compute (2)	Compute (2)	Compute (2)	Compute (2)	Compute (2)
Compute (2)	Compute (2)	Compute (2)	Compute (2)	Compute (2)
Compute (2)	Compute (2)	Compute (2)	Compute (2)	Compute (2)
Compute (2)	Compute (2)	Compute (2)	Compute (2)	Compute (2)
Storage	Storage	Storage	Storage	Storage
Control Spare	Control Spare	Control Spare	Control Spare	Control Spare
Control	Control	Control	Control	Control
IO	IO	IO	IO	IO

FIGURE 15-15   Example computing rack. A canonical signal processing rack. Each rack contains five chassis. Each chassis has 14 slots. Four of the slots may need to be reserved for IO, control (and spare), and storage. The result is that 100 processors can be fit into the entire rack.

Ambient Air. Minimal difference.

IO Channels. There are precisely four input channels $\Rightarrow$ four processors must be used in the first processing step.

Processor Selection

Once it has been decided to go ahead and build a signal processing system, then it is necessary to select the physical hardware to use. Often it is the case that the above form factor requirements entirely dictate this choice. [An extreme case is when an existing signal processor is already in place and a new application or mode must be added to it.] For example, we may decide that there is room for a total of five chassis with all the required power, cooling, and vibration isolation requirements. Let's say each chassis has 14 slots. In each chassis, we need one input buffer board, one master control computer (and a spare), and a central storage device. This leaves ten slots, each capable of holding a dual processor node (see Figure 15-15). The result is

$$N_P^{real} = (5 \text{ chassis}) \ (10 \text{ slots/chassis}) \ (2 \text{ processors/slot})) = 100 .$$

At this point, the die is cast, and it will be up to the implementors of the application to make the required functionality "fit" on the selected processor. The procedure for doing this usually consists of first providing an initial software implementation with some optimization on the hardware. If this implementation is unable to meet the performance requirements, then usually a trade-off is done to see if scaling back some of the algorithm parameters (e.g., the amount of data to be processed) can meet the performance goals. Ultimately, a fundamentally different algorithm may be required, combined with heroic efforts by the programmers to get every last bit of performance out of the system.

15.6.3 EFFICIENCY

Efficiency is the fraction of the peak capability of the system the application achieves. The value of $T_{comp}(1)$ implies a certain efficiency factor on one processor relative to the theoretical peak performance (e.g., $\varepsilon_{comp} \approx 0.2$). There are similar efficiencies associated with bandwidth (e.g., $\varepsilon_{comm} \approx 0.5$) and the memory (e.g., $\varepsilon_{mem} \approx 0.5$). There are two principal implications of these efficiencies. If the required efficiency is much higher than these values, then it may mean that different hardware must be selected (e.g., nonprogrammable hardware, higher bandwidth networking, or higher density memory). If the required efficiency is well below these values, then it means that more flexible, higher level programming environments can be used, which can greatly reduce schedule and cost.

The implementation of the software is usually factored into two pieces. First is how the code is implemented on each individual processor. Second is how the communication among the different processors is implemented. The typical categories for the serial implementation environments follow:

Machine Assembly. Such as the instruction set of the specific processor selected. This provides the highest performance, but requires enormous effort and expertise and offers no software portability.

Procedural Languages with Optimized Libraries. Such as C used in conjunction with the Vector, Signal, and Image Processing Library (VSIPL) standard. This approach still produces efficient code, with less effort and expertise, and is as portable as the underlying library.

Object-Oriented Languages with Optimized Libraries. Such as C++ used in conjunction with the VSIPL++ standard. This approach can produce performance comparable to procedural languages with comparable expertise and is usually significantly less effort. Portability may be either more or less portable than procedural approaches depending upon the specifics of the hardware.

High-Level Domain-Specific Languages. Such as MATLAB, IDL, and Mathematica. Performance is usually significantly less than procedural languages, but generally requires far less effort. Portability is limited to the processors supported by the supplier of the language.

The typical categories for the parallel implementation environment are the following:

Direct Memory Access (DMA). This is usually a processor and network-specific protocol for allowing one processor to write into the memory of another processor. It delivers the highest performance, but requires enormous effort and expertise, and offers no software portability.

Message Passing. Such as the MPI. This is a protocol for sending messages between processors. It produces efficient code, with less effort and expertise, and is as portable as the underlying library.

Threading. Such as OpenMP or pthreads.

Parallel Arrays. Such as those found in Unified Parallel C (UPC), Co-Array Fortran (CAF), and Parallel VSIPL++. This approach creates parallel arrays using PGAS, which allow complex data movements to be written succinctly.

Very rough quantitative estimates for the performance efficiencies of the above approaches are given in Table 15-2 (Kepner 2004; Kepner 2006). The first column (labeled ε_{comp}) gives a very rough relative performance efficiency of a single-processor implementation using the approach specified in the first column. The second row (labeled ε_{comm}) gives a very rough relative bandwidth efficiency using the approach specified in the first row. The interior matrix of rows and columns shows the combined product of these two efficiencies and reflects the range of performance that can be impacted by the implementation of the software.

The significance of the product $\varepsilon_{comp}\varepsilon_{comm}$ can be illustrated as follows. The overall rate of work can be written as

$$R(N_P) = W/T = W/(T_{comp}(N_P) + T_{comm}(N_P)).$$

Substituting $T_{comp}(N_P) = W/\varepsilon_{comp}R_{peak}N_P$ and $T_{comm} = D/\varepsilon_{comm}BW_{peak}N_P$ gives

$$R(N_P) = \frac{\varepsilon_{comp}\varepsilon_{comm}R_{peak}N_P}{\varepsilon_{comm} + \varepsilon_{comm}(D/W)(R_{peak}/BW_{peak})},$$

TABLE 15-2

Software Implementation Efficiency Estimates

Serial Code	Serial $\in_{comp}$	Communication Model			
		DMA	Messaging	Threads	PGAS
$\in_{comm}$	—	0.8	0.5	0.4	0.5
Assembly	0.4	0.36			
Procedural	0.2	0.16	0.1	0.08	0.1
Object Oriented	0.18	0.14	0.09	0.07	0.09
High Level	0.04		0.02	0.016	0.02

Note: The first column lists the serial coding approach. The second column shows a rough estimate for the serial efficiency ($\in_{comp}$) of the serial approach. The first row of columns 3, 4, 5, and 6 lists the different communication models for a parallel implementation. The second row of these columns is a rough estimate of the communication efficiency ($\in_{comm}$) for these different models. The remaining entries in the table show the combined efficiencies ($\in_{comp}\in_{comm}$). Blank entries are given for serial coding and communication models that are rarely used together.

where D/W is the inherent communication-to-computation ratio of the application and R_{peak}/BW_{peak} is the computation-to-communication ratio of the computer. Both ratios are fixed for a given problem and architecture. Thus, the principal "knob" available to the programmer for effecting the overall rate of computation is the combined efficiency of the serial coding approach and the communication model.

15.6.4 SOFTWARE COST

Software cost is typically the dominant cost of implementing embedded applications and can easily be 10× the cost of the hardware. There are many approaches to implementing a parallel software system and they differ in performance, effort, and portability. The most basic approach to modeling software cost is provided by the Code and Cost Modeling (CoCoMo) framework (Boehm et al. 1995):

$$\text{Programmer effort [Days]} \approx$$

$$(\text{Total SLOC}) \frac{(\text{New code fraction}) + 0.05(\text{Reused code fraction})}{\text{SLOC/Day}}$$

This formula says that the effort is approximately linear in the total number of software lines of code (SLOC) written. It shows that there are three obvious ways to decrease the effort associated with implementing a program:

Increased Reuse. Including code that has already been written is much cheaper than writing it from scratch.

Higher Abstraction. If the same functionality can be written using fewer lines of code, this will cost less.

Increased Coding Rate. If the environment allows for more lines of code to be written in a given period of time, this will also reduce code cost.

TABLE 15-3

Software Coding Rate Estimates

Serial Code	Relative SLOC	SLOC/Day	Serial	Communication Model			
				DMA	Messaging	Threads	PGAS
Expansion Factor	—	—	1	2	1.5	1.05	1.05
Effort Factor	—	—	1	2	2	2	1.5
Assembly	3	5	1.6	0.5			
Procedural	1	15	15	5	8	14	14
Object Oriented	1/2	25	50	17	25	45	47
High Level	1/4	40	160		80	140	148

Note: The first column gives the code size relative to the equivalent code written in a procedural language (e.g., C). The next column gives the typical rate (SLOC/day) at which
lines are written in that environment. The column labeled "Serial" is the rate divided
by the relative code size and gives the effective relative rate of work done normalized
to a procedural environment. The row labeled "Expansion Factor" gives the estimated
increase in the size of the code when going from serial to parallel for the various parallel programming approaches. The row labeled "Effort Factor" shows the relative
increase in effort associated with each of these lines of parallel lines of code. The interior matrix combines all of these to give an effective rate of effort for each serial programming environment and each parallel programming environment given by
(SLOC/day)/(Relative SLOC)/(1 + (Expansion Factor – 1)(Effort Factor)).

Very rough quantitative estimates for the programming impacts of the above approaches are
given in Table 15-3. The first column gives the code size relative to the equivalent code written in
a procedural language (e.g., C). The next column gives the typical rate (SLOC/day) at which lines
are written in that environment. The column labeled "Serial" is the rate divided by the relative code
size and gives the effective relative rate of work done normalized to a procedural environment. The
row labeled "Expansion Factor" gives the estimated increase in the size of the code when going
from a serial code to a parallel code for the various parallel programming approaches. The row
labeled "Effort Factor" shows the relative increase in effort associated with each of these lines of
parallel lines of code. The interior matrix combines all of these to give an effective rate of effort for
each serial programming environment and each parallel programming environment. For example,
in the case of an object-oriented environment, on average each line does the work of two lines in
a procedural language. In addition, a typical programmer can code these lines in a serial environment at rate of 25 lines per day. If a code written in this environment is made parallel using message
passing, we would expect the total code size to increase by a factor of 1.5. Furthermore, the rate at
which these additional lines are coded will be decreased by a factor of two because they are more
difficult to write. The result is that the overall rate of the parallel implementation would be 25 effective procedural (i.e., C) lines per day.

Figure 15-16 notionally combines the data in Tables 15-2 and 15-3 for a hypothetical 100-pro-
cessor system and illustrates the various performance and effort trade-offs associated with different
programming models.

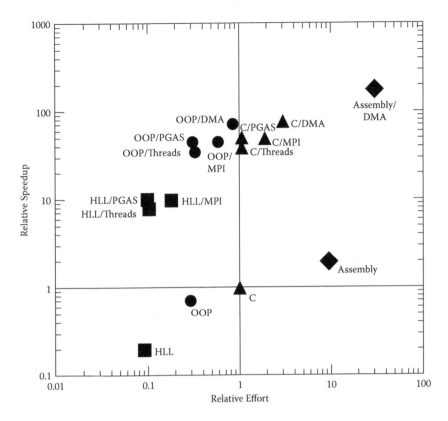

FIGURE 15-16   Speedup versus effort. Estimated relative speedup (compared to serial C) on a hypothetical 100-processor system plotted against estimated relative effort (compared to serial C).

REFERENCES

Bader, D.A., K. Madduri, J.R. Gilbert, V. Shah, J. Kepner, T. Meuse, and A. Krishnamurthy. 2006. Designing scalable synthetic compact applications for benchmarking high productivity computing systems. *CTWatch Quarterly* 2(4B).

Boehm, B., B. Clark, E. Horowitz, R. Madachy, R. Shelby, and C. Westland. 1995. Cost models for future software life cycle processes: COCOMO 2.0. *Annals of Software Engineering* 1: 57–94.

DeLuca, C.M., C.W. Heisey, R.A. Bond, and J.M. Daly. 1997. A portable object-based parallel library and layered framework for real-time radar signal processing. *Proceedings of the First Conference on International Scientific Computing in Object-Oriented Parallel Environments* 241–248.

Kepner, J. ed. 2004. Special issue on HPC productivity. *International Journal of High Performance Computing Applications* 18(4).

Kepner, J. ed. 2006. High productivity computing systems and the path towards usable petascale computing: user productivity challenges. *CTWatch Quarterly* 2(4A).

Lebak, J., J. Kepner, H. Hoffmann, and E. Rutledge. 2005. Parallel VSIPL++: an open standard software library for high-performance parallel signal processing. *Proceedings of the IEEE* 93(2): 313–330.

Loveman, D.B. 1993. High performance Fortran. *IEEE Parallel and Distributed Technology: Systems and Applications* 1(1): 25–42.

Luszczek, P., J. Dongarra, and J. Kepner. 2006. Design and implementation of the HPC Challenge Benchmark Suite. *CTWatch Quarterly* 2(4A).

Soumekh, M. 1999. *Synthetic Aperture Radar Signal Processing with MATLAB Algorithms*. New York: Wiley-Interscience.

Zosel, M.E. 1993. High performance Fortran: an overview. *Compcon Spring 93 Digest of Papers* 132–136.

APPENDIX A: A SYNTHETIC APERTURE RADAR ALGORITHM

This appendix provides the algorithmic details of the HPEC Challenge SAR benchmark.

A.1 SCALABLE DATA GENERATOR

The mathematical details of the Scalable Data Generator (SDG) are beyond the scope of this chapter and not relevant to the subsequent processing. The SDG simulates the response of a real sensor by producing a stream of $m_c \times n$ single-precision complex data matrices X, where

n = range (fast-time) samples, which roughly corresponds to the effective bandwidth times the duration of the transmitted pulses.

m_c = cross range (slow-time) samples, which roughly corresponds to the number of pulses sent.

In a real sensor, these input matrices arrive at a predetermined period T_{input}, which translates into an input data bandwidth

$$BW_{input} = (8 \text{ byte}) \, n \, m_c / T_{input} \, .$$

The processing challenge is to transform the raw data matrix into a sharp image before the next image arrives.

A.2 STAGE 1: FRONT-END SENSOR PROCESSING

This stage reads in the raw SAR data (either from a file or directly from an input buffer) and forms an image. The computations in this stage represent the core of a SAR processing system. The most compute-intensive steps involved in this transformation can be summarized as follows:

Matched Filtering. This step converts the raw data along the range dimension from the time domain to the frequency domain and multiplies the result by the shape of the transmitted pulse.

Digital Spotlighting. Performs the geometric transformations for combining the multiple views.

Interpolation. Converts the data from a polar coordinate system to a rectangular coordinate system.

The result of these processing steps is a $n_x \times m$ single precision image I. The core of the matched-filtering step (labeled "1a") consists of performing an FFT on each column of the input matrix X and multiplying it by a set of precomputed coefficients:

for $j = 1 : m_c$
    X(:,j) = FFT(X(:,j)) .* c $_{fast}$ (:,1) .* c$_1$ (:,j)
end

where

X = $n \times m_c$ complex single-precision matrix.
(:,j) = jth column of a matrix.
.* elementwise multiplication.

FFT() performs complex-to-complex one-dimensional FFT.

$c_{fast} = n \times 1$ complex vector of precomputed coefficients describing the shape of the transmitted pulse.

$c_1 = n \times m_c$ complex single-precision matrix of precomputed coefficients.

The computational complexity of this step is dominated by the FFT and is given by

$$W_{stage}^{1a} = 5\, n\, m_c\, (2 + log_2(n))\ \text{[FLOPS]}.$$

The parallelism in this step is principally that each column can be processed independently, which implies that there are m_c degrees of parallelism (DOP). Additional parallelism can be found by running each FFT in parallel. However, this requires significant communication. At the completion of this step, the amount of data sent to the next step is given by

$$D_{stage}^{1a \to 1b} = 8\, n\, m_c\ \text{[bytes]}.$$

The digital spotlighting step (labeled "1b") consists of performing an FFT on each row of X, which is then copied and offset into each row of a larger matrix Y. An inverse FFT of each row of Y is then multiplied by a set of precomputed coefficients and a final FFT is performed. Finally, the upper and lower halves of each row are swapped via the FFT_{shift} command. The result of this algorithm is to interpolate X onto the larger matrix Y.

```
for i = 1 : n
    X(i,:) = FFT(X(i,:))
    Y(i,1: m_c/2)) = m/m_c .* X(i,1: m_c/2)
    Y(i, m_c/2 + m_z + 1:m) = (m/mc) .* X(i, 1 + m_c/2:m_c)
    Y(i,:) = FFT_shift (FFT(FFT^{-1} (Y(i,:)) .* c_2 (i,:) ))
end
```

where

$m_z = m - m_c$.

$Y = n \times m$ complex single-precision matrix.

$(i,:) = i$th row of a matrix.

$(i_1 : i_2 ,:) = $ sub-matrix consisting of all rows i_1 to i_2.

$FFT_{shift}()$ swaps upper and lower halves of a vector.

$FFT^{-1}()$ performs complex-to-complex one-dimensional inverse FFT.

$c_2 = n \times m$ complex single-precision matrix of precomputed coefficients.

The computational complexity of this step is dominated by the FFTs and is given by

$$W_{stage}^{1b} = 5\, n\, (m_c(1 + log_2(m_c)) + m(1 + 2log_2(m))).$$

The parallelism in this step is principally that each row can be processed independently, which implies that there are n degrees of parallelism. At the completion of this step, the amount of data sent to the next step is given by

$$D_{stage}^{1b \to 1c} = 8\, n\, m\ \text{[bytes]}.$$

The backprojection step (labeled "1c") begins by completing the two-dimensional FFT_{shift} operation from step 1b. This consists of performing an FFT_{shift} operation on each column, which is then multiplied by a precomputed coefficient. The core of the interpolation step involves summing the projection of each element of $Y(i,j)$ over a range of values in $Z(i - i_K : i + i_K)$ weighted by the $sinc()$ and $cos()$ functions. The result of this algorithm is to project Y onto the larger matrix Z.

> **for** $j = 1:m$
> $Y(:,j)) = \text{FFT}_{shift} (Y(:,j)) .* c_2^{-1} (:,j)$
> **end**
>
> **for** $i = 1:n$
>   **for** $i_K = -n_K : n_K$
>   **for** $j = 1:m$
> $Z(i_{KX}(j) + i_K ,j) \mathrel{+}= Y(i,j) .*$
> $sinc(f_1(i,i_k,j)) .* (0.54 + 0.46 \cos(f_2(i,k_k,j)))$
>   **end**
>   **end**
> **end**

where

$Z = n_x \times m$ complex single-precision matrix.
n_K = half-width of projection region.
$f_{1,2}$ are functions that map indices into coordinates to be used by sinc() and cos() functions.

The computational complexity of this step is dominated by sinc() and cos() functions used in the the backprojection step,

$$W_{stage}^{1c} = 2\,n\,m\,n_K\,(O(sinc) + O(cos)) \approx 40\,n\,m\,n_K .$$

The parallelism in this step is principally that each column can be processed independently, which implies that there are n degrees of parallelism. This dimension is preferred because the interpolation step spreads values across each column. If this step were made parallel in the row dimension, this would require communication between neighboring processors. The parallelism in this step is the same as the beginning of the next step, so no communication is required.

The goal of the frequency to spatial conversion step (labeled "1d" and "1e") is to convert the data from the frequency domain to the spatial domain and to reorder the data so that it is spatially contiguous in memory. This begins by performing an FFT on each column, multiplying by a precomputed coefficient, and then circularly shifting the data. Next an FFT is performed on each row, multiplied by a precomputed coefficient and circularly shifted. Finally, the matrix is transposed and the absolute magnitude is taken. The resulting image is then passed on to the next stage.

> **for** $j = 1:m$
> $Z(:,j) = \text{cshift}(\text{FFT}^{-1} (Z(:,j)) .* c_3 (j), \text{ciel}(n_x /2))$
> **end**
>
> **for** $i = 1:n_x$
> $Z(i,:) = \text{cshift}(\text{FFT}^{-1} (Z(i,:) .* c_4 (i), -\text{ciel}(m/2))$
> **end**
> $I = |Z|^T$

where

cshift(,n) circular shifts a vector by n places.

$c_3 = m$ complex single-precision vector of precomputed coefficients.

$c_4 = n_x$ complex single-precision vector of precomputed coefficients.

$I = n_x \times m$ real single-precision matrix.

The computational complexity of these steps is dominated by the FFTs and is given by

$$W_{stage}^{1d} = 5\, m\, (n_x(1 + log_2(m))$$

and

$$W_{stage}^{1e} = 5\, n_x\, (m(1 + log_2(n_x)) \,.$$

The parallelism for step 1d is principally that each column can be processed independently, which implies m degrees of parallelism. For step 1e, the parallelism is principally that each row can be processed independently, which implies n_x degrees of parallelism. The amount of data sent between these steps is given by

$$D_{stage}^{1d \to 1e} = 8\, n_x\, m\ [\text{bytes}] \,.$$

The above steps complete the image formation stage of the application. One final step, that is a negligible untimed part of the benchmark, is the insertion of templates into the image. These templates are used by the next stage. Each template is a $n_{font} \times n_{font}$ matrix containing an image of a rotated capital letter. The total number of different templates is given by $n_{let}n_{rot}$. The templates are distributed on a regular grid in the image and with an occupation fraction of 1/2. The grid spacing is given by $4n_{font}$, so that the total number of templates in an image is

$$n_{templates} = floor(m/(4n_{font}))floor(n_x/(4n_{font})) \,.$$

A.3   STAGE 2: BACK-END KNOWLEDGE FORMATION

This stage reads in two images (I_1 and I_2) of the same region on the ground and differences them to find the changes. The differenced image is thresholded to find all changed pixels. The changed pixels are grouped to find a region of interest that is then passed into a classifier. The classifier convolves each region of interest with all the templates to determine which template has a best match.

$I_\Delta = \max(I_2 - I_1, 0)$
$I_{mask} = I_\Delta > c_{thresh}$
$i = 1$
while(NonZeros(I_{mask}))   ROI(i,1:4) = PopROI(I_{mask}, n_{font})
      $I_{sub} = I_\Delta$ (ROI(i,1):ROI(i,2),ROI(i,3):ROI(i,4))
      ROI(i,5:6) = MaxCorr(T,I_{sub})
end

where

$I_\Delta = m \times n_x$ single-precision matrix containing the positive difference between sequential images I_1 and I_2.

c_{thresh} = precomputed constant that sets the threshold of positive differences to consider for additional processing.

$I_{mask} = m \times n_x$ logical matrix with a 1 wherever the difference matrix exceeds c_{thresh}.

NonZeros() = returns the number of nonzeros entries in a matrix.

ROI = $n_{templates} \times 6$ integer matrix. First four values hold the coordinates marking the ROI. The final two values hold the letter and rotation index of the template that has the highest correlation with the ROI.

PopROI(I_{mask}, n_{font}) selects the "first" nonzero pixel in I_{mask} and returns four values denoting the n_{font} by n_{font} region of interest around this pixel. Also sets these locations in I_{mask} to zero.

$I_{sub} = n_{font} \times n_{font}$ single-precision matrix containing a subregion of I_Δ.

$T = n_{let} \times n_{rot} \times n_{font} \times n_{font}$ single-precision array containing all the letter templates.

MaxCorr(T,I_{sub}) correlates I_{sub} with every template in T and returns the indices corresponding to the letter and rotation with the highest correlation.

The computational complexity of this stage is dominated by computing the correlations. Each correlation with each template requires two n_{font}^4 operations. The total computational complexity of this stage is

$$W_{stage}^2 = 2\, n_{template}\, n_{let}\, n_{rot}\, n_{font}^4.$$

In the previous stage, at each point in the processing, either a row or a column could be processed independently of the others. In this stage, two-dimensional regions are involved in the calculation. Assuming the preferred parallel direction is that coming out of the previous stage (i.e., the second dimension), then this would imply n_x degrees of parallelism. However, the computation of column depends upon n_{font} neighboring columns. To effect this computation requires that overlapping data is stored on each processor. This effectively limits the degrees of parallelism to n_x/n_{font}. The amount of data sent between Stage 1 and Stage 2 is what is needed to implement this overlap:

$$D_{stage}^{1 \to 2} = 8\, m\, N_P\, n_{font} \text{ [bytes]}.$$

An alternative to the above approach is to break up the data in both dimensions, which exposes more parallelism ($m\, n_x / n_{font}^2$). However, this requires more communication to set up ($8\, m\, n_x$ [bytes]).

16 Programming Languages

James M. Lebak, The MathWorks

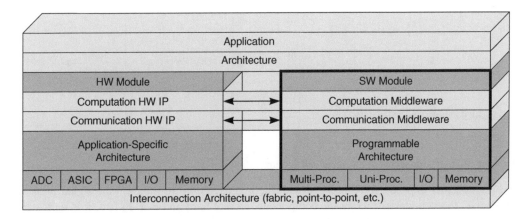

In this chapter, programming languages for high performance embedded computing are considered. First, principles of programming embedded systems are discussed, followed by a review of the evolution of programming languages. Specific languages used in HPEC systems are described. The chapter concludes with a comparison of the features and popularity of the various languages.

"I wish life was not so short," he thought. "Languages take such a long time, and so do all the things one wants to know about."

— **J.R.R. Tolkien,** *The Lost Road*

16.1 INTRODUCTION

This chapter considers the state of programming languages for high performance embedded systems. To set the appropriate scope for the discussion, consider the embedded system shown in Figure 16-1. This system consists of one or more signal processors that take data from a sensor and operate on it to reconstruct a desired signal. The data processor extracts knowledge from the processed signals: for example, in a radar system, it might try to identify targets given the processed signal. The data processor displays these products in some form that the user can understand. The control processor is responsible for telling the signal processors what operations to perform based on user inputs and the knowledge extracted by the data processor.

Each component performs a different type of processing. The signal processor primarily performs high-intensity floating-point calculations in which high throughput and low latency are required. The data processor may perform a lower number of operations that center more on testing and matching than on floating-point computation. The control processor is typically more complicated in terms of number of lines of code, but again performs fewer operations than the signal processor.

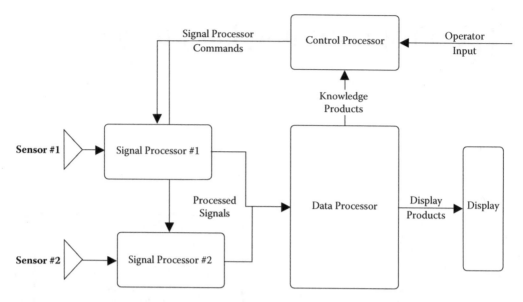

FIGURE 16-1 Components of an embedded system.

This chapter is primarily concerned with the language used to program the signal processor. Of course, the boundaries among the components are not hard and fast. In fact, it is increasingly common for a so-called "signal processor" to incorporate data processing and control processing in a single package. Furthermore, many of the languages mentioned in this chapter could be applied to all three types of processing. However, the focus of this chapter is on languages for the typically high-throughput, math-intensive operations performed by a signal processor.

16.2 PRINCIPLES OF PROGRAMMING EMBEDDED SIGNAL PROCESSING SYSTEMS

Signal processing systems are characterized by streams of data that arrive at a regular rate. In some radar systems under consideration in 2005, new datasets may arrive on a timescale of tens or hundreds of milliseconds, and each data product may take hundreds of millions to billions of floating-point operations (gigaFLOPs, GFLOPs) to process. Overall, these requirements lead to signal processor throughput requirements on the order of tens to hundreds of GFLOPs/s. While compute performance of this magnitude is easily within the theoretical capabilities of modern processors, embedded processors have additional constraints in terms of form factor, such as size, weight, or power. Designing the system to meet the required performance and fit the given form factor is a difficult task, and providing margin to allow for inefficient use of computer hardware adds to system cost. Further, the constant arrival of input data requires the system to have resources available to process them on a set schedule. Consequently, embedded system programming places a greater emphasis on achieving repeatable high performance than is typically found in desktop or scientific computer programming.

This emphasis on performance, together with the fact that many operations are repeated, leads to the first of three important principles for programming embedded systems: *do as much processing as possible beforehand*. Many embedded programs are separated into a "setup phase" that is performed before datasets begin to arrive and an "execute phase" that happens when the datasets are arriving. The term for performing part of an operation in a separate setup phase is *early binding*.

Memory allocation is a classic example of an operation that should be done in the setup phase. The amount of memory needed by an application is typically bounded, but memory allocation can take an unpredictable amount of time. Consequently, many embedded programs allocate their memory ahead of time and find ways to reuse buffers.

A second classic example of an operation performed at setup time is the weight computation for a fast Fourier transform (FFT). Weights are needed at every phase of the computation, and their values can be computed independent of the data being transformed. If multiple FFT operations are being performed, as is nearly always the case in a signal processing application, considerable speedup can be achieved by computing the weights during the setup phase. Obviously, the disadvantage to computing weights at setup time is that memory is required to store them. The cost of the workspace must be balanced against the performance improvement. Van Loan (1992) discusses some of the trade-offs involved.

The emphasis on repeatable performance has the potential to lead to a huge amount of work in optimizing applications. In many embedded signal processing applications, part of the application will be written in assembly language or using assembly-level instructions. A recent article emphasizes that this is often done for the purpose of guaranteeing deterministic execution time as much as for efficiency (Lee 2005). Assembly-level code is obviously nonportable, and so effort is expended to confine such optimizations to a limited number of places in the program. Fortunately, signal processing applications use many of the same operations over and over again; therefore, hardware vendors typically provide optimized libraries to perform these operations. This is the second principle for embedded system programming: *use optimized libraries to decrease the amount of work for the application programmer*. Hand-optimization need only be used when libraries are not available.

The emphasis on low-level programming in embedded systems does not solely arise from the need to achieve performance. As Figure 16-1 shows, the signal processor is dealing with a data stream coming directly from a sensor. Data from a sensor may be in a custom format and may include embedded control information. For these reasons, embedded system programming places a greater premium on data size and bit-level manipulation than does desktop or workstation programming.

In general, as systems have grown more complex, computer science has responded with the introduction of increased levels of abstraction. These levels are introduced to allow system programmers to manage complex systems more easily. However, as noted in Lee (2005), the abstractions used in today's programming languages grant ease of management at the expense of both performance and repeatability. This tension between the level of abstraction required to manage system complexity and that needed to interface to low-level devices and ensure repeatable performance is a key challenge for embedded system programming. This tension may be expressed as a third principle of embedded systems programming: *minimize the performance cost of the abstractions used*.

16.3 EVOLUTION OF PROGRAMMING LANGUAGES

Computer science classifies programming languages into five "generations." The first generation of programming languages is considered to be machine language, that is, coding the actual numbers the machine understands. This was common on the earliest computer systems. The second generation of programming languages is assembly language. The programmer uses human-understandable mnemonics to represent instructions and an external program, called an assembler, performs the translation into the machine codes. An early form of assembly language was available on the UNIVAC in the early 1950s. On that computer, programmers used an instruction such as "A100" to add the contents of memory location 100 to the "add register." Koss characterizes such instructions as "… a major programming advance over the coding for any machine developed before that time" (2003). For a signal processing application, assembly language offers the additional advantage that the machine resources are explicitly exposed to the programmer, enabling a more certain prediction of execution time and resource availability. A common example of assembly-level programming in recent systems is the use of Altivec code that takes advantage of the vector instructions in Motorola's PowerPC family of microprocessors (Altivec 1999).

Today's familiar *high-level programming languages* are actually the third generation of programming languages. Ada, C, C++, Java, and Fortran are all third-generation languages, in which a compiler translates code written in the language into sequences of machine-readable instructions. Fourth-generation languages attempt to raise the level of abstraction at which a program is written. Sometimes such languages serve only a limited purpose. The Structured Query Language (SQL) is an example of a fifth-generation language targeted at database applications. Fifth-generation languages are targeted at artificial intelligence. Due to their application-specific nature and the high levels of abstraction at which they operate, fourth- and fifth-generation languages do not see much use in high performance embedded systems, and so no more will be said of them here.

16.4 FEATURES OF THIRD-GENERATION PROGRAMMING LANGUAGES

In general, all third-generation languages provide users with constructs to express iteration or looping, in which a given sequence of code is executed multiple times. They also allow the conditional execution of particular statements, calls to subroutines, and operations on variables of different abstract types, such as integer, floating-point, and character data. The major differences from assembly-language programming are that (1) the variables used in high-level language programming are not associated directly with machine registers and (2) management of those registers is done by the compiler.

The role of the compiler in third-generation languages has evolved over time. The original Fortran compiler translated code directly into machine language. In later compilers, such as those associated with the UNIX operating systems and the C language, programs are translated by the compiler into *object files* that are at the level of machine language but cannot be executed. Multiple object files are joined by a program called a *linker* to form an executable program. The link step may also introduce precompiled object files provided as part of a library. Linking can be *static*, meaning that the executable program uses the specific version of the library available at link time, or *dynamic*, meaning that the executable code uses the version of the library available at runtime.

The compile and link process in Java is different from that of other languages. In the Java language, the output of the compiler is Java byte code, targeted at a portable abstraction called the Java virtual machine (JVM). At execution time, Java byte code is actually executed by an interpreter specialized for the platform, which also at that time performs functions that other languages perform with a linker. The version of libraries used is, therefore, determined at runtime, similar to dynamic linking.

Languages are evolving to meet the needs of complex systems. Many languages are adding support for multiprocessing, but they tend to support it in different ways. However, recent languages do have several common features that we briefly consider here.

16.4.1 OBJECT-ORIENTED PROGRAMMING

Object-oriented programming packages data and the operations on those data into a single entity called an *object*. The implementation of the object's operations and exact storage of its data are hidden from the outside world. This allows the object to present a consistent interface across different platforms while being internally optimized for each platform. Libraries of objects thus can provide portability across platforms and code reuse among different applications. Object-oriented programming is a methodology that can be applied in any language; however, language support of this methodology makes writing applications easier.

16.4.2 EXCEPTION HANDLING

Exceptions are errors experienced by programs during execution. Some languages provide features that allow the application programmer to explicitly define and group categories of exceptions. Such languages also provide a way to specify areas of a program where exceptions could be expected to occur and the methods to be used to handle them when they occur.

16.4.3 GENERIC PROGRAMMING

Some paradigms in computer science are commonly applied to a variety of different types of data. Examples include stacks and queues. It is wasteful to have to write an implementation of such a construct multiple times for every conceivable data type in a program, as the details of the implementation typically do not depend on the type. To eliminate this redundant work, some languages provide support for *generic programming*, in which a construct can be programmed independent of type. The compiler can instantiate the construct for any type needed by the application program.

16.5 USE OF SPECIFIC LANGUAGES IN HIGH PERFORMANCE EMBEDDED COMPUTING

This section presents the characteristics of five specific high-level languages in the context of high performance embedded computing (HPEC): C, Fortran, C++, Ada, and Java. Although other languages are certainly in use in this field, we discuss these five for two reasons. First, they are all historically important, either in embedded systems or in general programming. Second, the five languages together constitute a large enough sample to demonstrate the state of programming in this field.

16.5.1 C

The ANSI/ISO standard C programming language was invented at Bell Laboratories in the 1970s. The standard reference for the language is the book by Kernighan and Ritchie (1988). An update to the language, referred to as C99, was issued by the ISO and adds features such as complex arithmetic to the language (British Standards Institute 2002). However, the original standard version of the language (C89) is still the most commonly used in HPEC as of this writing.

The C programming language is traditionally the most widely used language in the embedded space and remains one of the most popular languages in the world. As one indicator of its general popularity, consider that a survey of open-source programming projects on the website freshmeat.net indicated that C was used for projects on that site about twice as often as the next closest competitor, Java (Welton 2004). A reason for C's popularity is its use as the basis for the UNIX and Linux operating systems, and the degree to which these operating systems have in turn been implemented on a wide variety of architectures. A second reason is that implementing a C compiler is a relatively easy task, due to the low level of abstraction in the language and the use of external libraries to keep the language itself small.

In the embedded space, C's popularity undoubtedly comes from the degree to which its programming model matches the programming model of embedded systems. Details like pointers, word size, memory allocation, registers, and "volatile" locations (that change outside of program control and that are already part of the C language) are frequently necessary in an embedded application. Insertion of assembly language into a C program is easy by design. The constructs provided by the C language are a good match to the principles of embedded programming previously cited. C's malloc and free functions for memory management allow the user explicit control over the timing of memory allocation, enabling the user to set up resources ahead of time in accord with the first principle of embedded system programming. C's reliance on external libraries rather than on language features matches the second principle of embedded system programming. C's low level of abstraction relative to the machine matches the third principle of embedded system programming.

As good as C is for embedded system programming, it is not perfect. C's low-level programming model, which is a major strength for small systems, does not always extend well to large multiprocessor systems. To perform collective computation in such systems, users must send messages between processors and manage individual computation on a per-processor basis in order to achieve performance. This distributed-memory programming model is in general hard to manage and may need to be done differently on each system or even for each size of system, thereby leading to brittle, nonportable code.

A further criticism of C is that the use of pointers can make it difficult for compilers to optimize complex programs. This problem comes from the fact that in some cases it is impossible to decide at compile time whether two pointers point to the same quantity in memory or not. This problem is made harder in complex programs where pointers are passed into a function from outside (Hennessy and Patterson 2003).

16.5.2 Fortran

Fortran was the original high-level language, invented for the IBM 704 system in 1954. The name of the language is an abbreviation of the phrase *formula translation*. According to Malik, "The overall success of Fortran is difficult to overstate: it dramatically changed, forever, the way computers are used" (1998). The language, which was intended to ease the implementation of mathematical formulas, provided capability to write expressions using variables. The language itself was so simple that it was easy for a compiler to optimize these expressions and achieve performance near that of assembly code (Chapman 1998; Metcalf and Reid 1999). The language has evolved through several versions, which have introduced such features as support for complex arithmetic (in FORTRAN 77), operations on whole arrays rather than individual elements, dynamically allocatable arrays (Fortran 90), full support for the exceptions defined by the IEEE arithmetic standard, and standard ways to interact with C (Fortran 2000) (ANSI 1978; ISO 1991; Loveman 1996; Reid 2003). The language is a favorite in the scientific computing space. Many scientific math libraries, such as the linear algebra package LAPACK, as well as many scientific applications, are written in Fortran (Anderson et al. 1999).

At first, one might consider Fortran to be a natural fit for the embedded space because of the emphasis on mathematical programming. Fortran compilers generally produce very efficient code. Complex data types, which are often used in signal processing, were directly supported as early as FORTRAN 77, whereas they were not added to C until C99. Fortran 95 supports distribution of data over multiple processors and global array operations, which are not present in C.

However, several shortcomings of FORTRAN 77 seem to have kept it from seeing heavy use in HPEC at a crucial time when the field was growing. First, there was not a standardized way for FORTRAN 77 to access operating system quantities: access to command-line arguments, for example, was only introduced in Fortran 2000. Second, interaction with other languages and libraries was not part of the standard and was dependent on the compiler and operating system [see Lebak (1997) for a discussion of the issues involved in cross-platform, cross-language portability]. Finally, FORTRAN 77 lacked support for many of the low-level operations necessary in embedded systems. An example is operations on bit-level quantities: these were added as an extension to the standard (MIL-STD 1753 in 1978) and became part of Fortran 95. Another example is dynamic memory allocation, which was added in Fortran 90 (Dedo 1999).

In summary, the choice of C over FORTRAN 77 in the HPEC space can be seen as a consequence of FORTRAN 77 being seen as existing at the wrong abstraction level for an embedded system. Nonetheless, it should be noted that all of the shortcomings of FORTRAN 77 listed here have been corrected in later versions of the standard.

16.5.3 Ada

In 1975, the U.S. Department of Defense (DoD) formed a "higher-order language working group" to develop a single language for programming DoD applications (Malik 1998). This effort resulted in the development of the Ada language specification in 1983. The intent was that this language would continue to evolve and support all DoD program needs, including those in the embedded space. Booch and Bryan (1994) list the design goals of the Ada language team as (1) recognition of the importance of program reliability and maintainability, (2) concern for programming as a human activity, and (3) efficiency.

Ada has many advantages for embedded system programming. It includes structures to implement task parallelism inside the language, a feature that must be supported in C++ by external libraries such as Parallel VSIPL++ (Lebak et al. 2005). It includes capabilities for object-oriented programming, generic programming, and exception handling, all capabilities that are lacking in C and Fortran.

The nature of the Ada language makes it easier for the compiler to find errors in programs. In one case study, a class in embedded systems programming taught at the University of Northern Iowa used C and Ada in successive years to control a model railroad. In semesters in which Ada was used, better than 50% of the students finished the project: in semesters in which C was used, none of the students finished the project. The instructor, John McCormick, concluded from examination of the student projects that strong typing played an important role in this success rate (McCormick 2000). A market survey done in the United Kingdom found that software managers and developers regarded the language as "safe, reliable, and robust" and gave it at least partial credit for their ability to successfully deliver products (Gilchrist 1999).

Nonetheless, Ada has fallen out of favor not just in embedded system programming but everywhere since the DoD mandate for the use of the language expired in 1997. The major reason for this appears to be the lack of commercial use of the language, leading to the decreased availability of tools for the language and a dearth of programmers trained in its use. A pair of studies by Venture Development Corporation show that while the market for all embedded system development tools is projected to grow from $250 million in 2003 to $300 million in 2007, the market for Ada development tools has remained approximately constant (at about $49 million) (Lanfear 2004; Lanfear and Balacco 2004). A white paper by the Software Engineering Institute noted that some important embedded programs like the Joint Strike Fighter program use Ada; however, development of Ada tools is at this point largely sustained by such projects rather than by any commercial use of the language. The paper concluded that despite the advantages of Ada for maintenance, implementation, and teaching, "...Ada is a programming language with a dubious or nonexistent future" (Smith 2003).

16.5.4 C++

The C++ programming language began as a set of extensions to the 1989 ANSI C standard to allow object-oriented programming. The ISO C++ standard was completed in 1998 and includes object-oriented programming, exception handling, and generic programming, while retaining compatibility with C. The C++ standard template library (STL), which is part of the standard, includes many useful type-independent constructs, including containers such as lists and queues, iterators to extract data from containers in a generic way, and input and output streams (Stroustrup 1997). This is a powerful combination of features that has led to widespread use of the language. Unlike Ada, Java, or Fortran 90 and its successors, C++ has no explicit support for multiprocessing or multithreading, preferring to rely on libraries for these features.

C++ has proved to be more extensible than C and has been shown to be able to achieve both high performance and productivity. For example, the parallel object-oriented methods and algorithms (POOMA) library demonstrated the use of C++ to implement global array handling features on distributed arrays while also achieving performance comparable to Fortran 90 (Cummings et al. 1998). Veldhuizen and Jernigan (1997) hypothesized that properly written C++ libraries could achieve performance better than Fortran.

A C++ compiler is much more complex than a C compiler, and, in fact, fully conforming compilers did not appear for years after the ratification of the standard. Early concerns about the size of the language, the predictability of exception handling, and the tendency of the use of templates to increase program size led to the development of a so-called "Embedded C++" subset of the standard by a consortium of (mostly Japanese) companies (Embedded C++ Technical Committee 1999). Those involved with the development of this subset engaged the ISO C++ standard committee directly to address their concerns; the result was a technical report on C++ performance that describes ways that implementations can avoid these problems (ISO Working Group 21 2003).

With these concerns addressed, C++ seems to be a logical heir-apparent to C in the embedded space. It still has some of the same shortcomings as C in the areas of memory management and multiprocessor programming; however, many of these limitations can be addressed by external libraries that are enabled by the extensibility of the language.

16.5.5 Java

The Java language, developed at Sun Microsystems, is focused on improving application programmer productivity. The language includes features aimed at raising the level of abstraction at which machines are programmed, including object-oriented programming and exception handling. Java also aims to improve productivity by removing the need for the programmer to directly manage memory. Finally, the language is designed to be portable to a variety of platforms. This portability is achieved by requiring programmers to write applications targeted to the Java virtual machine (JVM). The JVM itself provides a standard set of features for the application programmer; it is ported to and optimized for different platforms. Multithreading is a well-defined feature of the language and the JVM, so that parallel programs written in Java can be portable. There are different versions of the JVM defined and provided for different classes of devices: a Java Micro Edition targets cell phones and similar consumer devices, while a Java Enterprise Edition is targeted at business applications that run on large clusters of computers (Flanagan 1997; Sun Microsystems 2005).

The elimination of the need to manage memory in Java has two important effects. First, Java does not include pointers or bit-level manipulation capability. Thus, device drivers must be written in a lower-level language like C. Second, Java introduces a *garbage collection* mechanism that is responsible for de-allocating memory that is no longer needed by the application. This garbage collection mechanism may run at unpredictable times and can, therefore, be a detriment to obtaining repeatable performance. Sun's real-time specification for Java corrects this by providing threads that cannot be interrupted by garbage collection (Bollella et al. 2000). The combination of high-productivity features and real-time performance makes the real-time Java specification an attractive possibility for high performance embedded systems.

16.6   FUTURE DEVELOPMENT OF PROGRAMMING LANGUAGES

One important criticism of third-generation programming languages is that they mostly use the abstraction of globally accessible memory. This abstraction is becoming less and less sustainable as computers move toward a chip multiprocessor model, in which multiple processors and associated memories are implemented on the same chip and exposed to the programmer. Computer system designers are being driven in this direction by two factors, an increase in propagation time for large wires and a need to manage the increased number of transistors that can be accommodated on a chip. Refer to Ho, Mai, and Horowitz (2001), Burger et al. (2004), and Taylor et al. (2002) for more details on these trends. Alternative computer architectures developed under the Defense Advanced Research Projects Agency (DARPA) Polymorphous Computing Architecture (PCA) Program can be programmed in a *streaming* mode, in which the output of a computation serves as the direct input to the next computation without being stored in main memory. This mode of operation can increase application efficiency but is not well supported by the languages presently available in the embedded space. For a discussion of the performance available from a streaming programming model on a chip multiprocessor, see Hoffmann's (2003) work on the Massachusetts Institute of Technology (MIT) Raw chip. Languages being developed to support streaming include the Brook

TABLE 16-1

Programming Language Feature Summary

Language	Object-Oriented Programming	Exception Handling	Generic Programming	Parallelism Support?
C	No	No	No	None
Fortran	Yes	No	Yes	Data-parallel
Ada	Yes	Yes	Yes	Task-parallel
C++	Yes	Yes	Yes	None
Java	Yes	Yes	No	Threading

language developed at Stanford (Buck 2003) and the StreamIt language developed at MIT (Theis et al. 2002).

The DARPA High Productivity Computing Systems (HPCS) Program is developing systems and languages to increase both the performance of future applications and the productivity of future programmers. The languages—Fortress, Chapel, and X10—all provide abstractions to manage parallelism and memory hierarchies (Allen et al. 2005; Callahan, Chamberlain, and Zima 2004; Sarkar 2004). At present, these languages support different paradigms for managing parallelism and are all research languages. However, it is likely that one or more of these languages will be used to program future high performance computing systems. They all, therefore, have the potential to migrate into the embedded computing space.

16.7   SUMMARY: FEATURES OF CURRENT PROGRAMMING LANGUAGES

Table 16-1 summarizes the features of the programming languages discussed here. Interestingly, Ada, which arguably has a feature set that is nearly a superset of those of the other languages, is seen to have a small user base and to be declining in popularity. On the other hand, C, which lacks most features that more recent languages include, remains one of the most popular languages in the world, especially in embedded systems. There are many potential explanations for this apparently contradictory pattern of usage. This author agrees with Lee that predictability of execution time plays a leading role in selection of language for embedded system projects (Lee 2005). Additional levels of abstraction in current languages obscure this predictability. In this view, C's success may largely be a consequence of the ability the language gives the programmer to easily understand and predict the execution time of the compiler's output.

REFERENCES

Allen, E., D. Chase, V. Luchangco, J.-W. Maessen, S. Ryu, G.L. Steele, Jr., and S. Tobin-Hochstadt. 2005. *The Fortress Language Specification*. Santa Clara, Calif.: Sun Microsystems.

AltiVec Technology Programming Interface Manual. 1999. Motorola Semiconductor Products.

American National Standards Institute, Inc. 1978. *American National Standard Programming Language FORTRAN*. New York: American National Standards Institute, Inc. Document ANSI X3.9-1978.

Anderson, E., Z. Bai, C. Bischof, S. Blackford, J. Demmel, J. Dongarra, J. Du Croz, A. Greenbaum, S. Hammarling, A. McKenney, and D. Sorensen. 1999. *LAPACK User's Guide*, 3rd edition. Philadelphia: Society for Industrial and Applied Mathematics Press.

Booch, G. and D. Bryan. 1994. *Software Engineering with Ada*, 3rd edition. Redwood City, Calif.: Benjamin Cummings.

British Standards Institute. 2002. *The C Language Standard: Incorporating Technical Corrigendum 1*. Hoboken, N.J.: John Wiley & Sons.

Brosgol, B., J. Gosling, P. Dibble, S. Furr, and M. Turnbull. G. Bollella, ed. 2000. *The Real-Time Specification for Java™*. Boston: Addison-Wesley.

Buck, I. 2003. *Brook Specification v.0.2.* Technical Report CSTR 2003-04. Stanford, Calif.: Stanford University.

Burger, D., S.W. Keckler, K.S. McKinley, M. Dahlin, L.K. John, C. Lin, C.R. Moore, J. Burrill, R.G. McDonald, and W. Yoder. 2004. Scaling to the end of silicon with EDGE architectures. *IEEE Computer* 37(7): 44–55.

Callahan, D., B.L. Chamberlain, and H.P. Zima. 2004. The Cascade high productivity language. *Proceedings of the Ninth International Workshop on High-Level Parallel Programming Models and Supportive Environments* 52–60.

Chapman, S.J. 1998. *Introduction to Fortran 90/95.* New York: McGraw-Hill.

Cummings, J.C., J.A. Crotinger, S.W. Haney, W.F. Humphrey, S.R. Karmesin, J.V.W. Reynders, S.A. Smith, and T.J. Williams. 1998. Rapid application development and enhanced code interoperability using the POOMA framework. Presented at the SIAM Workshop on Object-Oriented Methods for Interoperable Scientific and Engineering Computing. Yorktown Heights, N.Y.

Dedo, C.T. 1999. Debunking the myths about Fortran. *ACM SIGPLAN Fortran Forum* 18(2): 12–21.

The Embedded C++ Technical Committee. 1999. *The embedded C++ specification.* Available online at http://www.caravan.net/ec2plus.

Flanagan, D. 1997. *Java in a Nutshell.* Sebastopol, Calif.: O'Reilly and Associates.

Gilchrist, I. 1999. Attitudes toward Ada—a market survey. *Proceedings of ACM SIGAda Annual International Conference* 229–252.

Hennessy, J.L. and D.A. Patterson. 2003. *Computer Architecture: A Quantitative Approach*, 3rd edition. San Francisco: Morgan Kaufmann.

Ho, R., K.W. Mai, and M.A. Horowitz. 2001. The future of wires. *Proceedings of the IEEE* 89(4): 490–504.

Hoffmann, H. 2003. Stream algorithms and architecture. Master's thesis. Massachusetts Institute of Technology, Cambridge, Mass.

ISO. 1991. *Fortran 90.* Documents ISO/IEC 1539: 1991 (E) and ANSI X3. 198–1992.

ISO Working Group 21. 2003. *Technical Report on C++ Performance.* ISO/IEC PDTR 18015.

Kernighan, B.W. and D.M. Ritchie. 1988. *The C Programming Language*, 2nd edition. Englewood Cliffs, N.J.: Prentice Hall.

Koss, A.M. 2003. Programming on the UNIVAC 1: a woman's account. *IEEE Annals of the History of Computing* 28(1): 48–59.

Lanfear, C. 2004. Ada in embedded systems. *The Embedded Software Strategic Market Intelligence Program 2001–2002* vol. 5. Natick, Mass.: Venture Development Corporation.

Lanfear, C. and S. Balacco. 2004. *The Embedded Software Strategic Market Intelligence Program 2004.* Natick, Mass.: Venture Development Corporation.

Lebak, J.M. 1997. Portable parallel subroutines for space-time adaptive processing, Ph.D. thesis. Cornell University, Ithaca, N.Y.

Lebak, J.M., J. Kepner, H. Hoffmann, and E. Rutledge. 2005. Parallel VSIPL++: an open standard software library for high performance parallel signal processing. *Proceedings of the IEEE* 93(2): 313–330.

Lee, E.A. 2005. Absolutely positively on time: what would it take? *IEEE Computer* 38(7): 85–87.

Loveman, D.B. 1996. Fortran: a modern standard programming language for parallel scalable high performance technical computing. *Proceedings of International Conference on Parallel Processing* 140–148.

Malik, M.A. 1998. Evolution of the high level programming languages: a critical perspective. *ACM SIGPLAN Notices* 33(12): 72–80.

McCormick, J.W. 2000. Software engineering education: on the right track. *Crosstalk* 13(8).

Metcalf, M. and J. Reid. 1999. *Fortran 90/95 Explained*, 2nd edition. Oxford, U.K.: Oxford University Press.

Reid, J. 2003. The future of Fortran. *IEEE Computing in Science and Engineering* 5(4): 59–67.

Sarkar, V. 2004. Language and virtual machine challenges for large-scale parallel systems. Presented at the Workshop on the Future of Virtual Execution Environments. Armonk, N.Y.

Smith, J. 2003. What about Ada? The state of the technology in 2003. Technical Note CMU/SEI-2003-TN-021. Pittsburgh: Carnegie Mellon University/Software Engineering Institute.

Stroustrup, B. 1997. *The C++ Programming Language*, 3rd edition. Boston: Addison-Wesley.

Sun Microsystems. 2005. *CDC: Java™ Platform Technology for Connected Devices.* White paper.

Taylor, M.B., J. Kim, J. Miller, D. Wentzlaff, F. Ghodrat, B. Greenwald, H. Hoffmann, P. Johnson, J.-W. Lee, W. Lee, A. Ma, A. Saraf, M. Seneski, N. Shnidman, V. Strumpen, M. Frank, S. Amarasinghe, and A. Agarwal. 2002. The Raw microprocessor: a computational fabric for software circuits and general-purpose programs. *IEEE Micro* 22(2): 25–36.

Theis, W., M. Karczmarek, M. Gordon, D. Maze, J. Wong, H. Hoffmann, M. Brown, and S. Amarasinghe. 2002. *StreamIt: A Compiler for Streaming Applications.* MIT/LCS Technical Memo LCS-TM-622. Cambridge: Massachusetts Institute of Technology.

Van Loan, C. 1992. *Computational Frameworks for the Fast Fourier Transform.* Philadelphia: Society for Industrial and Applied Mathematics.

Veldhuizen, T.L. and M.E. Jernigan. 1997. Will C++ be faster than Fortran? Presented at the 1997 Workshop on International Scientific Computing in Object-Oriented Parallel Environments. Marina del Rey, Calif.

Welton, D.A. 2004. Programming language popularity. Available online at http://www.dedasys.com/articles/language_popularity.html.

17 Portable Software Technology

James M. Lebak, The MathWorks

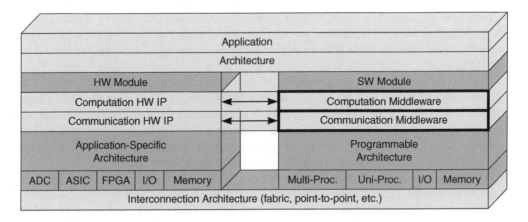

This chapter discusses software technologies that support the creation of portable embedded software applications. First, the concept of portability is explored, and the state of the art in portable middleware technology is surveyed. Then, middleware that supports portable parallel and distributed programming is discussed, and advanced techniques for program optimization are presented.

A library is not a luxury but one of the necessities of life.

— **Henry Ward Beecher**

17.1 INTRODUCTION

A program is said to be *portable* if it may be run without alteration on another computer system. In the high performance embedded computing (HPEC) field, which makes use of many different computer architectures, portability takes on special importance. In the period 1995–2005, high performance embedded systems have made use of a number of different computer architectures, including digital signal processors such as Analog Devices' SHARC processor and general-purpose processors such as the Intel i860 and Motorola PowerPC G3 and G4. These systems may now move in the direction of new architectures such as the Cell processor developed by IBM, Sony, and Toshiba (Cico, Greene, and Cooper 2005). By way of contrast, portability has not been a primary concern for developers writing programs for the Intel-based PC. Intel's chips have retained compatibility with a particular instruction set for over 25 years, making portability between generations of technology almost automatic.

For embedded system developers, then, portability of software is a vital concern simply because of the wide range of possible target architectures. This is particularly true for military signal processing systems that are required to implement complex algorithms on complex hardware. These systems are expected to remain in service for decades and undergo several "technology refresh" cycles in which hardware and software are replaced. Portability enables easier technology refresh and preserves the investment in the application software.

Portability can take two major forms. *Source-code portability* means that source code may be compiled and run on multiple systems. *Object-code portability* means that after the source is compiled into object code, some further tool is used on the object code to make it run on a different target system.

Source-code portability has been a goal of programming languages since the design of Fortran and the first compilers. Indeed, source-code portability is an important advantage of programming in a high-level language. The C programming language enhanced portability by standardizing not only the language, but a set of libraries for common tasks and an interface to the operating system. Source-code portability for C programs was enabled by the simplicity of the language, the compiler, and the library.

Object-code portability involves a translation from either a particular machine's instruction set or an intermediate form to the executing machine's instruction set. There are two broad classes of techniques for achieving object-code portability. Static techniques translate the instructions before the program is run, while dynamic techniques perform the translation as the code is running. Implementations of the Java language typically employ a compiler to translate source code to Java byte code, which is then executed by an interpreter: this is a dynamic technique.

For high performance embedded signal processors, we strive for application source-code portability for two reasons. First, if we desire to build a system to achieve high performance, we generally have the source code available or are willing to re-create it. Object-code portability is used when the source code is not available. A second reason to strive for source-code portability in a high performance situation is that a compiler typically performs platform-specific optimization as part of the translation of source code into object code. It is more difficult to undo these optimizations and reoptimize for a new target platform than it is to optimize for the new platform beginning with the source code.

Portability is often at odds with achieving high performance. In general, achieving high performance requires tuning code for a particular platform to take account of its unique register set, memory hierarchy, and platform-specific features. Therefore, code that performs well on one platform may not perform well for another without extensive modification. This is true either at the source-code or object-code level. Developers speak of a program as having *performance portability* if it achieves similar high performance on multiple platforms (portable low performance being easy to achieve and generally of little interest). If a program is not fully portable, we may measure portability by the number of lines of source code that must be changed. We may similarly use the number of lines of code that must be changed to achieve high performance as a measure of performance portability.

Precision is an important aspect of portability that is less often considered. We can say that a portable application should achieve the same answer on different machines. For floating-point calculations, when we say "the same answer," we actually mean that the answer is within the error bounds for the problem, given the numerical precision of the quantities involved. Since the IEEE Standard 754 for floating-point arithmetic was published in 1985, there has been widespread agreement about the definitions of *single-precision* and *double-precision* floating-point formats (IEEE 1985). However, some machines do not fully implement the standard, while others provide a mode in which certain aspects of the standard may be turned off to achieve higher floating-point performance at the expense of accuracy. Furthermore, even if the same formats are used, different implementations of an algorithm may accumulate error differently. The system designer must have an understanding of both the algorithm and the floating-point format to properly bound the expected error from a computation. Bounds may be portable between machines, given the same format; actual error quantities should not be expected to be. A good overview of these issues is provided by Goldberg (1991).

17.2 LIBRARIES

An important principle of embedded system programming is to use optimized libraries to decrease the amount of work for the application programmer. In the early 1990s, there were many different embedded signal processing platforms, each with its own library. This situation required application developers to create nonportable code, as the library calls were different on different platforms. Developers could insulate themselves from these effects by creating common "portability layers," useful for writing applications. However, this approach was costly and represented a potential performance overhead.

Application portability can be preserved by the use of standard libraries for particular tasks. The standard defines an interface. Implementors are free to implement the routines as they choose, as long as the interface is maintained. Therefore, platform-specific optimizations can be made by the implementor, leading to good performance on a variety of platforms. The ANSI C standard library is an example of such an interface.

In general, there is a trade-off in system programming between implementation using libraries and using source code. Libraries embody platform-specific knowledge that can be used to optimize code, while application-specific knowledge could possibly be used to optimize code in other ways. Applications that do not use libraries may have more opportunity to optimize source code as more of it is visible. On the other hand, if the application does not use a library, then the application programmer must perform all the necessary platform optimizations.

In response to this trade-off, libraries are becoming less and less a collection of generic processing routines. Increasingly, they are active participants in the tuning of application programs. Providing the library with more information about the application can make the application simpler and more portable by moving the responsibility for platform-specific optimizations into the library.

17.2.1 DISTRIBUTED AND PARALLEL PROGRAMMING

The use of multiple processors in an application adds a new dimension to the problem of portable software technology. For the purposes of this discussion, we consider *distributed* programs to run locally on a single machine but make use of services provided by other machines. We consider *parallel* programs to consist of a tightly coupled set of computing resources working on parts of the same problem. Informally, we are considering the primary difference between the two types of programs to be the degree to which they assume coupling to other components of the system. Typically, we might expect to find distributed programming used in communicating among the control, data, and signal processing components of an embedded system, and parallel programming used within the signal processor and possibly the data processor. See Chapter 18 for more discussion of these topics.

Distributed programs require two pieces of information in order to make use of a service: the location of the machine that provides the service and the protocol required to use the service. They also require a communication interface to exchange data with the remote service; this interface may have to take account of the topology and hardware of the network that connects the machines. For a distributed program to be truly portable, that is, independent of the network and the locations of particular services, all of these aspects of the program must be isolated from the application programmer.

Parallel programs may make use of two types of parallelism: task parallelism, in which different processing stages communicate data, and data parallelism, in which a mathematical object such as a matrix or vector is shared among many processors. In each of these cases, full portability can only be achieved when the program is independent of the number of processors on which it is to be run. This quality of a program is known as *map independence*. Without map independence, a program will have to be rewritten to take advantage of more processors becoming available or be moved to a new platform. Map independence implies that the program uses a source of information outside of the program source code to determine the number of processors available.

17.2.2 Surveying the State of Portable Software Technology

In this section, the state of portable software technology is surveyed. First, portable interfaces for computation and the evolution of technology to gain portable performance using those interfaces are considered. Following this is a description of portable software technology as it applies to distributed and parallel systems.

17.2.2.1 Portable Math Libraries

Portable math library interfaces have existed for some time. The basic linear algebra subroutines (BLAS), which are used by the LAPACK package, are perhaps the oldest example in current use (Anderson et al. 1992). Originally, the BLAS were a set of simple function calls for performing vector operations such as vector element-wise add and multiply operations, plane rotations, and vector dot-products. These vector routines are referred to as level-1 BLAS routines. Level-2 BLAS were later defined that extended the interface to matrix-vector operations, including matrix-vector multiply, rank-one updates to a matrix, and solution of triangular systems with a vector right-hand side. Level-3 BLAS include matrix-matrix operations, both element-wise and product operations, and extensions to the level-2 BLAS, including generalized updates and solution of triangular systems with multiple right-hand sides. The major motivation for the level-2 and level-3 BLAS is to aggregate operations so that the number of function calls required to implement a given function is greatly reduced.

The BLAS were originally defined with only a Fortran interface. Since there was until recently no standardized way for Fortran subroutines and C programs to interact, C programmers who wished to make use of the BLAS had to write a portability layer or make their programs nonportable. See Lebak (1997) for an example of the difficulties involved. In 1999, this situation was rectified by the creation of a standard interface to the BLAS from C by a group known as the BLAS Technical Forum (Basic Linear Algebra Subprograms Technical Forum 2002).

Despite the fact that the BLAS are a well-established portable interface, they are not widely used on high performance embedded systems. Historically, this has to do with the rejection of Fortran by the embedded community (see Chapter 16). It also reflects the fact that the BLAS do not provide standard interfaces to useful signal processing functions such as the fast Fourier transform (FFT). Until the late 1990s, embedded vendors provided such functions using their own proprietary library interfaces. Today, embedded system vendors have available a standard library, the Vector, Signal, and Image Processing Library (VSIPL), which provides the routines needed by high performance embedded computing. VSIPL includes not only matrix operations like the BLAS, but traditional signal processing routines such as convolution, correlation, FFTs, and filters. The complete standard is very large (Schwartz et al. 2002); an overview is provided by Janka et al. (2001). A C++ interface to VSIPL, called VSIPL++, was defined in 2005 (CodeSourcery 2005).

VSIPL is designed from the ground up to be implemented on embedded systems. Three major features are provided to support such systems. First, VSIPL separates the concepts of storage and computation into the *block* and *view* objects, respectively. Multiple views may reference the same block of data or different subsets of the block, avoiding the need for costly copy operations. Second, VSIPL supports early binding on most of its major computational routines, allowing the user to set up ahead of time for an operation that is to be repeatedly executed. Examples of such routines include the FFT, filters, and linear algebra decompositions. Third, VSIPL makes data blocks *opaque*, allowing vendors to optimize the underlying memory layout in a way that is optimal for their machines.

17.2.2.2 Portable Performance Using Math Libraries

In general, there are two techniques to achieving portable high performance with math libraries. The first technique is to have an external party, usually the library or platform vendor, tune the library.

The second technique is to provide mechanisms for either the user or the software package itself to tune the library. The first technique is that taken by VSIPL and by the original implementation of the BLAS. In the remainder of this section, mechanisms for the second technique are considered.

An early example of allowing the user to tune the code is provided by the linear algebra package LAPACK. Built on top of the BLAS, LAPACK allows the user to select a parameter called the *block size* when the library is compiled. This block size parameter, which is related to the machine's memory hierarchy, is used as the input size to the level-3 BLAS operations performed by the library. Provision of this parameter allows the library to achieve high performance on a platform-specific basis and retain source-code portability. Other parameters are also provided that can be tuned for the individual platform (Anderson et al. 1992).

The tuning technique in LAPACK requires the user to properly pick parameters. A more automatic method is provided by the automatically tuned linear algebra subprograms, or ATLAS. This is a set of routines that generate and measure the performance of different implementations of the BLAS on a given machine. The results are used to select the optimal version of the BLAS and of some LAPACK routines for that machine (Demmel et al. 2005). A similar technique is used to optimize the FFT by the software package known as the "fastest Fourier transform in the West," or FFTW, and to optimize general digital signal processor transforms by a package known as SPIRAL (Frigo and Johnson 2005; Püschel et al. 2005). In all of these approaches, a special version of the library tuned for performance on the platform of interest is generated and used.

Most libraries such as LAPACK, BLAS, and VSIPL provide not only basic matrix operations such as add, subtract, and multiply, but also combinations of two or three of these element-wise operations. The reason for this is simple: on modern microprocessors, moving data between the processor, the various levels of cache, and main memory is frequently the bottleneck. To overcome this bottleneck, it is important to perform as many computations as possible for each access to memory. If a simple matrix expression such as $Y = aX + B$ must be computed in two steps, $C = aX$ and $Y = C + B$, then elements of the matrix C must be fetched from memory into cache twice. Combining these operations and properly blocking the memory accesses can make the implementation of this operation much more efficient.

However, libraries are obviously limited in the number of calls that they can provide. If an expression required by an application is not provided directly by a library, it must be created from a set of intermediate calls. In such a case, intermediate variables must be generated that are cycled through the cache twice. This is inherently inefficient. In the C++ language, a technique known as *expression templates* has been developed to deal with this problem. Expression templates allow the generation of efficient element-wise loops for an expression from the high-level code for that expression. The element-wise nature of the loops preserves good performance in the presence of a cache. Expression template techniques are described by Veldhuizen (1995) and were used in the Los Alamos POOMA library (Cummings et al. 1998; Haney et al. 1999). An example of the use of expression templates in the MIT Lincoln Laboratory Parallel Vector Library (PVL) library is given at the end of this chapter.

17.2.3 PARALLEL AND DISTRIBUTED LIBRARIES

Distributed programming frameworks provide location independence in a network. An example of such a framework is the Common Object Request Broker Architecture (CORBA), maintained by the Object Management Group (OMG). In CORBA, which comes from the world of distributed business computing, services are provided by an Object Request Broker (ORB) object. The ORB knows how to identify and find services in the system and isolates the client from the location of the service. While this approach is very flexible, it has the potential to be very time-consuming. Therefore, CORBA is most useful in HPEC systems for communication of low-volume control information. More discussion of CORBA is provided in Chapter 18.

```
/* Setup phase */
Vector<> w(M);
Matrix<> A(M, N);
qrd qrdObject(M, N,
              QRD_SAVEQ);
/* end of setup phase */
/* Generate or read A & w here */
/* Compute phase */
qrdObject.decompose(A);
qrdObject.prodq<VSIP_MAT_HERM,
              VSIP_MAT_LSIDE>(w);
/* end of compute phase */
```

(a)

```
/* Setup phase */
Vector<> w(M, wMap);
Matrix<> A(M, N, aMap);
qrd qrdObject(M, N,
              QRD_SAVEQ);
/* end of setup phase */
/* Generate or read A & w here */
/* Compute phase */
qrdObject.decompose(A);
qrdObject.prodq<VSIP_MAT_HERM,
              VSIP_MAT_LSIDE>(w);
/* end of compute phase */
```

(b)

FIGURE 17-1 Sample VSIPL++ (a) and parallel VSIPL++ (b) code to factor an $M \times N$ matrix A into orthogonal factor Q and triangular factor R and compute the product $Q^H w$. In both examples, a compute object called qrdObject is created to provide storage for the operation. In (b), the names wMap and aMap refer to maps defined in an outside file. (From Lebak, J.M. et al., Parallel VSIPL++, *Proceedings of the IEEE* 93(2): 318, 2005. With permission. © 2005 IEEE.)

Embedded systems typically consist of multiple nodes connected by a high-bandwidth communication fabric. The nodes communicate using some sort of direct memory access (DMA) engine. The communication software is typically customized for each embedded platform. In the last several years, embedded systems have begun to make use of the message-passing interface (MPI) standard that evolved in the high performance computing (HPC) world. Use of MPI makes code that performs communication operations portable. However, MPI implements a two-sided communication protocol that may not be optimal for DMA engines, which typically only require the sender or the receiver to initiate communication. Therefore, although MPI is in wide use in the HPC space, it is used less often in the HPEC space. More discussion of MPI can be found in Chapter 18.

A limitation of using MPI or DMA engine calls in a parallel program is that care must be taken to avoid writing code that depends specifically on the number of nodes in the system. Such code must obviously be rewritten when the number of nodes changes. Avoiding the need to change code in this way is the motivation for the map independence feature in PVL and in the parallel VSIPL++ (pVSIPL++) standard.

An example of the use of map independence in a pVSIPL++ program is shown in Figure 17-1. The single-processor code is shown on the left, and the multiprocessor code is shown on the right. Notice that the only difference between the single-processor and multiprocessor code is the presence of the map object, highlighted in bold on the right.

A map object must contain information about the processors that the data object is mapped to and the type of distribution used. Any block-cyclic distribution is allowable, including the block and cyclic subtypes: see Chapter 18 for a description of different distribution types. The key point is that by encapsulating the distribution information in the map object, the application program can become independent of the way in which the data are distributed and the number of nodes.

Besides making a program portable, map independence enables automated determination of the optimal maps for parallel programs. An off-line program can generate maps for the application, and the application can be run and timed with successive sets of maps to determine the best mapping. Many examples of such programs exist. An early example of automatic mapping of signal processing flow graphs to a parallel machine was provided by Printz et al. (1989). Moore et al. (1997) demonstrated a graphical model-based approach that optimally maps a signal flow graph for an image processing application onto a parallel hardware architecture, given a set of constraints and benchmarks of the key components. Squyres et al. (1998) built software to transparently distribute processing operations, written in C, onto a cluster. The implementation described uses "worker nodes" to perform the parallel image processing (Squyres, Lumsdaine, and Stevenson 1998). The Parallel-Horus library is a portable image processing library including a framework that allows the implementation to self-optimize for different platforms. The library is based on the concept of "user transparency," which is very similar to map independence (Seinstra and Koelma 2004). Hoffmann and Kepner developed a framework called Self-Optimizing Software for Signal Processing (S³P), which demonstrates generation of maps for a simple application in a cluster context (Lebak et al. 2005). S³P requires the application to be broken up into large blocks called tasks. An approach invented by Travinin et al. (2005) for the parallel MATLAB library, called pMapper, is able to generate the distributions of individual data objects in a parallel application.

17.2.4 EXAMPLE: EXPRESSION TEMPLATE USE IN THE MIT LINCOLN LABORATORY PARALLEL VECTOR LIBRARY

Both PVL and pVSIPL++ are designed to enable the use of expression templates in C++ to give high performance. This approach enables the library to generate efficient code without requiring the implementor to have coded every possible expression ahead of time. This approach was used in the POOMA library (Cummings et al. 1998).

Other, similar approaches exist in the literature. To cite just one example, Morrow et al. (1999) use a *delayed execution* model to construct programs that are passed to an offline optimizer. The optimizer in turn builds implementations to run on a "parallel co-processor."

This section shows an example of using the PVL approach to obtain high performance. The following material originally appeared in Lebak et al., "Parallel VSIPL++," *Proceedings of the IEEE* 93(2): 313–330, and is used by permission (© 2005 IEEE).

PVL is implemented using the C++ programming language, which allows the user to write programs using high-level mathematical constructs such as

$$A = B + C * D,$$

where A, B, C, and D are all distributed vectors or matrices. Such expressions are enabled by the *operator overloading* feature of C++ (Stroustrup 1997). A naive implementation of operator overloading in C++ will result in the creation of complete copies of the data for each substep of the expression, such as the intermediate multiply C*D, which can result in a significant performance penalty. This penalty can be avoided by the use of *expression templates*, which enable the construction of a chained expression in such a way as to eliminate the temporary variables (see Figure 17-2). In many instances, it is possible to achieve better performance with expression templates than using standard C-based libraries because the C++ expression-template code can achieve superior cache performance for long expressions (Haney et al. 1999).

Consider the PVL code to perform a simple add, shown in Figure 17-3. Intuitively, it is easy to understand that this code fragment adds the vectors b and c to produce a. Vectors a, b, and c may use any block-cyclic distribution. The mechanism used to effect this addition is not obvious from the code. We use the portable expression template engine [PETE (Haney et al. 1999)] to define operations on vectors and matrices. PETE operators capture the structure of an expression in a PETE expression and defer execution of that expression until after all parts of the expression are known. This has the potential to allow optimized evaluation of expressions by eliminating temporary objects. The operations contained in a PETE expression are processed by a PVL object called an *evaluator*.

Add, subtract, multiply, and divide operations in PVL are performed by a binary element-wise computation object (which operates on the BinaryNode in Figure 17-2), internal to the evaluator. Such an object is used whenever an optimized vector routine, such as a VSIPL vector add, is desired for performance reasons. This object contains code to move the input data to the working locations, actually perform the add, and move the sum to the desired output location.

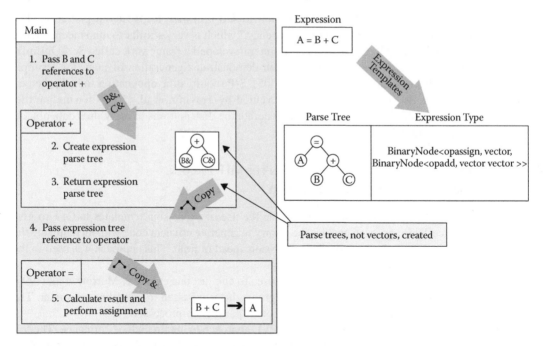

FIGURE 17-2 C++ expression templates. Obtaining high performance from C++ requires technology that eliminates the normal creation of temporary variables in an expression. (From Lebak, J.M. et al., Parallel VSIPL++, *Proceedings of the IEEE* 93(2): 322, 2005. With permission. © 2005 IEEE.)

```
void addVectors(const int vecLength)
    {
    Vector< Complex<Float> > a("a", vecLength, aMap);
    Vector< Complex<Float> > b("b", vecLength, bMap);
    Vector< Complex<Float> > a("c", vecLength, cMap);
    // Fill the vectors with data
    generateVectors(a,b,c);

    a=b+c;
    // Check results and end
    }
```

FIGURE 17-3 PVL vector add. PVL code to add two vectors. The distribution of each vector is described by its map object. (From Lebak, J.M. et al., Parallel VSIPL++, *Proceedings of the IEEE* 93(2): 322, 2005. With permission. © 2005 IEEE.)

Thus, the simple statement a=b+c actually triggers the following sequence of events:

1. A PETE expression for the operation is created that references b and c and records that the operation is an add;
2. The assignment operator (operator=) for vector uses an evaluator to interpret the expression; and
3. The evaluator calls its internal binary element-wise computation object to perform the add and assign its results to a.

Creating the evaluator object is a time-intensive operation and one that can be optimized using early binding. Therefore, the PVL library provides an optimization method that creates the evaluator for the particular expression and stores it for future reference. When the assignment operator for a distributed view is called, it checks whether an evaluator has been stored for this particular expression, and if so, uses that evaluator to perform the expression rather than creating one. If the

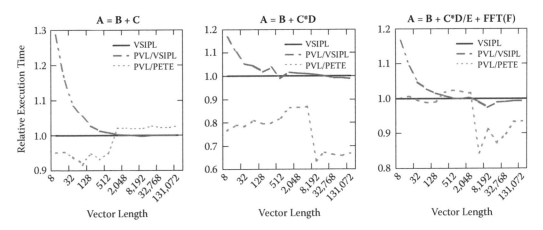

FIGURE 17-4 Single-processor performance. Comparison of the single-processor performance of VSIPL (C), PVL (C++) on top of VSIPL (C), and PVL (C++) on top of PETE (C++), for different expressions with different vector lengths. PVL with VSIPL or PETE is able to equal or improve upon the performance of VSIPL. (From Lebak, J.M. et al., Parallel VSIPL++, *Proceedings of the IEEE* 93(2): 323, 2005. With permission. © 2005 IEEE.)

evaluator has not been stored, one is created, stored, and associated with a view for future use. This approach provides early binding for system deployment without requiring it in system prototyping.

Figure 17-4 shows the performance achieved using templated expressions in PVL on a single Linux PC. We compare expressions of increasing length [A = B + C, A = B + C*D, and A = B + C*D/E + FFT(F)] and examine the performance of three different approaches: the VSIPL reference implementation, PVL layered over the VSIPL reference implementation, and PVL implemented using expression templates (bypassing VSIPL for all the operations except the FFT). Notice that layering PVL over VSIPL can introduce considerable overhead for short vector lengths; this overhead is eliminated by the expression template approach. For long expressions, code that uses templated expressions is able to equal or exceed the performance of VSIPL.

Figures 17-5 and 17-6 show the performance achieved using templated expressions on a four-node cluster of workstations, connected using gigabit Ethernet. The expressions used are the same as in Figure 17-4, and the approaches are the same, except that the basic approach uses a

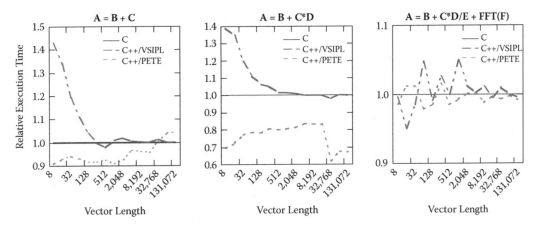

FIGURE 17-5 Multiprocessor (no communication). Comparison of the multiprocessor (no communication) performance of VSIPL (C), PVL (C++) on top of VSIPL (C), and PVL (C++) on top of PETE (C++), for different expressions with different vector lengths. (From Lebak, J.M. et al., Parallel VSIPL++, *Proceedings of the IEEE* 93(2): 323, 2005. With permission. © 2005 IEEE.)

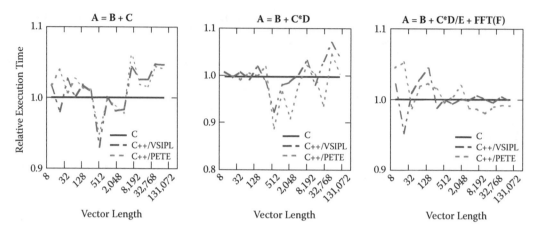

FIGURE 17-6   Multiprocessor (with communication). Comparison of the multiprocessor (with communication) performance of VSIPL (C), PVL (C++) on top of VSIPL (C), and PVL (C++) on top of PETE (C++), for different expressions with different vector lengths. (From Lebak, J.M. et al., Parallel VSIPL++, *Proceedings of the IEEE* 93(2): 324, 2005. With permission. © 2005 IEEE.)

combination of VSIPL and MPI in C. In Figure 17-5, all the vectors are identically distributed, so no communication needs to be performed except by the FFT operation. Therefore, the first two expressions show results similar to the single-processor results. For the third expression, the communication cost of the FFT dominates and all approaches are roughly comparable. In Figure 17-6, vector A is distributed over two nodes, and the remaining vectors are distributed over four nodes each. This introduces the requirement to communicate the results from one set of nodes to another, and this communication cost dominates the computation time so that all the approaches have similar performance.

Figures 17-4 through 17-6 show a comparison of expression templates with the unoptimized, VSIPL reference implementation as a proof of concept. However, it is obviously important for the library to be able to leverage all the features of a hardware architecture. An example of such a feature is the Altivec extensions provided by the PowerPC G4 processor, described in Chapter 13. Franchetti and Püschel (2003) have shown that SPIRAL can generate code that uses such extensions and have implemented it using the similar SSE and SSE-2 extensions for Intel architectures. Similarly, Rutledge (2002) demonstrated that PETE can make use of Altivec extensions directly and achieve comparable or better performance to optimized implementations of VSIPL by doing so. A summary of his results is shown in Figure 17-7. He compared a hand-generated Altivec loop (assuming unit stride) with an Altivec-optimized VSIPL implementation provided by MPI Software Technology, Inc., and PVL using PETE (again assuming unit stride) to generate Altivec instructions, for a series of expressions. The VSIPL implementation achieves lower performance on average. At least in part, this probably has to do with the requirement that the VSIPL implementation support non-unit strides, but it has more to do with the necessity in C-based VSIPL to perform multiple function calls to evaluate long expressions. The most encouraging result, however, is that PVL using PETE is able to achieve performance comparable to handwritten Altivec code.

17.3   SUMMARY

The foundation of portable application software is the concept of a library of subroutines. In the beginning, portable software technology consisted of well-defined interfaces to commonly used routines. As libraries have evolved, techniques such as self-optimization and expression templates have increased the amount of optimization possible within the library. These techniques have increased the performance portability of applications. As embedded applications make increased

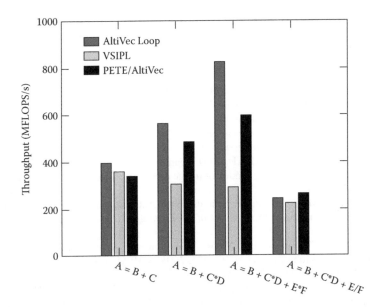

FIGURE 17-7 Combining PETE with Altivec. Comparison of peak throughput achieved by a hand-generated Altivec loop, an optimized VSIPL implementation, and PETE modified to use Altivec instructions. (From Lebak, J.M. et al., Parallel VSIPL++, *Proc. IEEE* 93(2): 324, 2005. With permission. © 2005 IEEE.)

use of parallelism, libraries that allow these applications to become independent of the mapping to processors will enable greater portability among parallel systems.

REFERENCES

Anderson, E., Z. Bai, C. Bischof, S. Blackford, J. Demmel, J. Dongarra, J. Du Croz, A. Greenbaum, S. Hammarling, A. McKenney, and D. Sorensen. 1992. *LAPACK User's Guide*. Philadelphia: Society for Industrial and Applied Mathematics Press.

Basic Linear Algebra Subprograms Technical Forum. 2002. Basic Linear Algebra Subprograms Technical (BLAST) Forum Standard. *International Journal of High Performance Applications and Supercomputing* 16(1).

Cico, L., J. Greene, and R. Cooper. 2005. Performance estimates of a STAP benchmark on the IBM Cell processor. *Proceedings of the Ninth Annual High Performance Embedded Computing Workshop*. Available online at http://www.ll.mit.edu/HPEC/agendas/proc05/agenda.html.

CodeSourcery LLC. 2005. VSIPL++ Specification 1.0. Available online at http://www.hpec-si.org.

Cummings, J., J. Crotinger, S. Haney, W. Humphrey, S. Karmesin, J. Reynders, S. Smith, and T. Williams. 1998. Rapid application development and enhanced code portability using the POOMA framework. Presented at the SIAM Workshop on Object-Oriented Methods for Interoperable Scientific and Engineering Computing. Yorktown Heights, NY.

Demmel, J., J. Dongarra, V. Eijkhout, E. Fuentes, A. Petitet, R. Vuduc, R.C. Whaley, and K. Yelick. 2005. Self-adapting linear algebra algorithms and software. *Proceedings of the IEEE* 93(2): 293–312.

Franchetti, F. and M. Püschel. 2003. Short vector code generation for the discrete Fourier transform. *Proceedings of the 17th International Parallel and Distributed Processing Symposium* 58–67.

Frigo, M. and S.G. Johnson. 2005. The design and implementation of FFTW3. *Proceedings of the IEEE* 93(2): 216–231.

Goldberg, D. 1991. What every computer scientist should know about floating-point arithmetic. *ACM Computing Surveys* 23(1): 5–48.

Haney, S., J. Crotinger, S. Karmesin, and S. Smith. 1999. Easy expression templates using PETE, the portable expression template engine. *Dr. Dobb's Journal* 23(10): 89–95.

Institute of Electrical and Electronics Engineers. 1985. IEEE Standard 754-1985 for binary floating-point arithmetic. Reprinted in *SIGPLAN* 22(2): 9–25.

Janka, R., R. Judd, J. Lebak, M. Richards, and D. Campbell. 2001. VSIPL: an object-based open standard API for vector, signal, and image processing. *Proceedings of the 2001 IEEE International Conference on Acoustics, Speech, and Signal Processing* 2: 949–952.

Lebak, J.M. 1997. Portable parallel subroutines for space-time adaptive processing, Ph.D. thesis. Cornell University, Ithaca, N.Y.

Lebak, J.M., J. Kepner, H. Hoffmann, and E. Rutledge. 2005. Parallel VSIPL++: an open standard software library for high-performance parallel signal processing. *Proceedings of the IEEE* 93(2): 313–330.

Moore, M.S., J. Sztipanovitza, G. Karsaia, and J. Nicholsa. 1997. A model-integrated program synthesis environment for parallel/real-time image processing. *Proceedings of SPIE: Volume 3166—Parallel and Distributed Methods for Image Processing* 31–45.

Morrow, P.J., D. Crookes, J. Brown, G. McAleese, D. Roantree, and I. Spence. 1999. Efficient implementation of a portable parallel programming model for image processing. *Concurrency: Practice and Experience* 11(11): 671–685.

Printz, H., H.T. Kung, T. Mummert, and P. Scherer. 1989. Automatic mapping of large signal processing systems to a parallel machine. *Proceedings of SPIE: Volume 1154—Real-Time Signal Processing XII* 2–16.

Püschel, M., J.R. Johnson, D. Padua, M.M. Veloso, B.W. Singer, J. Xiong, F. Franchetti, Y. Voronenko, K. Chen, A. Gacic, R.W. Johnson, and N. Rizzolo. 2005. SPIRAL: code generation for DSP transforms. *Proceedings of the IEEE* 93(2): 232–275.

Rutledge, E. 2002. Altivec extensions to the portable expression template engine (PETE). *Proceedings of the Sixth Annual High Performance Embedded Computing Workshop.* MIT Lincoln Laboratory, Lexington, Mass. Available online at http://www.ll.mit.edu/HPEC/agendas/agenda02.html.

Schwartz, D.A., R.R. Judd, W.S. Harrod, and D.P. Manley. 2002. Vector, Signal, and Image Processing Library (VSIPL) 1.1 application programmer's interface. Georgia Tech Research Corporation. Available online at http://www.vsipl.org.

Seinstra, F. and D. Koelma. 2004. User transparency: a fully sequential programming model for efficient data parallel image processing. *Concurrency and Computation: Practice and Experience* 16(7).

Squyres, J.M., A. Lumsdaine, and R.L. Stevenson. 1998. A toolkit for parallel image processing. *Proceedings of SPIE:Volume 3452—Parallel and Distributed Methods for Image Processing II* 69–80.

Stroustrup, B. 1997. *The C++ Programming Language*, 3rd edition. Reading, Mass.: Addison-Wesley.

Travinin, N., H. Hoffmann, R. Bond, H. Chan, J. Kepner, and E. Wong. 2005. pMapper: automatic mapping of parallel MATLAB programs. *Proceedings of the Ninth Annual High Performance Embedded Computing Workshop.* MIT Lincoln Laboratory, Lexington, Mass. Available online at http://www.ll.mit.edu/HPEC/agendas/proc05/agenda.html.

Veldhuizen, T. 1995. Expression templates. *C++ Report* 7(5): 26–31.

18 Parallel and Distributed Processing

Albert I. Reuther and Hahn G. Kim, MIT Lincoln Laboratory

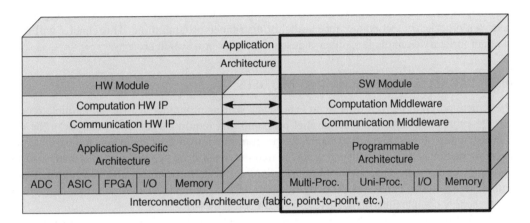

This chapter discusses parallel and distributed programming technologies for high performance embedded systems. Parallel programming models are reviewed, and a description of supporting technologies follows. Illustrative benchmark applications are presented. Distributed computing is distinguished from parallel computing, and distributed computing models are reviewed, followed by a description of supporting technologies and application examples.

18.1 INTRODUCTION

This chapter discusses parallel and distributed programming technologies for high performance embedded systems. Despite continual advances in microprocessor technology, many embedded computing systems have processing requirements beyond the capabilities of a single microprocessor. In some cases, a single processor is unable to satisfy the computational or memory requirements of the embedded application. In other cases, the embedded system has a large number of devices, for example, devices that need to be controlled, others that acquire data, and others that process the data. A single processor may not be able to execute all the tasks needed to control the entire system. Increasingly, embedded systems employ multiple processors to achieve these stringent performance requirements.

Computational or memory constraints can be overcome with parallel processing. The primary goal of parallel processing is to improve performance by distributing computation across multiple processors or increasing dataset sizes by distributing data across multiple processors' memory. Managing multiple devices can be addressed with distributed processing. Distributed processing connects multiple devices—often physically separate, each with its own processor—that communicate with each other but execute different tasks.

Let us more formally define the differences between parallel and distributed processing. The most popular model for parallel processing is the single-program multiple-data (SPMD) model. In SPMD, the same program is executed on multiple processors but each instantiation of the program processes different data. The processors often share data and communicate with each other in a tightly coordinated manner. Distributed processing is best described with the multiple-program multiple-data (MPMD) model. Each processor executes a different program, each processing different data. Programs often communicate with each other, sending and receiving data and signals, often causing new activity to occur within the distributed system.

Armed with a networking library, any adept programmer can write parallel or distributed programs, but this is a cumbersome approach. The purpose of this chapter is to describe various structured models for writing parallel and distributed programs and to introduce various popular technologies that implement these models. Note that vendors of embedded computers may supply their own proprietary technology for parallel or distributed programming. This chapter will focus on nonproprietary technologies, either industry standards or research projects.

18.2   PARALLEL PROGRAMMING MODELS

The typical programmer has little to no experience writing programs that run on multiple processors. The transition from serial to parallel programming requires significant changes in the programmer's way of thinking. For example, the programmer must worry about how to distribute data and computation across multiple processors to maximize performance and how to synchronize and communicate between processors. Many standard serial data structures and algorithms require significant modification to work properly in a parallel environment. The programmer must ensure that these new complexities do not affect program correctness.

Parallel programming is becoming a necessary technique for achieving high performance in embedded systems, especially as embedded-computer vendors build multiprocessor systems and microprocessor manufacturers increasingly employ multiple cores in their processors. Parallel programming models are key to writing parallel programs. They provide a structured approach to parallel programming, reducing its complexity. While nearly all parallel programming technologies are SPMD, the SPMD model can be broken down into several subcategories. The following sections will describe the threaded, message-passing, and partitioned global address space models and briefly introduce various technologies that implement each model. Each of the technologies presented has its advantages and disadvantages, a number of which are summarized in Table 18-1. Clearly, none of these technologies are suitable for all problems requiring parallel computation and care must be taken when selecting one for a specific application.

To gain greater understanding and appreciation of the concepts discussed in the following sections, read about the trade-offs of computation versus communication in parallel and distributed computing in Chapter 5.

18.2.1   THREADS

Although most programmers will likely admit to having no experience with parallel programming, many have indeed had exposure to a rudimentary type in the form of threads. In serial programs, threads execute multiple sections of code "simultaneously" on a single processor. This is not true parallelism, however. In reality, the processor rapidly context-switches between threads to maintain the illusion that multiple threads are executing simultaneously.*

---

* Recently, symmetric multithreading (SMT) technology has been finding its way into processor architectures. SMT allows instructions from multiple threads to be dispatched simultaneously. An example of this is Intel's Hyper-Threading Technology (see website: http://www.intel.com/technology/hyperthread).

TABLE 18-1

Advantages and Disadvantages of Various Parallel Programming Technologies and the Languages Supported by Each

Technology	Advantages	Disadvantages	Languages
Pthreads	Widely available	Not designed for parallel programming; extremely difficult to use	C, C++
OpenMP	Allows incremental parallelization	Limited to shared-memory architectures; requires specialized compilers	Fortran, C, C++
PVM	Supports heterogeneous clusters	Not well suited for embedded computing systems	Fortran, C, C++
MPI	Supports wide range of parallel processor architectures; implementations available for nearly every system	Difficult to use; lack of support for incremental parallelization	Fortran, C, C++
UPC	Allows incremental parallelization; supports wide range of architectures	Requires specialized compilers	C
VSIPL++	Targets the signal and image processing domain; supports wide range of architectures	Not a general-purpose parallel programming library	C++

True parallelism can be achieved by executing multiple threads on multiple processors. The threaded parallel programming model is targeted at shared-memory processor architectures. In shared-memory architectures, all processors access the same physical memory, resulting in a *global address space* shared by all processors. Figure 18-1 depicts a shared-memory architecture. In a shared-memory architecture, multiple memory banks are logically grouped together to appear as a single address space. Every processor has direct access to the entire address space. Many embedded computing platforms, such as multiprocessor single-board computers, use shared memory.

A typical threaded program starts execution as a single thread. When the program reaches a parallel section of code, multiple threads are spawned onto multiple processors and each thread processes a different section of the data. Upon completion, each thread terminates, leaving behind the original thread to continue until it reaches the next parallel section.

The threaded programming model is well suited for incremental parallelization, i.e., gradually adding parallelism to existing serial applications without rewriting the entire program. One common example of code that is amenable to incremental parallelization is data parallel computation. In data parallel applications, the same set of operations is applied to each data element. Consequently, the dataset can be divided into blocks that can be simultaneously processed. Consider the example of adding two vectors, A and B, and storing the result in another vector, C. A and B can be added using a simple for loop. If there are four processors, A, B, and C can be divided such that each

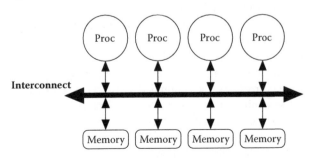

FIGURE 18-1 A shared-memory architecture.

TABLE 18-2

One Example of How a For Loop Would Execute in Parallel by Four Threads

Serial	Thread 0	Thread 1	Thread 2	Thread 3
for i = 0 to 1023	for i = 0 to 255	for i = 256 to 511	for i = 512 to 767	for i = 768 to 1023
C[i] = A[i]+B[i]	C[i] = A[i] + B[i]	C[i] = A[i] + B[i]	C[i] = A[i] + B[i]	C[i] = A[i] + B[i]
end	end	end	end	end

processor processes one-fourth of the loop iterations. Table 18-2 shows how this for loop would be executed in parallel by four threads.

18.2.1.1 Pthreads

The most well-known thread technology is the Pthreads library. Pthreads is a part of POSIX, an IEEE standard for operating system application program interfaces (APIs) initially created for Unix (IEEE 1988). POSIX has spread beyond Unix to a number of operating systems, making Pthreads a widely available parallel programming technology (Nichols, Buttlar, and Proulx Farrell 1996). Pthreads provides mechanisms for standard thread operations, e.g., creating and destroying threads and coordinating interthread activities, such as accessing shared variables using memory-locking mechanisms such as mutexes and condition variables.

The wide availability of Pthreads can make it an attractive option for parallel programming. Pthreads was not explicitly designed for parallel programming, however, but rather was designed to provide a general-purpose thread capability. This flexibility results in a lack of structure in managing threads, thus making parallel programming using Pthreads very complex. Because the programmer is responsible for explicitly creating and destroying threads, partitioning data between threads, and coordinating access to shared data, Pthreads and other general-purpose thread technologies are seldom used to write parallel programs. Consequently, other thread technologies designed explicitly for parallel programming, such as OpenMP, have been developed to address these issues.

18.2.1.2 OpenMP

Open Multiprocessing (OpenMP) is a set of language extensions and library functions for creating and managing threads for Fortran, C, and C++ (Chandra et al. 2002; OpenMP webpage). Language extensions are implemented as compiler directives, thus requiring specialized compilers. These directives support concurrency, thread synchronization, and data handling between threads.

The programmer uses directives to mark the beginning and end of parallel regions, such as the for loop described earlier. The compiler automatically inserts operations to spawn and terminate threads at the beginning and end of parallel regions, respectively. Additional directives specify which variables should be parallelized. For example, in the for loop, OpenMP computes which loop iterations should be executed on each processor. In Pthreads, the programmer is responsible for writing the code to perform all these operations.

OpenMP programs can be compiled for both serial and parallel computers. Serial compilers ignore the directives and generate a serial program, while OpenMP compilers recognize the directives and generate a parallel program; therefore, OpenMP can support incremental parallelization. OpenMP directives can be added to an existing serial program. The program is compiled and executed on a single processor to ensure correctness, then compiled and executed on a parallel processor to achieve performance. This process significantly simplifies the development of parallel applications. Nevertheless, due to complexities such as race conditions between iterations in a parallel for loop, it is possible for programmers to produce incorrect parallel programs even if the serial version is correct.

OpenMP was developed through a collaboration among industry, government, and academia and is widely used for implementing parallel scientific codes on shared-memory systems. Its use in

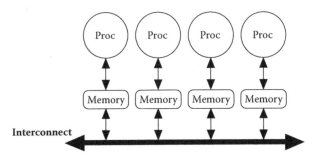

FIGURE 18-2 A distributed-memory architecture.

embedded applications is limited but growing, as more processor manufacturers are moving toward multicore architectures.

18.2.2 MESSAGE PASSING

Messaging passing is arguably the most popular parallel programming method used today. Unlike threads, message passing presents a *partitioned address space*. The parallel computer is viewed as a set of processors, each with its own private address space, connected by a network. As a result, each data element is located in a single processor's local memory.

Previously, we described data parallel applications. Unfortunately, data parallel applications make up only a portion of all parallel applications. Due to data dependencies, it is often necessary for processors to exchange data. A partitioned address space requires interprocessor communication in order to share data. Message passing provides mechanisms to send and receive messages between processors.

A drawback of message passing is that it can significantly increase programming complexity. Unlike technologies such as OpenMP, incremental parallelization of existing serial programs is difficult using message passing. Nevertheless, it has several compelling advantages. First, message passing accommodates a wide range of parallel architectures, including both shared- and distributed-memory machines. Figure 18-2 depicts a distributed-memory architecture. In distributed-memory architectures, each processor has its own local memory that is typically accessed over a high-speed bus. Memory for all processors is connected via a separate, usually slower, interconnect, allowing processors to access another processor's memory. Second, message passing is very flexible. It can be used to implement nearly any kind of application, from data parallel computations to problems with complex interprocessor communication patterns. Third, a partitioned address space allows message-passing technologies to achieve very high performance. A processor accessing its own local memory is considerably faster than one accessing another processor's memory.* Consequently, distributed-memory architectures typically achieve higher performance than do shared-memory systems. Finally, shared-memory architectures are limited in scalability. Distributed-memory architectures use scalable interprocessor networks to build large computing systems, including embedded systems, with tens to hundreds of processors. Message passing's support for distributed memory allows programs to scale to much larger numbers of processors than can be scaled with threads.

18.2.2.1 Parallel Virtual Machine

The Parallel Virtual Machine (PVM) is an outgrowth of a research project to design a solution for heterogeneous parallel computing (Geist et al. 1994; Parallel Virtual Machine webpage). The PVM library and software tools present the programmer with an abstract "parallel virtual machine," which

---

* This is true for even some types of shared-memory architectures.

TABLE 18-3

Pseudocode for an Example of Using Ranks to Perform Communication

Rank 0	Rank 1
if (my_rank = = 0)	if (my_rank = = 0)
Send message to rank 1	Send message to rank 1
else if (my_rank = = 1)	**else if (my_rank = = 1)**
Receive message from rank 0	**Receive message from rank 0**
end	**end**

```
int N = 1024;
int blockSize = N/numProcs;
int start = blockSize * my_rank;
int end = start + blockSize - 1;

for i = start to end
   C[i] = A[i] + B[i];
end
```

FIGURE 18-3 Pseudocode for an example of data-parallel computation with MPI.

represents a cluster of homogeneous or heterogeneous computers connected by a network as a single parallel computer. Machines can enter or leave the virtual machine at runtime. PVM uses a leader-worker paradigm; programs start as a single-leader process, which dynamically spawns worker processes onto multiple processors, all executing the same program. All processes are assigned a unique identification (ID), known as a *task ID*, used for communicating. PVM supplies functions for message passing: adding and removing processors to and from the virtual machine; managing and monitoring processes executing on remote machines.

Despite the fact that PVM was the first widely used message-passing programming technology supporting Fortran, C, and C++, MPI has become the most popular message-passing technology in both the high performance computing (HPC) and high performance embedded computing (HPEC) domains. PVM's ability to run on heterogeneous computers and dynamically resize the virtual machine, however, makes it popular for parallel programming on clusters.

18.2.2.2 Message Passing Interface

The Message Passing Interface (MPI) was created by a consortium of industry vendors, application developers, and researchers to create a standard parallel programming interface that would be portable across a wide range of machines (Gropp 1999). Since the formation of its specification, implementations of MPI in C and Fortran have been developed for nearly every type of system, including free (see MPICH at http://www-unix.mcs.anl.gov/mpi/mpich and LAM at http://www.lam-mpi.org) and commercial implementations optimized for various platforms.

When an MPI program is launched, the program starts on each processor. The core MPI functions revolve around communication, i.e., send and receive. Each process is assigned an ID, known as a rank. Ranks are used to identify source and destination processors when performing communication. Table 18-3 contains MPI code in which rank 0 sends a message to rank 1. Due to MPI's SPMD nature, the exact same program runs on both processors. The *if-else* statement distinguishes which sections of code run on each processor. The table shows the code that executes for each rank in bold. Ranks are also used to partition data across multiple processors. For example, consider the vector add from Table 18-2. Figure 18-3 contains pseudocode for MPI that calculates which indices in A and B from Table 18-2 the local processor should add based on the processor's rank.

A new version of MPI, called MPI-2, has been developed to address many of the limitations of MPI (Gropp, Lusk, and Thakur 1999). For example, in MPI the number of processors is fixed for the duration of the program; MPI-2 allows for dynamic resizing of the number of processors, similar to PVM.

MPI is one of the most popular technologies for implementing parallel embedded applications. Implementations are available for embedded processing platforms either directly from the vendor or third-party developers. MPI is often used with math and scientific processing libraries to implement high performance programs. One such example of these libraries is the Vector, Signal, and Image Processing Library (VSIPL), a C library for high performance embedded signal and image processing (see VSIPL website at http://www.vsipl.org).

On the surface, MPI and PVM look similar and are often compared, but they are actually two very different implementations of the message-passing model and may be used to solve different problems. The differences between MPI and PVM are discussed in Gropp and Lusk (1997).

18.2.3 PARTITIONED GLOBAL ADDRESS SPACE

All of the technologies discussed thus far have their advantages and limitations. Threads benefit from a global address space, which is much easier to program and enables technologies like OpenMP to support incremental parallelization of existing serial applications. Thread technologies are thus restricted to shared-memory architectures and, consequently, are limited in scalability.

Because message passing supports nearly any type of parallel processor architecture, it has become the most widely used programming model for large-scale applications. Messaging passing's partitioned address space, though, places the burden of partitioning and communicating data between processors on the user. Programmers naturally think of a dataset as a single array rather than multiple subarrays. Transitioning from a global address space to a partitioned address space is not trivial.

The partitioned global address space (PGAS) model combines the advantages of both models by abstracting the partitioned address space of distributed memory architectures as a global address space. Additionally, PGAS provides a means to express how to parallelize and manipulate arrays at a high level. The programmer implicitly specifies parallelism using high-level constructs, rather than explicitly writing code to distribute data. It is the responsibility of the underlying technology to distribute the data. Some PGAS technologies implement these constructs using language extensions; others use library functions. Each of the technologies presented here, however, uses a similar vocabulary to describe data parallelization.

Consider the problem parallelizing a two-dimensional array. First, the programmer specifies the number of processors across which to parallelize each dimension of the array. Second, the programmer specifies how to distribute each dimension across the processors, i.e., using block, cyclic, or block-cyclic distribution. In the block distribution, a processor owns a contiguous set of indices. In the cyclic distribution, data elements are distributed across processors in a cyclic, or round-robin, manner. In the block-cyclic distribution, each processor owns multiple blocks of contiguous indices, which are distributed across processors in a cyclic manner.* Finally, the programmer specifies the processor on which each section of the array resides. For example, global array semantics allows the programmer to specify that the columns of a 4 × 16 matrix should be parallelized with a block distribution across processors 0 to 3. Figure 18-4 graphically depicts this array. The global array is shown at the top of the figure. Each processor stores a subset of the global array such that the columns of the global array are distributed using a block distribution across the four processors. Unlike threaded or message-passing paradigms, each processor has knowledge of the entire global array, such that each processor knows which elements of the global array it owns and which elements are owned by other processors.

PGAS enables the programmer to focus on writing application code rather than parallel code. Placing the responsibility of partitioning data and computation on the technology instead of the

---

* Note that block and cyclic are special cases of block-cyclic, but block and cyclic are so commonly used that they are often categorized separately.

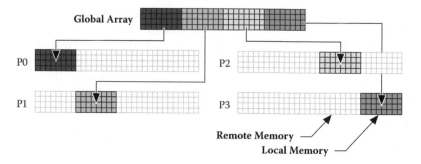

FIGURE 18-4   Example of parallelizing a matrix using global array semantics.

programmer results in fewer errors (e.g., incorrectly partitioning indices between processors), as well as in code that is easier to write, more robust, and easier to debug.

PGAS is emerging as an accepted programming model, though not yet as widely used as threading and message passing. Note that PGAS is well suited for applications that have regular data structures, e.g., vectors, matrices, and tensors. This makes PGAS less flexible than message passing for certain applications. However, HPEC systems are often built for applications, such as signal and image processing, that process regular data structures.

18.2.3.1   Unified Parallel C

Unified Parallel C (UPC) is an extension of the C programming language for parallel programming, jointly developed by a consortium of industry, government, and academia (see UPC website at http://upc.gwu.edu). UPC supports both shared- and distributed-memory architectures. Programs are modeled as a collection of independent threads executing on multiple processors. Memory is logically partitioned between the threads. Each thread's memory is divided into a private space and shared space. Variables can be allocated in either space. Private variables can be accessed only by the local thread, while shared variables can be accessed by any thread. Collectively, the shared address space represents a single, unified address space.*

UPC adds a new `shared` qualifier that allows the programmer to declare variables as shared in the variable declaration. Shared arrays can be declared, also; block, cyclic, or block-cyclic data distributions can also be concisely described in the variable declaration. When a thread accesses shared data allocated in another thread's memory, the data are communicated transparently. UPC also contains syntax for supporting shared pointers, parallel loops, thread synchronization, and dynamic memory allocation.

UPC is a formal superset of C; compiling a C program with a UPC compiler results in a valid program that executes N copies of the programs in N threads. This capability enables the incremental parallelization of existing C programs using UPC.

A number of vendors, including IBM and Hewlett-Packard, offer UPC compilers, but targeted at their HPC systems (see IBM UPC website at http://www.alphaworks.ibm.com/tech/upccompiler and the HP UPC website at http://h30097.www3.hp.com/upc). Several UPC research groups have developed free compilers that can be used with a range of processor architectures and operating systems, including embedded computing platforms (see the George Washington UPC website at http://upc.gwu.edu).

18.2.3.2   VSIPL++

Like OpenMP, MPI, and UPC, the VSIPL specification was developed through a collaboration among industry, government, and academia (see the HPEC Software Initiative website at http://

---

* Co-array Fortran and Titanium are extensions to Fortran and Java, respectively, that apply concepts similar to UPC.

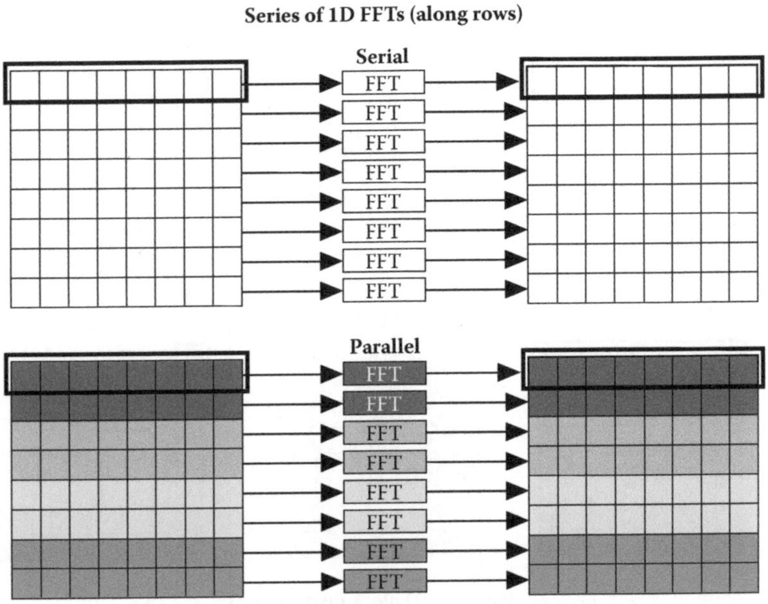

Series of 1D FFTs (along rows)

FIGURE 18-5   Mapping the input and output of a sequence of one-dimensional FFT operations. The heavily outlined boxes indicate the input data required to generate the corresponding output data.

www.hpec-si.org). After the establishment of VSIPL, work began on a C++ successor to VSIPL called VSIPL++ (Mitchell 2006). Both free and commercial implementations of VSIPL++ are available (see CodeSourcery VSIPL++ website at http://www.codesourcery.com/vsiplplusplus).

VSIPL++ provides high-level data structures (e.g., vectors and matrices) and functions [e.g., matrix multiplication and fast Fourier transform (FFT)] that are common in signal and image processing (see HPC Challenge website at http://www.hpcchallenge.org and HPEC Challenge website at http://www.ll.mit.edu/hpecchallenge). Unlike VSIPL, VSIPL++ provides direct support for parallel processing. The specification includes functionality that easily enables VSIPL++ to allocate parallel data structures and execute parallel functions through the use of maps.

VSIPL++ does not support incremental parallelization of existing code as OpenMP or UPC do. Rather, programmers can develop a serial application first using VSIPL++, then parallelize the application by adding maps. Maps are objects that concisely describe how to parallelize data structures. A matrix can be parallelized by adding the map object as an additional argument to the matrix constructor. VSIPL++ detects the map and constructs a parallel matrix accordingly. Any operation supported by the library, e.g., FFT, that is applied to the matrix will execute in parallel.

Different algorithms have different mappings that maximize the computation-to-communication ratio. Figures 18-5 and 18-6 show how different mappings are required to parallelize a series of one-dimensional FFT operations and a matrix multiply.

Consider a set of vectors that must be converted from the time domain to frequency domain. The vectors can be organized as the rows of a matrix, then a series of one-dimensional FFT operations is applied to the rows. Since each FFT requires an entire row as input, the matrix can be broken into blocks of rows, with each block assigned to a different processor. This is shown in Figure 18-5.

In matrix multiplication, each element of the output matrix requires an entire row from the first input matrix and an entire column from the second input matrix. Unlike parallelizing a series of one-dimensional FFTs, no mapping of matrix multiplication eliminates all communication. Figure 18-6 shows two different mappings of matrix multiplication. The first is a one-dimensional block distribution, in which only one dimension of each matrix is distributed. The second is a one-dimensional block distribution, in which both dimensions of each matrix are distributed. In

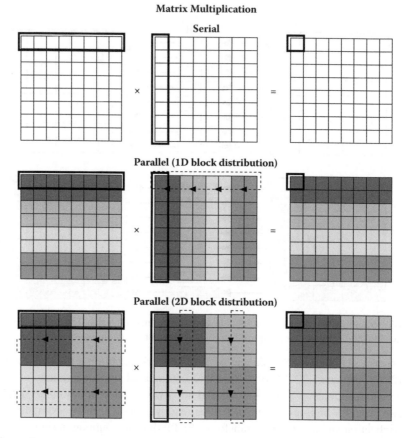

FIGURE 18-6 Two different methods of mapping the inputs and output of a matrix multiply. The heavily outlined boxes indicate the input data for each output element. The dashed arrows indicate the directions blocks must be shifted during computation.

each mapping, blocks of the parallel matrix must be shifted between processors during computation in order to compute all values of the output matrix. The one-dimensional and two-dimensional block distributions each have advantages. More about parallel matrix multiplication can be found in Grama et al. (2003).

Often, a chain of operations will require different mappings for each operation. VSIPL++ can automatically redistribute data structures from one distribution to another. This removes the need for the application programmer to write the code necessary to redistribute data among processors. Maps are more thoroughly discussed in Chapters 17 and 19.

Unlike the other technologies described in this chapter, VSIPL++ is targeted at a specific domain, i.e., signal and image processing. Targeting a specific domain relieves the programmer from integrating a parallel programming library with scientific and math libraries, but also means VSIPL++ may not be well suited for other types of applications.

18.2.4 APPLICATIONS

The Defense Advanced Research Projects Agency's (DARPA's) High Productivity Computing Systems (HPCS) program created the HPC Challenge benchmark suite in an effort to redefine how to measure performance, programmability, portability, robustness, and productivity in the HPC domain (HPC Challenge website; Luszczek et al. 2005). The benchmarks measure various aspects of HPC system performance, including computational performance, memory-access performance, and input/output (I/O). Along similar lines, DARPA's HPCS and Polymorphous Computing Architecture (PCA) programs created the HPEC Challenge benchmark suite to evaluate the performance of high performance embedded systems (HPEC Challenge website; Haney et al. 2005). The benchmark suite consists of a number of computational kernels and a synthetic aperture radar (SAR) application benchmark.

This section will briefly introduce parallelization strategies for the algorithms employed in the HPC Challenge FFT benchmark and the HPEC Challenge SAR benchmark.

18.2.4.1   Fast Fourier Transform

The HPC Challenge FFT benchmark measures performance of a vital signal processing operation that is often parallelized for two reasons: higher performance or larger data sizes. Figure 18-7 depicts the algorithm for the parallel one-dimensional FFT. First, the input vector is divided into subvectors of equal length, which are organized as the rows of a matrix and distributed among the processors (step 1). The distribution of vectors across processors is indicated by the different gray shades. Each processor applies a serial one-dimensional FFT to its rows (step 2) and multiplies its rows' elements by a set of weights, W, known as *twiddle factors* (step 3). The matrix is reorganized such that the columns are distributed equally among processors (step 4). Each processor applies a serial one-dimensional FFT to its columns (step 5). Finally, the output vector is constructed by concatenating the rows of the matrix (step 6). This algorithm is mathematically equivalent to applying a serial one-dimensional FFT to the entire vector (Grama et al. 2003).

The reorganization operation is commonly known as a *corner turn* and requires all processors to communicate with all other processors. The corner turn typically dominates the runtime of parallel FFTs; consequently, minimizing corner-turn latency is key to optimizing parallel FFT performance.

As mentioned before, threaded programming technologies are designed for shared-memory architectures. Thus, the corner turn requires no actual communication since all processors can access any address in memory. As a result, the corner-turn operation simply requires each processor to compute the column indices for which it is responsible and apply FFTs to those columns, providing a significant performance improvement. Technologies like OpenMP compute indices automatically, greatly simplifying index computation. The maximum size of the vectors that can be processed, however, is limited; shared-memory embedded computers typically employ no more than a handful of processors.

Message-passing applications can process larger data sizes since message passing supports distributed-memory systems. Technologies such as MPI, however, place the burden of calculating how array indices are partitioned on the programmer. To perform the corner turn, each processor must compute the column indices owned by every other processor. Each processor then sends the values in its rows that intersect with every other processor's columns. This index computation can

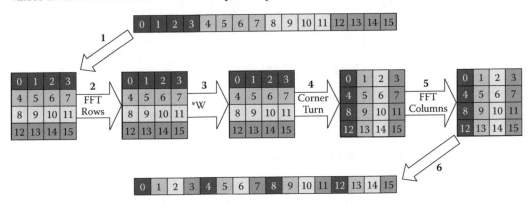

FIGURE 18-7   Parallel one-dimensional FFT algorithm.

be complicated, especially for matrices with dimension lengths that are not evenly divisible by the number of processors.

The emergence of PGAS technologies has provided programmers with a means of implementing parallel FFT algorithms that are scalable and easy to implement and that achieve high performance. This is especially true with VSIPL++, which includes a built-in parallel FFT operation. Two parallel matrices are defined, one distributed row-wise and the other distributed column-wise. The row-wise parallel matrix contains the input, and the column-wise parallel matrix contains the output. The FFT and weight multiplication are applied to the input matrix. To perform the corner turn, the input matrix is simply assigned to the output matrix. The library will automatically redistribute the data. Finally, the FFT operation is applied to the output matrix.

18.2.4.2　Synthetic Aperture Radar

The HPEC Challenge SAR benchmark contains an end-to-end example of a SAR processing system consisting of two major computational stages: front-end sensor processing and back-end knowledge formation, shown in Figure 18-8. In stage 1, the raw data are read, an image is formed, targets from a set of templates are inserted, and an image is written to disk for a specified number of images. Stage 2 consists of a similar loop in which a specified number of image sequences are read. For each chosen image sequence, the difference between pairs of sequential images is computed, determining regions of the images that have changed. These changed regions are then identified and stored in subimages. The goal of the benchmark is to represent the computation and I/O requirements commonly found in embedded systems that are utilized in a broad range of military and commercial image processing applications.

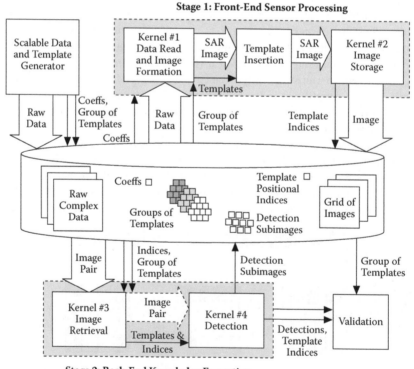

FIGURE 18-8　HPEC Challenge SAR benchmark. (From Bader et al., Designing scalable synthetic compact applications for benchmarking high productivity computing systems, *CTWatch Quarterly* 2(4B), 2006. With permission.)

This section focuses on the image-formation kernel in the front-end sensor processing stage. This kernel iterates over raw SAR data generated by the data generator to produce images. Unlike the FFT benchmark, the image-formation kernel consists of a sequence of operations that can be parallelized in either a *coarse-grained* or *fine-grained* manner (Mullen, Meuse, and Kepner 2006).

In the coarse-grained approach, each image can be processed independently. A common name for this type of parallelism is *embarrassingly parallel*. Since images are independent from each other, no interprocessor communication is necessary. Each processor loads data for a single image and executes the image-formation kernel to produce an entire image. In general, finding independence between iterations in the outermost loop of the application is the best approach to improving performance in an application. It is easy to implement and requires no communication. This approach comes with some caveats, however. First, it results in higher latency in the time to generate a single image. In a real-time system, depending on the performance requirements, higher latency may be an issue. Second, the data for each image must fit within a single processor's memory.

The data generator in the SAR benchmark can produce data larger than memory on a single processor. In this case, a fine-grained approach is necessary. The data and computation for a single image are distributed across multiple processors. The image-processing kernel consists of several steps, with each successive step processing data along different dimensions. Naive serial implementations of the image-processing kernel may result in a large number of corner turns when the code is parallelized. To reduce the number of corner turns, the processing chain may have to be modified in order to place computations along the same dimensions next to each other, thus removing the need for a corner turn between those two steps. The canonical parallelized SAR image-formation kernel requires three corner turns.

The fine-grained approach allows larger images to be processed and can reduce the latency of generating a single image, but is more complex than the coarse-grained approach. Additionally, the fine-grained approach requires several corner turns that must be optimized to maximize performance.

18.3   DISTRIBUTED COMPUTING MODELS

The three aforementioned parallel programming models can be used to architect and solve myriad high performance embedded computing problems; however, there are situations in which different application codes need to execute on different embedded processors. As mentioned in the introduction of this chapter, this distributed programming model is often described as a multiple-program multiple-data model. A few reasons for choosing a distributed programming model include differences in the capabilities of the processors and associated peripherals within a multiprocessor embedded system, the potential simplicity with which each of the application code components can be programmed, and the ability to satisfy timing restraints of each component of the overall system without using data parallelism. Similar to parallel programming, there are multiple distinct models for distributed programming: client-server and data driven. A subset of the client-server model is the peer-to-peer model.

When one compares these two models and their corresponding middleware libraries, there are many factors to consider. These factors include the communications patterns between the processes; the interoperability between processes, different operating systems, and platforms; the fault tolerance of the system; and the ease of use of the library. To a great extent, these factors depend on how the client code and the client stub interact and how the server code and server stub interact. One can think of the client stub (or proxy) and the server stub (or skeleton) as the translator layer between the client-server applications and the middleware that connects heterogeneous operating environments. Table 18-4 summarizes the advantages and disadvantages of the distributed programming technologies discussed here.

TABLE 18-4

Advantages and Disadvantages of Various Distributed Programming Technologies and the Languages Supported by Each

Technology	Advantages	Disadvantages	Languages
SOAP	Interoperability with many languages; client stub code can be autogenerated using WSDL (Web Services Description Language)	Encoding and decoding XML (eXtensible Markup Language) can be slow; generally not for use with hard real-time systems	C, C++, Java, Perl, Python, PHP, Ruby, JavaScript
Java RMI	Integrated in Java language	Synchronization and blocking of remote objects is difficult; only server can clone objects	Java
CORBA	Mature, robust, distributed client-server middleware library	Not all CORBA implementations are interoperable	C, C++, Java, COBOL, Smalltalk, Ada, Lisp, Python, Perl
JMS	Robust set of publish-subscribe features	JMS (Java Messaging Service) central servers are single points of failure	Java
DDS	Distributed publish-subscribe middleware library (no central server); quality-of-service features	Open-source version currently lacks performance	C, C++, Java

18.3.1 CLIENT-SERVER

The simpler of the two distributed computing models is the client-server model. The client is a requester of services, while the server is a provider of services (Tanenbaum and van Steen 2002). A conceptual diagram of the client-server model is depicted in Figure 18-9. This figure shows several computers that are taking on the role of either a client or a server (but not both simultaneously). Though the client requests services, the client process is executing an application; that is, it is not relying on the server for all of the processing capability, as is the case in mainframe computing. The clients use remote procedure calls (RPCs), remote method invocation (RMI), standard query language (SQL), or similar mechanisms to communicate with the server. Many client-server systems are familiar processes; for instance, requesting web pages from a web server with a browser client constitutes one of these familiar capabilities. The client browser sends service requests to one or more web servers to retrieve documents and other data from the server(s). Each of the web servers, executing file-retrieval software along with server-side scripts (written in Java, Perl, PHP,

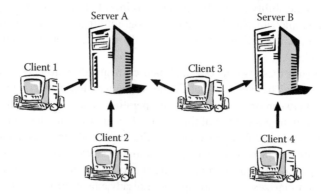

FIGURE 18-9 A conceptual diagram of the client-server model.

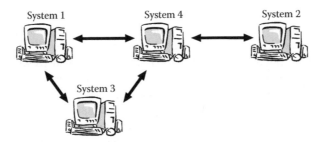

FIGURE 18-10 A conceptual diagram of the peer-to-peer model.

Ruby, etc.) and databases, receives requests and returns the requested documents and data to the client browser. Upon receipt of the documents and data, the client browser renders the data into the web pages with which we are familiar and executes client-side applets (written in JavaScript, Java, etc.) and helper applications (such as Acrobat Reader and QuickTime) to further display the received documents and data.

When clients require services, they request the services from servers (often termed *communication by invocation*), and they generally wait for the reply before continuing local application execution. The model is procedure/method-centric since it centers around choosing services that are executed on remote computers (referenced by procedure/method identifier). Computers in the system are made aware of each other's service and server names either directly (hardcoded) or through a service broker (Tanenbaum and van Steen 2002). This model works best for one-to-one or many-to-one communication patterns with central information such as databases, transaction processing systems, and file servers (Vaughan-Nichols 2002). Some common client-server technologies are the remote shell (rsh) and secure shell (ssh) programs, Simple Object Access Protocol (SOAP), Java RMI, and Common Object Request Broker Architecture (CORBA).

One potential challenge with this model is that the names of the servers and the services generally must be known in advance. This requirement implies that whenever a failure occurs on one of the servers, a change must be made either at the clients or at the broker that redirects requests (from the broker to the server) that will prevent clients or the broker from invoking services on the failed server. This results in a disruption in services to the clients. Furthermore, this requirement could introduce challenges in scaling the application since either the broker or the server providing services can become overloaded if too many requests arrive in a short time interval. Another challenge is that the client usually waits and does nothing until the server replies or a timeout occurs; while this response latency could be hidden by writing multithreaded applications (thereby introducing more complexity and timing challenges), valuable computation time is being spent waiting. However, the remote call may be entirely necessary because the server may be the only place certain data or services may be available.

A variation on the client-server model is the peer-to-peer model. In this model, processor nodes are simultaneously clients and servers. A conceptual diagram of the peer-to-peer model is depicted in Figure 18-10. Though peer-to-peer systems are fairly easy to configure, they are not used much in embedded systems because of the potential conflict for processing time between the server process(es) and the client process(es) running on each processor. That is, in most embedded systems, a more deterministic capability is required of the system, and such determinism can be undermined by having both client and server processes running on individual processors. If one chooses to run processors in a peer-to-peer manner, one must be certain that the nodes will not be overwhelmed by peak load instances.

18.3.1.1 SOAP

Recently, significant interest has been paid to Web Services, which are a set of cross-platform technologies that deliver network services across the Internet (Vaughan-Nichols 2002). The key enablers for Web Services are the eXtensible Markup Language (XML) and specific schema used for SOAP calls. (The acronym used to mean Simple Object Access Protocol, but it has been redefined to not stand for anything.) SOAP is one of three Web Services XML standards along with the Web Services Description Language (WSDL) and the Universal Description, Discovery, and Integration (UDDI) protocol. WSDL describes the services that a server offers as well as the interfaces to the services, while UDDI, itself a WSDL-defined service, is the lookup service for WSDL entries (Cerami 2002). SOAP implements the language- and platform-agnostic interface to the services by encoding the data delivery, object method invocation, and resultant service reply in XML, thus bridging heterogeneous systems and programming languages (Snell, Tidwell, and Kulchenko 2001). SOAP libraries have been written for a wide array of languages, including Java, C, C++, Perl, Python, PHP, Ruby, JavaScript, and many others. A client (but not server) capability can even be implemented in MATLAB. For many of these languages, the client stub object can be autogenerated at development time from the service WSDL. Due to several factors, including the relatively slow transaction rate with XML and the best-effort manner in which SOAP and Web Services operate, generally SOAP and Web Services are not well suited for hard real-time embedded systems. However, Web Services could be considered for soft real-time embedded systems.

18.3.1.2   Java Remote Method Invocation

While Java can be used extensively in building Web Services (Chappell and Jewell 2002), if the entire distributed system will be implemented in Java and remote interfaces are well defined and static, Java RMI could be a simpler solution (Grosso 2001). Since it is integrated into the language, Java RMI has a high degree of distribution transparency; that is, to the programmer, a remote object looks and behaves like a local object. A Java remote object resides on only one server (which is a fault-tolerance risk), while its client proxy resides in the address space of the client process (Tanenbaum and van Steen 2002). There are two differences between a local object and a remote object. First, synchronization and blocking of the remote object are difficult, so client-side blocking must be implemented by the proxy. Second, only the server can clone its objects; that is, a client cannot request that a copy of a remote object be made and located onto another server—only the originating server of the object can do it. Nevertheless, Java RMI is a viable option for a distributed embedded system if the entire system is written in Java.

18.3.1.3   Common Object Request Broker Architecture

Historically within embedded systems, the most commonly used distributed client-server middleware is CORBA. CORBA defines a distributed-object standard. The programmer describes the CORBA distributed objects in the CORBA interface definition language (IDL). The IDL code is then translated to build the client stub and server skeleton in the target language; these are used to interface the client and server code to the CORBA object request broker (ORB) (depicted in Figure 18-11). Thus, the programmer is not required to write these code stubs. The ORB is the translation (or marshalling) layer that enables heterogeneous platforms within the system to communicate. CORBA can be used in code written in a variety of languages, including C, C++, Java, COBOL, Smalltalk, Ada, Lisp, Python, and Perl, and some major interoperability problems were solved with the General Inter-ORB Protocol (see the Object Management Group's website at http://omg.org/gettingstarted/corba-faq.htm). By default, a client invokes an object synchronously. The request is sent to the corresponding object server, and the client blocks

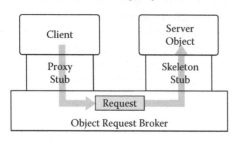

FIGURE 18-11   Components of a CORBA client-server interface.

the request until it receives a response. However, a client can also communicate with a CORBA object in a one-way, best-effort manner if it is not expecting a reply, or in an asynchronous request manner when it wants the reply to come back as a execution-interrupting callback (Tanenbaum and van Steen 2002).

Clients can look up object references by using a character-based name that is unique for each CORBA object on each server. Finally, fault tolerance is dealt with by replicating objects into object groups, with identical copies of the same objects residing on the same or different servers. References to object groups are made possible by the Interoperable Object Group Reference (IOGR) (Tanenbaum and van Steen 2002). CORBA provides a rich set of features for high performance embedded computing; a real-time version of CORBA called TAO is available based on the Adaptive Communication Environment (ACE), an open-source, object-oriented framework for high performance and real-time concurrent communications software (Schmidt 2007). While CORBA has many advantages, not all implementations of CORBA are equal; one cannot assume that two different implementations are interoperable. While the General Inter-ORB Protocol solved many interoperability issues, there are still incompatibilities between some implementations. Therefore, one must be sure that all processors in a network will be using interoperable CORBA libraries.

18.3.2 DATA DRIVEN

In the client-server model, the services and servers that provide services are named or have brokers that mediate access to the named services and servers. Having named services and servers is ideal for one-to-one and many-to-one communication patterns. When communication patterns become more complex, such as with one-to-many and many-to-many patterns, the data-driven model, also known as publish-subscribe, has advantages. In this model, server and client node names as well as service names are anonymous. Instead, requests are made for named data; hence, it is a data-centric paradigm. A conceptual diagram of the data-driven model is depicted in Figure 18-12. The data-driven model is similar to newspapers or magazines to which readers subscribe. At certain intervals, the publication is published and delivered to the subscribers. To continue with the newspaper metaphor, data streams (one or more time-oriented data items) are named as topics (referenced by data identifier), while each instance of a topic is called an issue (Schneider and Farabaugh 2004). Each topic is published by one or more data producers and subscribed to and consumed by one or more data consumers. Consumers can subscribe to any topic being published in the system, and every subscriber/consumer of a topic receives a copy of each message of that topic. The producers publish data when new data are available; the consumers receive the data and are notified via callbacks that a new message has arrived (communication by notification). Since data can be produced and consumed anywhere in the network of computers, the data-driven model is ideal for one-to-many and many-to-many communications and is particularly useful for data-flow applications. Though

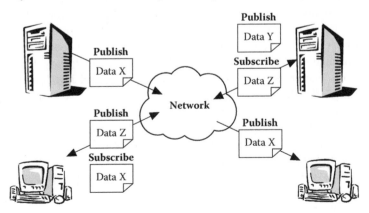

FIGURE 18-12   A conceptual diagram of the data-driven model.

there are many publish-subscribe middleware libraries available, two stand out from the rest: Java Messaging Service (JMS) and Data Distribution Service (DDS).

One challenge in using the data-driven model is that it is not well suited for distributing procedures/objects with which many of us are very familiar; it requires us to adopt an architecture methodology that centers on how the data moves in the system rather than on the procedures or methods that are called. Therefore, transaction processing and database accesses are usually best implemented with the client-server model (Schneider and Farabaugh 2004). Data reliability in data-driven models can be more challenging than in client-server systems, but data persistence (keeping data available for consumers that have not received it) and subscription acknowledgments overcome this hurdle (Monson-Haefel and Chappell 2001). Also, if a data producer fails, the data-driven model relies on mechanisms outside of its middleware specification to either revive the failed producer or start a replacement producer (Monson-Haefel and Chappell 2001). Finally, due to the dynamic nature of the publish-subscribe paradigm, nodes that appear and disappear dynamically are handled robustly because messages are sent to whichever computer is listening to a topic rather than being sent explicitly from one computer to another.

18.3.2.1 Java Messaging Service

Java Messaging Service is usually associated with enterprise business systems since it is a Sun Java standard and is included as an enterprise service with most Java 2 for Enterprise Edition (J2EE) application servers (Monson-Haefel and Chappell 2001). However, it is finding a variety of uses in embedded systems. As the name implies, it can only be used in Java. Though JMS also features point-to-point messaging, it is usually employed for its publish-subscribe capabilities. JMS uses one or more JMS servers to store and forward all of the issues that are published on each topic; this usage enables persistent messages, which are stored for a period of time in case some subscribers missed the message due to failure (called a durable subscription) or new consumers subscribe to the topic. Consumers can be selective of which messages they receive by setting message selectors. However, even with durable subscriptions and persistent messages, JMS is a best effort communication middleware library; it currently does not have any quality-of-service (QoS) features. Similar to Java RMI, JMS is a viable option for a distributed embedded system if the entire system is written in Java. A disadvantage of JMS is that since communication is centralized, it creates a single point of failure.

18.3.2.2 Data Distribution Service

The Data Distribution Service for Real-Time Systems standard was finalized in June 2004 (Object Management Group 2004). Four companies—Real-Time Innovations (RTI), Thales, MITRE, and Object Interface Systems (OIS)—submitted the standard jointly. Currently, there are four implementations of the DDS standard: RTI-DDS from RTI; Splice DDS from PrismTech and Thales; DDS for TAO from Object Computing, Inc.; and OpenDDS, an open-source project. In the United States, the most popular of these is RTI-DDS, which is a middleware library that can be used across many heterogeneous platforms simultaneously and has C, C++, and Java bindings (Schneider and Farabaugh 2004). RTI-DDS, like some versions of CORBA, is scalable to thousands of processes. Discovery of topics and publishers is distributed among all of the RTI-DDS domain participants, so it does not rely on a central message server. Beyond topics and issues, RTI-DDS provides message keys that can help in distinguishing message senders or subtopics.

RTI-DDS also provides many features for QoS and fault tolerance. Several of the QoS properties that can be set on topics, publishers, and subscriptions include, among many others,

- Reliability—guaranteed ordered delivery;
- Deadlines—delivering periodic issues on time or ahead of time;

- Time-based filtering—only receiving new issues after a certain time since last issue was received;
- Publisher priorities—priority of receiving issues from certain publisher of a topic;
- Issue liveliness—how long an issue is valid;
- Issue persistence and history—how many issues are saved for fault tolerance; and
- Topic durability—how long issues of a topic are saved.

18.3.3 APPLICATIONS

Two examples of how distributed computing models are being used at MIT Lincoln Laboratory are the Radar Open Systems Architecture (ROSA) (Rejto 2000) and Integrated Sensing and Decision Support (ISDS), a Lincoln Laboratory internally funded project with roots in the Silent Hammer sea trial experiment (Pomianowski et al. 2007). Unlike the parallel processing models, the choice of distributed computing model usually is easier to make because of the inherent communication patterns of the applications.

18.3.3.1 Radar Open Systems Architecture

The ROSA project is a multiphase project that is modernizing the computational signal processing facilities of radar systems owned and operated by MIT Lincoln Laboratory. The ROSA model decomposes the radar control and radar processing systems into individual, loosely coupled subsystems (Rejto 2000). Each of the subsystems, or radar peripherals, has openly defined interfaces so that any subsystem can easily be swapped out. Improvements to the architecture occur in phases; the first phase mainly standardized the radar subsystems while the second phase is standardizing the radar real-time processor. Figure 18-13 shows the phase 1 system diagram of ROSA. The architecture is broken into the radar front-end, the radar subsystems, and the radar real-time processor. The subsystems include (1) electromechanical controls for the radar dish and (2) analog beamforming and signal processing that feed into the real-time (digital) processor. The first phase included the design of many interoperating radar hardware subsystems using open VME and VXI board and bus standard technology, and the radar real-time processor (RTP) was an SGI shared-memory computer (Rejto 2000). Changes to the RTP included adding CORBA interfaces as part of the external communication capabilities of the system. This enabled external computer systems to control the radar as well as receive system telemetry and radar products, all through a standard CORBA interface. CORBA was chosen because the control of the radar system is inherently a remote procedure call, and CORBA provided implementations in a variety of languages.

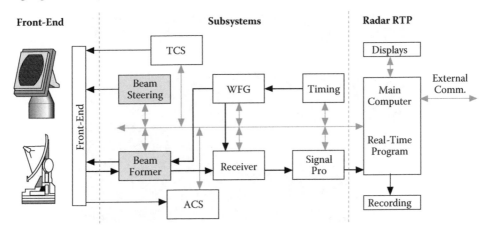

FIGURE 18-13   Radar Open Systems Architecture (ROSA) phase 1 system diagram.

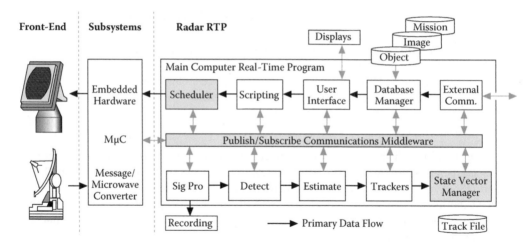

FIGURE 18-14   Radar Open Systems Architecture (ROSA) phase 2 system diagram.

The second phase is targeting improvements to the RTP. As shown in Figure 18-14, the various subsystems of the RTP are broken into separate processes, and communications between these subsystems are enabled by a publish-subscribe communication middleware like DDS. DDS was chosen because many of the communication patterns in the RTP are one-to-many and many-to-many. Also, DDS provides QoS parameters, which are utilized to guarantee that radar data are processed in required time intervals. Modularizing the components of the RTP enables the use of less expensive hardware (like a cluster with commodity networking) while still satisfying performance requirements. Furthermore, the external communication interface has been augmented with a publish-subscribe communication middleware so that multiple external computers can receive data from the radar as they are being received and processed by the RTP.

18.3.3.2 Integrated Sensing and Decision Support

The ISDS project brings together several areas of research that are making sensor intelligence gathering, processing, and analysis more efficient and effective. Research is directed at managing, sharing, discovering, and analyzing distributed intelligence data. These intelligence data are composed of the enhanced intelligence data (pictures, videos, target tracks, etc.) and metadata objects describing each of these data items. The data and metadata are gathered and stored within an intelligence analysis cell (IAC), where they can be searched and analyzed from within the IAC

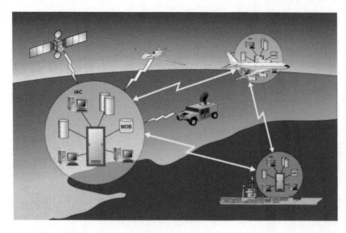

FIGURE 18-15   Conceptual diagram of internetworked IACs.

(see Figure 18-15). However, an IAC generally does not know what data other IACs have in their databases because of very low bandwidth between the IACs. The key to providing that knowledge is to intelligently distribute the metadata of the data items among the IACs. Since the IACs are based on J2EE servers and the network within an IAC is quite reliable, the interprocess communication is handled by the publish-subscribe and point-to-point capabilities of JMS. The inter-IAC communication is inherently unreliable and many-to-many in nature. Therefore, that traffic is handled by DDS-based publish-subscribe communication to take advantage of the QoS features in DDS. Each IAC can choose to receive various topics of metadata updates that other IACs are collecting. By using the QoS features of DDS, the link reliability issues can be overcome, while the lack of bandwidth is addressed by only sharing metadata, rather than moving much larger data products across the network, whether they are needed or not. When an IAC searches and finds a data item that it wants to acquire, it can then transfer only that data item.

18.4 SUMMARY

This chapter has presented three parallel programming models and two distributed processing models for high performance embedded computing. These five models were accompanied with brief overviews of current implementations. When comparing these models and their corresponding implementations, one must consider many factors, including the communications patterns between the processes; the interoperability between processes, different operating systems, and platforms; the fault tolerance of the system; and the programmability of the library. Though improvements to these models and implementations surely will come in the future, the underlying fundamentals and trade-offs remain the same and will drive the design decisions on which model and implementation to use in an HPEC system.

REFERENCES

Bader, D.A., K. Madduri, J.R. Gilbert, V. Shah, J. Kepner, T. Meuse, and A. Krishnamurthy. 2006. Designing scalable synthetic compact applications for benchmarking high productivity computing systems. *CTWatch Quarterly* 2(4B).

Cerami, E. 2002. *Web Services Essentials.* San Sebastopol, Calif.: O'Reilly.

Chandra, R., R. Menon, L. Dagum, D. Kohr, D. Maydan, and J. McDonald. 2002. *Parallel Programming in OpenMP.* San Francisco: Morgan Kaufman.

Chappell, D.A. and T. Jewell. 2002. *Java Web Services.* San Sebastopol, Calif.: O'Reilly.

Geist, A., A. Beguelin, J. Dongarra, W. Jiang, R. Manchek, and V.S. Sunderam. 1994. *PVM: Parallel Virtual Machine: A Users' Guide and Tutorial for Network Parallel Computing.* Cambridge, Mass.: MIT Press.

Grama, A., A. Gupta, G. Karypis, and V. Kumar. 2003. *Introduction to Parallel Computing.* Boston: Addison-Wesley.

Gropp, W. 1999. *Using MPI: Portable Parallel Programming with the Message-Passing Interface*, 2nd edition. Cambridge, Mass.: MIT Press.

Gropp, W. and E. Lusk. 1997. Why are PVM and MPI so different? *Proceedings of the Fourth European PVM/MPI Users' Group Meeting on Recent Advances in Parallel Virtual Machine and Message Passing Interface* 3–10.

Gropp, W., E. Lusk, and R. Thakur. 1999. *Using MPI-2: Advanced Features of the Message Passing Interface.* Cambridge, Mass.: MIT Press.

Grosso, W. 2001. *Java RMI.* San Sebastopol, Calif.: O'Reilly.

Haney, R., T. Meuse, J. Kepner, and J. Lebak. 2005. The HPEC Challenge benchmark suite. *Proceedings of the Ninth Annual High Performance Embedded Computing Workshop.* MIT Lincoln Laboratory, Lexington, Mass. Available online at http://www.ll.mit.edu/HPEC/agendas/proc05/agenda.html.

HPC Challenge webpage. Available online at http://www.hpcchallenge.org.

Institute of Electrical and Electronics Engineers. Standards interpretations for IEEE standard portable operating system interface for computer environments (IEEE Std 1003.1-1988). Available online at http://standards.ieee.org/regauth/posix/.

Intel Hyper-Threading website. Available online at http://www.intel.com/technology/hyperthread/.

Luszczek, P., J. Dongarra, D. Koester, R. Rabenseifner, B. Lucas, J. Kepner, J. McCalpin, D. Bailey, and D. Takahashi. 2005. *Introduction to the HPC Challenge Benchmark Suite.* Lawrence Berkeley National Laboratory, Berkeley, Calif.. Paper LBNL-57493. Available online at http://repositories.cdlib.org/lbnl/LBNL-57493/.

Mitchell, M. 2006. Inside the VSIPL++ API. *Dr. Dobb's Journal.* Available online at http://www.ddj.com/dept/cpp/192501827.

Monson-Haefel, R. and D.A. Chappell. 2001. *Java Message Service.* San Sebastopol, Calif.: O'Reilly.

Mullen, J., T. Meuse, and J. Kepner. 2006. HPEC Challenge SAR benchmark: pMatlab implementation and performance. *Proceedings of the Tenth Annual High Performance Embedded Computing Workshop.* MIT Lincoln Laboratory, Lexington, Mass. Available online at http://www.ll.mit.edu/HPEC/agendas/proc06/agenda.html.

Nichols, B., D. Buttlar, and J. Proulx Farrell. 1996. *PThreads Programming: A POSIX Standard for Better Multiprocessing.* San Sebastopol, Calif.: O'Reilly.

Object Management Group. 2004. Data distribution service for real-time systems. Available online at http://www.omg.org/technology/documents/formal/data_distribution.htm.

OpenMP webpage. Available online at http://www.openmp.org.

Parallel Virtual Machine webpage. Available online at http://www.csm.ornl.gov/pvm/pvm_home.html.

Pomianowski, P., R. Delanoy, J. Kurz, and G. Condon. 2007. Silent Hammer. *Lincoln Laboratory Journal* 16(2): 245–262.

Rejto, S. 2000. Radar open systems architecture and applications. *The Record of the IEEE 2000 International Radar Conference* 654–659.

Schmidt, D. 2007. Real-time CORBA with TAO (the ACE ORB). Available online at http://www.cs.wustl.edu/~schmidt/TAO.html.

Schneider, S. and B. Farabaugh. 2004. Is DDS for you? White paper by Real-Time Innovations. Available online at http://www.rti.com/resources.html.

Snell, J., D. Tidwell, and P. Kulchenko. 2001. *Programming Web Services with SOAP.* San Sebastopol, Calif.: O'Reilly.

Tanenbaum, A.S. and M. van Steen. 2002. *Distributed Systems: Principles and Paradigms.* Upper Saddle River, N.J.: Prentice Hall.

Vaughan-Nichols, S.J. 2002. Web services: beyond the hype. *IEEE Computer* 35(2): 18–21.

19 Automatic Code Parallelization and Optimization

Nadya T. Bliss, MIT Lincoln Laboratory

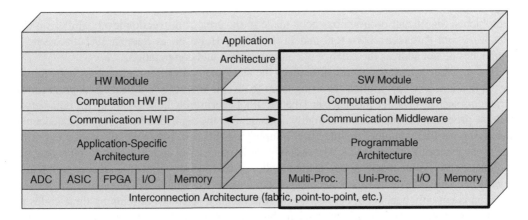

This chapter presents a high-level overview of automated technologies for taking an embedded program, parallelizing it, and mapping it to a parallel processor. The motivation and challenges of code parallelization and optimization are discussed. Instruction-level parallelism is contrasted to explicit parallelism. A taxonomy of automatic code optimization approaches is introduced. Three sample projects, each in a different area of the taxonomy, are highlighted.

19.1 INTRODUCTION

Over the past decade, parallel processing has become increasingly prevalent. Desktop processors are manufactured with multiple cores (Intel), and commodity cluster systems have become commonplace. The IBM Cell Broadband Engine architecture contains eight processors for computation and one general-purpose processor (IBM). The trend toward multicore processors, or multiple processing elements on a single chip, is growing as more hardware companies, research laboratories, and government organizations are investing in multicore processor development. As an example, in February 2007 Intel announced a prototype for an 80-core architecture (Markoff 2007). The motivation for these emerging processor architectures is that data sizes that need to be processed in industry, academia, and government are steadily increasing (Simon 2006). Consequently, with increasing data sizes, throughput requirements for real-time processing are increasing at similar rates. As radars move from analog to wideband digital arrays and image processing systems move toward gigapixel cameras, the need to process more data at a faster rate becomes particularly vital for the high performance embedded computing community.

As the hardware architecture community moves forward with processing capability, new programming models and tools are necessary to truly harness the power of these architectures. Software accounts for significantly more government spending than does hardware because as processor architectures change so must the software. Portability is becoming prevalent (see Chapter 17 for a detailed discussion of portability). In addition to accounting for different floating-point standards and instruction sets, one must also consider the concerns emerging with parallel architectures. In an ideal scenario, a program written to run on a single processor would run just as efficiently without modifications on a multiprocessor system. In reality, the situation is quite different. For the purpose of this discussion, let us consider applications that are data driven and require access to mathematical operations such as transforms, solvers, etc. These types of applications are relevant to both the scientific and embedded processing communities.

Taking a single-processor program and turning it into a multiprocessor program turns out to be a very difficult problem. Computer scientists have been working on it for over 20 years (Banerjee et al. 1993; Hurley 1993; Nikhil and Arvind 2001; Wolfe 1996; Zima 1991) and, in the general case, the problem of automatically parallelizing a serial program is NP-hard, or in the hardest class of problems. Significant progress has been made in the area of instruction-level parallelism (ILP) (Hennessy and Patterson 2006; Rau and Fisher 1993). However, as Section 19.2 discusses, ILP approaches do not address the balance of communication and computation.

The problem is made even more difficult by the fact that parallel programming is challenging for the application developers, often requiring them to switch languages and learn new programming paradigms such as message passing (see Chapter 18). Determining how to split up the data between multiple processors is difficult and is highly dependent on the algorithm being implemented and the underlying parallel computer architecture. Simplifying this task is beneficial for efficient utilization of the emerging architectures. The task of parallelizing serial code is a significant productivity bottleneck and is often a deterrent to using parallel systems. However, with multicore architectures quickly becoming the de facto architecture standard, closer examination of automatic program optimization is not only beneficial, but necessary. Additionally, scientists developing these codes tend not to be computer scientists; instead, they are biologists, physicists, mathematicians, and engineers with little programming experience who, therefore, face a significant learning curve. Often, they are required to learn a large number of new parallel programming concepts.

This chapter provides a high-level discussion on automatic code optimization techniques with an emphasis on automatic parallelization. First, the difference between explicit parallelization and instruction-level parallelism is presented. Then, a taxonomy of automatic code optimization approaches is introduced. Three sample projects that fall into different areas of the taxonomy are highlighted, followed by a summary of the chapter.

19.2   INSTRUCTION-LEVEL PARALLELISM VERSUS EXPLICIT-PROGRAM PARALLELISM

Parallel computers and the research of parallel languages, compilers, and distribution techniques first emerged in the 1970s (Kuck et al. 1974; Lamport 1974). One of the most successful research areas has been ILP. A large number of processors and compilers have successfully incorporated ILP techniques over the past few decades. While some ILP approaches can be applied to explicit-program parallelization, ILP techniques are not sufficient to take advantage of multicore and explicitly parallel processor architectures. This section introduces ILP concepts and compares the ILP problem with a more general problem of explicit-program parallelization.

Instruction-level parallelism refers to identifying instructions in the program that can be executed in parallel and/or out of order and scheduling them to reduce the computation time. Let us consider a simple example.

Figure 19-1 illustrates a simple program and an associated dependency graph or parse tree. Note that there are no dependencies between computation of C and F. This indicates to the compiler that these sets of two addition operations can be executed in parallel or out of order (for example, if changing the order speeds up another computation farther down the line). If the architecture allows for multiple instructions to be executed at once, this approach can greatly reduce execution time.

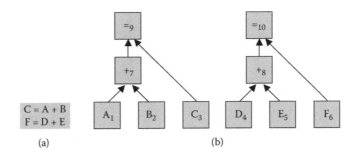

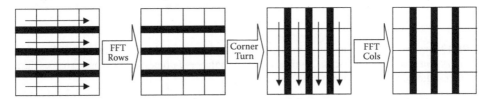

FIGURE 19-1 Simple program example to illustrate ILP; (a) presents a set of operations (or instructions), while (b) is the corresponding dependency graph or parse tree. Observe that there are no dependencies between computation of C and F. Thus, a compiler can execute the two addition operations in parallel or out of order.

FIGURE 19-2 Parallel implementation of an FFT on a large vector via two one-dimensional FFTs. The arrows inside the matrix indicate along which dimension the FFT is performed. The thick lines indicate the optimal distribution for the particular FFT. A corner turn is defined as an all-to-all data redistribution from row-distributed to column-distributed or vice versa.

This optimization technique at the instruction level has been incorporated into a number of commercial products. At a higher level, specifically at the kernel and function levels, this technique is referred to as *concurrency analysis* and has also been researched extensively and incorporated into parallel languages and compilers as discussed in a later section (Yelick et al. 2007).

A natural question is, why doesn't concurrency analysis or instruction-level parallelism solve the automatic parallelization problem? Why is it not the same as just building a dependency graph and understanding what nodes in the graph can be executed in parallel? Let us consider a simple example. A common implementation of a parallel fast Fourier transform (FFT) on a large vector is performed via two one-dimensional FFTs with a multiplication by twiddle factors (see Chapter 18). For the purpose of this discussion, let us ignore the multiplication step and simply consider the two FFTs. First, the vector is reshaped into a matrix. Then, one FFT is performed along rows and the second along columns, as illustrated by Figure 19-2.

First, consider the complexity of execution of this algorithm on a single processor. The time complexity of the operation is simply the computational complexity of the two FFTs, which is $2*5*N*\log_2(N)$, where N is the number of elements in the matrix. Second, consider the details of the parallel implementation on two processors, as illustrated in Figure 19-3(a). Here, the time complexity is equivalent to computational complexity divided by two (two processors) plus the additional time needed to redistribute the data from rows to columns. Third, consider the same operation but using four processors [Figure 19-3(b)]. Now the computation time is reduced by a factor of four, but there is the extra communication cost of the four processors communicating with each other.

This delicate balance of computation and communication is not captured through concurrency analysis. Specifically, the dependency graph of a serial program does not provide sufficient information to determine the optimal processor breakdown. This is true because the computation is no longer the only component, and the network architecture and topology influence the computation time significantly. For example, on a slow network it might only be beneficial to split the computation up between fewer nodes to reduce the communication cost, while on a faster network more processors would provide greater speedup.

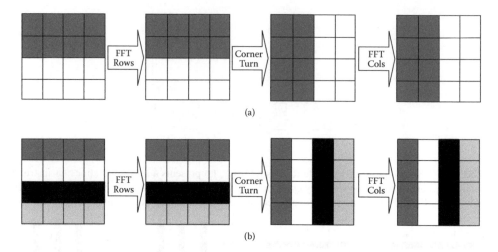

FIGURE 19-3   Two- and four-processor implementations of the parallel one-dimensional FFT. The different shades represent different processors. Observe that as the number of processors increases, so does the amount of communication necessary to perform the corner turn.

This section has distinguished ILP from explicit-program parallelization. In the following section, a taxonomy of program parallelization and optimization approaches is presented.

19.3   AUTOMATIC PARALLELIZATION APPROACHES: A TAXONOMY

Automatic code optimization is a rich area of research (Blume 1995; Carter and Ferrante 1999; Darte, Robert, and Vivien 2000; McKinley 1994). The current approaches can be classified according to a simple taxonomy described here. Automatic optimization approaches can be described by four characteristics. The first one is *concurrency*, or what memory hierarchy the approach is applied to. Concurrency can be either serial or parallel. If the concurrency is serial, as in Whaley, Petitet, and Dongarra (2001), then the approach is finding an efficient mapping into a memory hierarchy of a single machine. If the concurrency is parallel (Chen et al. 2003), the approach is optimizing the code for a parallel or distributed-memory hierarchy.

The second characteristic is *support layer*, or in which layer of software the automatic distribution and optimization are implemented. These approaches tend to be implemented either in the compiler or middleware layer. If the parallelization approach is implemented in the compiler (FFTW; Wolfe 1996), it does not have access to runtime information which could significantly influence the chosen mapping. On the other hand, if the approach is implemented in middleware (Hoffmann, Kepner, and Bond 2001; Whaley, Petitet, and Dongarra 2001) and invoked at runtime, it could incur a significant overhead. Balancing the amount of information available and the amount of time spent on the optimization is an important issue when designing an automatic parallelization capability.

The third characteristic is *code analysis*, or how the approach finds parallelism. Code analysis is *static* or *dynamic*. Static code analysis involves looking at the source code as text and trying to extract inherent parallelism based on how the program is written. Dynamic code analysis (University of Indiana Computer Science Dept.) involves analyzing the behavior of the code as it is running, thus allowing access to runtime information. Dynamic code analysis can only be implemented in middleware, as compilers do not provide access to runtime information.

Finally, the fourth characteristic is the *optimization window*, or at what scope the approach applies optimizations. Approaches could be local (peephole) or global (program flow). Local optimization approaches find optimal distribution of individual functions. Local optimization approaches (FFTW; Whaley, Petitet, and Dongarra 2001) have had the most success and are utilized by many parallel programmers. It is often true, however, that the best way to distribute individual functions is not the

TABLE 19-1

Automatic Program Optimization Taxonomy

Concurrency	Serial	Parallel
Support Layer	Compiler	Middleware
Code Analysis	Static	Dynamic
Optimization Window	Local (peephole)	Global (program flow)

best way to distribute the entire program or even a portion of the program. Global optimization (Kuo, Rabbah, and Amarasinghe 2005) addresses this issue by analyzing either the whole program or part of the program consisting of multiple functions. Table 19-1 summarizes the taxonomy.

19.4 MAPS AND MAP INDEPENDENCE

Although the concepts of maps and map independence are mentioned in Chapters 17 and 18, these concepts are vital to this chapter and are considered here in greater detail. The concept of using maps to describe array distributions has a long history. The concept of a map-like structure dates back to the High Performance Fortran (HPF) community (Loveman 1993; Zosel 1993). The concept has also been adapted and used by MIT Lincoln Laboratory's Space-Time Adaptive Processing Library (STAPL) (DeLuca et al. 1997), Parallel Vector Library (PVL) (Lebak et al. 2005), and pMatlab (Bliss and Kepner 2007) and is part of the VSIPL++ standard (Lebak et al. 2005). This chapter is limited to data parallel maps; however, the concepts can be applied to task parallelism. Observe that task parallelism can be achieved via careful manipulation of data parallel maps. A map for a numerical array defines how and where the array is distributed. Figure 19-4 presents an example of a map, mapA, and a mapped array, A.

Note that a map consists of three pieces of information: grid specification, distribution specification, and processor list. The grid defines how to break up the array between processors, e.g., column-wise, row-wise, on a two-by-two grid, etc. The distribution specification describes the pattern into which the data should be broken. In signal processing applications, support for block-cyclic distributions is usually sufficient. However, computations such as mesh partitioning often require irregular distributions. MIT Lincoln Laboratory software libraries only support the regular distribution sets; however, as more post-processing applications requiring sparse matrix computations and irregular data structures are moving to the sensor front-end, more research is necessary into additional distributions. Finally, the last argument in the map specification is the processor list. This is simply the list of ranks, or IDs, assigned to processors used for the computation by the underlying communication layer. See Chapter 18 for a more detailed discussion of processor ranks.

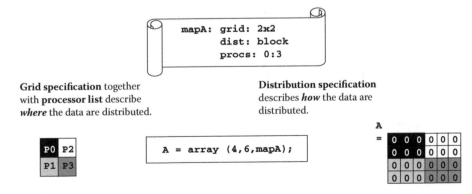

FIGURE 19-4   An example of a map and a mapped array.

```
%Create arrays
A = array(N,M, mapA);
B = array(N,M, mapB);
C = array(N,M, mapB);
%Perform FFT along the 2nd dimension (row)
A(:,:) = fft(A,2);
%Corner-turn the data
B(:,:) = A;
%Perform FFT along the 1st dimension (col)
C(:,:) = fft(B,1);
```

FIGURE 19-5 Parallel FFT pMatlab code. This is a common implementation of a parallel FFT on a large vector via two one-dimensional FFTs.

An important concept to note is that a map abstracts away a large number of details about underlying processing architecture. At this level, the developer simply has to understand how many processors are available. It is true that knowing which processor IDs (or ranks) bind to which underlying computational units could provide the user with more accurate information, but in the simplest case that information can be ignored. Maps essentially provide a layer of abstraction between the hardware and the application. Ideally, as the maps change, the application code should remain intact. It is not always the case, but this discussion will make that simplifying assumption. Essentially, the maps could be defined by one developer, while the algorithm development could be the task of another.

Nonetheless, usually an expert application mapper is not available and the same individual is both writing and mapping the application. Constructing an efficient map for a particular computation is difficult and requires knowledge of parallel algorithms and architectures. Usually programs consist of multiple computations, and then the task is made even more difficult as the locally optimal map is likely not to be the globally optimal one. The algorithm developer must take special care balancing computation and communication costs to efficiently map the program, as discussed in Section 19.2 and illustrated by the parallel one-dimensional FFT example in Figures 19-2 and 19-3. Figure 19-5 is the pMatlab code for the two FFTs and a corner turn (full all-to-all communication) between them. Note the map objects (mapA, mapB) passed into the arrays.

The locally optimal maps (Figure 19-3 in Section 19.2) for the two FFTs are trivial (map along the dimension on which the FFT is performed using the maximum number of processors) yielding an embarrassingly parallel operation, or an operation requiring no interprocessor communication. Yet, what about the corner turn? On a higher-latency single-path network, a more efficient mapping would be one with fewer processors. Additionally, if the FFT is performed multiple times, it could be worthwhile to pipeline it or use disjoint sets of processors to distribute the two FFTs. This is a very simple example, yet it highlights the numerous choices that are available.

Let us return to the concept of map independence. Earlier it was mentioned that using maps in the program allows for the separation of the tasks of mapping the program and ensuring program correctness. The map independence also allows the first party to be another program or an automatic mapper. While clearly a large number of parallel languages and language extensions exist, ones that allow for map independence allow for a cleaner approach to automatic parallelization. The rest of this chapter highlights some of the existing approaches that fall into different classes of the taxonomy. While specific research initiatives are highlighted, they are highlighted as examples of techniques that are effective in automatic code optimization. Research initiatives are highlighted, though it is important to note that this is an active area in industry as well.

19.5   LOCAL OPTIMIZATION IN AN AUTOMATICALLY TUNED LIBRARY

As discussed previously, one of the categories in the automatic optimization taxonomy is the layer in which the optimization is implemented. The two categories previously cited are compiler and middleware. Note that middleware approaches are often libraries (as discussed in Chapter 17) that provide high performance routines for certain key operations. It is common that libraries additionally provide constructs such as distributed arrays and maps, as in PVL and pMatlab. This discussion will concentrate on the libraries that provide routine optimizations.

Consider the Basic Linear Algebra Subprograms (BLAS) library that provides an application programmer interface (API) for a number of basic linear algebra operations (Dongarra et al. 1988; Hanson, Krogh, and Lawson 1973; Lawson et al. 1979). Numerous implementations of the BLAS routines exist, optimized for different architectures. Clearly, it is impractical to create new implementations of the routines for all possible emerging architectures. This is a time-consuming process requiring expertise in processor architectures and linear algebra. The ATLAS project provides a solution (Whaley, Petitet, and Dongarra 2001). ATLAS automatically creates BLAS routines optimized for processor architectures using empirical measurements. It is unfair to say that ATLAS is purely a library approach as some of the ATLAS techniques could be incorporated in and are beneficial to compilers. However, it is not performing compile time analysis of the code.

ATLAS uses source-code adaptation in order to generate efficient linear algebra routines on various architectures. In this type of optimization, two categories of automatic optimization techniques can be utilized: (1) multiple routines to test which routine performs better on a given architecture and (2) code generation. ATLAS uses both types of optimization techniques together, yielding a tractable approach that scales well to a variety of architectures.

Multiple-routine source-code adaptation is the simpler of the two approaches. Essentially, the optimizer is provided with multiple implementations of each routine. It then executes the various implementations on the target architecture and performs a search for the lowest execution time. This approach can be very effective if there exists a large community that provides a large set of specialized programs; however, this is not always the case.

On the other hand, the code-generation approach provides the ultimate flexibility. The code generator, or a program that writes other programs, is given a set of parameters that specify various optimizations that could be utilized. Examples of such parameters include cache sizes and length of pipelines. While very large search spaces of codes can be explored, the approach could require exponentially long search times to determine the best generated code. To get the best of both worlds, the ATLAS approach takes advantage of code generation but supplements it with multiple routines.

Note that ATLAS does not explicitly address parallelism; however, it could work well on architectures with multiple floating-point units by unrolling loops, thus allowing for analysis of what routines could be executed in parallel. Additionally, as ATLAS determines what array sizes could fit into the cache, the same technique could be applied in an increasingly hierarchical manner, specifically, determining what size operation fits in local memory and then in memory of a subprocessor (on an architecture such as the IBM Cell). Further, this can be extended to automatically determining proper block sizes for out-of-core applications as discussed in Kim, Kepner, and Kahn (2005).

Another important point to mention is that a similar approach can be utilized on parallel architectures, given basic implementations of parallel BLAS routines and/or guidelines for code generation for those routines. If there is sufficient architecture and function coverage, an approach of benchmarking multiple versions of routines and storing the optimal for later use could be applied to a parallel architecture. As a matter of fact, a later section on pMapper discusses an approach that does some parallel benchmarking up front and uses that data at a later time.

ATLAS and other approaches like it, such as the "Fastest Fourier Transform in the West" (FFTW), perform all of the work once when a library has to be defined for an architecture. Specifically, the user acquires a processor architecture and runs ATLAS on it, thereby creating an optimized set of routines for that architecture. The optimizations are performed during architecture initialization and not at runtime. There are certain benefits to this type of approach. First, highly optimized routines are produced, as the time to generate the library is not critical to overall computation. Second, there is no runtime overhead in using ATLAS—once routines are determined they are used until a user decides to rerun the library-tuning routine. On the other hand, no dynamic updates are performed to the library and the initial compute time could be rather long, often on the order of hours (Seymour, You, and Dongarra 2006).

An additional benefit of this type of approach includes its feasibility. ATLAS is used in various commercial products (ATLAS). This is a successful project that started as a small academic endeavor. One of the factors contributing to the success of projects such as FFTW and ATLAS is that the creators limit themselves to specific domains (FFTs and linear algebra routines, respectively) and concentrate on the optimization of those routines.

A potential drawback of an approach such as ATLAS is that it optimizes specific routines and does not take into account global optimization of the code. Specifically, consider the matrix multiply example. ATLAS is likely to find a very fast matrix multiply for various array sizes and data types. However, ATLAS does not consider the composition of those routines. Specifically, the optimal utilization of cache could be, and likely is, different for combination of the routines versus for each independent routine.

19.6 COMPILER AND LANGUAGE APPROACH

On the other end of the taxonomy spectrum are parallel languages and language extensions and compilers and compiler extensions. These approaches can often provide more general support for parallelism, but require significantly larger implementation effort and are harder to adapt by the programming community (either embedded or scientific). One such example is the Titanium language (Yelick et al. 2007). Titanium is not a completely new language but a set of extensions to the Java programming language, making it easier to adapt for application developers. Note that Java-like languages are not usually a good fit for embedded systems, but we are using Titanium as an example for concept illustration.

Titanium adds significant array support on top of the arrays present in Java languages. Specifically, while Java provides support for multidimensional arrays, it does not do so efficiently. Multidimensional arrays are stored as arrays of arrays, yielding an inefficient memory access pattern. Titanium provides an efficient array implementation, thus allowing this language dialect to be more suitable to signal processing algorithms and other application domains that are highly array based. Titanium arrays can also be distributed, providing support for the Partitioned Global Address Space (PGAS) programming model (see Chapter 18 for the definition of the PGAS model), which is a natural model for many array-based computations and is used by pMatlab, PVL, and Unified Parallel C (UPC Community Forum), to name a few examples.

Since Titanium is a dialect of Java and is, therefore, a language extension, compiler support needs to be implemented for it. In addition to compiler techniques, there are also runtime optimizations that are implemented. The Titanium code is first translated into C code by the Titanium translator. Before the C compiler is run, a Titanium front-end compiler performs some optimizations that can be fed to the C compiler. Let us consider one of these optimization techniques in detail—concurrency analysis. Concurrency analysis identifies parts of the code that can be run at the same time or have no interdependencies (see Section 19.2). Some of the Titanium language restrictions allow for stronger concurrency analysis.

Titanium requires that barriers in the code are textually aligned. A barrier in a parallel single-program multiple-data (SPMD) code (see Chapter 18 for more details on programming models) is a construct that requires all processors that are executing the code to get to the same place in the program prior to continuing. The code between barriers identifies phases of the execution. Since all processors must reach the barrier before continuing past it, no two phases can run concurrently. This provides information to the Titanium high-level compiler and can prevent race conditions, or critical dependence on timing of events.

Titanium uses *single* qualifiers to indicate that a value has to be the same across all the processors. The use of the single qualifier indicates to the compiler that conditionals guarded by the single variable cannot be executed concurrently since the variable has the same value everywhere. For more details on the Titanium concurrency analysis algorithm, see Kamil and Yelick (2005).

In addition to concurrency analysis, Titanium also uses alias analysis—a common technique in parallelizing compilers to determine whether pointer variables can access the same object—and local qualification inference which analyzes the code and determines pointers that are local and have not been marked as such. The local qualification inference allows for significant optimization as handling global pointers requires more memory and time because when referencing a global pointer, a check must be performed regarding whether network communication is necessary.

If we go back to our taxonomy, it is worthwhile to point out that this analysis is performed at compile time and not at runtime; therefore, significant benefits are gained as optimization can be performed at compile time and errors can be detected sooner. On the other hand, some information that could aid optimization might not be available at compile time. The final example, pMapper, performs the code analysis at runtime and is discussed in the next section.

19.7 DYNAMIC CODE ANALYSIS IN A MIDDLEWARE SYSTEM

pMapper (Travinin et al. 2005), developed at MIT Lincoln Laboratory, falls in between the two approaches discussed in the preceding two sections. pMapper was initially developed to map MATLAB programs onto clusters; however, it later became clear that the techniques are general enough for a variety of systems. To use pMapper, the following two conditions must be met:

1. Presence of an underlying parallel library
2. Map independence in the program

Specifically, we simply replace the distributed array maps with parallel tags to indicate to the system that the objects should be considered for distribution, and let the parallelization system figure out the maps.

Let us consider a concrete example. The pMatlab code in Figure 19-6 performs a row FFT, followed by a column FFT, followed by a matrix-matrix multiply.

Note the maps in the code. They are not trivial to define, and it is not often clear what performance will result from choosing a particular map. In Figure 19-7, all maps are replaced with parallel tags (*ptags*). That is the syntax adapted by pMapper. The tags indicate to the system that arrays A–E should be considered for distribution.

```
%Define maps
map1 = map([4 1],[0:3]);
map2 = map([1 4],[0:3]);
map3 = map([2 2],[0:3]);
map4 = map([2 2],[0 2 1 3]);
%Initialize arrays
A = array(N,N,map1);
B = array(N,N,map2);
C = array(N,N,map3);
D = array(N,N,map4);
E = array(N,N,map3);
%Perform computation
B(:,:) = fft(A,2);
C(:,:) = fft(B,1);
E(:,:) = C*D;
```

FIGURE 19-6 Simple processing chain consisting of two FFTs and a matrix-matrix multiply.

```
%Initialize arrays
A = array(N,N,ptag);
B = array(N,N,ptag);
C = array(N,N,ptag);
D = array(N,N,ptag);
E = array(N,N,ptag);
%Perform computation
B(:,:) = fft(A,2);
C(:,:) = fft(B,1);
E(:,:) = C*D;
```

FIGURE 19-7 Functionally the same code as in Figure 19-6, but the map definitions are removed and map argument in the array constructor is replaced with *ptag*.

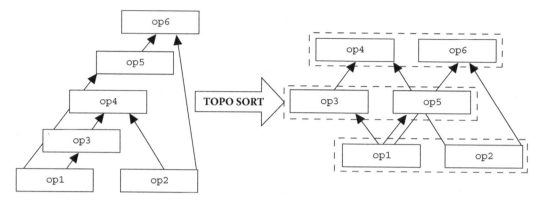

FIGURE 19-8 The dependency graph, or parse tree, on the left shows the operations in chronological order, while the tree on the right shows the operations in topological order. Note that after the topological sort, it becomes apparent that (op1, op2),(op3, op5) and (op4, op6) can be executed concurrently.

In order to provide efficient mappings for the application, the parallelization architecture has to either model or benchmark library kernels to get a sense for the underlying architecture performance. pMapper does both. If a parallel library and a parallel architecture exist, timing benchmarks are performed on a real system with computation and communication kernels. On the other hand, when the architecture is not available or a parallel library for the architecture currently does not exist, pMapper uses a machine model based on the architecture parameters to create a benchmark database. The collection of timing data on individual kernels is time-intensive and is done once when pMapper is initialized for the system. Collecting empirical information about the underlying architecture via benchmarking is similar to approaches used by local optimization systems such as ATLAS and FFTW. Once the timing information exists, pMapper uses a fast runtime method based on dynamic programming to generate mappings.

pMapper uses *lazy evaluation* to collect as much information about the program as possible prior to performing any program optimization. A lazy evaluation approach delays any computation until it is absolutely necessary, such as when the result is required by the user. Until the result is necessary, this approach simply builds up a dependency graph, or parse tree, of the program and stores information about array sizes and dependencies. Once a result is required, the relevant portion of the dependency graph is extracted and analyzed.

Some of the analysis techniques are similar to the approaches used by parallelizing compilers as discussed in Section 19.6. For example, pMapper performs concurrency analysis and figures out which portions of the code can be executed independently. This is done via topological sorting of the dependency graph. Nodes that end up in the same stage (as illustrated by Figure 19-8) could be executed in parallel.

Additionally, pMapper performs global optimization on the dependency graph by looking at multiple nodes at the same time when determining the maps. Global optimization allows for balancing communication and computation, and produces more efficient program mappings than does local per-kernel optimization.

Once the maps are determined (see Figure 19-9 for maps and speedup achieved for code in Figure 19-7), they can either be returned to the user for inspection and reuse, or the program can be executed using pMapper executor. Additionally, if the underlying architecture does not exist and a machine model simulation is used to determine the mappings, a simulator will produce the expected execution time of the program. The simulation capability allows suitability assessment of various architectures and architecture parameters for specific applications. See Bliss et al. (2006) for additional pMapper results both on large codes [synthetic aperture radar (SAR) benchmark discussed in Chapter 15] and as a predictor of performance on the Cell architecture.

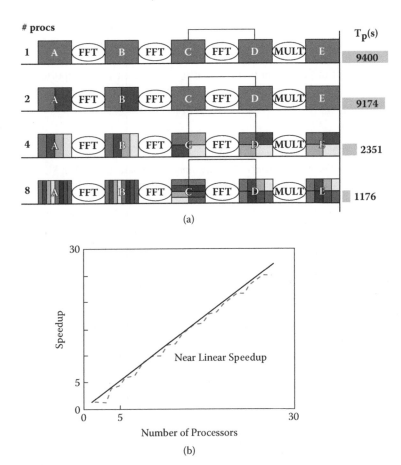

(a)

(b)

FIGURE 19-9 **(Color figure follows p. 278.)** pMapper mapping and speedup results. These results were obtained for a low-latency architecture that would be consistent with a real-time embedded processor. Note how pMapper chooses mapping to balance communication and computation. At two processors, only arrays A and B are distributed as there is no benefit to distributing the other arrays. At four processors, C is distributed to benefit the matrix multiple operation and not the FFT operation. (From Travinin, N. et al., pMapper: automatic mapping of parallel Matlab programs, *Proceedings of the IEEE Department of Defense High Performance Computing Modernization Program Users Group Conference*, pp. 254–261. © 2005 IEEE.)

19.8 SUMMARY

Automatic program optimization is an active area of research and significant progress is being made. While serial program optimization is important, optimizing applications for parallel architectures is becoming vital for utilizing emerging multicore processor architectures. As parallel processing becomes increasingly mainstream, parallelizing compilers and runtime systems have to keep up with the development.

It is not yet clear what the best approach to automatic parallelization is. Optimizing individual kernels in an application is certainly beneficial, particularly if the application is heavily dependent on a particular kernel. However, it does not help with the overall program optimization.

Specialized parallel languages such as Titanium and the set of HPCS languages are making significant strides in parallelizing compilers and runtime systems (Allen et al. 2005; Callahan, Chamberlain, and Zima 2004; Sarkar 2004). However, new language adoption is often a very slow process and legacy codes have to be rewritten.

Additionally, general-purpose program parallelizers are still out of reach. When one is choosing a runtime or compiler approach, it is important to choose one that performs well for a particular type of application. Limiting the optimization space to a particular type of processing—for example, linear algebra for signal processing—as done by pMapper, will usually yield the best results in the near term.

Automatic program optimization is a fascinating and increasingly important area. With even desktop computers shipping with multiple cores, improving performance without requiring significant changes to a program is paramount for future applications.

REFERENCES

Allen, E. et al. 2005. *The Fortress Language Specification*. Santa Clara, Calif.: Sun Microsystems.

Automatically Tuned Linear Algebra Software (ATLAS). Available online at http://math-atlas.sourceforge. net/.

Banerjee, U., R. Eisenmann, A. Nicolau, and D.A. Padua. 1993. Automatic program parallelization. *Proceedings of the IEEE* 81(2): 211–213.

Bliss, N.T., J. Dahlstrom, D. Jennings, and S. Mohindra. 2006. Automatic mapping of the HPEC challenge benchmarks. *Proceedings of the Tenth Annual High Performance Embedded Computing Workshop*. Available online at http://www.ll.mit.edu/HPEC.

Bliss, N.T. and J. Kepner. 2007. pMatlab parallel MATLAB library. Special issue on High-Productivity Programming Languages and Models, *International Journal of High Performance Computing Applications* 21(3): 336–359.

Blume, W. 1995. Symbolic Analysis Techniques for Effective Automatic Parallelization. Ph.D. thesis. University of Illinois, Urbana–Champaign.

Callahan, D., B.L. Chamberlain, and H. Zima. 2004. The cascade high-productivity language. *Proceedings of the Ninth International Workshop on High-Level Parallel Programming Models and Supportive Environments*. IEEE Computer Society 52–60.

Carter, L. and J. Ferrante. 1999. *Languages and Compilers for Parallel Computing: 12th International Workshop, LCPC '99 Proceedings*. Lecture Notes in Computer Science series. New York: Springer.

Chen, Z., J. Dongarra, P. Luszczek, and K. Roche. 2003. Self-adapting software for numerical linear algebra and LAPACK for clusters. *Parallel Computing* 29(11–12): 1723–1743.

Darte, A., Y. Robert, and F. Vivien. 2000. *Scheduling and Automatic Parallelization*. Boston: Birkhäuser.

DeLuca, C.M., C.W. Heisey, R.A. Bond, and J.M. Daly. 1997. A portable object-based parallel library and layered framework for real-time radar signal processing. *Proceedings of the International Scientific Computing in Object-Oriented Parallel Environment Conference* 241–248.

Dongarra, J., J. Du Croz, S. Hammarling, and R. Hanson. 1988. An extended set of FORTRAN basic linear algebra subprograms. *ACM Transactions on Mathematical Software* 14(1): 1–17.

FFTW site. Fastest Fourier transform in the west. Available online at http://www.fftw.org.

Hanson, R., F. Krogh, and C. Lawson. 1973. A proposal for standard linear algebra sub-programs. *ACM SIGNUM Newsletter* 8(16).

Hennessy, J.L. and D.A. Patterson. 2006. *Computer Architecture: A Quantitative Approach*, 4th edition. San Francisco: Morgan Kaufman.

Hoffmann, H., J. Kepner, and R. Bond. 2001. S3P: automatic optimized mapping of signal processing applications to parallel architectures. Given at the Fifth Annual High Performance Embedded Computing Workshop, MIT Lincoln Laboratory, Lexington, Mass.

Hurley, S. 1993. Taskgraph mapping using a genetic algorithm: a comparison of fitness functions. *Parallel Computing* 19: 1313–1317.

IBM. The Cell project at IBM research. Available online at http://www.research.ibm.com/cell/.

Intel. Intel Core Microarchitecture. Available online at http://www.intel.com/technology/architecture/coremicro.

Kamil, A. and K. Yelick. 2005. Concurrency analysis for parallel programs with textually aligned barriers. Presented at the 18th International Workshop on Languages and Compilers for Parallel Computing, Hawthorne, N.Y.

Kim, H., J. Kepner, and C. Kahn. 2005. Parallel MATLAB for eXtreme Virtual Memory. *Proceedings of the IEEE Department of Defense High Performance Computing Modernization Program Users Group Conference* 381–387.

Kuck, D., P. Budnik, S.-C. Chen, E. Davis, Jr., J. Han, P. Kraska, D. Lawrie, Y. Muraoka, R. Strebendt, and R. Towle. 1974. Measurements of parallelism in ordinary FORTRAN programs. *Computer* 7(1): 37–46.

Kuo, K., R. Rabbah, and S. Amarasinghe. 2005. A productive programming environment for stream computing. *Proceedings of the Second Workshop on Productivity and Performance in High-End Computing* 35–44.

Lamport, L. 1974. The parallel execution of DO loops. *Communications of the ACM* 17(2): 83–93.

Lawson, C., R. Hanson, D. Kincaid, and F. Krogh. 1979. Basic linear algebra subprograms for Fortran usage. *ACM Transactions on Mathematical Software* 5(3): 308–323.

Lebak, J., J. Kepner, H. Hoffmann, and E. Rutledge. 2005. Parallel VSIPL++: an open standard software library for high-performance parallel signal processing. *Proceedings of the IEEE* 93(2): 313–330.

Loveman, D.B. 1993. High performance Fortran. *Parallel and Distributed Technology: Systems and Applications.* IEEE 1(1).

Markoff, J. 2007. Intel prototype may herald a new age of processing. *The New York Times* February 12, Sec. C, 9.

McKinley, K. 1994. Automatic and Interactive Parallelization. Ph.D. thesis. Rice University, Houston, Tex.

Nikhil, R.S. and Arvind. 2001. *Implicit Parallel Programming in pH*. San Francisco: Morgan Kaufman Publishers.

Rau, B.R. and J.A. Fisher. 1993. Instruction-level parallel processing: history, overview, and perspective. *Journal of Supercomputing* 7(1): 9–50.

Sarkar, V. 2004. Language and virtual machine challenges for large-scale parallel systems. Presented at the Workshop on the Future of Virtual Execution Environments, Armonk, N.Y.

Seymour, K., H. You, and J. Dongarra. 2006. *ATLAS on the Blue/Gene/L—Preliminary Results*. University of Tennessee, Computer Science Department Technical Report, ICL-UT-06-10.

Simon Management Group for Interactive Supercomputing. 2006. The development of custom parallel computing applications.

Travinin, N., H. Hoffmann, R. Bond, H. Chan, J. Kepner, and E. Wong. 2005. pMapper: automatic mapping of parallel MATLAB programs. *Proceedings of the IEEE Department of Defense High Performance Computing Modernization Program Users Group Conference* 254–261.

University of Indiana Computer Science Dept. The Dynamo project: dynamic optimization via staged compilation. Available online at http://www.cs.indiana.edu/proglang/dynamo.

UPC Community Forum. Available online at http://upc.gwu.edu.

Whaley, R.C., A. Petitet, and J.J. Dongarra. 2001. Automated empirical optimizations of software and the ATLAB project. *Parallel Computing* 27(1–2): 3–35.

Wolfe, M. 1996. *High Performance Compilers for Parallel Computing*. Redwood City, Calif.: Benjamin Cummings.

Yelick, K., P. Hilfinger, S. Graham, D. Bonachea, J. Su, A. Kami, K. Datta, P. Colella, and T. Wen. 2007. Parallel languages and compilers: perspective from the Titanium experience. Special Issue on High Productivity Programming Languages and Models, *International Journal of High Performance Computing Applications* 21(3): 266–290.

Zima, H. 1991. *Supercompilers for Parallel and Vector Computers*. New York: ACM Press/Addison-Wesley.

Zosel, M.E. 1993. High performance Fortran: an overview. *Compcon Spring '93, Digest of Papers* 132–136.

Section V

High Performance Embedded Computing Application Examples

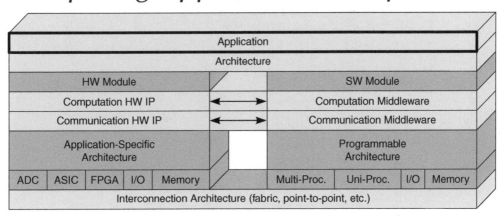

Chapter 20 Radar Applications
Kenneth Teitelbaum, MIT Lincoln Laboratory

This chapter explores the application of high performance embedded computing to radar systems, beginning with a high-level description of some basic radar principles of operation and fundamental signal processing techniques. This is followed by a discussion of the mapping of these techniques onto parallel computers. Some actual radar signal processors developed at MIT Lincoln Laboratory are presented.

Chapter 21 A Sonar Application
W. Robert Bernecky, Naval Undersea Warfare Center

This chapter introduces computational aspects pertaining to the design and implementation of a real-time sonar system. The chapter provides an example development implementation using state-of-the-art computational resources to meet design criteria.

Chapter 22 Communications Applications

Joel I. Goodman and Thomas G. Macdonald, MIT Lincoln Laboratory

This chapter discusses typical challenges in military communications applications. It then provides an overview of essential transmitter and receiver functionalities in communications applications and their signal processing requirements.

Chapter 23 Development of a Real-Time Electro-Optical Reconnaissance System

Robert A. Coury, MIT Lincoln Laboratory

This chapter describes the development of a real-time electro-optical reconnaissance system. The design methodology is illustrated by the development of a notional real-time system from a non-real-time desktop implementation and a prototype data-collection platform.

20 Radar Applications

Kenneth Teitelbaum, MIT Lincoln Laboratory

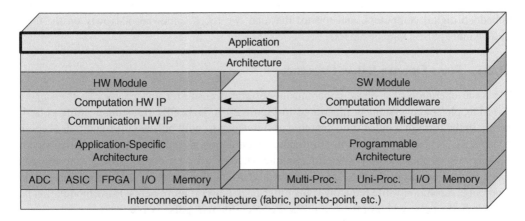

This chapter explores the application of high performance embedded computing to radar systems, beginning with a high-level description of some basic radar principles of operation and fundamental signal processing techniques. This is followed by a discussion of the mapping of these techniques onto parallel computers. Some actual radar signal processors developed at MIT Lincoln Laboratory are presented.

20.1 INTRODUCTION

RADAR (**RA**dio **D**etection **A**nd **R**anging; the acronym is now commonly used in lowercase), a technique for detecting objects via scattering of radio frequency (RF) electromagnetic energy, traces its history to the early days of World War II when systems such as the British Chain Home were employed successfully for the detection of aircraft (Britannica 2007). Early systems relied on operators monitoring oscilloscopes for target detection, and they suffered from the inability to distinguish energy reflected from targets from energy reflected from terrain (clutter) and were susceptible to jamming (deliberate transmission of high-power RF signals from an adversary intended to mask radar detection of targets). With the advent of the analog-to-digital converter and the digital computer, digital processing of radar signals became commonplace, resulting in the development of automatic detection techniques and improved target detection in clutter and jamming. As Moore's Law improvements in digital device technology have brought faster and more capable computers, radar signal processing has become increasingly sophisticated, and radar performance in difficult environments has steadily improved. High performance embedded computing (HPEC) has become an indispensable component of modern radar systems.

This chapter explores the application of high performance embedded computing to radar systems, beginning with a high-level description of some basic radar principles of operation and fundamental signal processing techniques. This is followed by a discussion of the mapping of these techniques onto parallel computers. Finally, some actual radar signal processors developed at MIT

Lincoln Laboratory are presented as examples of what can be accomplished in this domain and how capability has evolved with time.

20.2 BASIC RADAR CONCEPTS

A complete tutorial on the design, operating principles, and applications of radar systems is well beyond the scope of this chapter; however, some excellent texts that treat the subject well are Skolnik (1990) and Stimson (1998). The intent here is to provide sufficient background to introduce the ensuing discussion on implementing radar signal processing algorithms with high performance embedded digital computers, and toward that end, we focus somewhat narrowly on the class of multichannel pulse-Doppler radars.

20.2.1 PULSE-DOPPLER RADAR OPERATION

The basic operation of a phase-coherent, monostatic (transmitter and receiver at the same site) pulse-Doppler radar system is illustrated in Figure 20-1. The radar transmitter emits pulsed RF energy in the form of a pulse train centered at frequency f_0, with pulsewidth τ and pulse-repetition frequency f_{pr}. Typically some form of intrapulse modulation is applied to the transmitted signal for the purposes of pulse compression, a technique whereby a longer coded pulse is transmitted and match-filtered on receive to produce a narrow, higher amplitude compressed pulse, thus providing good range resolution while reducing transmitter peak power requirements. If the modulation bandwidth is B, the pulse width of the compressed (match-filtered) pulse is $1/B$.

Each transmitted pulse is reflected by objects illuminated by the antenna beam. The reflected pulse, substantially weaker than the transmitted pulse, is delayed in time by the round-trip delay between antenna and target (a function of the range to the target), and is Doppler-shifted by the motion of the target relative to the radar. The reflected energy is then amplified and downconverted to a low intermediate frequency (IF) by the receiver and digitized. If the receiver output is sampled at a rate of f_s samples/second, then there will be f_s/f_{pr} samples per pulse repetition interval (PRI). Usually only samples in the inter-pulse period (IPP) are collected, as the receiver is typically blanked for protection during transmit. This process results in the collection of $N_s = f_s(1/f_{pr} - \tau)$ samples per PRI. Since successive samples relate to targets at greater range (longer round-trip delay), these intra-PRI samples are typically referred to as *range samples*. In order to permit measurement of target Doppler shifts (and thus target radial velocities) and to permit separation of target and clutter in velocity space, phase coherence of transmit and receive signals is required over a group of N_{pr} pulses (at constant center frequency), known as a coherent processing interval (CPI).

The data collected from a single CPI can be organized as an $N_s \times N_{pr}$ matrix with columns consisting of the individual intra-PRI range samples from a single PRI. Processing of the CPI data matrix proceeds on a column-by-column basis, as shown in Figure 20-2, beginning with the completion of the downconversion process and resulting in baseband quadrature (real and imaginary) channels. This is followed by pulse compression (matched-filtering) of the receive signal, usually via fast Fourier transform (FFT)-based fast-convolution techniques (Oppenheim and Schafer 1975). Doppler processing proceeds on a range-sample–by–range-sample basis. The target echo in a par-

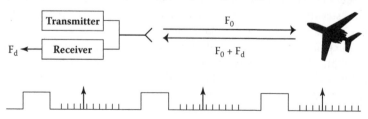

FIGURE 20-1   Pulse Doppler radar operation.

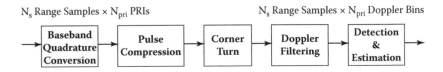

FIGURE 20-2   Pulse-Doppler signal processing.

ticular range sample is effectively sampled at the pulse repetition frequency (PRF), and an N_{pr}-point discrete Fourier transform (DFT) yields a bank of N_{pr} Doppler filters that span a range of target Dopplers from $-f_{pr}/2$ to $f_{pr}/2$. For the case of a stationary radar and a moving target, the return from the ground (clutter) falls in the zero-velocity Doppler bin while the target, because of its nonzero velocity, falls in a different bin, thus permitting effective suppression of clutter. For computational reasons, the DFT is usually replaced by an FFT of length an integer power of 2, and the data for each range sample are zero-padded to adjust the length. Some form of window function is typically applied to control the sidelobe response in the frequency domain in order to prevent clutter energy from leaking into the sidelobes of the target Doppler filter. Sometimes a moving target indicator (MTI) canceller is applied to attenuate clutter at zero Doppler in order to minimize the loss from the low-sidelobe taper (Skolnik 1990). Following Doppler processing, the columns of the data matrix now represent the individual Doppler filters rather than the PRI samples. Detection and estimation typically follow. For search radar operation, in which the target range and velocity are generally not known *a priori*, the entire range-Doppler space represented by the data matrix must be searched for the presence of targets. In order to accomplish this, a background noise level is estimated for each range-Doppler cell, and the amplitude in that cell is compared to a threshold, which is typically a multiple of the background noise level. Threshold crossings in groups of contiguous range-Doppler cells (clusters) are collapsed into a single report and adjacent range and Doppler cells are used to refine the estimates of target range and Doppler.

20.2.2   MULTICHANNEL PULSE-DOPPLER

Increasingly, radar systems have incorporated multichannel antenna designs in order to deal effectively with jamming and the problem of clutter suppression from a moving platform. A block diagram of such a multichannel pulse-Doppler radar system is shown in Figure 20-3. A uniformly spaced linear array of antennas is shown; however, these channels could be beamformed rows or columns of a two-dimensional phased array or some number of formed beams and auxiliary elements. By applying a complex (amplitude and phase) weight to each element, usually digitally, it is possible to shape the antenna pattern, placing high gain in the direction of a target of interest

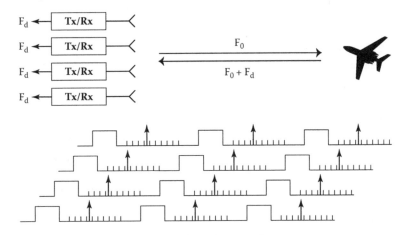

FIGURE 20-3   Typical multichannel pulse-Doppler radar operation.

and a null in the direction of interference—either jamming or clutter. The digitized data collected are now in the form of a data cube of dimension N_s range samples × N_{pr} PRIs × N_c antenna channels. Much of the processing along two dimensions (range, Doppler) of the cube is similar to the previous section, although processing the third cube dimension (the multiple antenna channels) warrants further discussion.

20.2.3 ADAPTIVE BEAMFORMING

In a typical radar scenario, target and jammers are spatially separated, and adapting the antenna pattern to place nulls on the jammers and high gain on the target can be an effective technique for mitigating jamming. Since the number, type, and location of jamming sources is not known *a priori*, it is necessary to adapt the antenna pattern based on received data containing jamming energy. One approach for this, known as *sample matrix inversion* (SMI), is illustrated in Figure 20-4.

In order to avoid nulling the target, the jamming environment is sampled prior to radar transmission, and time samples from each antenna channel are formed into a sample matrix X where the columns represent the individual antenna channels and the rows represent a particular snapshot in time. The sample covariance matrix R can be estimated as $R = X^H X$. The steering vector v is a complex vector that consists of the quiescent (nonadapted) weights required to steer the antenna beam to have gain on the target. The adapted weight vector $w = R^{-1}v$ will preserve the target gain and place the desired null on the jammer. In practice, numerical considerations in the presence of strong interference or jamming make direct computation of the sample covariance matrix impractical, and the solution of the desired weight vector follows from QR decomposition of the sample matrix and double back-substitution.

Often the interference to be nulled is wideband in nature, and the mismatch between channel transfer functions of real receivers and analog-to-digital converters has the potential to limit performance with a scheme, such as suggested in Figure 20-4, where there is but a single degree of freedom per channel. One way to avoid such difficulties is by the insertion of finite impulse response

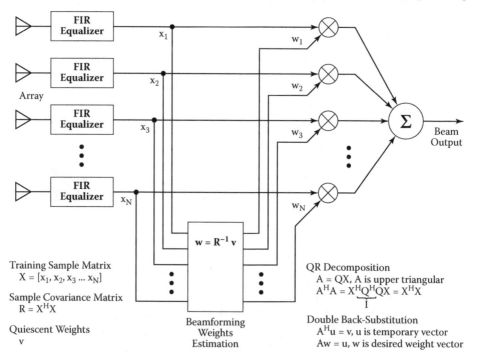

FIGURE 20-4   Adaptive beamforming.

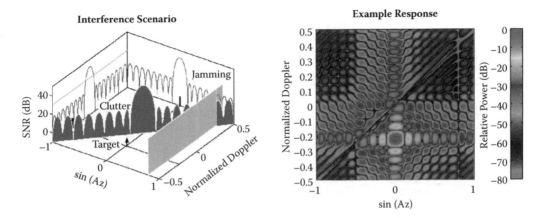

FIGURE 20-5   (**Color figure follows p. 278.**) Space-time adaptive processing to suppress airborne clutter. (Ward 1994, reprinted with permission of MIT Lincoln Laboratory.)

(FIR) equalizing filters in each channel designed to match channel transfer functions across all channels (Teitelbaum 1991). This technique, known as *channel equalization*, can substantially increase improvements in signal-to-interference-plus-noise ratios achievable via adaptive beamforming.

20.2.4   SPACE-TIME ADAPTIVE PROCESSING

To this point, we have considered the case of a stationary radar and moving target, where the clutter has zero Doppler and the target has nonzero Doppler, permitting effective separation of target from clutter. If the radar is moving, as in the case of airborne radar, the motion of the radar will spread the clutter return in Doppler between $\pm 2v/\lambda$, where v is the velocity of the radar.

It is still possible to separate the target and clutter in Doppler, but the Doppler cell(s) containing the clutter vary spatially (as a function of angle), and a two-dimensional, or space-time (angle-Doppler), filter is needed. When this two-dimensional filter is constructed in a data-adaptive manner, the technique is known as space-time adaptive processing, or STAP. An example is shown in Figure 20-5, where the interference consists of both clutter and jamming. The clutter is constrained by geometry to lie on a straight line in sin (angle)-Doppler space known as the clutter ridge, while jamming exists at a constant angle(s) for all Dopplers. It is possible to construct a two-dimensional filter that has a deep null along the clutter ridge and at the jamming angle(s), while maintaining high gain in hypothesized target locations in angle and Doppler.

There are many ways to construct these space-time filters (Ward 1994), but in essence, the process consists of solving for and applying an adaptive weight vector, as in Figure 20-4, with inputs (degrees of freedom) that could come from both antenna elements or beams, and time-delayed or Doppler-filtered versions of those elements or beams, depending on the application.

The problem of training the adaptive weights, that is, estimating the covariance of the interference (clutter, jamming), is more of an issue than for the jamming-only, spatially adaptive case in which we could sample the interference environment prior to transmission. For STAP, we must transmit in order to sample the clutter environment, meaning that we must avoid using the same range cells for covariance estimation and weight application in order to prevent target nulling. We must also contend with the nonstationary nature of the interference, with the power and Doppler of the clutter a strong function of range, angle, and terrain reflectivity. Several schemes have been devised for this purpose (Rabideau and Steinhardt 1999).

An example STAP algorithm is shown in Figure 20-6. It is representative of an element-space, multibin post-Doppler STAP processor. It operates on inputs which are taken from each element, row, column, or subarray of an antenna. Processing of each individual channel is similar to that of Figure 20-2 except that detection/estimation has been deferred until the STAP processing is

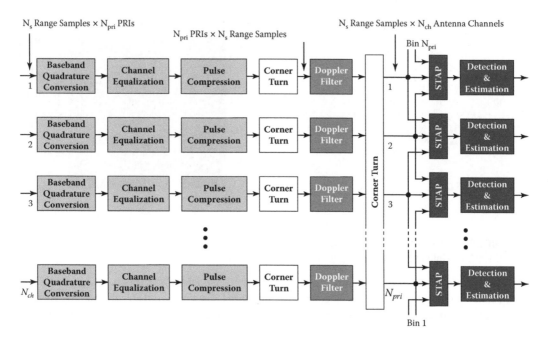

FIGURE 20-6 Adjacent-bin post-Doppler STAP algorithm.

completed and FIR channel equalizers have been added. The result of this processing is a set of N_{ch} matrices of dimension $N_s \times N_{pri}$ that contain range-Doppler data for each channel. The data are then rearranged (corner-turned) to produce a series of N_{pri} matrices of dimension $N_s \times N_{ch}$ that contain all of the range samples for each channel for one Doppler bin. The STAP algorithm is performed independently on each Doppler bin. For the adjacent-bin algorithm, $(k - 1)/2$ (k odd) adjacent bins on either side of the bin of interest are included in the STAP algorithm, resulting in $k \times N_{ch}$ inputs or degrees of freedom (DOF) for the adaptive beamformer shown in Figure 20-6. The weights are computed as before, but with the sample covariance matrix estimated from a sample matrix drawn from snapshots at multiple range gates. As a typical rule of thumb (Brennan's Rule), three times as many samples as DOF are required for good performance. These range gates may come from any ranges; however, the cells with the largest clutter-to-noise ratio are often selected (power-selected training) to prevent undernulling of clutter. The adaptive weights thus computed are applied to all range cells; however, care must be taken to prevent self-nulling of the target for range cells that are part of the training set. In this case, the sample data for the cell under test must be excised from the covariance estimate. There are computationally efficient down-dating algorithms (sliding hole) for this that minimize the associated computational burden.

Typically, multiple (N_b) beams are formed using multiple steering vectors, and the result of the STAP process is an $N_s \times N_b$ data matrix for each Doppler bin. This is then thresholded and clusters of threshold crossings in contiguous range/Doppler/angle cells must be collapsed (now in three dimensions). Comparison of amplitudes in adjacent cells (also in three dimensions) can now be used to estimate target range, Doppler, and angle.

20.3 MAPPING RADAR ALGORITHMS ONTO HPEC ARCHITECTURES

Successfully mapping this class of algorithms onto multiprocessor architectures requires careful consideration of two principal issues. First, it is necessary to ensure that we satisfy throughput requirements, processing datasets at least as quickly as they arrive. In the radar context, this means that we must, on average, process one CPI of data in one CPI or less, driving the requirement for the number of processors we must map onto. The second issue concerns latency—specifically, how long after a CPI of data is collected and input to the processor must the processed result be available? Latency requirements are highly dependent on the particular application. Some radar applications—missile guidance, for example, in which latent data can cause the missile to miss its intended target—have very tight latency requirements, typically measured in fractions of a second. Other applications, surveillance radar, for example, in which it may take the radar many seconds to search an area for targets, have less stringent latency requirements. Latency requirements will drive the mapping strategy—specifically, how do we allocate processing and/or datasets to the available processors?

In order to get a sense of the scale of the mapping problem, consider that adaptive radar signal processing algorithms typically have peak computation rates of hundreds of GFLOPS (gigaFLOPS, one billion floating-point operations per second) to a TFLOP (teraFLOP, one trillion floating-point operations per second) or two and might require hundreds of parallel processing nodes to satisfy the real-time throughput constraint. At the same time, typical latencies are measured in a handful of CPIs.

20.3.1 ROUND-ROBIN PARTITIONING

The simplest approach from a computational perspective is to have each processor in an N-processor circular queue process an entire CPI of data. The first CPI goes to the first processor, the second CPI to the second processor, and the Nth CPI to the Nth processor. CPI $N + 1$ will go to the first processor, and the number of processors is chosen such that the first processor will have just completed processing the first CPI when CPI $N + 1$ is ready. The latency of this approach, of course, is N CPIs, so if we need hundreds of processors to meet real-time throughput, we will have a latency of hundreds of CPIs also—not acceptable for typical radar signal processing applications.

20.3.2 FUNCTIONAL PIPELINING

Another simple approach would be to pipeline data between processors, with each processor implementing one function for all of the data. One processor could implement the baseband quadrature conversion, another pulse compression, another Doppler filtering, etc. Of course, it would be difficult to employ more than a handful of processors in this manner (one for every processing stage), and adding processors does nothing to decrease the latency.

20.3.3 COARSE-GRAIN DATA-PARALLEL PARTITIONING

The key to being able to employ hundreds of processors while maintaining low latency is to exploit the inherent parallelism of the data cube. Since baseband quadrature conversion and pulse compression operate in parallel on every channel and Doppler bin, we could employ up to $N_{ch} \times N_{pri}$ processors (one per channel/bin) in parallel without requiring interprocessor communication. Doppler processing operates in parallel on every channel and range cell, and we could employ up to $N_{ch} \times N_s$ processors for Doppler processing. Similarly, STAP operates in parallel on each range cell and Doppler bin, and $N_s \times N_{pri}$ processors could be employed. Effectively, each stage of the processing operates primarily along one dimension of the data cube and is parallelizable along the other two dimensions of the cube. Unfortunately, there is no single mapping that works for all stages of the computation without having to communicate data between processors. A typical approach, suggested by the grayscale coding in Figure 20-6, employs three pipelined stages (each represented by a different shade) that exploit data parallelism within each stage and rearrange the data between each stage. The STAP stage is slightly more complicated than the others since it is necessary to distribute multiple Doppler bins to each STAP processor (because of the adjacent-bin STAP algorithm). Also, while the STAP weight application parallelizes cleanly along the range dimension, the weight computation might require training data from range cells in different processors. The communication burden for this, however, is typically quite small.

The number of processors employed at each stage is typically chosen so that each stage processes a single CPI of data in a single CPI, and the communication between stages is typically overlapped with the computation, resulting in an overall latency of three CPIs.

20.3.4   Fine-Grain Data-Parallel Partitioning

In this context, we define *coarse-grain partitioning* to mean that each processor has all of the data it requires to execute multiple problem sets without interprocessor communication and has sufficient computation rate to execute those problem sets within an allotted time. We define *fine-grain partitioning* to mean that more than one processor must cooperate to compute even a single problem set within the allotted time. In applications in which latency constraints cannot be met with the coarse-grain data-parallel approach summarized in the previous section, fine-grain data-parallel partitioning approaches can be employed, at the cost of increased interprocessor communications. One area in which this sometimes becomes necessary involves the computation of STAP weights. As discussed previously, this computation is based on QR decomposition of the sample data matrix, and the required computation rate grows as the cube of the number of adaptive degrees of freedom. For systems with many DOF, the required computation rate can become quite large. A hybrid coarse-grain/fine-grain approach based on distributing the Householder QR algorithm is discussed for a single-instruction multiple-data (SIMD) processor array in McMahon and Teitelbaum (1996).

Systolic array architectures are also applicable to this problem and may be well suited to the emerging class of tiled processor architectures discussed in Chapter 13. A systolic STAP processor, based on the QR approach of McWhirter and Shepherd (1989), is shown in Figure 20-7. It consists of an array of systolic cells, each of which performs a specific function in cooperation with its nearest

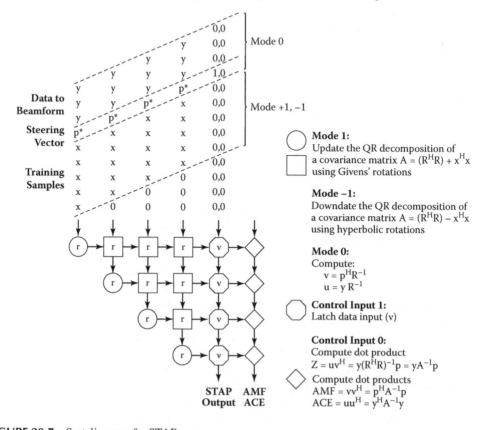

FIGURE 20-7   Systolic array for STAP.

neighbors. In Mode 1, training samples are input from the top of the array in time-skewed fashion. The circular cells compute Givens rotations, and the square cells apply those rotations, accumulating the triangular factor R of the estimated sample covariance matrix within the array of circular and square cells. In Mode 0, with the control input set to 1, the array steering vector is input from the top of the array, the circular and square cells compute the back-substitution $v = p^H R^{-1}$, and the octagonal cells latch the computed intermediate vector v. The diamond-shaped cells compute the adaptive matched-filter (AMF) normalization factor vv^H. Still in Mode 0, but with the control input set to 0, the data to be beamformed, y, is input from the top of the array and the circular and square cells compute the back-substitution $u = yR^{-1}$. The octagonal cells now compute the dot product uv^H, which is the STAP-beamformed result, output from the bottom of the array. The diamond-shaped cells compute the AMF and adaptive coherence estimator (ACE) normalizations to facilitate implementation of the adaptive sidelobe blanker, which has proven useful in mitigating false alarms in the presence of undernulled clutter (Richmond 2000). With the triangular factor R stored in the circular and square cells, rank-k modifications of the covariance estimate are easily accomplished. Updates are accomplished by input of new training data from the top in Mode 1, and down-dates are accomplished by input of the training data to be deleted from the top in Mode −1, which uses hyperbolic rotations. In this manner, sliding-window or sliding-hole training approaches can be used.

The mapping of systolic array cells to processors in a multiprocessor architecture varies with the application and the required degree of fine-grained parallelism. Typically, multiple systolic cells are mapped onto each processor. The array shown in Figure 20-9 can be folded onto a linear array in a manner similar to that in Liu et al. (2003), suitable for field programmable gate array (FPGA) implementation (Nguyen et al. 2005). Such an approach might be well suited to hybrid FPGA/programmable processor architectures.

20.4 IMPLEMENTATION EXAMPLES

Over the course of the last two decades, the development of real-time signal processors for radar systems at MIT Lincoln Laboratory has focused on the application of programmable HPEC techniques. Several of the systems built over this period are illustrated in Figure 20-8, and serve as system examples of how commercial off-the-shelf (COTS) products may be productively employed in this application area.

20.4.1 RADAR SURVEILLANCE PROCESSOR

During the early 1990s, as part of the Radar Surveillance Technology Program, MIT Lincoln Laboratory undertook the development of a ground-based, rotating ultrahigh frequency (UHF) planar array radar with the goal of demonstrating the practicality of real-time digital adaptive beamforming for the suppression of jamming. The real-time processing consisted of baseband quadrature sampling and channel equalization on each of 14 parallel channels (each corresponding to a single row of antenna elements), adaptive beamforming, and then pulse compression and Doppler filtering on each of two adapted beams. The total required computation rate was 5.5 GOPS (giga, or billion, operations per second), the bulk of which was consumed by the digital filtering operations on each channel prior to beamforming. The processor was composed from two custom module types, a processing element (PE), and a systolic processing node (SPN) (Teitelbaum 1991). The PE implemented the front-end digital filtering using a COTS FIR filter application-specific integrated circuit (ASIC), the INMOS A100. The SPN was a programmable processor, based on the AT&T DSP-32C [one of the first COTS floating-point digital signal processors (DSPs)], and had multiple custom input/output (I/O) ports to support inter-SPN communication in a nearest-neighbor grid. It was programmed in C and assembler, and a custom parallel debugger was developed to support software development. Mapping of the adaptive processing algorithm to the SPN array was accomplished in a fine-grained, data-parallel manner.

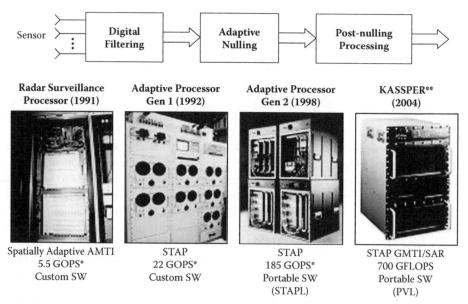

FIGURE 20-8   Adaptive radar signal processor development at MIT Lincoln Laboratory, 1991–present.

20.4.2 ADAPTIVE PROCESSOR (GENERATION 1)

Developed at approximately the same time as the Radar Surveillance Processor and using a similar architectural approach, the first-generation Adaptive Processor was designed for space-time adaptive processing. The total required computation rate was on the order of 22 GOPS, again dominated by the front-end digital filtering requirements. The processor comprised multiple custom board types, employing the INMOS A100 for baseband quadrature sampling, channel equalization, and pulse compression, and a programmable PE based on the TI TMS320C30 (another early floating-point DSP processor) for Doppler filtering and STAP processing. The STAP processing was mapped onto a 64-PE array in a coarse-grained data-parallel fashion, with each PE processing one or more Doppler bins.

20.4.3 ADAPTIVE PROCESSOR (GENERATION 2)

By the end of the decade, COTS offerings at the system level had begun to emerge and eliminated the need to design custom programmable processors based on COTS DSP chips for each application. MIT Lincoln Laboratory developed the second-generation Adaptive Processor based on a product from Mercury Computer Systems that featured Analog Devices' ADSP 20060 SHARC DSP processor in a scalable communications fabric Mercury Computer Systems called the RACE-WAY. The processor comprised 912 SHARC processors (73 GFLOPS) in four 9U-VME chassis.

The processing flow for the Generation 2 processor is shown in Figure 20-9. While the generation of baseband quadrature samples (digital I/Q data) was accomplished as before using custom hardware (developed around a new custom FIR filter integrated circuit developed at MIT Lincoln Laboratory), channel equalization and pulse compression were implemented in the programmable processor using fast convolution techniques. Doppler filtering is also implemented within the digital filtering subsystem (DFS). A two-step adaptive beamforming approach was used, first adaptively forming beams from element-level data while nulling jamming, and then employing a beamspace post-Doppler STAP technique to null clutter. Detection and parameter estimation were also implemented within the programmable processor.

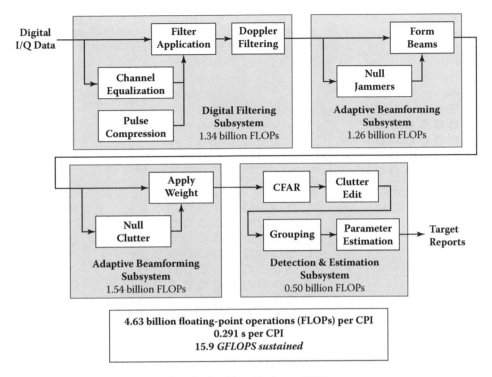

FIGURE 20-9 Adaptive Processor (Gen 2) algorithm (Arakawa 2001).

The mapping of the processing flow shown in Figure 20-9 onto the target hardware reflects a careful balance of real-time throughput, latency, and interprocessor communication. The mapping is complicated by the fact that each compute node comprises three SHARC processors that share a network interface and local DRAM. The digital filtering subsystem is mapped so that each input channel is processed by a single triple-SHARC compute node. For I/O bandwidth considerations, the 48 nodes are split between two chassis, DFS-1 and DFS-2, each of which processes 24 input channels. The two DFS chassis feed two adaptive beamforming subsystem (ABS)/detection and estimation subsystem (DET) chassis in round-robin fashion so that both DFS chassis write the first CPI to ABS/DET-1 and then write the second CPI to ABS/DET-2 while ABS/DET-1 processes CPI-1. Within each ABS/DET chassis, the ABS is mapped such that each ABS node processes one Doppler bin for all channels and all range gates. In order to support the different mappings in DFS and ABS, the data are globally corner-turned (rearranged) during the data transfer process. Each ABS/DET chassis has 96 compute nodes dedicated to ABS (one per Doppler) and 32 nodes dedicated to the DET (three Dopplers per compute node).

Software for the Generation 2 Adaptive Processor was written in C and featured the development of a portable library for STAP applications called the Space-Time Adaptive Processing Library (STAPL). The required total computation rate was on the order of 16 GFLOPS sustained, and the mapping shown in Figure 20-10 required a total of 304 triple-SHARC nodes—a total of 912 SHARC processors for a total peak computation rate of 73 GFLOPS. This amounted to an overall efficiency of about 22%, although the measured efficiency varied from about 50% for the FFT-intensive DFS to about 12% for the DET, which made heavy utilization of data-dependent operations.

20.4.4 KASSPER

Following the development of the Generation 2 Adaptive Processor, MIT Lincoln Laboratory embarked on the development of a new STAP processor for ground moving-target indication (GMTI) and synthetic aperture radar (SAR), sponsored by the Defense Advanced Research Proj-

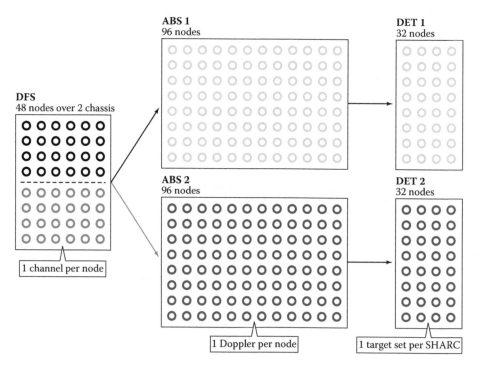

FIGURE 20-10   Adaptive Processor (Gen2) mapping (Arakawa 2001).

ects Agency (DARPA). The Knowledge-Aided Sensor Signal Processing and Expert Reasoning (KASSPER) project focused on the development of knowledge-aware STAP algorithms that could exploit *a priori* information to improve performance, development of a software infrastructure for knowledge storage and management, and the implementation of a real-time processor test bed (Schrader 2004).

Often in real-world applications of STAP, STAP performance can be degraded due to many factors related to STAP weight training, including heterogeneous clutter environments (e.g., land/water interfaces), clutter discretes (e.g., water towers, radio/TV antenna towers, etc.), and the presence of targets in the training sets. *A priori* information, such as digital terrain elevation database (DTED) and vector smart map (VMAP) data, could identify land/water interfaces and permit segmenting STAP training along regions of similar terrain. Similarly, known road locations could be omitted from training sets to avoid target nulling, and target reports in close proximity to known clutter discretes could be censored to reduce false alarms. Dynamic knowledge (i.e., knowledge that changes during the course of a mission) could be exploited as well. Ground targets in track, for example, could be censored from STAP training regions to prevent target nulling.

Like the Generation 2 Adaptive Processor, the KASSPER test bed was developed using COTS hardware from Mercury Computer Systems, although with the continuing improvement of technology over time, the KASSPER test bed was PowerPC-based and could support up to 180 PowerPC G4 processors running at 500 MHz for a peak computation rate of 720 GFLOPS. This represented a 10× improvement in computation rate, and a 5× reduction in the number of processors, although the efficiency achieved on the data-dependent knowledge-aware algorithms was somewhat poorer than what was demonstrated with the Generation 2 Adaptive Processor. Software for the KASSPER test bed was written in C++ and was based on the Parallel Vector Library (PVL), an object-oriented evolution of the STAPL library developed for the Generation 2 Adaptive Processor (Kepner and Lebak 2003). PVL facilitates the development of parallel signal processing applications in a portable manner using a standards-based approach.

20.5 SUMMARY

This chapter has discussed the application of high performance embedded computing to the radar domain, beginning with an introduction to radar operating principles and radar processing techniques, focusing on adaptive-antenna MTI and STAP systems. Also discussed were the mapping of these adaptive algorithms onto parallel processors and the evolution of several processors developed at MIT Lincoln Laboratory over the past 15 years. We have seen an evolution in the underlying hardware technology (migrating from custom multiprocessors based on early DSP chips to COTS multiprocessors based on the latest commodity microprocessors) and an evolution in the underlying software technology (from custom software to library-based portable software that facilitates migrating applications to newer and more capable hardware platforms) that has enabled an evolution in the sophistication and performance of radar signal processing algorithms.

REFERENCES

Arakawa, M. 2001. Private communication.

Encyclopædia Britannica. 2007. Online edition, s.v. "radar." Accessed 8 August 2007 at http://www.search. eb.com/eb/article-28737.

Kepner, J. and J. Lebak. 2003. Software technologies for high-performance parallel signal processing. *Lincoln Laboratory Journal* 14(2): 181–198.

Liu, Z., J.V. McCanny, G. Lightbody, and R. Walke. 2003. Generic SoC QR array processor for adaptive beamforming. *IEEE Transactions on Circuits and Systems—II: Analog and Digital Signal Processing* 50(4): 169–175.

McMahon, J.S. and K. Teitelbaum. 1996. Space-time adaptive processing on the Mesh Synchronous Processor. *Proceedings of the Tenth International Parallel Processing Symposium* 734.

McWhirter, J.G. and T.J. Shepherd. 1989. Systolic array processor for MVDR beamforming. *IEE Proceedings F: Radar and Signal Processing* 136(2): 75–80.

Nguyen, H., J. Haupt, M. Eskowitz, B. Bekirov, J. Scalera, T. Anderson, M. Vai, and K. Teitelbaum. 2005. High-performance FPGA-based QR decomposition. *Proceedings of the Ninth Annual High Performance Embedded Computing Workshop.* MIT Lincoln Laboratory, Lexington, Mass. Available online at http://www.ll.mit.edu/HPEC/agendas/proc05/agenda.html.

Oppenheim, A.V. and R.W. Schafer. 1975. *Digital Signal Processing.* Englewood Cliffs, N.J.: Prentice Hall.

Rabideau, D.J. and A.O. Steinhardt. 1999. Improved adaptive clutter cancellation through data-adaptive training. *IEEE Transactions on Aerospace and Electronic Systems* 35(3): 879.

Richmond, C.D. 2000. Performance of the adaptive sidelobe blanker detection algorithm in homogeneous environments. *IEEE Transactions on Signal Processing* 48(5): 1053.

Schrader, G.E. 2004. The Knowledge Aided Sensor Signal Processing and Expert Reasoning (KASSPER) real-time signal processing architecture. *Proceedings of the IEEE Radar Conference* 394–397.

Skolnik, M.I. 1990. *Radar Handbook, Second Edition.* New York: McGraw-Hill.

Stimson, G.W. 1998. *Introduction to Airborne Radar, Second Edition.* Mendham, N.J.: SciTech Publishing.

Teitelbaum, K. 1991. A flexible processor for a digital adaptive array radar. *Proceedings of the 1991 IEEE National RADAR Conference* 103–107.

Ward, J. 1994. *Space-Time Adaptive Processing for Airborne Radar,* MIT Lincoln Laboratory Technical Report 1015. Lexington, Mass.: MIT Lincoln Laboratory.

21 A Sonar Application

W. Robert Bernecky, Naval Undersea Warfare Center

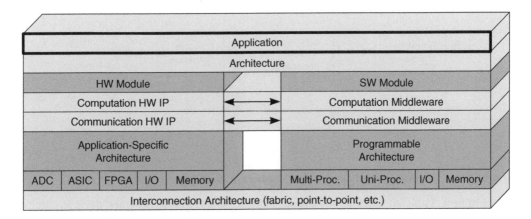

This chapter introduces the computational aspects pertaining to the design and implementation of a real-time sonar system. The chapter provides an example development implementation using state-of-the-art computational resources to meet design criteria.

21.1 INTRODUCTION

Sonar, an acronym for **so**und **n**avigation **a**nd **r**anging, is the engineering science of how to use underwater sound waves to detect, localize, and classify the sound emitted or echoed from an object (Nielsen 1991; Urick 1983).

A sonar system is an embodiment of sensor signal processing, designed at the abstract level to extract information from the real world. It has such characteristics as real-time performance, continuous input from a set of sensors, and a restricted footprint. It thus shares much with other embedded computing applications, including radar and automatic target recognition.

It is the purpose of this chapter to focus solely on the computational aspects pertaining to the design and implementation of a real-time sonar system, and take as given the sensor hardware and the corpus of existing sonar or signal processing algorithms. The chapter is not concerned with the development of new algorithms, nor in the derivation of existing algorithms. The task before us is to develop an implementation, using the state-of-the-art computational resources to meet our design criteria.

21.2 SONAR PROBLEM DESCRIPTION

In the following, we will explore one particular version of a sonar system known as a *passive sonar system*, which uses a linear arrangement of equally spaced hydrophones to listen for sound radiated by a target. Our goal is to design and build such a system to detect a quiet acoustic source amidst the loud noise sources of the ocean. In addition, the system will measure the arrival angle from this acoustic source to provide a degree of localization. The system must fit within prescribed physical limits (its footprint), consume a known quantity of electrical energy, and dissipate a corresponding

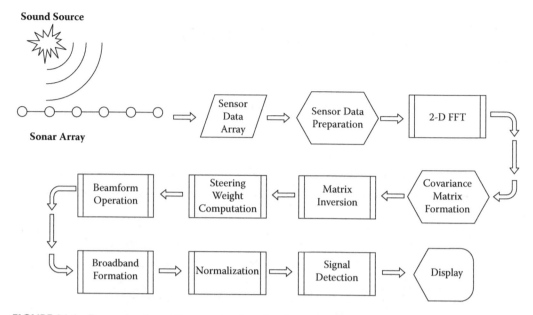

FIGURE 21-1   Processing thread for a passive, broadband sonar system.

degree of heat. These are the high-level requirements. For the system designer, these requirements are not flexible.

At a high level, the computation in a sonar system may be represented as a coarse-grained pipeline, from arrays to displays, as depicted in Figure 21-1. The main functions are sensor data collection, two-dimensional fast Fourier transform (FFT), covariance matrix formation, covariance matrix inversion, adaptive beamforming, broadband formation, normalization, detection, and display preparation and operator controls.

21.3   DESIGNING AN EMBEDDED SONAR SYSTEM

Once the sensor array has been chosen and the high-level performance goals specified, the sonar system engineer must design and implement the system. A proven methodology that we will use in this chapter is outlined in Figure 21-2. In overview, the engineer (1) defines the signal processing functions that compose the system, (2) implements a non-real-time prototype system, (3) identifies the computational requirements of each of the main functional components of the system, (4) analyzes the system for parallelism, (5) implements a real-time system, (6) instruments the real-time system to verify real-time performance, and (7) validates the output of the real-time system against the output of the prototype system.

In this methodology, it is crucial in step 2 to use a very-high-level signal processing development environment to ensure rapid and correct implementation of the system. If done carefully, the prototype that is developed can serve usefully as the "gold standard" baseline and the definition of correct output behavior.

21.3.1   THE SONAR PROCESSING THREAD

The main functional components of the system that we will design and implement are given in Figure 21-1. The sonar system engineer would, of course, have an understanding of the array gain (the ability to amplify a weak signal buried in noise), the bearing resolution (how accurately a direction of arrival can be measured), the details of exactly what frequency band will be processed, and so on. For our purposes, we may assume the role of implementer and rely on the sonar system engineer

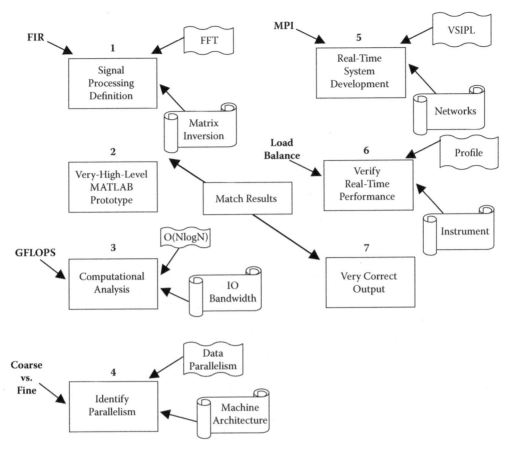

FIGURE 21-2   Development methodology for a complex, embedded sonar system.

for signal processing expertise. Together, the implementer and sonar system engineer will design a system that meets the requirement specifications.

21.3.2 PROTOTYPE DEVELOPMENT

The main functions of the system have been identified. The next step in developing the sonar application is to define and implement a baseline, non-real-time version of the system. The purpose of this prototype system is to process a known input data set and compute a known output, *without regard to executing in real time.* That is, the system engineer must build a system that operates correctly. However, it need not operate in real time, nor execute on specialized hardware. This initial system, along with an accompanying input dataset, becomes the baseline definition for the embedded, real-time system.

As a rule, the prototype system should be developed in a very-high-level programming language that enables quick development and easy verification of correct performance. For example, MATLAB (a language and development environment) is well suited for rapidly implementing a signal processing application (Kamen and Heck 1997).

It is possible (though unlikely, in the near term) that the prototype system, implemented in a high-level language, will provide the required real-time performance within the power and footprint constraints. (Ultimately, this happy result should occur with advancing computer science and improved parallelizing compilers.) Generally, though, the prototype system will only provide *correct* performance, but at a rate significantly less than real time. This takes us to step 3 in implementing the system.

21.3.3 Computational Requirements

The third step is to quantify the computational requirements for real-time behavior. To be specific, a particular function in the system will demand a certain number of multiplies and adds every second (floating-point operations per second, or FLOPS) to execute at a rate greater than the incoming data (the definition of real time, in this context). The system engineer should also evaluate the requirements for memory storage (main memory or disk), memory bandwidth, input/output (I/O) bandwidth, and communication bandwidth between the major functions.

The computational requirements for a given function can be determined either by theoretical analysis, e.g., an FFT algorithm is an $O(N \log(N))$ computation, or by instrumentation of the prototype system. However, there is often a variety of algorithms that one might choose in order to implement a particular function, and this choice of algorithm can profoundly affect the computation. This is what makes system design difficult. Understanding the trade-offs between different algorithmic approaches is crucial to success. Note that at this stage there may be feedback to the sonar system engineer as the implementer and system engineer work through the implications of various algorithm choices.

If there are significant effects that must be evaluated—for example, using fixed-point arithmetic instead of double-precision floating-point—the designers may wish to implement a lower-level prototype that incorporates these attributes. This latter prototype is meant to capture the effects that occur when modifications are made in the interest of reducing computation. The engineer will be interested in quantifying how the behavior of the resulting system deviates from the "gold standard."

The computational resources required to implement the system will determine, in a nonobvious way, the expected footprint, energy consumption, and cooling capacity of the physical system. If any of these constraints exceeds the design goals, the design work must iterate over the choice of algorithms, or (conceding defeat) relax the original goals.

Once the computational requirements have been defined, the system engineer is ready to tackle the difficult task of mapping the computation to a hardware suite that can support real-time execution. Though not necessarily so, this process generally involves *parallelizing* the major functions composing the system.

21.3.4 Parallelism

Parallelism is a technique that maps a sequence of computer instructions to a corresponding number of computers, so that the entire set of instructions can be executed simultaneously. As described, this is fine-grained parallelism, in which the computation has been decomposed into the smallest units, e.g., computer instructions. Aggregating instructions into functions and parallelizing at the level of functions is known as coarse-grained parallelism. Generally, fine-grained parallelism requires a specially designed computer architecture (note that some fine-grained parallelism is achieved in modern microprocessors, with deep instruction pipelines and multiple functional units). On the other hand, coarse-grained parallelism is a natural choice for applications that have large, difficult computations that can be mapped to powerful, general-purpose computers.

It is evident from Figure 21-1 that the sonar system can be implemented on a coarse-grained, pipeline architecture. Each of the major functions could execute on a dedicated processor, and the intermediate data sent through the parallel system via a simply connected communication scheme. However, let us suppose that the analysis from the previous step reveals that not all the functions require comparable FLOPS. In such a case, the developer might wish to consider load-balancing, a process that equalizes the computational workload assigned to each processor.

But to make significant speedups, the computation should be analyzed in terms of data parallelism. In a sensor processing system such as sonar or radar, the computation can often be decomposed into a set of computations applied to each data input. Thus, in principle, every datum can be sent to a separate processor, and the system can process the entire dataset in constant time. For large datasets (as provided by the constant stream of data from the sensors), the speedup in computation can be many orders of magnitude greater than for a simple coarse-grained pipeline. However, the requisite hardware must meet the footprint constraints. In the end, the specific implementation that the designer adopts will be driven by the constraints; the goal is to choose an approach that meets the constraints, but with the least cost and effort.

21.3.5   IMPLEMENTING THE REAL-TIME SYSTEM

Developing the real-time version of the embedded sonar application requires a deep understanding of the requirements, an accounting of where the FLOPS are being spent, the degree of parallelism required for real-time performance, and, finally, the proper choice of system architecture to implement the parallelized sonar processing thread.

System architecture is strongly determined by the desired FLOPS/unit volume, recurring and nonrecurring costs, and the details of the computation. At the simplest, the system could be a single general-purpose computer processing a small number of sensors. More realistically, the sonar community has gravitated to commodity processors, e.g., Intel microprocessors, interconnected with a commercial off-the-shelf (COTS) network. Higher-density computations have been realized using more specialized hardware such as field programmable gate arrays (FPGAs) or digital signal processors (DSPs). But beware: the system implementer must be cognizant of the costs associated with using these latter devices, as the software development costs are historically much greater than those of microprocessors.

In addition to the system architecture, the system software—i.e., operating system, communication network, I/O support, plus the development environment—plays a crucial role in implementing a real-time system. Software such as Message Passing Interface (MPI) (Gropp, Lusk, and Skjellum 1994) and standardized libraries such as the Vector, Signal, and Image Processing Library (VSIPL) [http://www.vsipl.org] should be adopted by the implementer to achieve efficient, portable code.

21.3.6   VERIFY REAL-TIME PERFORMANCE

Once the system, or each major component of the system, has been developed, the implementer must verify and validate its performance. Verifying that the system processes data in real time requires a real-time simulator and instrumented code that profiles the execution times of the system.

21.3.7   VERIFY CORRECT OUTPUT

The prototype baseline system serves the important role of defining the correct output for pre-defined input datasets. Once the system correctly generates outputs that match the baseline, sensor data (recorded from at-sea experiments, if available, or simulated, if need be) are used to exercise the system. These laboratory tests evaluate the system's robustness and mean time to failure. The final test of the system would be at sea, where carefully constructed scenarios demonstrate the correct operation of the complete sonar system.

21.4   AN EXAMPLE DEVELOPMENT

As noted in the introduction, sonar comprises a large variety of systems. Especially for the submarine community, there are a bewildering number of separate sonar systems on a single platform, with different performance metrics and functionality: low-frequency towed arrays, middle-frequency hull arrays, active (pinging) systems, high-frequency imaging systems, depth sensors, and communication systems. These same systems are also incorporated in unmanned undersea vehicles (UUVs), surface ships, and buoy systems. In every instance, the sonar design engineer had the task of implementing an embedded system that processed real-time data at rates as high as

Sonar Array Parameters

Number of hydrophones	100
Design frequency	1000 Hz
Design wavelength l	1.5 m
Array length	75 m
Sample rate	5000 Hz
Bits per sample	16

FIGURE 21-3 Array design parameters for a sonar example.

current technology would support. (As an aside, it is interesting to note that the sonar algorithms and functions have grown in complexity and sophistication to match the availability of ever-more computational resources.) At this time, it is easy to envision embedded systems that will consume teraFLOPS (trillions of FLOPS) of computation within the footprint of a small UUV device.

There is such a gamut of sensor systems and differing operational characteristics that no one system captures all the nuances of sonar system design and implementation. However, the passive towed array system that we will design touches on many of the most important components and clearly illustrates the methodology for implementing an embedded high performance sonar application.

21.4.1 System Attributes

We begin with a specification for the array of sensors that will be processed by the sonar system. These attributes strongly determine the computational requirements. Figure 21-3 summarizes the important features: the array length, the number of sensors, the design frequency, the rate at which the sensor data are sampled, and the number of bits per sensor sample from the analog-to-digital converters (ADCs).

As a rule of thumb, the number of beams (or *look directions*) that the system will form is a small multiple of the number of sensors. In this case, the system will compute 200 beams over a frequency range from DC to 1000 Hz. As another rule of thumb, the system will process data in blocks corresponding to 10 times the length of time it takes for sound to propagate down the array. From Figure 21-3, we have the array length as 75 m, so sound (at 1500 m/s) requires 50 ms to traverse the array. The block of time we will process should be about 0.5 s.

21.4.2 Sonar Processing Thread Computational Requirements

The computational requirements for the processing thread defined in Figure 21-1 can be estimated by counting the number of FLOPS required for each of the main functions. This initial analysis will identify the functions that dominate the computation.

21.4.3 Sensor Data Collection

At the indicated sample rate of 5000 Hz, each block of data from *one sensor* will be 2500 samples, or 5000 bytes. The dataset for one block of data from the entire array is 100 times greater, or 500 KB every 0.5 s. These data should be converted to single-precision floating point (but expect double-precision floating point to be the choice in the near future), increasing the data size to 1 MB, and the communication rate to 2 MB/s.

However, sensor data collection includes windowing techniques, multiplying each time sample by a Hamming window (Harris 1978), with a corresponding 50% overlap of data. This effectively doubles the communication rate to 4 MB/s, and incurs 2 MFLOPS (megaFLOPS, or one million FLOPS) of computation.

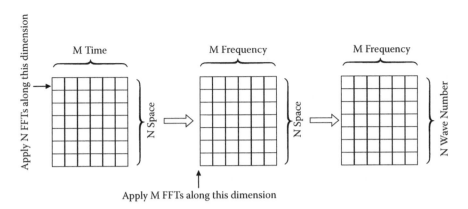

FIGURE 21-4   Two-dimensional fast Fourier transform of towed array data.

21.4.4   TWO-DIMENSIONAL FAST FOURIER TRANSFORM

Because the sonar system engineer realized that our system uses an equally spaced linear array of hydrophones and the array is assumed to be straight, the decision was made up front to use an FFT approach to translate the sensor data into the wave number/frequency domain (known as k-omega space). The important point is that the computation goes as $N\log_2 N$ when using an FFT as compared to a more general N^2 algorithm. For $N = 100$ (the number of sensors), the computation will be a factor of 14 less, due to this choice. This example clearly illustrates the effects of algorithm selection. The more general algorithm would compute the identical response, but would consume 14 times more computational resources. If the FFT approach requires one central processing unit (CPU), then the more general approach requires 14 CPUs.

The two-dimensional FFT can be implemented as 100 FFTs applied to the 2500 time samples in each of the 100 sensor data blocks, and then 2500 FFTs applied to the output of the temporal FFT (see Figure 21-4). This would transform the 100 sensors by 2500 time samples to 100 wave numbers by 2500 frequencies.

However, the system implementer may observe that a power of two number of samples might yield better efficiencies and suggest increasing the block size to 4096 or decreasing the block size to 2048. Either choice might simplify the implementation (perhaps through reusing previously developed software functions or taking advantage of an FFT library routine that does not support arbitrary sizes). The implementer and sonar system engineer must choose an FFT size and then accommodate it by zero padding (adding additional zeros to the data block) or by adding more data to the data block. The implications to the system of either strategy must be understood. Let us choose to reduce the FFT size to 2048, realizing that the original block size was chosen based on a rule of thumb. Experience informs us that this choice will be acceptable.

The implementer, being familiar with many signal processing algorithms (or, through diligence, having searched for various FFT algorithms), decides to use a variation of the generic FFT algorithm that efficiently computes the transform on purely real data (as opposed to complex data, which has an imaginary component). This choice effectively reduces the size of the FFT (for *computational purposes only*) by half, or about 1024 data samples.

The sonar system engineer is quick to point out to the implementer that the system will only process frequencies up to 1000 Hz, which is only a subset of the temporal FFT output, represented by the bins from 0 to about 420. Thus the spatial FFT need only be applied to 421 of the frequency bins.

The total computational cost of the two-dimensional FFT stage is approximately 8 (fixed factor) $\times$ 100 $\times$ 1024 $\times \log_2 (1024)$ + 8 $\times$ 421 $\times$ 100 $\times \log_2 (100)$ ~ 10.5 MFLOP per data block, or 10.5 MFLOP/0.482 s = 22 MFLOPS. Note that this estimate is well below the full-blown 100 $\times$ 2048 two-dimensional FFT that requires about 61 MFLOPS.

The implementer also estimates the memory requirements as 100×1024 single-precision complex numbers, or 800 KB per input buffer, and $421 \times 100 \times 8$ bytes, or 420 KB per output buffer. If the data are double-buffered, then the total memory storage—composed of input, working, and output data—would be 2*input (double buffers) + input + output + 2*output (again, double buffers), or 3*(800 KB + 420 KB) = 3.6 MB.

The data flow into and out of the two-dimensional FFT function is approximately 800 KB/0.482 s = 1.7 MB/s in, and less than 1 MB/s out.

The lesson to be extracted from this rendition of technical details is that an enormous amount of system engineering is devoted to minimizing the computational costs of the system. Implicitly, the designer or implementer is striving to reduce the footprint because an *embedded* system is almost always strongly constrained by rack space or energy consumption. Otherwise, a straightforward implementation would be the quickest and easiest path to an operational system.

21.4.5 COVARIANCE MATRIX FORMATION

The adaptive beamformer algorithm in the sonar processing thread requires an estimate of a covariance matrix, a mathematical representation of how signals arriving at the array interfere with one another. A covariance matrix is estimated by performing an outer product of one wave number vector (that is, at one frequency) to form a matrix. At each succeeding time epoch, a new wave number vector is added to the matrix to build up the required full-rank covariance estimate. However, a small subset of the full 100×100 covariance matrix will be required in this application. This is because, again due to our choice of working in the k-omega domain, the beamformer needs only a small number of adjacent wave numbers to generate its beams. In this case, five contiguous wave numbers will be used to form a beam, so a 5×5 covariance matrix may be extracted from a narrow diagonal band of the original matrix.

To ensure that a matrix is invertible, an $N \times N$ matrix should be formed from the average of $3N$ data updates. These required degrees of freedom will be achieved by averaging over both time and frequency. At each of ten time epochs (4.8 s of data), we add in the outer product for three adjacent frequency bins. This approach yields a matrix with 30 independent updates, twice the minimum 15 samples suggested for a 5×5 matrix.

The total computation requirement to estimate the covariance matrix is on the order of 12 MFLOP (double precision) per second. Note we have changed over to double-precision arithmetic for all the matrix algebra functions. The total working memory requirement is 6.7 MB. The I/O buffers, assuming double buffers, will aggregate to 20 MB.

21.4.6 COVARIANCE MATRIX INVERSION

There are about one hundred 5×5 matrices for each of 420 frequency bins, or 42,000 matrices that must be inverted every 5 s. Matrix inversion is $O(N^3)$, so the total computational cost is 8 (fixed factor) $\times 5^3 \times 42e3 = 42$ MFLOP (double precision) amortized over 4.82 s. This is about 9 MFLOP/s. The inverted matrices require 16.8 MB of storage.

21.4.7 ADAPTIVE BEAMFORMING

A beam is formed by summing the sensor outputs in such a manner that a signal at a given look direction is amplified relative to background noise and signals arriving from other directions. In the frequency domain, an $N \times 1$ steering vector is applied (via a dot product) to the $N \times 1$ sensor data vector to compute a beam response for one frequency bin. A different steering vector (for each frequency) is applied to the data, frequency by frequency, to generate a frequency-domain beam output. This output is often transformed via an inverse FFT to yield a digitized time series that can be fed into speakers for audio output and/or fed into a spectral estimation function for further analysis. In the k-omega domain, each wave number represents a look direction, so the steering vector is particularly simple; it selects the wave number bin of interest. A slightly more sophisticated steering vector interpolates to a look direction between two wave number bins, using a small number of bins centered about the look direction. We will use this latter method, with a 5×1 steering vector.

The discussion to this point has concerned a *conventional* steering vector, which is computed once at system startup and remains static, independent of the sensor data. For our system, there are 200 look directions and 420 frequency bins, thus requiring 84,000 steering vectors. Given that each vector is five double-precision complex numbers, the total memory required for the conventional steering vectors is 6.7 MB. However, there are techniques to reduce this to 400 KB (Bernecky 1998).

The adaptive beamforming (ABF) function of our system forms beams that exclude interference from other signals by using the input data, encoded in the covariance matrix, to compute an *adaptive* steering vector (Van Trees 2002). Here *adaptive* refers to the use of a steering vector that is adapted to the specific input data. The optimum adaptive steering vector $\mathbf{w}_f$ is given by

$$\mathbf{w}_f = \frac{\mathbf{R}_f^{-1}\mathbf{d}_f}{\mathbf{d}_f^{\mathrm{H}}\mathbf{R}_f^{-1}\mathbf{d}_f},$$

where $\mathbf{R}^{-1}$ is the inverse covariance matrix and $\mathbf{d}$ is the conventional steering vector, both defined for every frequency bin f. (The superscript H represents the complex conjugate transpose operator.)

The computational cost for generating the adaptive steering vectors is $O(N^2)$, or $8 \times 25 \times 84000$ FLOP = 16.8 MFLOP every 4.8 s. This translates to 3.5 MFLOPS. Storing the adaptive steering vectors requires 6.7 MB of memory. These steering vectors are used until a new set is regenerated in 5 s.

Finally, the ABF function must apply the adaptive steering vector to the input data at every epoch. This costs 84×10^3 complex vector dot products every 0.482 s, or 7 MFLOPS. At this stage, it is reasonable to convert down to single-precision floating-point output, especially if the system is not processing spectral data. Output size is 420 frequency bins $\times$ 200 beams $\times$ 8 bytes/single-precision complex = 0.6 MB.

The total cost for the ABF function, which encompasses the adaptive steering vector calculation and the application of the steering vector to compute the beams, is 10.5 MFLOPS, and 8 MB of memory (excluding I/O buffers).

21.4.8 BROADBAND FORMATION

The system can become very complex if we wish to process the individual frequencies. This so-called narrowband processing is composed of a multitude of analysis tools and approaches, including spectral estimation, harmonic analysis, and signal detection.

Our system will process broadband energy by summing the frequency components in a single beam. This enables the system to detect a signal and measure the signal's direction of arrival. Of course, this broadband approach precludes acoustic source discrimination based on its spectral content.

Generating the broadband output requires computing the power in each frequency bin and summing over all bins: 8 FLOP/datum $\times$ 420 frequency $\times$ 200 beams = 670 KFLOP every 0.482 s, or 1.4 MFLOPS. The memory requirements are very modest, 700 KB for the input data (assuming single-precision complex) and a mere 800 bytes for output.

The output of this function is generated every 0.482 seconds. This rate represents short time averaging (STA). Longer integration times allow lower signal-to-noise ratio (SNR) signals to be detected, so the system will also generate an intermediate time average (ITA) and a long time average (LTA) output by averaging successive STA outputs. For example, let ITA average five STA outputs to create an ITA with an output rate of 2.4 s. Similarly, let the LTA have an output rate of 9.6 s. Note that neither of these broadband outputs requires significant computation or memory.

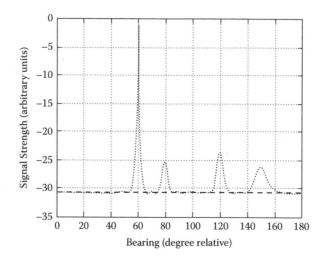

FIGURE 21-5 Example A-scan showing the broadband output of a sonar system as a function of bearing.

21.4.9 NORMALIZATION

The ABF output tends to preserve the strength of the signal source and represent noise as essentially a fixed-floor value. An "A-scan" of this output, depicted in Figure 21-5, clearly demarks the presence of a signal. However, the noise floor can vary with look direction, e.g., rain storms or shipping traffic, and it is useful to apply a normalization technique that removes this variation and thus emphasize signals. One simple algorithm subtracts the average of the beams in the immediate neighborhood from the center beam. For example, the average of the two beams to the right and to the left is subtracted from the center beam. Note that this process can be cast as a data-parallel operation. Indeed, it is embarrassingly parallel, as each beam can be normalized independently (though, of course, the scheme does require some local, nearest-neighbor communication). The computation cost for this process is only a few KFLOPS.

21.4.10 DETECTION

The system is required to automatically detect when a signal is in a beam. This is accomplished by comparing the output value of a beam (at each integration time, STA, ITA, and LTA) against a precomputed threshold that has been defined to achieve a given probability of false alarm (PFA). This comparison to a threshold requires less than 2 KFLOPS.

21.4.11 DISPLAY PREPARATION AND OPERATOR CONTROLS

The broadband data are typically displayed as a waterfall, with the newest data entering at the top of a screen and the oldest data scrolling off the bottom. The entire screen of data, time by bearing, enables the operator to visually identify the arrival directions of all the signals. In addition, any signals detected automatically by the system can be highlighted as a cue to the viewer.

There are many algorithms designed to optimize the gray levels used to paint the screen since a naive linear mapping of beam value to pixel intensity would quickly saturate. A typical scheme is to assign gray-scale values based on distance from the mean, as measured in standard deviations.

The man-machine interface (MMI) of the sonar system can take advantage of the commodity technologies developed by the commercial sector, which has invested heavily in display hardware and software. In fact, this area of system development may be considered its own specialty with its own development methodology, software tools, and hardware accelerators. As such, this function goes beyond high performance embedded computing.

TABLE 21-1

Summary of Computational Requirements

Function	MFLOPS	Memory (MB)	I/O Buffers (MB) In	I/O Buffers (MB) Out	I/O Rates (MB/s) In	I/O Rates (MB/s) Out
Sensor Data	2	1	1	3.2	1	1.6
2-D FFT	22	1.2	3.2	0.7	1.6	0.7
Cov Matrix	12	6.7	0.7	13.4	0.7	2.8
Matrix Inv	9	16.8	13.4	33.6	2.8	3.5
ABF	10.5	8	33.6	1.3	3.5	1.3

21.4.12 SUMMARY OF COMPUTATIONAL REQUIREMENTS

The computational requirements of the sonar processor being designed are summarized in Table 21-1.

21.4.13 PARALLELISM

The above analysis indicates that the embedded system, neglecting the display functions, requires about 60 MFLOPS sustained. This level of performance is easily achieved by a workstation executing a MATLAB program. Nevertheless, an embedded system constrained by footprint and power considerations may require a different target architecture that minimizes cost, power consumption, or footprint. In such instances, the design engineer must analyze the processing thread for parallelism and so take advantage of multiple CPUs (or DSPs) operating at slower clock speeds (lower cost, lower power consumption).

Section 21.4.12 summarizes the computational requirements. The four functions—two-dimensional FFT, covariance matrix estimation, covariance matrix inversion, and adaptive beamforming—are all of comparable FLOPS (within a factor of two), though the latter two functions require significantly more memory resources.

The first of the four functions, the FFT, is such a commodity algorithm that a deep analysis of it for parallelism is a redundant effort. There are excellent C or Fortran implementations available, e.g., "Fastest Fourier Transform in the West" (FFTW) (Frigo and Johnson 2005) and FFT chips (Baas 1999) optimized to efficiently compute standard-sized data blocks. However, it is possible the implementer may wish to develop a customized FPGA implementation of the FFT algorithm that more tightly integrates with the other components of the processing thread. In such instances, there are intellectual property (IP) cores that have been developed for specific FPGA chips, e.g., Altera, Xilinx, etc. In any event, current technology supports the two-dimensional FFT function operating in real time at less than one milliwatt.

The estimation of the covariance matrix is not a commodity algorithm, but it is embarrassingly parallel across frequencies. Each element of the band diagonal at a given frequency can be computed independently, so each estimate can proceed on a separate processor. However, because we are averaging over nearest-neighbor frequencies, the computations should take place in adjacent processors (as measured by the interconnect scheme) in order to preserve nearest-neighbor communication. A low-cost, multicore chip is a reasonable target for this function. In general, the implementer may use commercial compilers that automatically optimize the software to take advantage of the parallel processing features of these chips.

Examining the remaining functions, it is evident that a data-parallel approach, in which the problem is decomposed across frequency, will expose more than enough parallelism (say two orders of magnitude) to reduce the necessary CPU clock rates to power-conserving levels.

With careful engineering, this simple sonar processing thread can be reduced to a few milliwatts and occupy a footprint on the order of a few cubic inches.

On the other hand, the prototype system can be rapidly developed in MATLAB and tested on a desktop computer. We might also note that the desktop version of the system can often be tested in an at-sea configuration by relaxing the requirement for low-power execution. This desktop testing capability is useful because it enables the developer to design, test, and debug the signal processing aspects of the system before tackling the engineering complexities of implementing an embedded version of the system.

21.5   HARDWARE ARCHITECTURE

Let us examine the requirements for the system architecture that supports an embedded application, divorced from the specifics of a particular implementation. The motivating goal is to choose an architecture that achieves the FLOPS, power consumption, and footprint requirements by exploiting parallelism. The examples in this book all demonstrate that by departing from the von Neumann single-processor architecture, the implementer can gain FLOPS or reduce energy requirements, or both. Of course, these gains come at the expense of greater complexity, both in the hardware and software.

A parallel system that can deliver high performance has multiple processing units, local memory, and a communication network between the processing units. An excellent architecture is characterized by a proportionality between the three components of processing, memory, and communication. It is easy to see that a parallel computer is most similar to the familiar von Neumann machine in which any component processor can read data from anywhere in the system in constant time. In fact, shared-memory machines are designed to approach this level of performance, though it is very difficult to achieve success for more than a small number of processors. The critical point is that the communication bandwidth between processors should match the processing throughput, enabling the system (on average) to move data into a processor at the same rate it is being processed. Note that this communication rate is dependent on the algorithm being executed. An $O(N^3)$ algorithm requires much less I/O bandwidth than an $O(N \log N)$ process. A reasonable goal, without regard to a specific processing thread, is a system that matches MB/s to MFLOPS. If working with double precision, the communication bandwidth should be at least twice this rule of thumb. It is almost always a mistake to use a parallel system that has less communication bandwidth than this.

Similarly, the local memory size should dovetail to the processor's computational rate. Especially for coarse-grained applications, the input data flowing into the local memory will often represent a block of time corresponding to the system latency. At GFLOPS (giga, or one billion, FLOPS), or TFLOPS (tera, or trillion, FLOPS), this implies large local memories. If the system only supports small local memories, the developer must adopt a systolic algorithm, where data are read, processed, and sent along. Unfortunately, such algorithms are very specialized, are difficult or impossible for some applications, and often require specialized hardware support. To be safe, the hardware architecture should support large memories, as measured by the FLOP rating.

21.6   SOFTWARE CONSIDERATIONS

Current technology for high performance embedded computing relies on a message-passing paradigm that can be supported by distributed or shared-memory architectures. Even in circumstances in which the developer uses a very-high-level software programming language to implement a parallel application, the underlying software is most likely based on Message Passing Interface. Presumably, as the lessons of HPEC migrate into the mainstream of processors and systems, e.g., microprocessor chips with multiple functional units, the same software approach will also be adopted.

In the near term, the implementer of an HPEC application such as sonar will face the following software challenges: efficient implementations, portable software, and capture legacy software.

Enough discussion has taken place relative to point one. In summary, the implementer must work very hard at analyzing the application, choosing the appropriate hardware and architecture, and adopting software programming practices that provide efficient implementations based on message passing.

The implementer will also want to develop portable software that can be adapted to the next generation of hardware. Currently, using MPI and standardized signal and image processing libraries is the best approach. In addition, the implementer will wish to maintain a high-level prototype that captures the functionality of the embedded system. As technology makes obsolete specific hardware platforms, a port of an embedded application may entail mapping the prototype to the best existing hardware, using the latest software technology.

One of the greatest difficulties the implementer must surmount is the reuse of legacy software that operates correctly but does not easily port to new architectures. For example, in the sonar world, there are acoustic models written in FORTRAN that accurately predict how sound propagates through the ocean (Weinberg 1975). This software was developed well before the notion of parallelism was an important design attribute. Consequently, even with the aid of parallelizing compilers, such legacy code is not easily mapped to an embedded system. This is not a problem unique to sonar. To date, there are no completely satisfactory solutions.

21.7 EMBEDDED SONAR SYSTEMS OF THE FUTURE

This section is devoted to summarizing the future developments of complex sonar systems using HPEC technology (Ianniello 1998). The simple example application that we examined in the previous section highlighted the approach and the difficulties that arise when implementing a system. However, it falls far short of the systems now being developed. The curious reader may ask what drives the increase in the computational complexity of an embedded sonar system.

One important feature that impacts the computational requirements is sensor systems that involve many more hydrophones arrayed in an asymmetric manner. Loss of symmetry precludes the use of "nice" algorithms such as an FFT. Concurrently, the number of sensors in a sonar array has grown by a factor of ten. Together, asymmetry and more sensors lead to one or two orders of magnitude more computation.

A second major driver is the development of techniques that grid the ocean not simply in bearing (look direction) but also in range. This immediately makes the problem two-dimensional in nature. Still in their infancy are sonar techniques that grid the ocean in all three dimensions of range, bearing, and depth, thus extending the computation from, for example, 200 bearings to 200 bearings × 100 ranges × 100 depths. The problem grows from 200 beams to two million "beams." In addition, these advanced sonar system algorithms include the execution of ocean acoustic models, therefore adding additional computations to the system.

It is possible to tackle such difficult problems with current technology, but a complete system lies many years in the future.

REFERENCES

Baas, B.M. 1999. An approach to low-power, high-performance, fast Fourier transform processor design. Ph.D. dissertation. Stanford University, Stanford, Calif.

Bernecky, W.R. 1998. Range-focused k-w beam forming of a line array. NUWC-NPT Technical Memorandum 980118.

Frigo, M. and S.G. Johnson. 2005. The design and implementation of FFTW3. *Proceedings of the IEEE* 93(2): 216–231.

Gropp, W., E. Lusk, and A. Skjellum. 1994. *Using MPI.* Cambridge: MIT Press.

Harris, F.J. 1978. On the use of windows for harmonic analysis with the discrete Fourier transform. *Proceedings of the IEEE* 66(1): 51–83.

Ianniello, J.P. 1998. The past, present and future of underwater acoustic signal processing. *IEEE Signal Processing Magazine* 15(4): 27–40.

Kamen, E.W. and B.S. Heck. 1997. *Fundamentals of Signals and Systems Using MATLAB.* Upper Saddle River, N.J.: Prentice Hall, Inc.

Nielsen, R.O. 1991. *Sonar Signal Processing.* Boston: Artech House, Inc.

Urick, R.J. 1983. *Principles of Underwater Sound.* New York: McGraw-Hill.

Van Trees, H.L. 2002. *Optimum Array Processing.* Hoboken, N.J.: John Wiley & Sons.

Weinberg, H. 1975. Application of ray theory to acoustic propagation in horizontally stratified oceans. *Journal of the Acoustic Society of America* 58(1): 97–109.

22 Communications Applications

Joel I. Goodman and Thomas G. Macdonald,
MIT Lincoln Laboratory

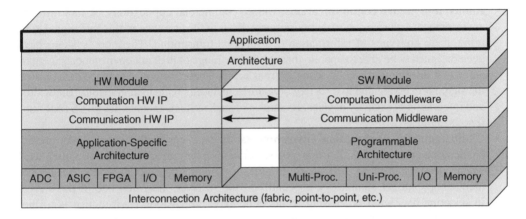

This chapter discusses typical challenges in military communications applications. It then provides an overview of essential transmitter and receiver functionalities in communications applications and their signal processing requirements.

22.1 INTRODUCTION

The objective of communications is to transfer information from one location to another. The field of communications is incredibly diverse as it spans a wide range of transport media such as spoken languages, the Internet, cell phones, etc. In order to focus the discussion, this chapter mainly pertains to wireless radio frequency (RF) communications for military applications. Establishing this focus area allows us to demonstrate how communications applications are implemented in embedded processing systems and how different communications applications place different demands on the processing systems. Many of these concepts are easily extended to commercial wireless systems. The relationship to wired and optical systems is similar but not as straightforward, and is beyond the scope of this chapter.

This chapter begins by discussing typical challenges in military communications applications. It then provides an overview of essential transmitter and receiver functionalities in communications applications and their signal processing requirements.

22.2 COMMUNICATIONS APPLICATION CHALLENGES

While in a simplistic view, communications is the process of moving information from one location to another, the type of information that is to be transferred will drive implementation choices. Most military communications applications that employ wireless RF transport are already employing embedded communications signal processing or are in the process of migrating to embedded communications signal processing. However, there is a wide range of signal processing implementations due to the variety of applications. A few example challenges in military communications applications are given below:

- **High data rate:** The military employs a number of different sensors (optical scanners, radars, etc.) that generate very high volumes of data. Examples of such applications include synthetic aperture radars (SAR) and hyperspectral imagers. These sensor outputs tend to contain large amounts of data and they generate these data at a continuous rate. The sheer amount and streaming nature of the data are a driving factor in the design of the embedded signal processing.

- **Robust and secure communications**: Unlike commercial applications, certain military environments experience hostile interference. Even in the presence of this intentional jamming, there are types of information that must be transmitted. Therefore, systems have been designed to be able to operate through a wide range of difficult environments to guarantee that information can be securely transferred. These systems can place demands on the signal processing due to a variety of features including wide instantaneous bandwidths and integration of security devices. Examples of such systems include the MILSTAR family of satellites.

- **Challenging RF propagation environments:** In one sense, all wireless links present challenging environments. For this reason, forward-error correcting codes are often used when operating in these difficult environments. The error-correcting codes, because of their computational complexity, especially at the receiver, are often one of the drivers in selecting the signal processing architecture. The propagation environment is further complicated by the mobility of the users in a wireless military network and the physical locations of these networks (e.g., urban or sea-based operations). The physical environment may motivate multiple processing chains to exploit multiple paths through the environment or motivate the processing need to control complicated antenna systems.

- **Time-critical data:** While this challenge also appears in commercial applications, military applications have apparent needs to move data with very low latency. For example, the information could be the location of an imminent threat so it has to be processed in real time. Typically, this type of information does not have large volume, but its unpredictable generation and latency demands drive up the embedded processing architecture complexity.

The diversity of military communications applications and their unique characteristics place different demands on communications signal processing requirements. Clearly, the burden of meeting different application needs does not fall solely on the processing elements (i.e., hardware), and, in fact, the design of firmware and software to meet these needs is an equally important challenge.

Much of the discussion above is on communications application features that would apply to the lower layer of the Open System Interconnection (OSI) seven-layer protocol stack [http://www.freesoft.org/CIE/Topics/15.htm]. Other aspects of networking features can also affect the selection and design of embedded processing elements.

In addition to the requirement of satisfying a specific communications application, there is a trend in both the military and commercial arenas to have a signal hardware platform (i.e., embedded signal processing system) that is able to be reconfigured to meet a variety of different communications applications. In the commercial world, this idea is manifested in cellular telephones that can connect to both time-division multiple-access (TDMA) and code-division multiple-access (CDMA) networks. Other efforts in the commercial sector are being headed up by the Software Defined Radio Forum [http://www.sdrforum.org]. For military applications, the Joint Tactical Radio System (JTRS) is one of the major driving forces behind reconfigurable hardware. Information on JTRS can be found at the website for the Program Office [http://jtrs.army.mil/]. The JTRS program is also promoting software that can be used on multiple processing solutions. In this vein, the software communications architecture (SCA) has been created. The motivation behind reconfigurable signal processing systems is that a single solution can be used for numerous applications, and there is an additional benefit in that it is much easier to insert new technology or evolving waveforms in these platforms.

As with any technology, there are a number of challenges and trade-offs in implementing communications applications. First, fundamental limits of individual components dictate overall system performance or processing architecture (see Chapter 8). As always, the cost of the processing system must be carefully weighed against the benefit of the solution. Finally, the physical implementation must take into consideration the form-factor and environmental requirements (i.e., size, weight, and power). These three attributes—performance, cost, and form factor—are applicable to almost any use of embedded processing. System restrictions or features that may be unique to wireless communications applications include a very limited amount of RF spectrum available and the large demands for this spectrum both by each potential user and by the large number of users clamoring for access to the scarce spectrum resource. This high demand results in specific solutions to allow for very efficient use of spectrum, but this elevated efficiency can also result in more complicated signal processing.

22.3 COMMUNICATIONS SIGNAL PROCESSING

Any communications system consists of both a transmitter and a receiver. The transmitter performs processing to prepare information for transmission over the channel medium (e.g., air, copper, or fiber). The receiver performs processing to extract the transmitted information from the received signal. The signal is subjected to channel effects that include, for example, energy lost along the propagation path, hostile interference, transmissions from other users in the system, and degradations due to mobility, such as Doppler shifts and spreads. The processing at both the transmitter and the receiver is intended to mitigate these channel effects. This section is broken into subsections that describe the different types of digital processing, often referred to as baseband processing, used in standard transmitters and receivers. The analog functions of a typical communications system (e.g., filtering, frequency upconversion, and amplification) are only given a cursory treatment in this chapter. The processing elements used in typical communications systems have evolved with technological advances. Early radios relied on analog processing. With the advent of more capable digital signal processing devices, more and more of the functions in the communications system are being converted to digital. In fact, some cutting-edge systems will directly sample the received signal and do all of the processing digitally. In general, the digital processors allow for much more capable systems than did their analog predecessors. An increasing trend is to use reconfigurable digital processors [such as field programmable gate arrays (FPGAs) or digital signal processors (DSPs)] to create systems that can be adapted to perform many functions (Tuttlebee 2003).

22.3.1 TRANSMITTER SIGNAL PROCESSING

A block diagram of a portion of an example baseband transmitter signal processing chain is illustrated in Figure 22-1. As mentioned earlier, the analog processing and the algorithms and functions pertaining to networking are not a focus of this chapter and, hence, are not included in Figure 22-1. Another critical area for any communications system, the control subsystem, is not shown in Figure 22-1. However, because the data path typically is the most computationally demanding and results in the most stressing processing requirements, the use of the functions in Figure 22-1 is the basis for discussion in the rest of this section.

For the system of Figure 22-1, the first digital processing element is a serial shift register with tapped delay lines and XOR logic to compute a cyclic redundancy check (CRC). The data input to the CRC processing is parsed in finite-length blocks and the CRC is appended to the tail-end of the data stream. The receiver uses the CRC to verify that the data received do not contain errors (Patapoutian, Ba-Zhong, and McEwen 2001). The CRC is not intended to identify where errors occur or

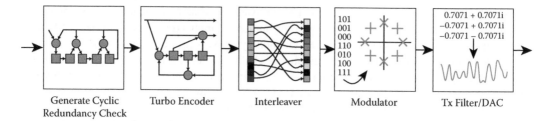

| Generate Cyclic Redundancy Check | Turbo Encoder | Interleaver | Modulator | Tx Filter/DAC |

FIGURE 22-1 Transmitter signal processing.

to correct them; rather, the CRC is a simple check over a block of data to provide some assurance at the output of the receiver that the data have not been corrupted. After completing the CRC processing in the transmitter, the digital stream is then sent to an encoder, adding redundancy and/or modifying the original information stream. This encoder is performing what is known as forward-error correction (FEC). That is, in anticipation of the hostile wireless communication channel, processing is being performed *a priori* to help find and correct errors.

There are many forms of coding such as linear block/cyclic codes, convolutional codes, low-density parity check (LDPC) codes, and turbo codes (Sklar and Harris 2004). These different types of codes each have strengths and weaknesses. Many of the most advanced codes, such as turbo codes, are very close to achieving the theoretical lower limit on energy required for error-free communication (Berrou and Glavieux 1996). Turbo codes are widely used today both in commercial systems (e.g., cell phones) and for military applications. However, because the more advanced FEC codes typically require more processing, they may not be good choices for very-high-rate systems or systems that are severely limited in processing power.

There are two common forms of turbo codes: parallel concatenated codes (PCC) and serial concatenated codes (SCC). A PCC encoder employs two recursive systematic convolutional (RSC) or block encoders separated by an interleaver as shown in Figure 22-2(a). An SCC has an outer code (e.g., Reed-Solomon) and inner code (e.g., RSC) separated by an interleaver as illustrated in Figure 22-2(b). As these examples illustrate, at the most fundamental level a turbo is simply a combination or concatenation of two simpler codes. The true power of turbo codes comes in the decoding of these codes, which will be described later. Separating the two constituent codes of a turbo code is an interleaver. An interleaver is a device that takes an input stream and outputs the data in a different, but deterministic, order. This function ensures that the data being input to each of the constituent encoders are highly uncorrelated. There are various forms of interleavers, such as block interleavers, random interleavers, and *s*-random interleavers. The choice of an interleaver impacts system performance; therefore, a search for a suitable interleaver is an important consideration in any communications system design (Yuan, Vucetic, and Feng 1999). The interleaving function can be memory intensive (storing the input sequence and reading it out in a different order) and can, therefore, be a driver in selecting processing elements both for memory sizing and interconnection considerations.

The forward-error correction coding processes introduce extra information or redundancy into the data stream. For a given type of code, the more redundancy that is added the better the performance will be. However, there is a penalty in adding redundancy in that the time and resources spent transmitting the redundant symbols are taken away from the amount of data being transmitted. This trade-off between having enough redundancy to guarantee high-quality transmission and not introducing redundancy to keep the data rate high is a key system design choice. In order to achieve the exact balance, certain redundancy bits are deleted from the transmitted stream; in communications, this process is known as puncturing. As an example, consider the PCC of Figure 22-2(a). For this PCC, each information bit generates two redundancy bits; thus, the overall rate of the code is said to be 1/3 (meaning one out of every three transmitted bits is a true information bit and the other two are redundant bits). A puncturing process could be performed at the output of this code by alternately puncturing (removing) every other redundant bit generated by the two

Data

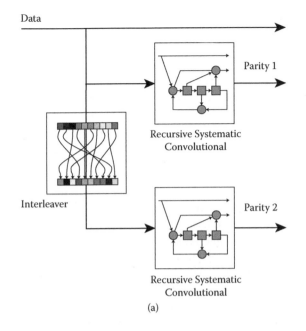

Recursive Systematic
Convolutional

Parity 1

Interleaver

Recursive Systematic
Convolutional

Parity 2

(a)

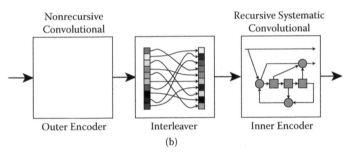

Nonrecursive
Convolutional

Recursive Systematic
Convolutional

Outer Encoder   Interleaver   Inner Encoder

(b)

FIGURE 22-2 Turbo encoder: (a) parallel concatenated coder; (b) serial concatenated coder.

constituent encoders. Some loss in decoding performance is incurred by puncturing, and this must be weighed against the attendant increase in data rate.

Referring back to Figure 22-1, one can see that the output of the encoder (which may possibly include puncturing) is input to another interleaver. Note that this is a separate interleaver from the one that may be used in the encoding process. This interleaver at the output of the encoder is used with almost all types of codes and is often called the channel interleaver. Its purpose is to provide a mechanism to break up bursts of errors at the receiver. These burst errors encountered during transmission can be from a variety of sources, including hostile interference or time correlated fades (Babich 2004). In order to break up these bursts of errors, the channel interleaver can span a very large number of data symbols and may require significant memory resources.

After interleaving, the next step in the transmission processing is modulation, which is the process of preparing the digital stream for transmission. Modulation uses the digital stream to alter the characteristics of the transmitted carrier waveform in amplitude and/or phase. There are many types of modulation techniques. Common choices for wireless communications include amplitude and phase modulation [e.g., quadrature amplitude modulation (QAM)], standard phase-only modulation only [e.g., phase shift keying (PSK)], or phase modulation with good spectral containment [e.g., continuous phase modulation (CPM)] (Proakis 2001). The different modulation choices offer different benefits to system performance. Some of the modulation formats are very efficient in their spectrum use (i.e., relatively high bits per second per unit of RF bandwidth); some of the modulation formats allow the receiver to operate with relatively poor estimates of the amplitude and phase; and some of the modulation formats are amenable to the effects of nonlinear power amplifiers. For example, QAM is very spectrally efficient but is sensitive to amplifier nonlinearities, while CPM is immune to the effects of the amplifier but is not as capable as QAM in delivering as high a throughput per unit energy. It is beyond the scope of this chapter to provide an in-depth listing of different modulation formats and their pros and cons. However, from a processing standpoint, all modulations parse the output of the channel interleaver into portions that are used to select one symbol from an amplitude/phase modulation constellation for transmission. For example, in 16-ary QAM, 4-bit chunks from the output of the interleaver are used to select one of 16 complex values (amplitude and phase) from the QAM constellation (Proakis 2001). All modulations perform similar mappings from bits to constellation symbols and so this processing must occur at the data rate. Certain modulations also require some memory in the process, although this is often implemented with delay lines rather than actual memory elements.

The last stage in Figure 22-1 is filtering prior to conversion to the analog domain. The modulated symbols are processed by a finite impulse response (FIR) filter to simultaneously limit out-of-band emission and shape the analog pulse. Limiting the out-of-band emissions is important for complying with regulatory agency policies and for allowing as many users as possible to access the frequency band allocated to the system. Pulse shaping is used in part to help proactively counteract some of the negative effects encountered during transmission, such as intersymbol interference and nonlinearities in the amplification process. The most commonly employed pulse-shaping filter is the (root-) raised cosine filter (Proakis 2001).

Once the filtered digital symbols are converted to an analog waveform, there are generally one or two stages of upconversion and then amplification before the modulated symbols are transmitted. When the signal is transmitted, it can take multiple pathways to reach the receiver. This is particularly relevant in wireless terrestrial communications in which many objects like buildings or trees can reflect additional signals toward the receiver (in addition to the line-of-sight path). Multiple pathways, or multipath, enable the signals to combine additively at the receiver. On pathways where the delay is less than a symbol interval, the signals add at the receiver constructively (multipath enhancement) or destructively (multipath fading). A path delay that is longer than a symbol interval causes intersymbol interference. It should be noted that as data rates increase, the likelihood of intersymbol interference increases since the symbol durations are getting shorter. The aggregate effects of complicated propagation environments and the transmission medium itself are often lumped together and called fading. Fading is a received signal that experiences fluctuations that were not part of the transmitted signal and, therefore, are often detrimental to performance. The fading is often correlated in time and/or frequency. Traditionally, fading has been mitigated by interleaving and/or by selecting or combining the output of multiple antennas at the receiver, while equalization at the receiver is used to combat intersymbol interference. Clearly, both of these techniques require more processing at the receiver and will be discussed in the next section. However, two relatively new approaches to mitigate intersymbol interference and leverage multipath fading can be applied on the transmission; they are orthogonal frequency division multiplexing (OFDM) and space-time coding, respectively.

OFDM mitigates the effect of multipath on the received signal by transmitting many slow-rate symbols in parallel (rather than serially transmitting high-rate symbols). Transmitter OFDM processing involves taking an inverse fast Fourier transform (IFFT) and adding a short prefix, which is typically the last samples from the IFFT. An FFT is taken at the receiver to recover the samples and convert the parallel streams back into a serial stream. With the addition of a cyclic prefix, intersymbol interference can be completely eliminated, given that the prefix is longer than the multipath delay spread (Morrison, Cimini, and Wilson 2001). OFDM is effective at mitigating multipath interference, but additional computational complexity is introduced at both the transmitter and receiver.

The transmitter can also explicitly exploit multipath by transmitting different streams of symbols from multiple antennas (where each antenna leads to a different path to the receiver). Space-

TABLE 22-1

Transmitter Signal Processing Operations

Function	Representation	Parameterized Operations
CRC	Bits	1 op/tap × 1 tap/bit × x bits/s × N_{crc} taps = $x \times N_{crc}$ ops
Encoder	Bits	1 op/tap × 1 tap/bit × x bits/s × N_{enc} taps = $x \times N_{enc}$ ops
Interleave	Bits	2 op/bit × x_{enc} bits/s = 2 x_{enc} ops
Modulate	Complex Words	2 op/M bits × x_{enc} bits/s = 2y ops
OFDM	Complex Words	$N \times \log_2 N$ op/N words × y words/s = $y \times \log_2 N$ ops
Pulse Shaping	Complex Words	2 op/tap × 1 tap/word × y words/s × N_{fir} taps = $2y \times N_{fir}$ ops
STC	All	$N_{antennas}$ × ops above (worst case)

Key: x: bit rate; x_{enc}: encoded bit rate; y: symbol rate; M: modulation efficiency; N_{crc}: number of taps in the CRC logic; N_{enc}: number of taps in encoder (feedforward/feedback); N_{fir}: number of taps in FIR filter; op: operations; and ops: operations per second.

time coding (STC) is a forward-error correction process in which different streams of data and redundant symbols are transmitted simultaneously (Tarokh, Seshadri, and Calderbank 1998). There are various forms of STC, such as block codes, (turbo) trellis codes, and layered codes. In multipath-rich environments, it has been shown that it is possible to achieve spectral efficiencies far in excess of any single transmitted signal system by employing numerous antennas at both the transmitter and receiver (Foschini 1996). The downside of such an approach is the extra hardware required to create multiple simultaneous transmitted signals and the associated processing.

22.3.2 TRANSMITTER PROCESSING REQUIREMENTS

Transmitter processing requirements are principally dominated by the data rate at which the communication system operates. As shown in Table 22-1, the processing requirements of the CRC, encoder, and pulse-shaping FIR filter are a function of both the data rate and the number of taps in their respective designs, while the interleaver and modulator are strictly a function of the data rate. The FFT-dominated OFDM processing requirements are, as expected, a base-2 logarithmic function of the block size, and the worst-case STC processing requirements are approximately equal to the aggregate of the aforementioned operations times the number of transmit antennas.

With an input data rate of 1 million bits per second (Mbps), a combined coding rate and modulation efficiency of 4 bits/symbol, a code rate of 1/2, a 12-tap CRC, a 14-tap encoder, and a 64-tap FIR filter, the aggregate processing requirements sans OFDM and STC are approximately on the order 100 MOPS (million operations per second). Transmitter operations scale linearly with data rate; in the example above, a 100 Mbps input data rate would correspond to a processing requirement of approximately 10 GOPS (giga-operations per second). Because processing at the transmitter is not iterative and can be easily pipelined, transmitter signal processing operations can be broken up among many general-purpose processing units (GPUs). In general, transmitter signal processing with data rates on the order of 10's of Mbps can be hosted on GPUs, while data rates in excess of 100 Mbps would require one or more FPGAs.

22.3.3 RECEIVER SIGNAL PROCESSING

A block diagram of an example receiver signal processing chain is illustrated in Figure 22-3. As with the previous discussion of the signal processing in the transmitter, the system of Figure 22-3 illustrates the critical processing components along the lower-layer processing of data, but does not illustrate analog processing, control subsystems, or processing associated with networking. In

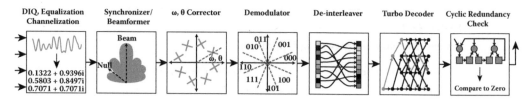

FIGURE 22-3   Receiver signal processing.

the first step of the receiver processing, which is represented in the left-hand side of Figure 22-3, digitized samples from analog-to-digital converters (ADCs) are channelized to produce in-phase (I) and quadrature (Q) samples. Channelization is the process by which a digitized wideband signal is filtered and converted to baseband to extract the signal in the band of interest. Note that although it is possible to employ two separate ADCs to generate I and Q data, many communications receivers jointly produce I and Q samples to avoid the deleterious effects of amplitude and phase differences that may arise between the I and Q samples prior to digitization.

To produce baseband I and Q data, the real-valued output of a single ADC is multiplied by a *sine* and *cosine* sequence with digital downconversion (i.e., low-pass FIR filtering). Note that digital downconversion is unnecessary if the signal consumes the entire band of interest. There may be multiple receiver chains (i.e., channels) and, hence, multiple digitized streams from multiple antennas. Multiple receivers enable diversity selection or diversity combining to mitigate multipath fading and adaptive beamforming (ABF) to spatially suppress interference. The processing burden on the receiver increases as the number of receiver chains increases.

A control system function that is critical for the receiver to perform is to synchronize the timing (and frequency) of its digital processing to the received signal. There are many ways this synchronization can occur from a separate control channel to special signals embedded in the data stream. As an illustrative example, consider a synchronization process that is intended for use in a packet-based system. A preamble is affixed to the beginning of the information bearing part of the packet and is used for synchronization and channel estimation. For multichannel synchronization, the process by which a receiver detects the onset of a transmitted packet and adjusts the sampling phase to maximize decoding performance is nearly identical to single-channel synchronization. In both the single-channel and multichannel cases, the preamble template is cross-correlated against the received samples to generate a peak; when above a given threshold, the peak indicates a successful detection and the start of a packet transmission. Both diversity selection and diversity combining are used to conduct synchronization on all channels. When a valid peak is detected, diversity selection chooses the stream with the highest signal-to-noise ratio for post-processing. In diversity combining, a weighted sum of the multiple streams is used in post-processing. Adaptive beamforming is a form of diversity combining that includes the capability of placing a null in the direction of an interfering source. The critical processing elements of this stage are conducted at the rate of the channel signaling and may involve parallel processing of numerous potential timing offsets. There is also the burden of coordinating with the control system and potentially combining or selecting a single signal from the multiple receive chains.

Because synchronization may need to be conducted in the presence of interference, combined adaptive beamforming and synchronization can be conducted to mitigate interference during the synchronization process. One approach is to first form a spatial correlation matrix from the received data symbols and then apply the root inverse to the received data to "whiten" the signal. Whitening the signal ensures that the power levels of the interference, noise, and signal of interest are brought to approximately the same level so that the integration gain during the correlation process enables robust synchronization. The next step in ABF is to generate a weight vector by estimating the spatial channel (i.e., the steering vector) via least-squares estimation (Paulraj and Papadias 1997), and then multiplying by the normalized inverse of the correlation matrix (Ward, Cox, and Kogon 2003).

With the possibility of 50 dB or more of interference suppression, ABF is an attractive technology for communication systems operating in an interference-laden environment. ABF can be combined with spread-spectrum techniques to significantly enhance interference suppression. Note that temporal channel estimation is conducted in an analogous manner to steering vector estimation with the exception that temporal samples rather than spatial samples are used.

Depending on the system requirements, a design having time synchronization may not be sufficient for acceptable performance. Further synchronization to accurately capture frequency and phase information may be needed (note that this is, in fact, just time synchronization at a much greater resolution) and may occur before the beamforming described above. A preamble can also be used as an example of one technique for frequency and phase offset estimation. Such a preamble consists of two or more repeated segments. The angle of the sum of the conjugate dot product across all pairs of preamble segments yields an estimate of the frequency offset. In a similar manner, the angle of the dot product of the preamble template with the received preamble yields an estimate of the phase offset. Frequency offset correction is conducted by applying a counter-rotating phase (rotating complex exponential) to the data prior to conducting phase-offset estimation and subsequent correction.

At the receiver, intersymbol interference can be suppressed by equalization. Adaptive equalization is either supervised or blind (Haykin 1996), and is implemented using a P-tap FIR filter in which P is chosen so that the aggregate of P-symbol periods is greater than or equal to the delay spread. In a multichannel receiver, temporal adaptive equalization can be combined with spatial adaptive beamforming to implement space-time adaptive processing (STAP) (Paulraj and Papadias 1997). STAP adaptively selects temporal and/or spatial degrees of freedom to mitigate interference or intersymbol, but this comes at the cost of increased computational complexity. The space-time correlation matrix now has P-times more rows and columns, and the steering vector is P-times longer than the ABF correlation matrix and steering vector, respectively. Mitigating intersymbol interference (ISI) with OFDM processing at the receiver involves taking the FFT of the received data sequence as discussed in the transmitter signal processing section above.

The synchronized, frequency/phase, offset-corrected, and equalized data stream is sent to the demodulator. In some cases, demodulation and decoding are a combined step, for example, in (turbo) trellis-coded modulation (TTCM) (Robertson and Worz 1998). Coherent demodulation generally involves the division of the complex amplitude and phase space into regions corresponding to the bits that were used to select the transmitted symbol (Wei and Mendel 2000). An example of the coherent demodulation of an 8-ary PSK symbol is illustrated in Figure 22-4. In Figure 22-4, the probabilities that the symbols (denoted by an x) are selected at the transmitter with the binary digits

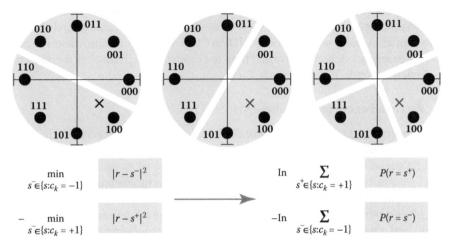

FIGURE 22-4 (Color figure follows p. 278.) 8-ary PSK maximum-likelihood demodulation.

1 or 0 are calculated using an L2-norm (Euclidean) metric for the most significant bit (left-most region), penultimate most significant bit (center region), and least significant bit (right-most region). In the case of 8-ary PSK, the log of the probability that the bit is a 1 is subtracted from the log of the probability the bit is a 0 so that for every received symbol three soft decisions (log-likelihood ratios) are generated. This fully generalizes to the M-ary modulation case in which $\log_2(M)$ soft decisions are generated for every symbol received. The soft decisions are next de-interleaved and presented to the decoder.

The decoder is responsible for reversing the operations of the encoder and correcting errors that may have occurred during transmission. In the case of either an SCC or PCC turbo decoder, this involves each constituent *a posteriori* probability (APP) decoder generating log-likelihood ratios (soft decisions) for every bit transmitted (Mansour and Shanbhag 2003). These soft decisions are then passed to the second constituent APP decoder (separated by an interleaver) after any prior likelihoods it may have received from the second APP have been subtracted. The second APP repeats the process of the first APP and so on until the decisions of the two APPs are highly correlated. It has been shown in the literature that the number of iterations needed is typically eight before any further iterations yield negligible improvement in decoding performance (Mansour and Shanbhag 2003). Unlike the demodulator, which generates soft decisions on individual symbols received in isolation, the decoder generates soft decisions on symbols based on not only the received symbol in question but also on the likelihood of all the prior and subsequent symbols. An efficient decoding algorithm known as BCJR (named after its inventors) is used to compute the soft decisions in the APP decoder (Bahl, Cocke, Jelinek, and Raviv 1974). APP decoding is the most computationally demanding procedure in any communication system. Computational complexity is a function of the number of states in the decoder and whether decoding is binary or multidimensional, as in the case of TTCM. APP decoding using the BCJR algorithm involves generating the probabilities of transitioning to every state from every previous state, as well as forward and backward metrics linking all the states. Log-likelihoods are formed from link-state probabilities and the forward and backward metrics (Bahl et al. 1974).

As mentioned previously, there are various forms of STC. One of the more computationally demanding forms of STC is a type of spatial multiplexing known as [horizontal, vertical, threaded] layered space-time (LST) coding (Foschini 1996). In LST, M separately encoded and modulated data streams are transmitted by M antennas. Digitized samples from N antennas at the receiver are used to recover the transmitted symbols. Because each of the transmitters is effectively an interferer, interference cancellation at the receiver is needed. One approach known as parallel interference cancellation (PIC) uses a bank of M decoders to recover the information transmitted (Sellathurai and Haykin 2003). PIC works by subtracting all of the previous estimates of the transmitted symbols from the current input with the exception of the input data stream currently being decoded. PIC is very similar to multiuser detection (MUD) in CDMA systems in which detected data streams from other users are successively subtracted from the received signal to isolate the data stream from the user of interest (Verdu 1998).

22.3.4 RECEIVER PROCESSING REQUIREMENTS

The receiver operations as tabulated in Table 22-2 are logically broken down into three categories: single-channel operation, multiple-channel operations with ABF, and multiple-channel operations with PIC-LST decoding. In both single-channel and multichannel modes, throughput requirements are driven by the turbo decoder; we will demonstrate this next by example.

With an input data rate of 500K symbols/s, for a combined coding rate and modulation efficiency of 4 bits/symbol, a code rate of 1/2, and a 32-tap channelizing filter, a 34-tap equalizer, a 12-tap CRC, and a 64-sample (tap) preamble, the aggregate processing throughput requirement is roughly 5 GOPS, of which 4 GOPS are consumed by the turbo decoder. If a special-purpose processor is used to host the turbo decoder, it may be possible to host the remaining operations on multiple

TABLE 22-2

Receiver Signal Processing Operations

Function	Representation	Parameterized Operations
DIQ	Real Words	1 op/word $\times$ 2y words/s = 2y ops
Channelization	Complex Words	8 op/tap $\times$ 1 tap/word $\times$ y words/s $\times N_{dwn} = 8y \times N_{dwn}$ ops
Single-Sync.	Complex Words	8 op/tap $\times$1 tap/word $\times$ y words/s $\times N_{pre} = 8y \times N_{pre}$ ops
Multi-Sync.	Complex Words	$((8N_{spc}N^2_{pre}/N_{pck}) + (8N^2_{spc} + 8N_{pre})) \times y$ ops
Equalization	Complex Words	8 op/tap $\times$ 1 tap/word $\times$ y words/s $\times N_{eql} = 8y \times N_{eql}$ ops
Channel Est.	Complex Words	8 op/tap $\times$ 1 tap/word $\times$ y words/s $\times N_{pre} = 8y \times N_{pre}$ ops
ABF	Complex Words	$(((8N_{spc}N^2_{pre} + 8N_{pre})/N_{pck}) + 8N_{spc}) \times y$ ops
Φ/Freq. Offset	Complex Words	8 op/tap $\times$ 1 tap/word $\times N_{pre} \times y$ words/s $= 8y \times N_{pre}$ ops
OFDM	Complex Words	$N \times \log_2 N$ op/N word $\times$ y word/s $= y \times \log_2 N$ ops
Demodulation	Complex Words	$8 \times M \times 2^M$ op/word $\times$ y word/s $= 8y \times M \times 2^M$ ops
Deinterleave	Real Words	2 op/bit $\times x_{enc}$ bits/s $= 2x_{enc}$ ops
Turbo Decode	Real Words	$2^{2(K+1)}$ op/word $\times N_{ite} \times x_{enc} = x_{enc} \times N_{ite} \times 2^{2(K+1)}$ ops
PIC (LST)	All	$N_{spc} \times N_{lst} \times$ demod-through-decode ops
CRC	Bits	1 op/tap $\times$ 1 tap/bit $\times x$ bits/s $\times N_{crc}$ taps $= x \times N_{crc}$ ops

Key: x: bit rate; x_{enc}: encoded bit rate; y: symbol rate; M: modulation efficiency; N_{crc}, N_{dwn}, N_{pre}, N_{spc}, N_{eql}, and N_{ces}: numbers of taps in CRC, channelizing filter, preamble samples, spatial channels, equalizer taps, respectively; N_{ite}: number of turbo decoding iterations; N_{pck}, and N_{lst}: number of samples in the packet and iterations of the parallel interference cancellation, respectively; op and ops: operations and operations per second, respectively.

GPUs. For the single-channel case, any symbol rate greater than 50K symbol/s and less than 25M symbols/s requires at least one FPGA as host processor, and anything greater than 25M symbols/s requires at least one application-specific integrated circuit (ASIC) as host processor.

In the multichannel case in which ABF is employed and it is assumed that the correlation matrix is only formed once per packet for both whitening (synchronization) and for ABF, with eight receive antennas, a 64-sample preamble, a 4096-sample packet, and an input data rate of 500K samples/s, the total number of operations for multichannel synchronization and ABF operations is approximately 600 MOPS, which is dominated by whitening in synchronization (512 MOPS). Processing requirements increase geometrically with increasing numbers of receive antennas and preamble samples.

Finally, PIC-LST is the most computationally daunting receiver architectures described above. Parallel banks of turbo decoders are iteratively invoked in the process of canceling interference. Using the single-channel parameters from the example above and with eight receive antennas and four iterations of the PIC, the computational requirements are roughly 32 times higher than those of the single channel alone, or roughly 160 GOPS. A 5M symbol/s input symbol rate would have a processing requirement of over 1 TOPS (tera-operations per second)!

22.4 SUMMARY

The previous sections have described examples of typical functions in communications. When possible, the impacts of these functions on the design and selection of embedded processing elements have been highlighted in an attempt to provide an introduction to the topic. However, the treatment in this chapter is cursory at best and does not even begin to highlight the complexities of a true system design. Despite the fact that only a small fraction of the possible choices for each function are listed and that only digital functions are covered (networking, analog, and control functions are not), it is hoped that a general theme of a generic communications system is conveyed.

There are almost as many ways to build a communications system as there are communications systems. As always, processing elements designed specifically for a task (e.g., ASICs) are often a choice, particularly for systems with very restrictive form factors, large production runs, or the benefit of only ever performing one function. An example of such an application would be commercial cell phones. Military applications certainly share the challenge of restrictive form factors with their commercial counterparts, but the military has additional constraints that motivate the use of reconfigurable hardware. Part of this trend is due to the desire to have one radio platform interconnect with many systems and for this platform to be updated with new functionality in the field. There are military applications that can run on state-of-the-art general-purpose processors, but at the time of the writing of this chapter many more computationally intensive military communications applications require more capable devices (e.g., FPGAs). The computational burden for these applications comes not only from high data rates, but from military-specific functionality such as resistance to hostile jamming and encryption.

REFERENCES

Babich, F. 2004. On the performance of efficient coding techniques over fading channels. *IEEE Transactions on Wireless Communications* 3(1): 290–299.

Bahl, L., J. Cocke, F. Jelinek, and J. Raviv. 1974. Optimal decoding of linear codes for minimizing symbol error rate. *IEEE Transactions on Information Theory* 20: 284–287.

Berrou, C. and A. Glavieux. 1996. Near optimum error correcting coding and decoding: turbo codes. *IEEE Transactions on Communications* 44(10): 1261–1271.

Foschini, G. 1996. Layered space-time architecture for wireless communication in a fading environment when using multiple antennas. *Bell Labs Technical Journal* 1(2): 41–59.

Haykin, S. 1996. *Adaptive Filter Theory*, 3rd edition. Upper Saddle River, N.J.: Prentice Hall.

Jinhong Yuan, B. Vucetic, and Wen Feng. 1999. Combined turbo codes and interleaver design. *IEEE Transactions on Communications* 47(4): 484–487.

Mansour, M.M. and N.R. Shanbhag. 2003. VLSI architectures for SISO-APP decoders. *IEEE Transactions on VLSI* 11(4): 627–650.

Morrison, R., L.J. Cimini, Jr., and S.K. Wilson. 2001. On the use of a cyclic extension in OFDM systems. *Proc. of the IEEE 54th Vehicular Technology Conference* 2: 664–668.

Patapoutian, A., S. Ba-Zhong, and P.A. McEwen. 2001. Event error control codes and their application. *IEEE Transactions on Information Theory* 47(6): 2595–2603.

Paulraj, A.J. and C.B. Papadias. 1997. Space-time processing for wireless communications. *IEEE Signal Processing Magazine* 14(6): 49–84.

Proakis, J. 2001. *Digital Communications,* 4th edition. New York: McGraw-Hill.

Robertson, P. and T. Worz. 1998. Bandwidth-efficient turbo trellis-coded modulation using punctured component. *IEEE Journal on Selected Areas in Communications* 16(2): 206–218.

Sellathurai, M. and S. Haykin. 2003. Turbo-BLAST: performance evaluation in correlated Rayleigh-fading. *IEEE Journal on Selected Areas in Communications* 21(3): 340–349.

Sklar, B. and F.J. Harris. 2004. The ABCs of linear block codes. *IEEE Signal Processing Magazine* 21(4): 14–35.

Tarokh, V., N. Seshadri, and A.R. Calderbank. 1998. Space-time codes for high data rate wireless communication: performance criterion and code construction. *IEEE Transactions on Information Theory* 45(2): 744–765.

Tuttlebee, W.H.W. 2003. Advances in software-defined radio. *Electronics Systems and Software* 1(1): 26–31.

Verdu, S. 1998. *Multiuser Detection.* New York: Cambridge University Press.

Ward, J., H. Cox, and S.M. Kogon. 2003. A comparison of robust adaptive beamforming algorithms. *Proceedings of the 37th Asilomar Conference on Signals, Systems and Computers* 2: 1340–1344.

Wei, W. and J.M. Mendel. 2000. Maximum-likelihood classification for digital amplitude-phase modulations. *IEEE Transactions on Communications* 48(2): 189–193.

23 Development of a Real-Time Electro-Optical Reconnaissance System

Robert A. Coury, MIT Lincoln Laboratory

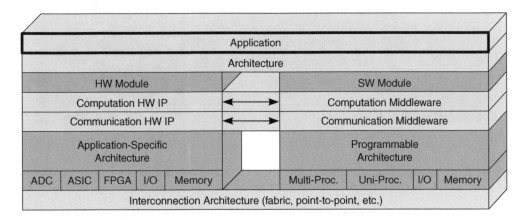

This chapter describes the development of a real-time electro-optical reconnaissance system. The design methodology is illustrated by the development of a notional real-time system from a non-real-time desktop implementation and a prototype data-collection platform.

23.1 INTRODUCTION

This chapter will describe the development of a real-time electro-optical (EO) reconnaissance system from an offline (i.e., non-real-time) capability. The chapter begins with a description of the problem and an offline solution that has been previously developed. The focus then shifts to study the transition from the desktop implementation of the algorithms, used with a prototype data collection platform, to a real-time implementation of the signal processing chain in an embedded platform. The selected methodology will be illustrated by the development of a notional real-time system.

23.2 AERIAL SURVEILLANCE BACKGROUND

Throughout history people have imagined how the world around them appeared from above. As an example, aerial, or bird's-eye, views of cities originated in Europe during the 16th century; this theme became popular again during the 19th century in the United States (see Figure 23-1). Not surprisingly, these prints were typically not actual views, but were renditions based upon observations from nearby points. The invention of the telescope allowed people to see scenes at greater distances

FIGURE 23-1	View of Boston, F. Fuchs, 1870. (Print Collection, Miriam and Ira D. Wallach Division of Art, Prints and Photographs, The New York Public Library, Astor, Lenox and Tilden Foundations.)

and with more detail than previously possible. With the development of hot air balloons in the late 18th century, people were able to ascend into the air and make observations of more terrain than could be seen from a tree or the top of a hill.

As photography was introduced and became more widespread in the early and mid-19th century, people worked to get the new technology into the air. The French artist Gaspard-Félix Tournachon (also known as Nadar) is credited with taking the first aerial photograph in 1858. One of the oldest surviving aerial photographs is a view of Boston taken by James Wallace Black at 1,200 feet from a captive balloon (see Figure 23-2). The first images from a free-flying balloon were captured by Triboulet over Paris in 1879. While balloons were the first platform for aerial photography, they were soon joined by others. Aerial photographs were taken from kites in the late 1880s; breast-mounted cameras for pigeons were patented in 1903; and in 1906 a photograph was taken from a compressed-air-powered rocket.

The strategic applications of such endeavors were realized early on as well. As early as 1859, Nadar met with the French military to discuss using aerial photographs for their campaigns. Balloons were also championed by Wallace for use as reconnaissance platforms during the American Civil War. Interestingly, a kite-mounted camera was used to survey the damage incurred by San Francisco in the 1906 earthquake. By World War I, aerial photography was a common reconnaissance practice.

As the fields of optics and photography developed, the quality of lenses and the devices into which they were placed grew ever higher. The optical elements' improved quality allowed for clearer images and higher-power telephoto lenses to be developed. The development of cameras that were able to take a sequence of images in quick succession made it possible to generate not only a single static image of a scene, but to capture the motion in the scene under observation.

Since they were first developed and introduced in the late 1960s and early 1970s, digital video cameras using charge-coupled device (CCD) imaging sensors have become increasingly common-

FIGURE 23-2　Balloon view of Boston, J.W. Black, 1860. (Courtesy of the Boston Public Library, Print Department.)

place. Once rare and expensive, now a wide variety of digital cameras can be purchased off the shelf by consumers. These range from small, inexpensive, relatively low-quality cameras, such as those integrated into cellular phones, to high-end single-lens reflex cameras that have, as of this writing, over 20 megapixels in their image sensor, with that number constantly growing. A complete discussion of the history and technology of digital photography is beyond the scope of this chapter, but many excellent references may be found in any library or on the Internet.

Over the years, high-quality photographic sensors, both digital and film-bearing, have been mounted in many platforms, including the ill-fated U-2 that was being flown by Francis Gary Powers when it was brought down in the Soviet Union on May 1, 1960. Like other high-altitude aircraft, the U-2 was designed to provide a deep-look capability while minimizing the risk to human life. However, regardless of design or method of operation, this risk can never be completely removed for any human-piloted aircraft. The mounting of video sensors on unmanned aerial vehicles (UAVs), like the Predator produced by General Atomics Aeronautical Systems, has opened up a new avenue for safe, reliable reconnaissance.

It is no understatement to say that a wide variety of UAVs are in production today. In the U.S. alone, 50 companies, government agencies, and academic institutions are working on developing more than 150 designs of UAVs (UAV Forum 2007). These include both fixed- and rotary-wing platforms and platforms flying at altitudes ranging from very low (10's of ft) to very high (55 kft, or higher), and they can be small enough to fit in the palm of your hand or large enough to carry a payload weighing over 1,000 kg. Although there is tremendous range in the scale of UAVs today, there is great interest in developing more and better capabilities for the smaller UAVs. These tend to be less costly and easier to deploy and maintain.

Aerial photography is not the only option available to those wishing to collect surveillance and reconnaissance information from a UAV. Since World War II, radar has been used with great success for airborne intelligence, surveillance, and reconnaissance applications. Radar systems have been placed on aircraft, satellites, and balloons. There is no doubt that radar has proved to be very successful in this field; however, as low-cost, high-quality video sensors become increasingly common, they are being progressively more utilized in surveillance and reconnaissance systems. There are various advantages and disadvantages to using either radar or video sensors; the trade-offs represent fundamental decisions that must be made early on when designing a new reconnaissance platform.

One of radar's primary features is that, unlike sensors that rely upon visible light, it is an all-weather, day-night capability. On the other hand, video sensors are comparatively inexpensive, are lightweight, and take up a small amount of space. Further, since video sensors are passive, they require only a minimal amount of power unless strobes are used. The passive nature of video surveillance and reconnaissance platforms is also of strategic interest since there is no way for areas under surveillance to detect that surveillance: they may detect the presence of the platform overhead, but there is no way of knowing whether it contains a sensor, if that sensor is turned on, or where it is looking. In addition, the imagery that is produced by a video sensor is easily and immediately understandable by operators, unlike radar imagery, which often must be interpreted by trained analysts, especially when looking at small regions, such as a single vehicle.

Video imagery does have its own set of challenges. It is not always easy to identify objects in video, especially in low light or shadow conditions or in the presence of obscuration such as clouds or rain. In actuality, it is not uncommon for UAVs to have sensors for multiple modalities. For example, the Predator UAV typically carries radar as well as EO and infrared (IR) cameras.

A variety of results can be generated by an airborne video sensor. These include developing three-dimensional terrain maps, detecting and tracking moving targets, and, of course, the video feed itself. In addition, there are a number of ways these capabilities can be exploited to extract additional information. For example, the frames of the video feed can be merged to form a large-scale video mosaic, images may be registered with and draped over preexisting digital elevation models (DEMs), and subsequent imagery of a location can be used with archived imagery to perform change detection. For a recent survey of aerial video surveillance platforms, the reader is referred to Kumar et al. (2001).

The remainder of this chapter will look at the example of developing a small stand-alone airborne reconnaissance platform that has as its goal the production of three-dimensional site maps with the ability to geolocate any points in the collected imagery without requiring any *a priori* knowledge of the underlying scene [cf. Collins et al. (2000), for example]. A standard problem in the area of computer vision is for a system with one or more cameras to develop a model of its surroundings so that it can maneuver. Such systems tend to focus on generating a relative model of the environment, not one that is registered to specific points on the Earth (Albus and Hong 1990; Cornall and Egan 2003; Davison 2003; Grossmann and Santos-Victor 2005; Huber and Hebert 1999). By utilizing its position and orientation information, our platform will provide geodetic coordinates (i.e., latitude, longitude, and altitude) for points in the scene. Having this information allows us to break the constraint of having prior knowledge introduced into our system, such as digital terrain elevation data (DTED) or geolocated reference imagery, before we can locate features of interest.

What follows will (1) detail a mathematical approach for generating three-dimensional surface models and performing geolocation, (2) describe a functioning test bed, and (3) develop a notional system. In addition, some of the key trade studies needing to be addressed will be discussed. The goal in this chapter is not to fully develop the end-product real-time system, but to illustrate the development process from design through the creation of a data-collection prototype test bed, which, in turn, sets the stage for the development of the final system.

23.3 METHODOLOGY

This section lays out an approach to generating three-dimensional site models and performing geolocation. To illustrate the method, consider the simple scenario of an aircraft flying over a scene containing a single point, as shown in Figure 23-3. For each frame in the video sequence, we need to know the location of the feature in the imagery, as well as the location and orientation of the camera when that frame was captured. The position and attitude of the camera are typically given by an inertial measurement unit (IMU) and global positioning system (GPS) information.

Having this information for each frame of imagery allows us to draw a ray in a geodetic coordinate system from the camera center through the feature location on the image plane. Since we have multiple looks at the feature, we can draw this ray for each frame. In the absence of error, these rays all intersect at a single point—the location of the feature on the ground.

In general, however, there is uncertainty, or error, in all of our information. The location and orientation of the camera are not known exactly, and even the location of the feature in the image plane may not be known precisely, but only to within a pixel or so. In this case, the rays do not all intersect at the actual feature location, or indeed at any single point, as is shown in Figure 23-4. However, since we do have multiple views of the same feature from different geometries, we can use this information to improve our estimate and lower the impact of the error in any particular measurement. We do this by formulating a least-squares problem that will provide the location of the feature by minimizing the distance from the estimated position to each of the rays and using that as the estimated location of the feature. In particular, if we have m frames in a video sequence, then the location of the tracked feature, $\hat{x}_T$, is given by

$$\hat{x}_T = \left[\sum_{i=1}^{m} \left[I_m - \hat{n}_i \hat{n}_i^T \right] \right]^{-1} \left[\sum_{i=1}^{m} \left[I_m - \hat{n}_i \hat{n}_i^T \right] \hat{x}_i \right], \tag{23.1}$$

where $\hat{x}_i$ is the estimated position of the plane at the ith frame in the video sequence, $\hat{n}_i$ is the estimate of the vector from the camera center to the tracked feature, and I_m is the $m \times m$ identity

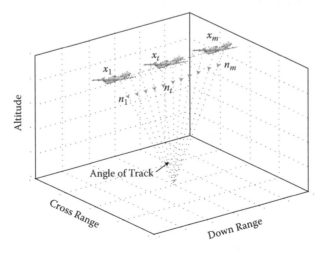

FIGURE 23-3   Geometry of a flyover.

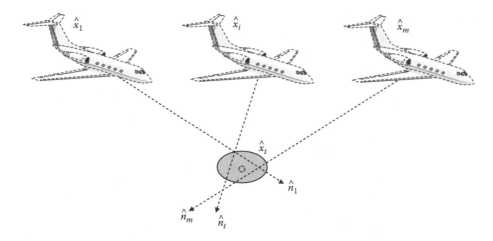

FIGURE 23-4   Single-feature location with noisy measurements.

matrix. We use the hat notation over the vectors to indicate the presence of uncertainty error in the quantity. Typically, Earth-centered, Earth-fixed (ECEF) coordinates are used for convenience in this processing.

By applying this technique to a number of features in the imagery, a three-dimensional point cloud that estimates the scene in the imagery can be developed. Information about points not tracked and estimated directly may be obtained by linearly interpolating between nearby points.

23.3.1 Performance Modeling

We have outlined our method and know that in the presence of error we will not have our rays converging at a single point. We know that we can formulate a least-squares problem that will minimize the distance from our estimated location to each of our rays, but we need to know how error in our location and pointing information will translate to error in our estimate of geodetic coordinates.

There are a variety of ways that one can determine these impacts. One might perform a bias and covariance analysis of the least-squares position estimator (Fessler 1996; Kragh 2004), or one may develop a model that simulates the performance of the system. While it is often faster and easier to develop a model than to perform the required mathematical analyses, both processes are useful in understanding system performance and may be used to perform a consistency check among results. The analytic results often can provide a different insight than can simulations. The development of the bias and covariance analysis is beyond the scope of this chapter, but we will briefly discuss how to develop a performance model and examine some of the results we can extract from it.

Since we are primarily interested in modeling the impact of random errors on our end product, we will use a Monte Carlo simulation in our performance model. Our model simulates the situation illustrated in Figure 23-4, namely, examining the behavior as our sensor flies over a single point on the ground. The parameters used by the simulation include

- target location (ECEF),
- starting camera position (ECEF),
- velocity of the camera,
- camera field of view,
- frame rate,
- imaging-plane dimensions (pixels),
- focal length,

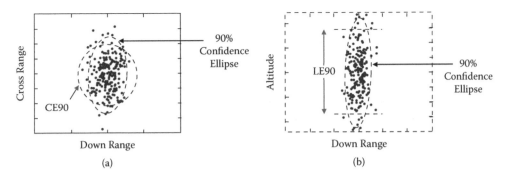

FIGURE 23-5 Scatter plots illustrating (a) circular and (b) linear error of lateral and vertical estimates, respectively.

- GPS/IMU bias and errors, and
- feature tracker error.

All of the parameters can, and should, be varied so that a complete understanding of the system requirements, and limitations, is obtained.

At each frame time, the location of the sensor and the pointing vector to the target is calculated. Uncertainty appropriate to our GPS/IMU specification is added in terms of a bias (fixed during a flight) and error (fluctuates frame to frame). Once all of the pointing vectors have been calculated, we estimate the target location using Equation (23.1). We run the same flight simulation multiple times so that the bias is randomly drawn for each flight.

This model can be used to reveal the expected performance of the system under a number of conditions. As shown in Figure 23-5, one can examine the impact of position and orientation error on our ability to geolocate a point. The two plots in this figure show two common measures of location accuracy. Horizontal accuracy is often measured by the radius of the circle within which a known object should be found. For CE90 (circular error 90%), the probability of the point being within a circle of this radius is 90%. Vertical accuracy is measured in a slightly different fashion: the LE90 (linear error 90%) provides the accuracy of the altitude estimate. Other measures such as CE50 and LE50 [also referred to as CEP (circular error probable) and LEP (linear error probable), respectively] are also common quantities to examine.

The results shown in Figure 23-5 are for a single-platform configuration: i.e., one set of aircraft, camera, and GPS/IMU parameters were used to estimate the target location. Now that we have our model we would like to examine the trade space available to us to see how various system design concepts perform. We can examine the performance of a variety of GPS/IMU units at different altitudes and for different angles of track, as illustrated in Figure 23-6.

These results are for a notional system flying at 100 ft/s (60 knots) with a frame rate of 24 fps and a feature tracker error of 1 pixel. Various pieces of information may be gleaned from plots such as these. First, from Figure 23-6(a), it is clear that all three systems have a "sweet spot," that is, an altitude at which the CE50 is minimized, and for all three systems this altitude is under 1.5 kft. It is also evident that the GPS/IMU represented by the solid curve has superior performance as we increase the altitude of our platform. If one delves a little deeper into the models, one finds that at lower altitudes the primary component of location error comes from the GPS measurements, while at higher altitudes the IMU pointing errors dominate the result.

From Figure 23-6(b) one can see that as the angle of track decreases, the ability to accurately estimate the altitude of a feature in the image drops dramatically, particularly for the system represented by the dotted curve. In any event, if we only track features for 10°, we expect our vertical accuracy to be 20 m or higher. We can use curves such as these to set specifications on components of our system, including GPS/IMU accuracy (in terms of both bias and error), lens focal length, gimbal precision, etc.

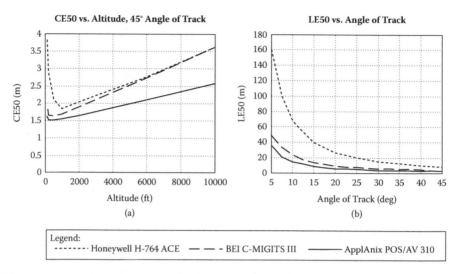

FIGURE 23-6 (a) CE50 and (b) LE50 versus altitude for three different GPS/IMU units.

23.3.2 Feature Tracking and Optic Flow

As mentioned previously, our goal is to determine the geodetic coordinates of features in imagery, i.e., estimate the latitude, longitude, and altitude of the point of interest. Unlike radar, the products produced by a video sensor are inherently bearing-only—no range information is contained in the images. In addition, the ground points are not measured directly by the sensor, but indirectly through the amount of reflected light. The imaging plane of the camera captures the brightness of each point on the ground. If we assume the brightness of each is slowly varying over the time interval between frames, we can associate from frame to frame. Thus, as the aircraft flies over a scene, we can observe the motion of points through the frames, and the distance to each point from the sensor can be inferred based on the rate at which the points transit through the imagery: points closer to the camera will move through the field of view faster than will points that are farther away.

The challenge of getting information from video has been a subject of much research by the computer vision community for many years. In particular, there are a number of optic flow techniques that capitalize upon the motion in imagery—whether the motion is caused by the sensor, the movement of objects in the scene, or both. Clearly, an in-depth discussion of this field is beyond the scope of this chapter, and one should consult the extensive literature for more details. For a good summary and comparison of optical flow techniques, the reader is referred to Barron, Fleet, and Beauchemin (1994) and Beauchemin and Barron (1995).

Broadly speaking, there are two general classes of optic flow techniques. Feature-based optic flow algorithms rely upon identifying features in the scene and then tracking them from image to image. These features are points at which there are significant brightness gradients, such as at corners or edges of buildings. An obvious limitation of feature-based optic flow techniques is that performance is based on the number of features contained, and detected, in the imagery. If a scene contains relatively few features, then the information about the flow through the scene will be sparse, and our resulting site model will have only a few points in it. Of course, the number of features tracked will directly impact the execution time of the system, which scales linearly with the number of features.

The other broad class of optic flow techniques is differential optic flow. Rather than tracking individual features, differential optic flow algorithms try to find a function that describes the change in brightness over the whole image in both time and space. Differential optic flow has the potential of generating a complete position map, but, in general, it is difficult to generate a good approximating function since most real-world brightness patterns are not smooth or even continuous. For the purposes of discussion, we will focus on using a feature-based optic flow technique for our application.

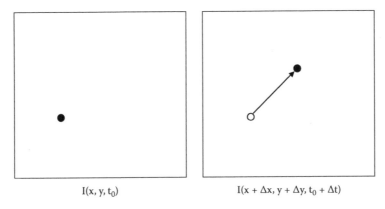

$$I(x, y, t_0) \qquad\qquad\qquad I(x + \Delta x, y + \Delta y, t_0 + \Delta t)$$

FIGURE 23-7 Illustration of pixel motion. Between time t_0 and $t_0 + \Delta t$, the pixel location has changed from (x, y) to $(x + \Delta x, y + \Delta y)$, where it is assumed that Δx and Δy are small.

The motion of an individual feature from frame to frame in an image sequence is illustrated in Figure 23-7. If we assume that the time between successive frames, Δt, is small enough that the motion of the pixel in the image, given by the vector $(\Delta x, \Delta y)$, is small, then we can take the Taylor series expansion of the image intensity, I, about a point

$$I\left(x+\Delta x,y+\Delta y,t\right)=I\left(x,y,t\right)+\frac{\partial I}{\partial x}\Delta x+\frac{\partial I}{\partial y}\Delta y+O\left(\Delta x^2,\Delta y^2\right), \tag{23.2}$$

and also use the fact that

$$I\left(x,y,t_0\right)\approx I\left(x+\Delta x,y+\Delta y,t_0+\Delta t\right). \tag{23.3}$$

Then, ignoring terms higher than linear in the Taylor series, we have that

$$\left(I\left(x,y,t_0+\Delta t\right)-I\left(x,y,t_0\right)\right)+\frac{\partial I}{\partial x}\Delta x+\frac{\partial I}{\partial y}\Delta y=0. \tag{23.4}$$

In the limit that $\Delta t \to 0$, we can write Equation (23.4) as

$$\frac{\partial I}{\partial t}+\nabla I\cdot\left[\frac{\partial x}{\partial t}\,\frac{\partial y}{\partial t}\right]=0. \tag{23.5}$$

As there are two unknown quantities, Δx and Δy, in Equation (23.5), we have an underdetermined system. To calculate the actual flow of a feature, we must develop more equations. One of the most common methods to do this is to assume that the flow in an image is locally smooth and further assume that the neighbors of a pixel have the same flow vector, $(\Delta x, \Delta y)$. For example, if a 5×5 window is used around every feature we wish to track, then we would have 25 equations of the form of Equation (23.5) for each feature. We can write these equations in the form

$$A\begin{bmatrix}\Delta x\\ \Delta y\end{bmatrix}=b, \tag{23.6}$$

where A is an $N \times 2$ matrix and b is an $N \times 1$ vector, where N is the number of pixels in the window we have selected. We can solve this overdetermined system of equations by formulating the least-squares problem

$$A^T A \begin{bmatrix} \Delta x \\ \Delta y \end{bmatrix} = A^T b \,, \tag{23.7}$$

where

$$A^T A = \begin{bmatrix} \sum \left(\dfrac{\partial I}{\partial x} \right)^2 & \sum \dfrac{\partial I}{\partial x} \dfrac{\partial I}{\partial y} \\[2ex] \sum \dfrac{\partial I}{\partial x} \dfrac{\partial I}{\partial t} & \sum \left(\dfrac{\partial I}{\partial y} \right)^2 \end{bmatrix} \tag{23.8}$$

$$A^T b = \begin{bmatrix} \sum \dfrac{\partial I}{\partial x} \dfrac{\partial I}{\partial t} \\[2ex] \sum \dfrac{\partial I}{\partial y} \dfrac{\partial I}{\partial t} \end{bmatrix} \tag{23.9}$$

and the summations in Equations (23.8) and (23.9) are over all of the pixels in the window chosen.

The preceding discussion was merely to give an overview of the fundamental concepts and equations of feature-based optic flow. Clearly, there are many more effects that need to be taken into account when attempting to track features through actual imagery. One algorithm commonly used for this purpose is the Kanade-Lucas-Tomasi feature tracking algorithm, or KLT (Lucas and Kanade 1981; Shi and Tomasi 1994; Tomasi and Kanade 1991). The KLT algorithm functions by looking at subsets, or windows, of each image and extracting features from these windows. These features may include regions where the intensity changes strongly in more than a single direction. The algorithm ranks the features and then attempts to correlate the strongest features to those in subsequent frames, subject to an unknown affine warping; the affine changes are determined by a Newton-Raphson minimization procedure. Features are tracked until the correlation becomes too low and is indistinct from noise. A good description of the KLT algorithm can be found in Jung and Sukhatme (2005).

23.3.3 THREE-DIMENSIONAL SITE MODEL GENERATION

Once the features have been detected and tracked, the rays from the sensor through the feature location in the image plane are calculated, and the locations of the features are estimated by using the least-squares method shown in Equation (23.1). Once the point cloud has been generated, a surface can be produced by using a method like Delaunay triangulation, which is a method that connects each point to its natural neighbors to create a triangular irregular network (TIN). Once the TIN has been created, then the geodetic coordinates of any point in the scene can be determined by using the TIN to interpolate the location of the desired point. In fact, once the TIN has been generated, we can resample the surface over a uniform grid to create a digital elevation model (DEM) that can be disseminated.

Clearly, the number of features used in creating the TIN will drive how accurately geolocation can be performed. In regions where few features are tracked, our TIN may have large triangles that do not accurately represent the underlying terrain.

In order to create a TIN that is a reasonable approximation to the actual surface, we need to remove any points that are statistically improbable. As shown in Figure 23-8, the generated

FIGURE 23-8 **(Color figure follows p. 278.)** Feature point cloud with outlier points flagged in red.

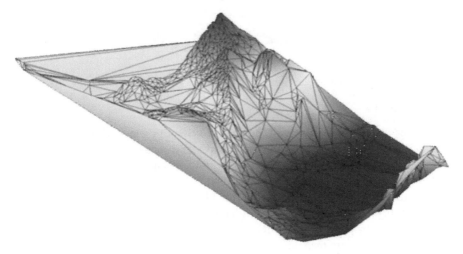

FIGURE 23-9 Terrain map with outliers removed.

point cloud has a number of outliers, indicated in red. These outlier points are due to a number of causes. These outliers could be due to system errors, but objects such as tall trees or flagpoles also may cause outlier points that are correctly estimated. By calculating gradients or using other more sophisticated techniques to detect the outliers, they can be removed and a smoother surface map, which is more representative of the underlying scene, can be generated, as shown in Figure 23-9.

In addition, moving targets also may appear as outlier points. This fact is illustrated in Figure 23-10. Since there is no range information in the video information, there is no way to discriminate between a static target at one range and a moving target at different range. The fact that moving targets show up as outliers provides one mechanism for performing moving-target indication (MTI). If the amount of vertical displacement can be determined, then this may be used to estimate the velocity of the target, v_T,

$$v_T = v_A \frac{\Delta h}{h + \Delta h}, \tag{23.10}$$

where v_A is the velocity of the aircraft, h is the altitude of the aircraft, and Δh is the height discrepancy. It turns out that there are more robust ways of performing MTI and velocity estimation based on registering subsequent frames and looking at the differences between them, but that problem is beyond the scope of this work.

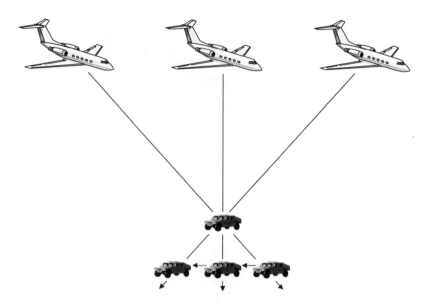

FIGURE 23-10 Moving targets appear as stationary targets with a height discrepancy.

23.3.4 CHALLENGES

Many challenges are involved in actually fielding a system to generate a three-dimensional site map using the method described previously. Obviously, the content of the scene in view can have a dramatic impact on the performance of our algorithm. When examining an urban environment, one sees there is a significant amount of obscuration that can occur behind buildings as the platform transits across the scene. If the feature density is not high enough, the features that are likely to be selected for tracking are corners at the tops of buildings, which can provide an incorrect estimate of the location of nearby points. Urban environments may also produce a disproportionate number of outlier points in the terrain map due to the presence of moving vehicles, as well as the presence of "point" features with strong height discontinuities. Another problem that may be encountered by all but the lowest-flying air platforms is partial obscuration of the scene due to clouds as they pass through the imagery.

The impact that moving vehicles can have on the generation of the point cloud has already been mentioned. When we are reconnoitering an area such as a highway interchange, as is illustrated in Figure 23-11, the moving vehicles' impact is something to be aware of and plan for accordingly. When, however, we wish to apply our technology to a scene in which the entire background is in motion, such as if we were flying over a body of water, then all of the features we wish to track would be in motion and the problem requires a different approach. There are many ways one can handle a case such as this, but this problem is beyond the scope of this work.

23.3.5 CAMERA MODEL

The ability of our system to geolocate points depends on having an accurate camera model. A commonly applied model is the linear camera model, which maps three-dimensional world coordinates, (x_w, y_w, z_w), into two-dimensional image coordinates, (u, v). This transformation is illustrated in Figure 23-12. The mapping from world coordinates to camera coordinates, (x_c, y_c, z_c), is given by

$$
\begin{bmatrix} x_c \\ y_c \\ z_c \end{bmatrix} = \begin{bmatrix} R_{11} & R_{12} & R_{13} \\ R_{21} & R_{22} & R_{23} \\ R_{31} & R_{32} & R_{33} \end{bmatrix} \begin{bmatrix} x_w \\ y_w \\ z_w \end{bmatrix} + \begin{bmatrix} T_x \\ T_y \\ T_z \end{bmatrix},
\tag{23.11}
$$

FIGURE 23-11 (**Color figure follows p. 278.**) Highway interchange; tracked features are shown in red, moving targets are shown in green.

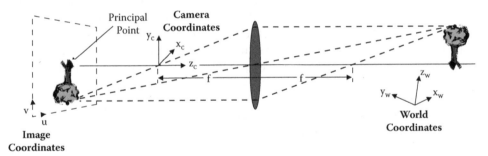

FIGURE 23-12 Linear camera model.

where the vector T and matrix R represent the translation and rotation between the world coordinate system (as measured by the GPS/IMU) and the camera coordinate system, respectively. The three translation components together with the three Euler angles that define the rotation matrix are collectively referred to as the extrinsic parameters of the camera; they are illustrated in Figure 23-13. Any uncertainty in the GPS/IMU information manifests itself in this transformation.

We can now use a perspective projection with a pinhole-camera geometry to transform from camera coordinates to image coordinates, (u, v),

$$\begin{bmatrix} u \\ v \end{bmatrix} = \begin{bmatrix} sf & \gamma \\ 0 & f \end{bmatrix} \begin{bmatrix} x_c / z_c \\ y_c / z_c \end{bmatrix} + \begin{bmatrix} u_0 \\ v_0 \end{bmatrix}, \tag{23.12}$$

where s is the pixel aspect ratio, f is the focal length, γ is the coefficient of skew, and the vector $[u_0 \, v_0]^T$ contains the location of the principal point. These five parameters of Equation (23.12) are often referred to as the *intrinsic parameters* of the camera.

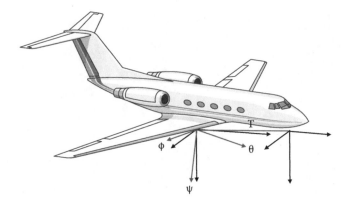

FIGURE 23-13 Extrinsic parameters: translation between sensor and GPS, T; rotation of sensor relative to IMU, $R(\theta, \phi, \psi)$.

23.3.6 DISTORTION

Simply, distortion is defined as the relative change in the location of a pixel compared to its "correct" location. If r is the distance of a pixel from the principal point of the image in the undistorted case and r' is the distance in the distorted image, then the relative distortion, D, is given by

$$D = \frac{r' - r}{r}.$$

(23.13)

While Equation (23.12) allows for the introduction of some distortion into the image from the horizontal scale factor and skew terms, there often are other sources of distortion present in images that are not accounted for in our linear model. These can broadly be bundled as being either radial, decentering, or thin prism distortions. Each of these different distortions is centered about a point called the *center of distortion*, where they have no effect. These centers of distortion for the various effects may or may not be collocated with each other or the principal point of the image.

Radial distortion, often the most noticeable type of distortion, changes the distance between the pixel and the principal point, but not the angle relative to the image sides. Radial distortion may be the result of one or more of the lens elements having a flawed radial curvature, or it may be a feature of the design of a fish-eye lens. Examples of radial distortion are illustrated in Figure 23-14.

Decentering distortion has not only a radial term, but also a tangential component and results from having a sequence of lenses whose optical axes are not aligned. Thin prism distortion, which also has both radial and tangential components, arises when there are imperfections in lens design or manufacturing, or when the sensor array has been misaligned so that it is not orthogonal to the optical axis.

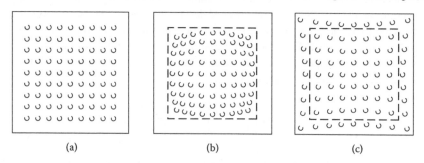

(a)   (b)   (c)

FIGURE 23-14 Sample radial distortion of a square grid: (a) undistorted grid, (b) barrel distortion, and (c) pincushion distortion. The outline of the original grid is shown in dashed lines in the distorted images.

A full discussion of distortion models is beyond the scope of this chapter [see, for example, Devernay and Faugeras (2001), Horn (2000), Tsai, (1987), and Weng, Cohen, and Herniou (1992)], but once a distortion model has been developed and the appropriate coefficients are determined by way of a calibration procedure, the location of the tracked feature in image space must be "undistorted" to compensate for the distortion in the system. The development of a good camera-calibration scheme is crucial since distortion from these parameters represents the most significant correctable source of error in our system.

It is important to note that when using a feature-based optic flow technique, one does not need to undistort each frame in its entirety; it is sufficient to merely apply the undistortion mapping to the locations of interest. In the result is a dramatic reduction in computational effort since the number of features tracked in an image is typically fewer than 1% of the number of pixels.

23.4 SYSTEM DESIGN CONSIDERATIONS

Many components of a surveillance and reconnaissance system must be examined when one wishes to start developing a prototype. Since we are working with an airborne platform, perhaps the first question that springs to mind is that of operating altitude. This choice alone can drive many other system design parameters.

23.4.1 ALTITUDE

From our Monte Carlo simulations, we know that as our altitude increases, we need to pay particular attention to the accuracy of our IMU so that the horizontal location errors do not grow too quickly. Furthermore, the amount of ground subtended by each pixel in the image frame increases with altitude, reducing the resolution of our image for a fixed sensor. Thus, if a specific spatial resolution is required, the altitude may impact the choice of image plane. On the other hand, if the altitude chosen is very low, the amount of area in the field of view can be quite small. If we have a requirement to cover a certain amount of area in a given time, this may drive us to a higher altitude or a shorter focal-length lens. As the focal length of the lens is reduced, we tend to introduce more distortion that must be compensated for, and while a small amount of distortion can effectively be undone, when we are dealing with severe fish-eye distortion, information at the edge of the scene may become corrupted.

23.4.2 SENSOR

Even the decision to use an optical sensor, as opposed to radar, for example, opens up many questions to be answered before making a selection of a particular sensor. In addition to the option of operating in the visible portion of the spectrum, there is also the option of using cameras that operate in the infrared band. A variety of sensors and filters exist for imaging in various portions of the IR band, each having a different phenomenology that must be understood and matched to the application under consideration. Polarization is another system consideration that, depending on the application, may be an avenue to pursue. If using the visible spectrum, one also must choose between a camera that uses CCD technology or one that is built around complementary metal oxide semiconductor (CMOS) imaging sensors (IR sensors use different detectors such as microbolometers or InGaAs detector arrays). There is also the question of whether to choose a color sensor as opposed to one that generates monochromatic (grayscale) imagery. One also must select the pixel depth, or how many bytes per pixel should be collected. The frame rate of the camera also plays an important role in developing a system.

After a camera is chosen, a type of lens (or possibly multiple lenses) must now be determined. There are many important issues associated with choosing lenses, not the least of which is the type of lens. While most common lenses are spherical, there are other types, such as anamorphic, that can have uses in certain applications. The focal length of the lens together with the size of the image sensor will determine the field of view of the sensor. That, coupled with the resolution of the image sensor, will determine the spatial coverage of a pixel in the image. The aperture range of a lens, which is the supported range of f-stops, also can have an impact on system performance. If the lens does not open wide enough, you may not be able to get the proper exposure for low-light scenes. Similarly, if the lens does not close down enough, you may have images that are saturated under some conditions. Obviously, the quality of the optical elements in a lens can have a dramatic impact on the imagery. For example, the lens contrast may be affected by the lens coatings. If a zoom lens is desired, then often some controller to change the focal length of the lens while in flight may be desired. This device must be calibrated in conjunction with the lens so that an accurate focal length is known. For some lenses, changing the focal length may require a refocusing of the lens. Often the controller that controls the zoom setting can also adjust the focus, but the focusing of a lens must be carefully done to achieve the sharpest imagery.

Some cameras may be mounted in gimbals, instead of in a fixed orientation to the aircraft, to provide additional flexibility to the system. In addition to providing a steering mechanism for the sensor, the gimbals can act as a source of stabilization, for example, isolating the sensor from the vibrations caused by the aircraft. If a gimbal is chosen, then the Euler angles associated with the rotation that the gimbal provides the sensor must be known well enough that they do not introduce too much error into our geolocation estimates. Further, if the rotation of the camera by the gimbal causes the translation between the GPS/IMU and the camera to change, then this change must be known and integrated into the processing chain. The decision to use a gimbal opens up a large set of questions and system design considerations that must be examined. This topic is beyond the scope of this work, but the manufacturers and vendors of gimbals for UAVs can often help with the selection process.

23.4.3 GPS/IMU

An inertial measurement unit is typically a self-contained unit that consists of sensors that measure linear and rotational accelerations. It is important to note that since the IMU measures accelerations, the IMU itself cannot produce absolute position or orientation; other information is required and a significant amount of processing must be completed to convert the collected IMU data into geodetic locations. Often, GPS information is blended with the IMU data to generate the position and attitude information. In addition to providing the IMU with a starting location, updates from the GPS unit can be used to control any drift that results from long-term integration of errors in the IMU. A full discussion of the processing required by an IMU is beyond the scope of this work, but a nice introduction can be found in Zywiel (2003).

Often accurately measuring the angles of rotation between the sensor and the IMU is quite difficult, especially if the IMU and the video sensor are not located in proximity to each other. Even at a relatively low height of 3 kft, one degree of bias in the orientation of the camera can result in over 15 m of error in geolocating a point on the ground. Unlike errors in location or orientation that arise during measurements, these biases cannot be reduced to zero by using multiple looks at the scene of interest—their impact is to determine the limit on how well the system may perform.

23.4.4 PROCESSING AND STORAGE

Often it is desirable to perform some amount of processing on board the system. For an initial test bed, often the most efficient, and inexpensive, option to pursue is that of rack-mounted, PC-quality computers. In addition to their reasonable cost, these systems are generally very flexible, supporting a wide variety of operating systems, hardware interfaces, and software development environments. It is important to keep in mind that the amount of hardware that one can put on board an aircraft is typically limited by space and power constraints; it is important to know those restrictions in as much detail as possible. For example, there may not be any power available at the location where the hardware is installed. Cooling is another fact to consider, but tends only to be an issue for very large processing systems.

The space and power constraints apply equally to storage hardware. This is often a concern if a high-capacity RAID (redundant array of inexpensive/independent disks/drives) system is required for data recording. The performance of the storage system under the environmental conditions of the aircraft also must be examined. The vibration, temperature, and other factors may serve to limit the effective data-recording rate that is achieved in flight. If data must be recorded and transferred simultaneously, either for processing or dissemination, then an analysis must be done to ensure that there is enough capacity in the data channel to support this. For some examples, a simple SCSI (small computer system interface) RAID system may suffice; for others, a more complicated system with multiple fibre channel host bus adapters (HBAs) may be required.

Initial prototyping of software is often done using C or C++, but for some applications, languages like MATLAB or Java may be appropriate, particularly if speed is not a concern. The ability to rapidly develop software in these languages often outweighs any performance penalties, at least at the outset. These languages also have the benefit of being platform-independent, so if one changes from Windows to Linux, for example, the code does not have to be rewritten. Of course, the interface and control of certain hardware devices often must be done in C or C++. One thing to always keep in mind when writing software is the eventual transition to an embedded form factor. When possible, it is always a good idea to write code that can be ported to the final system with a minimal effort. Using calls or software packages designed specifically for one platform can help speed up the initial development of a test bed system, but sometimes can lead to long development times later.

23.4.5 COMMUNICATIONS

The ability for the platform to communicate with a ground station is often a desired quality, if not a requirement. There are many different types of communications links available, operating at different frequencies and supporting different data rates. These can range from commercial standards like the various 802.11 wireless networks to the Common Data Link (CDL) that is installed on a variety of surveillance and reconnaissance platforms.

23.4.6 COST

In all of this, cost is often the driving factor in developing a test bed or prototype system. Having a robust and accurate performance model, as well as a good understanding of the problem being solved, will allow substitutions to be made that will allow development to proceed while producing data that can be used as emulations for the final product. Often it is a good practice to design the initial test bed or prototype system in such a way as to facilitate the scaling of results to the final system design.

23.4.7 TEST PLATFORM

An airborne test bed was developed to serve as a data-collection platform for the development of a small UAV. In this case, we mounted a camera to look down from the belly of a Sabreliner jet, a small, low-flying aircraft. We used a JAI CV-M4+CL camera with a 1 megapixel (1392 × 1040 pixels) 2/3 in. CCD that captures 8-bit grayscale frames 24 times a second. The camera weighs 250 g (without lens), occupies just under 180 cm^3, draws a maximum of 4.5 W on 12 V DC, and operates down to –5°C. We chose a Fujinon H16×10A-M41 zoom lens that could be adjusted to give us a (horizontal) field of view between 3° and 47.5° (10–160 mm focal length) and that weighs 900 g. Our system also included a DALSA Coreco X64-CL frame grabber, and a V1LC lens controller from Image Labs International was used to control the aperture during flight (typically the zoom and focus are not adjusted during a collection).

FIGURE 23-15   Sample image from test platform taken over Chelmsford, Massachusetts.

When the Sabreliner is flying at an altitude of 3,000 ft, its velocity is typically in the area of 150 m/s, or approximately 300 knots. At that altitude when the lens is completely zoomed out, we have images that are approximately 700 m on a side, meaning that each pixel is slightly under 1 m in each direction. Of course, the actual roll and pitch of the platform at any given time will change those numbers. That geometry also meant that, barring any dramatic changes in aircraft attitude, we would expect features in the scene to be in view for approximately 5 s, providing ample opportunity to track them and get robust estimates of their positions.

A sample frame of a video sequence from our test platform appears in Figure 23-15. It shows a view over Chelmsford, Massachusetts, at an altitude of just under 2,200 ft. In image space, the platform is flying to the left, and north is indicated on the image. The dark bands in the corners and at the right of the image are due to the mounting of the camera in a tube that partially obscured the field of view.

In addition to the video acquisition system, a Honeywell H-764 ACE GPS/IMU unit was set to collect information about the platform at a 50 Hz rate. This unit weighs 10.7 kg, occupies slightly over 9,300 cm³, and operates down to –54°C. Since the samples from this unit are being collected at a different sampling rate than those from the video camera, we have to filter the position and attitude to the image-collection times before we can merge the GPS/IMU information with the video frames to create the three-dimensional site map.

The data transfer and processing requirements on the GPS/IMU are very modest; the amount of data collected at any one time can be measured in bytes. With our relatively low sampling rate of 50 Hz, the amount of data generated by the GPS/IMU is dwarfed in comparison to that of the sensor.

The Sabreliner was also equipped with 802.11 g capability to disseminate data to people on the ground. Two classes of data products were sent down from the Sabreliner on this 54 Mbps channel. In addition to the traditional streaming video and telemetry feed, some high-resolution still imagery compressed using JPEG2000 and the three-dimensional surface model are also disseminated. The lossless compression ratio achieved using the JPEG2000 encoder was typically around 50%. With this compression ratio, a single frame from the video sequence could be sent to the ground in approximately one-tenth of a second. The streaming video sequence typically has reduced resolution, both in terms of number of pixels as well as frame rate, so when this fact is combined with efficient video coding techniques, there is plenty of room in the channel to support all of the required products and information.

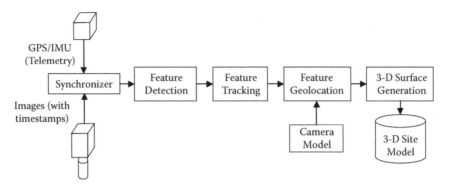

FIGURE 23-16 Test platform processing chain.

On board the Sabreliner were three rack-mounted PCs, each with dual Pentium-class processors. The first PC was connected to the GPS/IMU and was used to process the data to provide the location and orientation information. The second computer was connected to the camera and controlled the collection and storage of the data, while the third computer performed the optic flow calculations, detecting and tracking features, generating the site model, and encoding the requested JPEG2000 imagery.

Clearly, the bulk of the processing was performed by the third PC, and the main data-traffic path was between PCs 2 and 3. With our choice of CCD and frame rate, we collect 265 Mbits/s, which must be stored and served to the third PC for processing. This data rate is fairly modest and fits well within SCSI disk and standard gigabit Ethernet limits for storage and transfer, respectively.

The processing chain used on board the Sabreliner is illustrated in Figure 23-16. The various processing blocks were implemented in Visual C++ and employed an implementation of the KLT tracker written in C that is in the public domain (Birchfield 2007). To improve the performance of this implementation of the KLT tracker, the code was modified to use routines from the Intel Integrated Performance Primitives (IPP) to speed up the calculation of convolutions in both row and column space. The IPP was also used to perform the matrix manipulations required to estimate the location of the tracked features using Equation (23.1). To demonstrate the possibility of migrating to a real-time system without making substantial modifications to the code, it was necessary to reduce the video stream, both in terms of frame rate and image size. For ease of implementation, as well as to not introduce any additional processing burden, we reduced the frame rate to 12 fps and cut the resolution of the images by 1/2 in each dimension. Having done this, we were able to detect and track features through our stream in real time.

Efficient calculation of Delaunay triangulation for the feature locations on the ground plane can be accomplished using the public domain Qhull package, which uses the quickhull algorithm (Barber, Dobkin, and Huhdanpaa 1996). It should be noted that Qhull does not handle constrained Delaunay triangulation problems, such as the triangulation of non-convex surfaces, so the triangulation must be performed on the ground positions, applying the height information after the triangulation has been completed.

23.5 TRANSITION TO TARGET PLATFORM

Our goal in this section is not to actually design the real-time system, but show examples of commercial off-the-shelf (COTS) hardware that currently exists and may be used in the further development of the system. No endorsement is stated or implied for the products and vendors used as examples in this section.

From our demonstration platform that used COTS software with a few modifications, we know that we are able to produce a near-real-time system using three Pentium-class machines and throttling back on the image size and frame rate. After we proved out our concept on a test bed, the next step in the development of our system is to transition the technology so that we can operate from aboard a small UAV. For our small UAV, we would like to mimic the setup that was demonstrated on our test bed and select a camera with a roughly 0.25 megapixel imaging plane that collects 8-bit grayscale imagery at a rate of 12 fps. As when we were developing our test platform, the same system considerations must be taken into account, but now the constraints are much more severe.

23.5.1 Payload

Although some UAVs, such as the Predator, can carry internal payloads that are several hundred kilograms, many UAVs can carry only 10 kg or even less. When one is trying to develop a very small and inexpensive UAV, such as a UAV that may be carried and launched by an individual, the amount of the weight budget dedicated to the payload may put a severe restriction on the equipment that makes up the sensor and processing package.

In addition to the weight restriction, the interior space of a UAV is often quite limited. A UAV may have 5 ft^3 of interior space, as is the case with the SAIC Vigilante, or it may be 0.5 ft^3, as it is with the Arcturus T-15. It is important to understand that not only must the payload fit in the designated volume, conforming to the shape of the space, but the equipment must be adequately cooled as well.

23.5.2 GPS/IMU

The platform must be able to support a GPS/IMU system that will provide accurate enough measurements for geolocation to be performed to within the desired specification. As was shown previously, the quality of the IMU measurements is the primary driver of system performance, particularly at higher altitudes. For a small UAV flying at a low altitude, a smaller, relatively low-performance IMU may be sufficient to produce the accuracy desired. For some UAVs, an autopilot system is desired, and in that case the GPS/IMU is typically an integrated part of the autopilot system.

Several vendors are producing microelectromechanical systems (MEMS) IMUs, which have the advantages of being small and light, as well as consuming little power. As an example, the Crista OEM Sensor Head IMU from Cloud Cap Technology weighs 7 g and occupies just over 12.2 cm^3 of space. This unit draws 30 mA at 6.5 V and can operate at temperatures as low as –40°C. There are also very small GPS receivers such as the Garmin GPS 25LP, which weighs 38 g and occupies 37 cm^3 of space. This unit typically draws 155 mA at 5.0 V and can operate down to –30°C. Whichever specific GPS/IMU models are to be considered, the performance prediction model should be run using their specifications and the expected specifications of the airframe (altitude and velocity) to get an estimate of how their performance will limit the end product.

23.5.3 Sensor

The size of the UAV will determine the altitude and speed at which the platform can fly. This must be traded off between the sensor parameters (particularly field of view and number of pixels of the imaging plane) to determine whether sufficient resolution can be had to track features of interest. The power consumption of the payload may lead to the choice of a sensor based on CMOS technology rather than on the more common CCD sensors. However, the quality of many CMOS cameras is often not as good as that of their CCD counterparts. The CMOS cameras tend to be less sensitive and may be more prone to fixed pattern noise due to inequalities amongst the amplifiers. In the end, cost, as always, may be the most prominent factor in making a decision as to which technology to employ.

One camera that may warrant consideration as a sensor is the Cohu 7200 Series. This camera has a 659 × 494 pixel, 8-bit grayscale, CMOS 1/2 in. sensor (which actually has dimensions 6.4 mm × 4.8 mm) that can support frame rates of up to 50 fps. The camera occupies 93.5 cm^3 and weighs less than 100 g without a lens. It draws less than 2 W of power on 12 V and includes a FireWire (IEEE-1394A) connection. It is rated for operating temperatures as low as –5°C and has shock and vibration specs of 40 G and 12 G, respectively. It can support a variety of lenses with focal lengths ranging from 6.5–50 mm; the 8 mm lens option would provide a 53° diagonal field of view. With that lens, we would expect spatial resolution of about 1 m/pixel at a height of just over 2.7 kft (800 m).

For very small UAVs with no onboard processing capability (except for the GPS/IMU), one might be interested in examining the EO5-380 from Hi Cam. This 380 line 1/4 in. CCD camera weighs less than 30 g and occupies 21.5 cm^3 when equipped with a 73° lens (a 38° lens is also available). The transmitter for the camera is similarly small, weighing about 30 g and occupying 12 cm^3. A two-hour battery for both the camera and the transmitter weighs 54 g and requires 18 cm^3. All of this equipment can operate at temperatures as low as –10°C.

Regardless of the camera and lens chosen for the system, a calibration procedure must be executed to determine the intrinsic parameters of the system so that images may be successfully undistorted. Since the sensor may undergo small physical changes as a result of vibration, temperature changes, etc., it is important to calibrate the camera often so that accurate parameters are available for use. After the initial preflight calibration has been performed, one may want to consider updating the camera parameters in flight using a technique known as *bundle adjustment*. A full discussion of bundle adjustment and the various ways to implement it are beyond the scope of this work; the reader is referred to Triggs et al. (2000).

Since uncertainty in the orientation of the camera can yield large geolocation errors, it is important that the specifications of any gimbal system that is selected do not lead to uncertainty that cannot support the accuracy of the end product. On the other hand, the cost of very accurate gimbals is such that one does not want to purchase one that is unnecessarily precise. Indeed, in some circumstances, the price of adding a gimbaled mount to a system, when one considers all of the costs, such as increased complexity in the camera control system, may outweigh the advantages it offers.

23.5.4 Processing

In addition to the performance limitations of the GPS/IMU, the amount of processing power one can host on board can place severe restrictions on how well the end system performs. The throughput of the system will determine the frame rate that can be supported, as well as the size of the images that can be processed in that time step. Using one or more single-board computers (SBCs) is often a quick and inexpensive way to develop a prototype system. The SBCs often have multiple RS232, (gigabit) Ethernet, and other I/O ports that can be used to connect them to each other as well as to devices such as the GPS/IMU and the camera. The processors hosted on these SBCs can range from Intel to AMD to PowerPC. The amount of random access memory (RAM) on these boards likewise can range from 16 MB to 3 GB or more. The amount of space available may dictate a form factor that may reduce the available choices. A real-time operating system must also be chosen to run on the SBC. Again, a large number of choices are available, and the best one will depend on the specific requirements, as well as the specific hardware, chosen for the system. For the smallest UAVs, one may have to use field programmable gate arrays (FPGAs) or some other technology. For example, chips like the Acadia I Vision Accelerator may be able to handle much of the desired processing before downlinking to a ground station for further processing and exploitation.

Our software was written in C/C++ to support offline processing and analysis on typical desktop machines. While some amount of modification was done to get the code to run faster, more work needs to be done to achieve real-time performance, especially if we are not using a general-purpose processor. Issues concerning both memory management and execution speed need to be examined. Memory-management issues include reducing the footprint of the code so that it takes up less memory, reducing the amount of calls to dynamic memory allocation by reusing space when possible, and freeing allocated memory when it is no longer in use.

We previously mentioned that the IPP library was used to speed up the calculation of convolutions in the KLT code that we were using on our test bed. Investigating the parallelization of the code stream, if the system architecture supports parallel processing, may also be an area where performance gains could be made. For example, since the KLT algorithm iterates over multiple feature windows while performing the selection and tracking of features, one might consider parallelizing the code in such a way that this task is broken out over multiple processors. Obviously, one would need to examine the cost-benefit ratio incurred by separating out these tasks. In the end, the extent to which we pursue any of these avenues will depend largely on the final processing architecture. For example, there may be enough memory available so that small gains in the footprint of the code will not have a measurable impact on system performance.

The runtime performance of the KLT implementation can also be strongly influenced by the choice of some parameters that are input to the code. For example, the number of features to track, the size of the windows to examine, the number of pixels between tracked features, the maximum number of iterations to use while tracking, etc., all drive execution time. Careful attention must be paid to the selection of these, and other, parameters to ensure that the performance of the KLT algorithm does not diminish below what is required as the performance is tuned.

Work has also been done to implement other optic flow techniques in real time, studying the trade-offs between accuracy and efficiency and estimating their performance (Liu et al. 1998). Some of these techniques combine aspects of the KLT algorithm with other variational optic flow techniques and have been demonstrated to run on a Pentium-class PC at a frame rate of 18 fps with small frames (316×252 pixels) (Bruhn et al. 2005). This technology may offer the ability to generate more complete and complex terrain models from our video sensor.

23.5.5 COMMUNICATIONS AND STORAGE

Finally, attention needs to be paid to storage and communication systems. If there is not enough bandwidth to downlink all of the information desired, for instance over an 802.11 wireless network, then some amount of storage must be made available. In recent years, the capacity of solid-state storage devices as well as very small hard drives has been increasing dramatically. Clearly, the amount of storage required will be determined by the product of data-acquisition rate and the expected time of operation for the platform. New storage devices, such as Seagate's EE25 series of 2.5 in. hard drives with a capacity up to 40 GB, are designed to resist vibration and operate in temperatures down to –30°C. Solid-state storage devices are also an option to consider. For example, Memtech has a line of 2.5 in. solid-state flash drives that occupy 66 cm^3 with IDE, SATA, and SCSI interfaces that are as large as 64 GB with sustained transfer rates of up to 40 MB/s. They can operate at temperatures as low as –40°C, at altitudes up to 80 kft, and draw less than 750 mA at 5 V. With the Cohu camera described previously, a 64 GB drive is enough to store over 200,000 frames, which is over an hour of collection time if we capture frames at the maximum rate of 50 fps, or over 4.5 hours at our notional rate of 12 fps.

If the individual frames do not need to be stored, then just enough storage is required to act as a buffer for the communications link when requests for still images are received faster than they can be completed. Of course, it is possible for any buffer, no matter how large, to eventually fill up. At this point, a decision must be made about how to handle this situation. Typical ways range from halting the addition of new material to the buffer until it has been emptied, either completely or partially, to preempting the oldest material in the buffer to make room for the new requests. Buffer management often goes hand in hand with the development of the priority scheme for the communication link, which is needed to ensure that high-priority products (such as the streaming video might be) do not suffer additional latency to service less important requests (such as still imagery might be).

To improve the utilization of the communications channel, image compression schemes are typically used for both still imagery and video feeds. These schemes may be either lossy (i.e., may introduce noise) or lossless (such as JPEG2000). Schemes such as JPEG2000 can preserve information at the cost of additional processing requirements. Fortunately, there are hardware solutions that can perform this compression for real-time systems, such as the intellectual property (IP) cores from Barco Silex and Cast, Inc.

23.5.6 ALTITUDE

Of the potential components identified so far, the most restrictive in terms of operating temperature is the camera, which is rated for temperatures down to only −5°C. We can use a standard atmosphere model to find an estimate of the maximum altitude at which our system will be able to operate (NASA 1976). The lapse rate of temperature in the troposphere is typically considered to be −6.5°K/km, so if we use a nominal surface temperature of 15°C, then the maximum altitude we can fly and still have the ambient temperature be above that required for the camera is just over 3 km, or 10 kft. Of course, the actual weather conditions may vary, and other factors may limit our altitude, but this limit is far above that which we would fly in order to have pixels that were on the order of 1 m.

Many other aspects of a system such as this must be carefully examined: for example, the control system which will have to directly interact with most, if not all, of the other subsystems, such as the autopilot, gimbal control, camera adjuster, etc. We have not touched on any aerodynamic aspects such as the power supply, airframe design, or propulsion system.

23.6 SUMMARY

This chapter has presented an example of how one might develop a small UAV reconnaissance platform from an idea through the implementation of a data-collection test bed on a small aircraft. The tasks of site model generation and geolocation were used for the purpose of illustration. Clearly, many extensions to this work can be considered. What benefit could be had by changing the CCD imager to one that provides either color imagery or imagery in the IR portion of the spectrum? What other optic flow techniques could be implemented in this system to provide improved site models? How does one handle complicated and difficult environments, such as reconnaissance over water or when clouds are between the sensor and the scene?

By using some of the techniques in the other chapters of this book, we hope to demonstrate that developing something that initially seems a very daunting task is quite within reach.

ACKNOWLEDGMENTS

The author would like to acknowledge the work of a number of people at MIT Lincoln Laboratory including Michael Braun, Jason Cardema, Thomas Kragh, William Ross, and Randy Scott.

REFERENCES

Albus, J.S. and T.H. Hong. 1990. Motion, depth, and image flow. *Proceedings of the IEEE International Conference on Robotics and Automation* 2: 1161–1170.

Barber, C.B., D.P. Dobkin, and H.T. Huhdanpaa. 1996. The Quickhull algorithm for convex hulls. *Association for Computing Machinery Transactions on Mathematical Software* 22(4): 469–483.

Barron, J.L., D.J. Fleet, and S.S. Beauchemin. 1994. Performance of optical flow techniques. *International Journal of Computer Vision* 12(1): 43–77.

Beauchemin, S.S. and J.L. Barron. 1995. The computation of optical flow. *Association for Computing Machinery Computing Surveys* 27(3): 433–467.

Birchfield, B. 2007. KLT: Kanade-Lucas-Tomasi Feature Tracker. Available online at http://www.ces.clemson.edu/~stb/klt/ hosted by Clemson University.

Bruhn, A., J. Weickert, C. Feddern, T. Kohlberger, and C. Schnörr. 2005. Variational optical flow computation in real time. *IEEE Transactions on Image Processing* 14(5): 608–615.

Collins, R.T., A. Lipton, T. Kanade, H. Fujiyoshi, D. Duggins, Y. Tsin, D. Tolliver, N. Enomoto, and O. Hasegawa. 2000. *A System for Video Surveillance and Monitoring.* Tech. Rep. CMU-RI-TR-00-12. Robotics Institute, Carnegie Mellon University, Pittsburgh, Penn.

Cornall, T. and G. Egan. 2003. Optic flow methods applied to unmanned air vehicles. Department of Electrical and Computer Systems Engineering, Academic Research Forum, Monash University, Victoria, Australia. Available online at http://www.ctie.monash.edu.au/hargrave/TC_no_16_forum.pdf.

Davison, A.J. 2003. Real-time simultaneous localisation and mapping with a single camera. *Proceedings of the Ninth IEEE International Conference on Computer Vision* 2: 1403–1410.

Devernay, F. and O. Faugeras. 2001. Straight lines have to be straight. *Machine Vision and Applications* 13: 14–24.

Fessler, J.A. 1996. Mean and variance of implicitly defined biased estimators (such as penalized maximum likelihood): applications to tomography. *IEEE Transactions on Image Processing* 5(3): 493–506.

Grossmann, E. and J. Santos-Victor. 2005. Least-squares 3D reconstruction from one or more views and geometric clues. *Computer Vision and Understanding* 99(2): 151–174.

Horn, B.K.P. 2000. Tsai's camera calibration method revisited. Report available online at http://people.csail.mit.edu/bkph/articles/Tsai_Revisited.pdf.

Huber, D.F. and M. Hebert. 1999. A new approach to 3-D terrain mapping. *Proceedings of the IEEE International Conference on Intelligent Robots and Systems* 2: 1121–1127.

Jung, B. and G. Sukhatme. 2005. *Real-time Motion Tracking from a Mobile Robot*. Tech. Rep. CRES-05-008, Center for Robotics and Embedded Systems, University of Southern California, Los Angeles.

Kragh, T. 2004. MIT Lincoln Laboratory, private communications.

Kumar, R., H. Sawhney, S. Samarasekera, S. Hsu, Hai Tao, Yanlin Guo, K. Hanna, A. Pope, R. Wildes, D. Hirvonen, M. Hansen, and P. Burt. 2001. Aerial video surveillance and exploitation. *Proceedings of the IEEE* 89(10): 1518–1539.

Liu, H., T.-H. Hong, M. Herman, T. Camus, and R. Chellappa. 1998. Accuracy vs. efficiency trade-offs in optical flow algorithms. *Computer Vision and Image Understanding* 72(3): 271–286.

Lucas, B. and T. Kanade. 1981. An iterative image registration technique with an application to stereo vision. *Proceedings of the International Joint Conference on Artificial Intelligence* 674–679.

National Aeronautics and Space Administration. 1976. U.S. Standard Atmosphere, 1976. Washington, D.C.: U.S. Government Printing Office.

Shi, J. and C. Tomasi. 1994. Good features to track. *Proceedings of the IEEE Computer Society Conference on Computer Vision and Pattern Recognition* 593–600.

Tomasi, C. and T. Kanade. 1991. *Detection and Tracking of Point Features*. Tech. Rep. CMU-CS-91-132, Carnegie Mellon University, Pittsburgh, Penn.

Triggs, B., P. McLauchlan, R. Hartley, and A. Fitzgibbon. 2000. Bundle adjustment—a modern synthesis. In *Vision Algorithms: Theory and Practice*. B. Triggs, A. Zisserman, and R. Szeliski, eds. Lecture Notes in Computer Science, vol. 1883: 298–372. New York: Springer-Verlag.

Tsai, R.Y. 1987. A versatile camera calibration technique for high-accuracy 3D machine vision metrology using off-the-shelf TV cameras and lenses. *IEEE Journal of Robotics and Automation* RA-3(4): 323–344.

UAV Forum: Vehicles. Available online at http://www.uavforum.com/vehicles/vehicles.htm.

Weng, J., P. Cohen, and M. Herniou. 1992. Camera calibration with distortion models and accuracy evaluation. *IEEE Transactions on Pattern Analysis and Machine Intelligence* 14(10): 965–980.

Zywiel, M. 2003. Aided intertial technology: more than IMU + GPS. *Earth Observation Magazine* 12(4). Available online at http://www.eomonline.com/Common/Archives/2003jun/03jun_zywiel.html.

Section VI

Future Trends

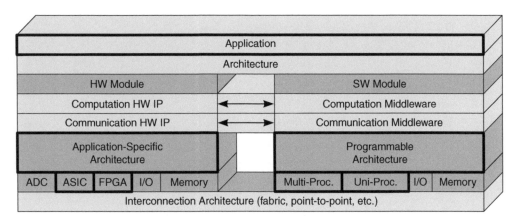

Chapter 24 Application and HPEC System Trends
David R. Martinez, MIT Lincoln Laboratory

This chapter looks at applications developed in the last decade and projects forward from the lessons learned to capabilities desired in the next decade. The advances in high performance embedded computing capabilities will depend on technologies spanning a balanced architecture between high performance hardware, cost-effective software, and high-speed memories and interconnects.

**Chapter 25 A Review on Probabilistic CMOS (PCMOS) Technology: From Device
Characteristics to Ultra-Low-Energy SOC Architectures**
*Krishna V. Palem, Lakshmi N. Chakrapani, Bilge E. S. Akgul, and
Pinar Korkmaz, Georgia Institute of Technology*

This chapter presents a novel technology to overcome energy consumption by CMOS chips and accompanying heat dissipation. This technology can have a major impact on chip performance (including better management of circuit noise) as devices scale into the nanometer regime.

Chapter 26 Advanced Microprocessor Architectures
   Janice McMahon and Stephen Crago, University of Southern California,
   * Information Sciences Institute*
   Donald Yeung, University of Maryland

This chapter focuses on microprocessor technologies. It begins with a high-level summary of processor trends over the recent past and projects ahead based on a wide range of architectures with a primary focus on on-chip parallelism.

24 Application and HPEC System Trends

David R. Martinez, MIT Lincoln Laboratory

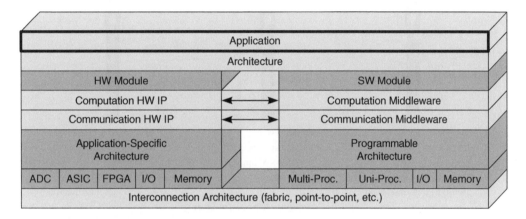

This chapter looks at applications developed in the last decade and projects forward from the lessons learned to capabilities desired in the next decade. The advances in high performance embedded computing capabilities will depend on technologies spanning a balanced architecture between high performance hardware, cost-effective software, and high-speed memories and interconnects.

24.1 INTRODUCTION

High performance embedded computing (HPEC) systems have become more and more complex as the application domains demand more of these systems. As described in previous chapters, earlier systems had to implement what today we consider to be very simple functions. Many of these systems developed in the late 1960s and 1970s were primarily *single-function centric*, that is, devoting all their computing resources to performing a single function in real time (e.g., the use of circuitry to compute a fast Fourier transform). This function was often implemented using dedicated hardware throughout the HPEC system.

Concurrently, applications evolved to a level of complexity that demanded multiple functions be implemented in real time (e.g., filtering, target detection, and tracking). In the late 1980s and early 1990s, the HPEC community continued to use custom designs but utilized commercial off-the-shelf (COTS) parts. Systems were built from COTS memories and microprocessors but were designed using custom high performance boards and unique custom software. In contrast to previous decades, when the systems were more single-function/single-hardware class, later systems were multiple-function/multiple-hardware class but limited to only a few channels of data streams to meet the size, weight, and power (SWAP) available.

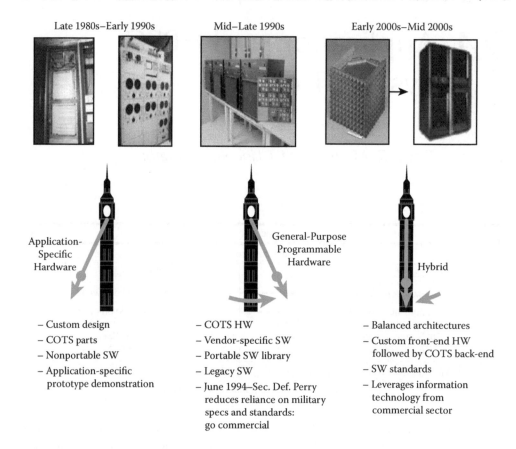

FIGURE 24-1 A decade of architectural changes.

The middle to late 1990s were heavily influenced by the Department of Defense (DoD) mandate to use COTS hardware and to minimize or eliminate requirements to comply with military hardware standards. As shown in Figure 24-1, the architecture pendulum swung from a custom solution to a fully COTS-compliant solution. HPEC vendors began offering hardware systems bundled with vendor software so that the user only needed to develop the application layer.

During this decade, the HPEC community was dominated by processing engines developed for applications driven by the non-DoD market. Some of these microprocessor engines were found in medical, automotive, industrial test and instrumentation, and other commercial applications that demanded real-time performance but were able to tolerate and be adjusted to increases in power usage. Many of the HPEC systems developed during the mid to late 1990s were based on general-purpose COTS programmable hardware but implemented in a parallel processing architecture. The systems were multifunctional and massively parallel. During this period, the HPEC community faced the dilemma of wanting large computation throughput but having power limitations dictated by the complexity of systems implemented in highly constrained platform environments. This dilemma led to a hybrid architecture that used a mix of both custom and general-purpose hardware.

The early 2000s to mid-2000s demanded a more balanced architecture requiring very high throughput hardware for the front-end processing [often designed using either application-specific integrated circuits (ASICs) or field programmable gate arrays (FPGAs)] followed by COTS back-end processors. The front-end designs leveraged significant advances in design automation. These front-ends were also closely coupled with the front-end of the sensor system. The antenna, transmitters, receivers, analog-to-digital converters (ADCs), and front-end processors were all designed to meet the most stringent SWAP requirements of the platform. The back-end processors were

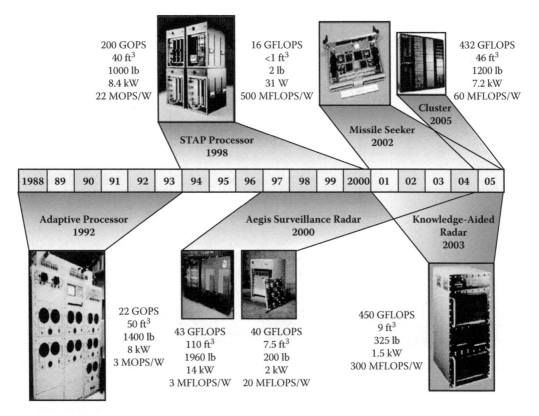

FIGURE 24-2 HPEC signal processor systems.

designed based on COTS hardware and allowed for more flexibility in the algorithms. The information technology from the commercial sector also contributed to a wide range of interconnection topologies and technologies (e.g., 10-Gbit Ethernet, rapid IO, InfiniBand, to name a few).

Most recently, this hybrid architecture has been incorporated into the back-end clusters of computers based on general-purpose PCs but ganged together to meet real-time performance. This cluster of computers can be assembled at a cost commensurate with the market cost of PCs. Currently, these systems are most prevalent in ground-based real-time systems. For highly constrained SWAP applications, the HPEC systems are assembled from commercially available processing engines, in some cases the same engines used in PCs (e.g., the PowerPC), and commercially available interconnections, all put together in an architecture with careful attention to managing size, weight, and power. These architectures can be unique, but not because of hardware components but because of how the thermal cooling, power consumption, and assembly are accomplished to meet the available SWAP of the platform (e.g., submarines, surface ships, airborne and spaceborne systems). These systems can be categorized as ones in which the hardware boards are either the same boards found in PCs or boards developed to meet small form factors (with the requisite cooling environment) assembled using commercially available processing engines and interconnection hardware. Figure 24-2 illustrates examples of complex signal processor systems, progressing from an application-specific system of the early 1990s, to general-purpose programmable systems of the mid- to late 1990s, and finally to hybrid systems based on a hybrid architecture using both custom hardware and general-purpose back-ends and to systems developed from clusters of PC computers. Figure 24-2 also shows the changes in SWAP and throughput performance dictated by the type of architecture employed and hardware used contemporary to the respective time period.

One important factor dominating the design of many HPEC systems today is the need to maintain low cost. More and more of the system is now dominated by the software cost, and as the ADCs

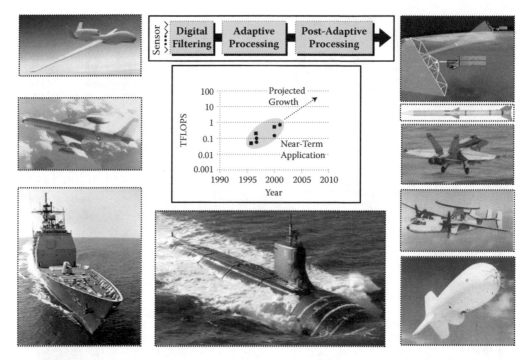

FIGURE 24-3 Projected embedded signal processing capabilities. (From Lebak, J.M. et al., Parallel VSIPL++, *Proc. IEEE* 93(2): 314, 2005. With permission. © 2005 IEEE.)

get closer to the antenna, more of the functionality is performed in the digital signal processor. The trend for the next decade will be toward minimizing the cost of both hardware and software. Systems will be highly distributed, leveraging very-high-speed interconnections across platforms (e.g., military platforms connected via the transformational communication architecture) and across wide-area nets (e.g., Future Defense Information Systems Network), while being compliant with a service-oriented architecture (SOA). The conceptual details of an SOA architecture are described later in this chapter.

Applications over the next decade will continue to demand multifunctional processing, but many of these DoD systems will also be multimodal. For example, with the advances experienced with active electronically scanned antennas (AESAs), the same antenna will be able to perform active surveillance [synthetic aperture radar (SAR) and ground moving-target indication (GMTI)], electronics intelligence (ELINT) operations, and direct communications. Furthermore, the onboard processing will be performed to a point sufficient enough to post the information onto a network. However, the large number of digital channels and high sampling rates will require computation throughputs exceeding several TeraOps (trillions of operations per second). As shown in Figure 24-3, the last decade has been dominated by capabilities ranging from over 10s to 100s of GigaOps (billions of operations per second). For the next decade, the applications will continue to grow in complexity, demanding aggregate computations between the front-end processing and back-end processing to several TeraOps, while still limited to very stringent SWAP requirements.

The advent of high-speed communication networks will introduce another dimension in the design of future architecture systems. As shown in Figure 24-1, the hybrid architecture incorporates both custom hardware in the front-end and general-purpose processing in the back-end. The extra dimension will be the need for determining how much onboard processing is necessary before the data are posted onto the net. If the communication networks connecting to the platforms are infinitely fast (which they are not), then all the processing could be done at a remote location on the ground. However, since the communication networks are limited in bandwidth throughput, the onboard real-time processing will be needed to transform the sensor data into useful bit streams

FIGURE 24-4 A system-of-systems net-centric architecture.

compatible with the available interconnection network bandwidths. The data posted onto the net must comply with a net-centric and distributed architecture. This means moving from a paradigm based on *need-to-know* to a new paradigm focused on *need-to-share*.

As an example, a Global Hawk unmanned aerial vehicle (UAV) might generate electro-optical/infrared (EO/IR), SAR, or GMTI data from the same platform (multimodalities) and interface these data onto a communication fabric able to format the data to comply with a service-oriented architecture. This type of system will be interfaced to the network using a unique Internet protocol (IP) address, just as today's computers are identified within an area network. The system will be a node among many posting data to the Global Information Grid (GIG) (National Security Agency 2007). The processing will be done on board with sufficient computation throughput to meet the size of the communication pipes and to maintain real-time performance. From this node in the network, users will be able to subscribe to data, and then the node containing the requisite data—or in a position to collect the data—will be able to post the data. This architecture will conform to what is now commonly referred to as a *publish-subscribe* architecture.

The sensor data must still be fused, registered, and mensurated, requiring additional processing off-board the platform. Many additional functions performed on the ground will be best characterized as converting information into knowledge. One can envision multiple sensors coming onto the network and leaving the network as they perform their required tasks. The resulting data will be processed at multiple sites across the world by those who need the information. This highly ubiquitous computing will be distributed both in time and location, and these systems will comply with a ubiquitous and distributed computing architecture (UDCA). The UDCA topology will significantly leverage standards that are becoming most prevalent in the commercial sector and that conform to a service-oriented architecture.

The next sections present more details about these predicted system trends. Addressed first is the sensor architecture, including trends in both hardware and software. In this discussion, a node will be a sensor embedded in a network consisting of multiple nodes. Figure 24-4 illustrates this topology in which the sensor has an onboard processing system and its outputs are posted onto a network joining these nodes together. These nodes will publish their data to a network. The subsequent sections address architecture-level trends and the hardware and software supporting the net-centric architecture. This type of UDCA architecture will have the benefit of riding the fast evolution in computing technology, Internet standards, and software protocols while at the same time meeting the requirements of defense systems for both strategic and tactical applications.

24.1.1 SENSOR NODE ARCHITECTURE TRENDS

Important considerations for future systems are the military's inventory of a large number of platforms and its imminent acquisition of new systems (e.g., additional UAVs, the future combat system,

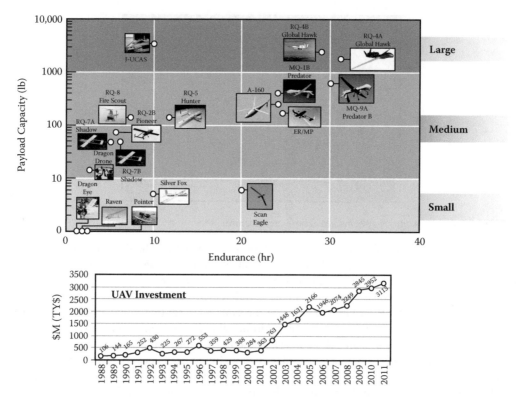

FIGURE 24-5   Types of UAVs and projected inventory growth (Office of the Secretary of Defense 2005).

the space radar, additional surface ships, etc.). Therefore, HPEC advances will impact existing systems by providing more capability and future systems by enabling capabilities not available to date. Figure 24-5 shows the payload capacity of existing or planned UAVs and the growth in the number of expected UAVs by the year 2011 (Office of the Secretary of Defense 2005). A UAV's total payload capacity ranges from small to large. For example, the Global Hawk has a total payload capacity of over 1000 lb. The signal processor alone weighs approximately 175 lb with a computation throughput of 12 gigaFLOPS in about 2 kW. So even for this relatively large platform, there is not a lot of available size, weight, or power if the future system capabilities demand TeraOps of computing in equivalent SWAP constraints. The capabilities of the processor are dramatically worsened if the signal processor must fit in either a medium- or small-sized UAV.

It is clear from Figure 24-5 that UAVs will continue for the foreseeable future to afford a significant opportunity for the insertion of HPEC systems. In addition to UAVs, the military is also investing in unmanned ground systems (UGS), unmanned underwater vehicles (UUVs), as well as in a large number of existing or planned platforms for land, air, sea, and space deployments. Figure 24-6 depicts the functions that these HPEC systems will perform, the sensor types, and the classes of platforms requiring onboard HPEC systems. Multiple sensors of different types will be housed in a single platform to avoid the proliferation of platforms with a single sensor capability. For every platform deployed, a large supporting "tail" accompanies it. Therefore, it is always desirable to minimize the number of platforms deployed to theater. A platform with multiple sensor modalities on board provides more flexibility.

Furthermore, the challenge today is not in the detection of suspected targets but in their proper identification with a very low probability of false alarms (PFA). Thus multiple modalities integrated into a single platform will allow for cross-referencing targets detected using different sensing modalities while in real-time providing accurate identification with a low PFA. The enabler for the functions and platforms shown in Figure 24-6 will be the ability of a single HPEC system to

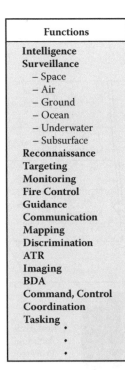
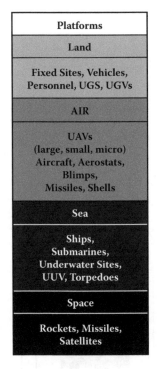
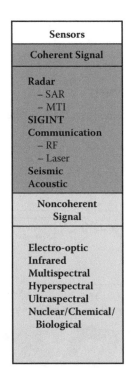

FIGURE 24-6 Functions, platforms, and sensor types.

perform the requisite processing of different sensor modalities. The prevailing architecture will be a hybrid that has for each respective sensor modality a dedicated front-end with real-time and high performance capabilities specific to the data and sensing modality. However, the back-end will be general purpose to accommodate variations across modalities and flexibility in algorithm types. It is commonly true that the front-end dominates the computational complexity but the operations are very well structured. Therefore, the use of dedicated hardware for these functions will lead to significant reductions in overall SWAP dearly needed for highly constrained platforms. However, the back-end is much more complex in types of operations, and it is where typically the architecture designers and users demand flexibility. In the subsequent sections, the example of a surveillance sensor system is used to illustrate the classes of technologies best matched to these different computing functions.

24.2 HARDWARE TRENDS

The best way to illustrate hardware trends projected to impact the systems shown in Figures 24-4 through 24-6 is through an example such as the one shown in Figure 24-7. The architecture consists of an active electronically scanned antenna used to send and receive signals. Advances in waveform generators have led to designs that permit very complex waveforms for specific applications. The reflected waveforms are received via highly compact receive modules encapsulated into compact transmit/receive (T/R) cavities. In some instances, the electronics for the receiver can be integrated into a chip. In other cases, the full chip will contain the T/R modules, the receive electronics, the ADC, and portions of the front-end processor. The specific design depends on the amount of power, levels of integration, and cost available for this part of the architecture. In Figure 24-7, the T/R modules, the analog receiver hardware, and the ADC are illustrated in separate hardware subsystems.

Architectures have been proposed with all the functions shown prior to the back-end signal processing integrated into a single chip device. The system applications suited for this type of

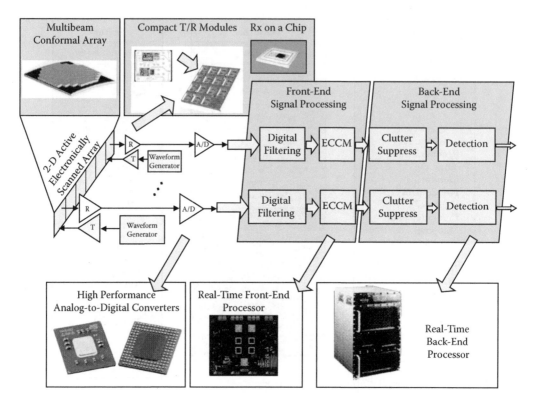

FIGURE 24-7   Sensor architecture for a representative surveillance system.

integration are either in millimeter-wave designs for the detection of very high resolution and close range signals (Kim et al. 2007) or in communication applications at about 24 GHz (Hajimiri et al. 2005). In both cases, one important enabler is silicon germanium (SiGe), needed as the device substrate technology. SiGe has become a prominent enabler as the transistor speed reaches several 100s of GHz (Rieh et al. 2005). Commercial applications of SiGe in high-frequency radio systems for either communication or automotive radar applications will drive the ability to integrate a complete system on a chip (SoC). The defense systems will also benefit from these advances driven by the commercial sector.

The hardware illustrated in Figure 24-7, preceding the back-end signal processor, can leverage from many of the commercial advances in silicon process technology. The SiGe example shown earlier is best matched to high-frequency (millimeter-wave) applications. However, the trend is either to integrate all mixed-signal components on a single device or to integrate several components into a single chip and then incorporate these multiple SoC onto a single highly integrated package. Recent publications describe the advances provided by integrating mixed-signal hardware onto a single device (Claasen 2006). There have also been approaches recommending the integration of a broad range of components—optical, analog, and digital electronics—into a system on package (SoP). The final level of integration will depend heavily on the amount of power dissipated, current and voltage leakages, and component-to-component communication bandwidths (Tummala 2006). Again, defense applications will be the main beneficiaries of these commercial technology advances. As described in earlier chapters in the handbook, several prototypes are under development to demonstrate the value and feasibility of integrating mixed-signal technologies into a SoC for several important defense applications.

The real-time front-end processor can be integrated as part of the previously discussed SoC and/or SoP if the digital filtering functions are simple and regular in structure. However, the inclusion of electronic countercountermeasures (ECCM) adds a high degree of complexity, leading to a high performance processor developed at the board level. It has not yet been demonstrated that it is feasible to incorporate ECCM functions plus all the front-end mixed-signal hardware on a SoC or SoP. There have been demonstrations of critical adaptive beamforming functions on a single device as needed for the ECCM portion of the processing stream shown in Figure 24-7 (Ku et al. 1988; Song 1994). The ECCM step requires matrix inversions and vector-vector multiplies that for the typical data rates of either airborne early warning systems or ground surveillance systems are too demandingly complex to integrate all the front-end hardware into a SoC or SoP. This part of the system is best delegated to a mix of dedicated hardware and FPGAs. The dedicated hardware can be used in the adaptive weight application and the FPGAs can be used in the weight computation used in ECCM algorithms (Nguyen 2004). Again, no general implementation approach fits all platforms for the architecture shown in Figure 24-7. The major determining constraint is the SWAP available in the chosen platform. For small platforms, a high level of integration and simplicity in front-end hardware is necessary to meet the capabilities suitable for real-time implementation.

As in the example shown in Figure 24-7, from the front-end analog and digital system to the back-end signal processor, the implementation choices change, thereby demanding more flexibility in the design. Typically, the real-time back-end signal processor involves a wide range of functions, such as clutter suppression for both air and surface surveillance systems and target detections. The types of operations performed require operating in different dimensions of the incoming data. For example, clutter suppression, as shown in Figure 24-7, would be done on a set of parallel radar beams. On the other hand, detections are often done on independent beams. These changes in data inputs cause what is commonly referred to as *corner turns*, as described in earlier chapters, and present the need to balance computation with communication to feed the signal processor with the right data and, therefore, achieve high levels of efficiency.

For the last several years, PowerPCs (IBM 2007) have been the predominant choice as the general-purpose CPU used in the design of HPEC systems. These systems have also leveraged advances in commercial interconnects integrated into a ruggedized chassis meeting very stringent SWAP requirements. Some of the systems shown in Figure 24-2, particularly those systems developed in the late 1990s and early 2000, were dominated by massively parallel architectures with 100s of PowerPC chips operating concurrently. The challenge has been in maintaining this large number of parallel processors working efficiently.

The newer dominants in the computing industry are emerging from the game industry. IBM, Sony, and Toshiba have formed a partnership to develop the Cell processor (Halfhill 2006; Kahle et al. 2005; Ruff 2005). This processor system is more than just a fast general-purpose processor. As shown in Figure 24-8, the Cell processor consists of a general-purpose controller based on the PowerPC core and a bank of eight synergistic processor elements (SPEs) used to achieve very high computational throughputs. The architecture of this chip is an example of a SoC built under a 90 nm complementary metal oxide semiconductor (CMOS) manufacturing process line.

Sony plans to use the Cell processor with the Nvidea graphics processor as the main horsepower in the PlayStation 3 (Guizzo and Goldstein 2006). The impressive performance comes from an architecture very well balanced between computation and communication. For example, running at 3.2 GHz, the Cell processor can communicate to Rambus DRAM via a bus running at 25.6 Gbytes/s. If the chip can achieve a computation throughput of 200 GFLOPS, this represents a ratio of about 7–8 ops/byte, which is necessary to keep the microprocessor busy (Krewell 2005).

The HPEC community will benefit for many years to come from the advances experienced in the game industry. Figure 24-9 shows a number of critical computing enablers that will be well matched to the back-end signal processor needs of systems such as the one described in Figure 24-7. However, several challenges will need to be resolved in order to properly integrate these enablers into a working system. One of these challenges is meeting the SWAP available (particularly the power and heat dissipation requirements). Another challenge is the development of software and programming approaches to efficiently program what amounts to a SoC with multiple single-

- Total peak performance over 200 GFLOPS running at 3.2 GHz
- 90 nm CMOS process
- 25.6 GBytes/s to Rambus DRAM
- Up to 20 GBytes/s communication bandwidth to graphics processor

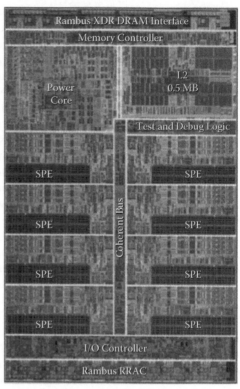

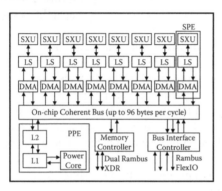

FIGURE 24-8   Cell processor system [adapted from Kahle et al. (2005)].

Cell Processor

- 3.2 GHz clock
- 200+ GFLOP/s peak
- Estimated 30–60 W

Graphics/Video Processors

- ATI
- NVidia
- 256–512 bit memory bus
- 1–2 TeraOps

FPGAs

- Xilinx
  - 330K logic elements
  - 550 MHz clock
- Altera
  - 622 user I/O pins

Chassis

- ATR
- VME/VME64/VME64x

- Air, conduction, or liquid cooling
- Shock isolated vs. hard mounted

Interconnects

- 10 Gbit Ethernet
- 3GIO
- HyperTransport/infinipath
- InfiniBand
- Myrinet
- QsNet
- RapidIO
- SCI
- StarFabric

Software

- VSIPL/VSIPL++
- MPI
- CORBA

- VxWorks
- Linux/rt Linux

- Service-oriented architecture
  - XML
  - WSDL
  - SOAP
  - UDDI

FIGURE 24-9   Critical HPEC enablers.

instruction/multiple-data stream SPEs running in parallel with multiple Cell processors operating together to meet several TeraOps of computational throughput capability.

24.3   SOFTWARE TRENDS

Software for HPEC systems continues to be a major concern because system costs are now dominated by software development. This is a particularly serious concern for defense applications because, as illustrated in Figure 24-10, there is an exponential growth in the complexity (measured as 1000s lines of source code) in many of today's military platforms (Nielsen 2002). The exponential growth shown in Figure 24-10 includes more than just the HPEC system. However, as the HPEC systems become more complex, requiring many processors operating in parallel, rigorous and disciplined software development approaches are needed. As shown in the right-hand side of Figure 24-10, in contrast to the early 1990s, the software now must comply with a layered approach separating the vendor-specific software from the application software. This approach leads to better reuse and a higher likelihood of portability of the application from one HPEC system to another. This portability challenge is important since the hardware refresh evolves more rapidly than do the changes in application requirements. As new hardware becomes available, the user would like to transition existing software to new hardware to minimize size, weight, and power and to better utilize premium space in the embedded platforms.

In the past several years, the HPEC community [supported by the Office of the Secretary of Defense, Director of Defense Research and Engineering (DDR&E)] has striven to standardize the software development for HPEC systems. DDR&E helped in the development of the object-based Vector, Signal, and Image Processing Library (VSIPL++), and several other organizations (e.g., the Defense Advanced Research Projects Agency, National Science Foundation, Department of Energy, Navy) also contributed to earlier instantiations of these signal processing libraries. As shown in Figure 24-11, the evolution of these libraries included the adaptation of middleware libraries developed for non-real-time scientific computing but modified to meet real-time signal processing applications. The development of these libraries also involved the evolution of communication libraries designed for massively parallel systems, such as the Message Passing Interface (MPI) library.

The trend in software development for HPEC systems is (1) to acquire a parallel hardware system based on the critical enablers shown in Figure 24-9, including vendor-specific software, and (2) to develop the application on top of this software but compliant with embedded software standards

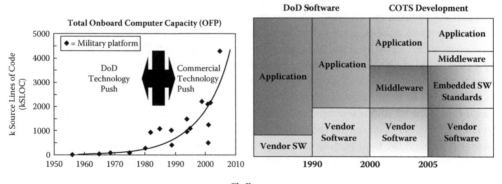

FIGURE 24-10   Embedded software challenges.

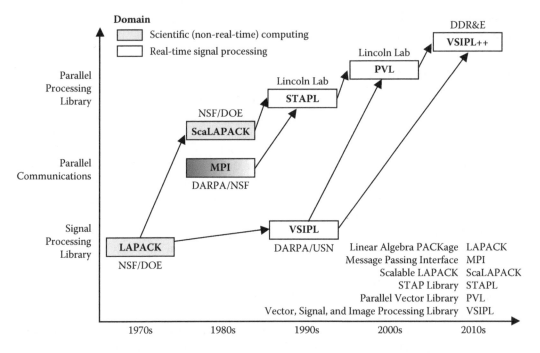

FIGURE 24-11　Evolution of modern signal processing middleware libraries.

(e.g., VSIPL++). Since existing and/or future defense platforms will likely demand a broad range of functions (shown in Figure 24-6), this software development approach will permit rapid software development for different types of applications resident on the same underlying hardware system while reusing a lot of the same vendor-specific software and standard middleware libraries. This approach will result in lower software development costs than if every new application function were developed from unique HPEC software.

In addition to the cost incurred in software development, another significant driver in the cost of developing HPEC systems is the integration, code validation, and verification to make sure that the whole system works as designed. One approach that will be more prevalent in the future is to verify the software code for functionality on a single workstation. The advent of MATLAB (MathWorks 2007) permits direct verification of the real-time algorithm code against an equivalent version of the algorithm running in MATLAB. As shown in Figure 24-12, once the code is verified for functionality, the same code is ported over to a cluster of computers to begin parallelizing the code. If the code were originally developed using the layered approach illustrated in Figure 24-10 and implemented based on middleware libraries shown in Figure 24-11, the porting to a cluster of computers to verify that the code is properly working in a parallel architecture can be done expediently. Once the code has been parallelized properly, it can be ported over to the final real-time platform to ascertain latencies and throughput performance. All along, a set of "gold standard" vectors and algorithm code developed during the earlier stages using the single workstation and MATLAB can be carried through the final porting to the real-time system to verify proper algorithm functionality while running in real time. This approach has been successfully implemented in many of the systems illustrated in Figure 24-2.

This section has addressed hardware and software technologies relevant to sensor node architectures. A sensor node is any type of platform shown in Figure 24-4. In the future, these platforms will be nodes in a network conforming to the DoD-sponsored developments referred to as the Global Information Grid. The sensor node will post its processed data (output of the HPEC system) to the network and will be identified by a unique IP address. The next section addresses the system complexity and technologies of a net-centric architecture consisting of multiple sensor nodes.

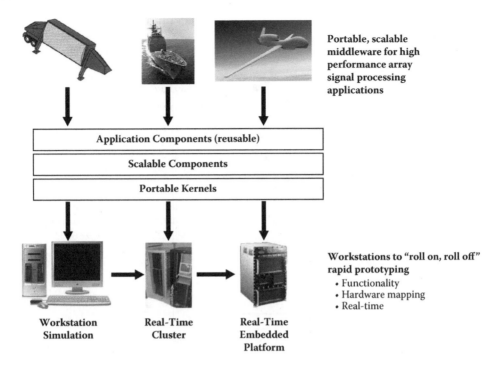

Portable, scalable
middleware for high
performance array
signal processing
applications

Application Components (reusable)

Scalable Components

Portable Kernels

Workstations to "roll on, roll off"
rapid prototyping
- Functionality
- Hardware mapping
- Real-time

Workstation	Real-Time	Real-Time
Simulation	Cluster	Embedded
		Platform

FIGURE 24-12   HPEC software development methodology.

24.4   DISTRIBUTED NET-CENTRIC ARCHITECTURE

Military systems are evolving toward a more net-centric architecture (Wilson 2000). A net-centric architecture consists of multiple nodes distributed worldwide and connected by way of an architecture composed of multiple levels of networks. The networks include terrestrial networks, wireless networks connecting ground systems to air breathers and satellites, and satellite-to-satellite high-speed communications. Ultimately, the goal is to have node systems (sensor nodes described earlier are an example of this class of systems) able to communicate irrespective of their space and time location across the earth. The enabler will be an Internet-like infrastructure with equivalent protocols and interconnection standards leveraging the commercial advances.

This formidable goal will be enabled by the development of the GIG, which not only encompasses communications links but also includes theater and worldwide communications plus the supporting infrastructure. Figure 24-13 illustrates the GIG's principal elements. In addition to the communications infrastructure (across satellites, airborne platforms, and terrestrial links), Net-Centric Enterprise Services (NCES) will provide the network services necessary to make the GIG a useful architecture (Defense Information Systems Agency 2007).

HPEC systems have a unique opportunity to play a significant role in the future as high performance computing can complement high-speed networks being developed under the GIG. There will be a need to deploy worldwide high-speed processing systems operating in a distributed fashion. Distributed storage centers and distributed compute centers will be interconnected via the GIG infrastructure. This distributed net-centric architecture will enable the military to rapidly transform sensor data to information and actionable knowledge. As illustrated in Figure 24-14, sensor nodes (with capabilities previously described) will be attached to a network (shown as the ring on the left image of Figure 24-14). This ring of sensors located across the globe will post as well as receive data on a need-to-share basis.

How much is done on board the platform will depend on the availability of SWAP inside the platform, system complexity, and cost. In some instances, information, knowledge, and intelligence,

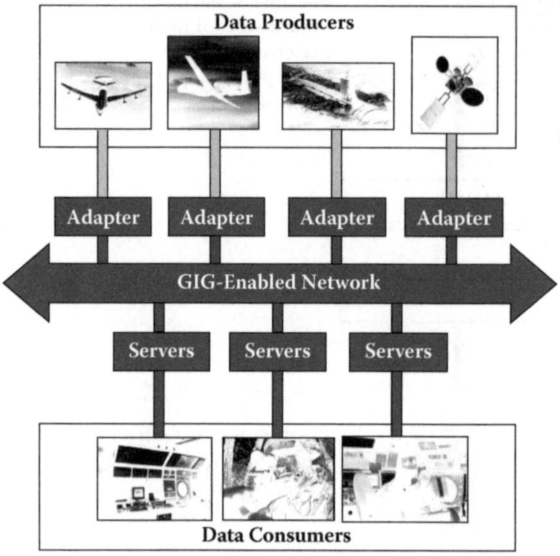

Global Information Sharing

Data Producers

Adapter   Adapter   Adapter   Adapter

GIG-Enabled Network

Servers   Servers   Servers

Data Consumers

Net-Centric Technologies

- **Commercial sector**
  - Service-oriented architectures
  - Web services
  - Terrestrial and wireless comms
  - Semantic Web

- **Military sector**
  - Global Information Grid
    - GIG-Bandwidth Expansion (terrestrial)
    - Joint Tactical Radio System
    - Transformational Satellite (TSAT)
    - Network-Centric Enterprise Services (NCES) (core enterprise services to the GIG)
    - Information assurance (encryption)
    - Teleport (theater, reach-back)
    - Joint network management system (joint network management tools)

FIGURE 24-13   The Global Information Grid (GIG) principal elements.

shown on the right-hand side of Figure 24-14, might best be done on the ground across multiple processing centers. The sensor node might have a single modality or it might span a broader range of modalities (e.g., infrared, RF, multispectral, etc.). However, the transformation of data from these sensor nodes into actionable intelligence might be made most cost-effective by leveraging the high-speed networks and exploiting data across multiple receive centers located on the ground in the theater of operations or back to the Continental U.S. (CONUS via "reachback" links).

HPEC advances in computing technology are very well matched to the needs of a distributed net-centric architecture. Ground centers will consist of terabytes to petabytes of storage in massively parallel processing centers with the ability to reach across the network in a distributed architecture. The massively parallel processing is necessary to (1) rapidly process the large amount of data incoming from different and distributed sensors, as well as from other processing centers, (2) share the information, and (3) deliver products to the users in real-time.

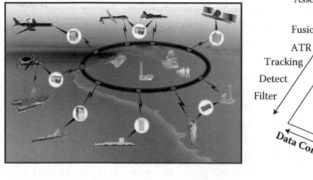

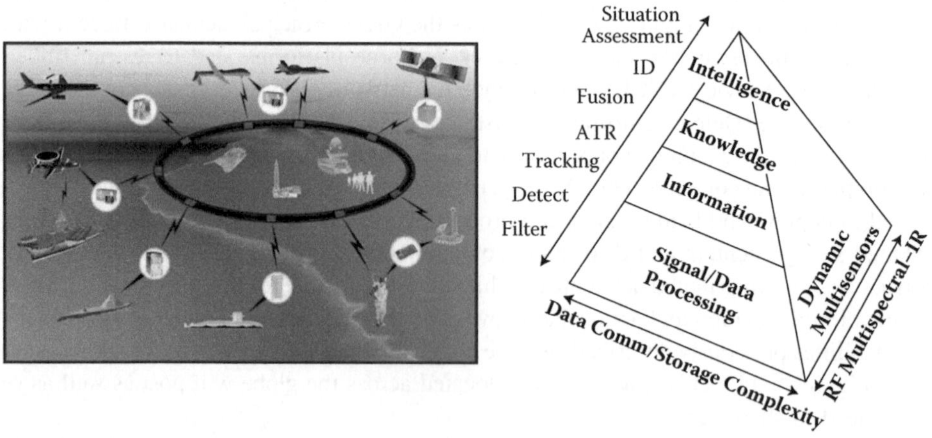

FIGURE 24-14   Distributed net-centric architecture.

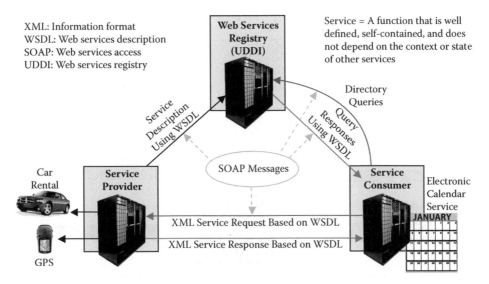

FIGURE 24-15 Web-services basics [adapted from D.K. Barry (2003)].

Commercial enterprises, such as banking, finance, and travel, are already utilizing terrestrial high-speed networks to more effectively provide services to the consumers. They are adhering to what is referred to as a service-oriented architecture. The SOA enables disparate enterprises to utilize and develop capabilities based on a common set of standards (Barry 2003). The SOA operates in a web-based infrastructure. The web services allow consumers to receive the best available services and enable service providers to offer the most competitive services without having to completely instantiate new web-based protocols for each respective consumer-provider set of transactions. Figure 24-15 illustrates a simple example, adapted from Barry (2003), describing the interface of a consumer electronic calendar with a car rental provider. The car rental provider posts its services to a service registry. Its services are described via the Universal Description, Discovery and Integration (UDDI) standard. A consumer performs a query via the web services registry and receives a response that a car rental provider has the services queried; the consumer and the service provider carry out their transactions across their respective computer systems, utilizing standards developed under an SOA architecture (e.g., XML, WSDL, SOAP).

An analogous web-based architecture for a military distributed net-centric architecture is illustrated in Figure 24-16. The service provider is analogous to one or more sensor nodes connected to the GIG network; the consumers are the ground processing centers receiving the data. Consumers query for the type of data needed, via a web-services registry, and once the services are identified, the consumers and service provider perform a binding and transactions start to occur to meet the transformation of sensor node data into actionable intelligence. These requisite standards and protocols, as well as the overall infrastructure, are under development within the rubric of the NCES (Lewis 2006).

An SOA architecture for use by the military, developed under the GIG infrastructure, must leverage in a significant way the advances ongoing in the commercial sector. Figure 24-16 illustrates the analogy between the commercial enterprise and the military enterprise. The commercial sector has a community of interest (COI) employing the SOA architecture, such as shown on the left side of Figure 24-16. This is at the application layer. The analogous COIs in the military are the space control, theater ballistic missile defense, intelligence-surveillance-reconnaissance applications, to name a few. The NCES is adapting many of the SOA network services from the commercial sector. Similarly, transformational communications efforts are leveraging many of the commercial advances in global networks and wireless communications at the physical and transport layers. This utilization of commercial advances will make the availability of these capabilities feasible and

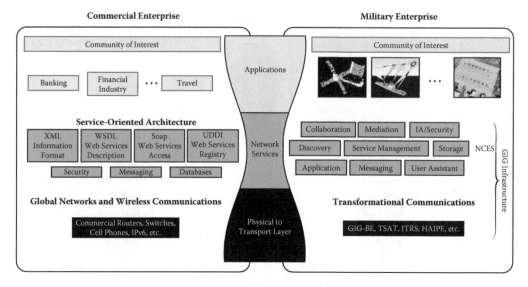

FIGURE 24-16 Building from a commercial service-oriented architecture.

affordable for the military. HPEC systems will be at the center of these capabilities by providing the computational engines that will enable the data information/knowledge processing to take place.

24.5 SUMMARY

Applications utilizing complex HPEC systems continue to proliferate. It is becoming more common to employ fewer platforms in theater, each with a much more capable system on board. Future systems will employ multiple modalities for target identification with the lowest possible false alarm. This combination of capabilities will lead to HPEC systems exceeding several TeraOps in computational throughput.

The advent of high-speed communication networks will also permit a trade-off between what computation is done on board the platform versus what computation is best done at a ground computing center. The DoD is investing significantly in the GIG, endeavoring to tie sensor nodes, processing centers, weapon systems, and users together irrespective of time or location across the globe. This linked capability is crucial to make the net-centric architecture a reality.

HPEC systems will be deployed in highly constrained platforms. Examples of these platforms are the full class of unmanned systems, such as UAVs, UGS, and UUVs. Since these platforms are very constrained in size, weight, and power, a hybrid architecture will be used to implement a mix of custom hardware for compute-intensive functions close to the front-end of the system and general-purpose COTS-based hardware for the back-end of the system. The back-end of the system will be fully programmable to permit the implementation of different classes of algorithms commensurate with the different classes of modalities available in a single platform.

The availability of high-speed networks will permit a ubiquitous and distributed computing architecture in which a platform performing a given set of functions will come into the network via a unique IP address and post its data for the user to exploit on a need-to-share basis. Users will subscribe to the type of data needed, and platforms able to produce the desired data will publish the results. This data interchange, referred to as a publish-subscribe architecture, should guarantee that no one in the GIG network will be hindered in the ability to share information.

The net-centric architecture will leverage architectures and standards becoming available through the commercial sector. One important enabler is the service-oriented architecture evolving in the commercial sector that includes banking, financial, travel, and other industries seeking to more easily provide a wide range of services to consumers. The military is adopting this SOA to permit service providers such as sensing platforms, intelligence centers, and weapons systems to more rapidly and effectively communicate and share information. These service providers and consumers will demand complex HPEC systems deployed in highly constrained platforms, as well as processing centers to rapidly process data emanating from the sensing systems and/or available from large archived databases.

The evolution in Internet standards combined with the exponential evolution in computing technologies will present a rich set of enablers crucial to the HPEC systems of the future. Therefore, the DoD's goal of achieving a transformation in military operations will be possible through the confluence of the evolving commercial capabilities in computing, communication networks, Internet protocol standards, and software methodologies, all adapted to the significant and unique demanding requirements of military systems.

REFERENCES

Barry, D.K. 2003. *Web Services and Service-Oriented Architectures*. San Francisco: Morgan Kaufmann.

Claasen, T. 2006. An industry perspective on current and future state of the art in system-on-chip (SoC) technology. *Proceedings of the IEEE* 94(6): 1121–1137.

Defense Information Systems Agency. 2007. Net-Centric Enterprise Services (NCES). Available online at http://www.disa.mil/nces/index.html.

Guizzo, E. and H. Goldstein. 2006. Expressway to your skull. *IEEE Spectrum* 43(8): 34–39. Also available on *IEEE Spectrum* online at http://www.spectrum.ieee.org/aug06/4256.

Hajimiri, A., H. Hashemi, A. Natarajan, X. Guan, and A. Komijani. 2005. Integrated phased array systems in silicon. *Proceedings of the IEEE* 93(9): 1637–1655.

Halfhill, T. 2006. Cell processor isn't just for games. *Microprocessor Report* 20(1).

International Business Machines. Accessed 2007. IBM PowerPC 7XX and 6XX Microprocessors. Available online at http://www-306.ibm.com/chips/techlib/techlib.nsf/products/PowerPC_7XX_and_6XX_Microprocessors.

Kahle, J.A., M.N. Day, H.P. Hofstee, C.R. Johns, and T.R. Maeurer. 2005. Introduction to the Cell processor. *IBM Journal of Research and Development* 49(4/5). Available online at http://www.research.ibm.com/journal/rd/494/kahle.html.

Kim, H., S. Duffy, J. Herd, and C. Sodini. 2007. SiGe IC-based mm-wave imager. *IEEE International Symposium on Circuits and Systems* 1: 1975–1978.

Krewell, K. 2005. Powering next-gen game consoles. *Microprocessor Report* 19(7): 26–29.

Ku, W.H., R.W. Linderman, P.M. Chau, and P.P. Reusens. 1988. High Performance Signal Processor, U.S. Patent 4,791,590, filed November 19, 1985, and issued December 13, 1988.

Lebak, J.M., J. Kepner, H. Hoffman, and E. Rutledge. 2005. Parallel VSIPL++: an open standard software library for high-performance parallel signal processing. *Proceedings of the IEEE* 93(2): 313–330.

Lewis, D. 2006. NCES SOA Foundation. Defense Information Systems Agency.

The MathWorks. MATLAB details can be found online at http://www.mathworks.com/.

National Security Agency. Accessed 2007. The GIG vision enabled by information assurance. Available online at http://www.nsa.gov/ia/industry/gig.cfm?MenuID=10.3.2.2.

Nielsen, P. 2002. Perspective on embedded computing, keynote address. *Proceedings of the Sixth Annual High Performance Embedded Computing Workshop*, MIT Lincoln Laboratory, Lexington, Mass. Available online at http://www.ll.mit.edu/HPEC/agendas/agenda02.html.

Nguyen, H. 2004. 130-GOPS radar front-end processor for wideband channelization and adaptive beamforming. Division Seminar. Lexington, Mass.: MIT Lincoln Laboratory.

Office of the Secretary of Defense. 2005. *Unmanned Aircraft Systems Roadmap 2005–2030*.

Rieh, J., D. Greenberg, A. Stricker, and G. Freeman. 2005. Scaling of SiGe heterojunction bipolar transistors. *Proceedings of the IEEE* 93(9): 1522–1535.

Ruff, J. 2005. Cell Broadband Engine Architecture and Processor. IBM Systems and Technology Group.

Song, W. 1994. VLSI bit-level systolic array for radar front-end signal processing. *Conference Record of the 28th Asilomar Conference on Signals, Systems, and Computers* 2: 1407–1411.

Tummala, R. 2006. Moore's law meets its match. *IEEE Spectrum* 43(6): 44–49. Also available on *IEEE Spectrum* online at http://www.spectrum.ieee.org/jun06/3649.

Wilson, J. 2000. Network-centric warfare 21st century. Military & Aerospace Electronics. Available online at http://mae.pennnet.com/articles/article_display.cfm?Section=ARCHI&C=Feat&ARTICLE_ID=66864 &KEYWORDS=command-and-control&p=32.

25 A Review on Probabilistic CMOS (PCMOS) Technology: From Device Characteristics to Ultra-Low-Energy SOC Architectures

Krishna V. Palem, Lakshmi N. Chakrapani, Bilge E. S. Akgul, and Pinar Korkmaz, Georgia Institute of Technology

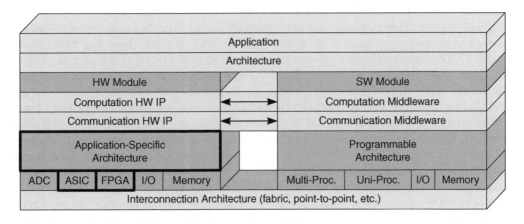

This chapter presents a novel technology to overcome energy consumption by CMOS chips and accompanying heat dissipation. This technology can have a major impact on chip performance (including better management of circuit noise) as devices scale into the nanometer regime.

25.1 INTRODUCTION

Complementary metal oxide semiconductor (CMOS) technology scaling, primarily driven by Moore's Law (Intel), faces serious obstacles as it continues into the nano-regime. The challenges posed by effects such as deep submicron noise (Kish 2002; Sano 2000), parameter variations (Borkar et al. 2003; Bowman et al. 2000), energy consumption (Hegde and Shanbhag 2000; ITRS 2005), and the associated heat dissipation of these devices (Kim et al. 2003; Mudge 2001, 2004), emerge as major challenges in the International Technology Roadmap for Semiconductors. The twin obstacles—noise and energy consumption—to technology scaling have been the subjects of extensive studies. A few studies worth highlighting are the following: Stein (1977) has studied the limitations imposed by thermal noise on the reduction of supply voltage, and hence switching energy, of CMOS circuits; similarly, Natori and Sano (1998) have studied the relationship between reliability and the energy consumption of CMOS circuits; and Kish (2002) has studied the limitations imposed by thermal noise on CMOS device scaling. These studies indicate that scaling down feature sizes and lowering energy consumption, while preserving reliable operation of CMOS devices in the presence of noise, are difficult challenges.

This chapter reviews a novel technique, referred to as *probabilistic* CMOS or PCMOS technology, as a promising approach to addressing the aforementioned challenges and as a dramatic shift from previous work. Devices based on this technology, in which noise is harnessed as a resource to implement CMOS devices exhibiting probabilistic behavior, are guaranteed to compute correctly with a probability p. Here, p is a design parameter; and by design, the devices are expected to compute incorrectly with a probability $(1 - p)$. The foundations of PCMOS technology are rooted in the physics of computation, algorithms, and information theory. Earlier, using techniques derived from the physics of computation and information theory, Palem (2003) showed that *the thermodynamic cost of computing a bit of information is directly related to its probability p of being correct.* This proof uses purely entropic arguments derived from the second law of thermodynamics (Palem 2005). Further, using an abstract model of computation, the randomized bit-level random access machine (RaBRAM) (Palem 2003), and the example of the *distinct vector problem* (Palem 2003), it has been demonstrated that such energy savings at the switching level can be harnessed at the application level to construct a probabilistic algorithm [the value amplification algorithm (Palem 2003)] that is more (energy) efficient than is the *best possible deterministic algorithm.*

While the work mentioned above demonstrates the energy efficacy and utility of probabilistic behavior in abstract computational models, the foundational principles have also been extended into the CMOS domain through a systematic characterization of a PCMOS device (Cheemalavagu et al. 2004, 2005; Korkmaz, Akgul, and Palem 2006; Korkmaz et al. 2006). This characterization involves the study of the relationship between the probability p of reliable operation, the noise level, and the switching energy. Further, this relationship (captured by the two PCMOS laws) is verified through simulations of PCMOS devices implemented in TSMC 0.25 μm and AMI 0.5 μm processes. Through analytical modeling and simulations, it is shown that, while devices based on PCMOS technology exhibit probabilistic behavior due to low-voltage operation and noise susceptibility, they achieve extreme energy savings in return. Further, we demonstrate the utility of PCMOS technology to computing platforms by studying application-specific architectures that cannot only harness PCMOS technology to implement real-world applications, but also can lead to extremely efficient implementations—both in terms of energy measured in Joules as well as performance (running time) measured in seconds, simultaneously captured by the *energy × performance* metric—when compared to conventional CMOS-based designs. The device studied is a PCMOS-based inverter, ubiquitous to digital design and a key building block used in probabilistic applications such as hyperencryption (Ding and Rabin 2002). The PCMOS-based implementation achieves up to an order of magnitude savings (quantified through the *energy × performance* metric) when compared to a competing custom application-specific integrated circuit (ASIC) realization of the hyperencryption application. The utility and benefits of such PCMOS-based architectures in implementing probabilistic applications such as pattern recognition, optimization, classification, patient monitoring, and Windows printer troubleshooting through randomized neural networks (Gelenbe 1991; Gelenbe and Batty 1992), probabilistic cellular automata (Fuks 2002), and Bayesian inferencing (Beinlich et al. 1989; Pavlovic et al. 2000; Rehg, Murphy, and Fieguth 1999) are detailed in our previous work in the context of a probabilistic system on a chip (PSOC) (Chakrapani et al. 2006).

Besides the aforementioned probabilistic applications, we have also shown the utility and benefits of PCMOS-based probabilistic computing in the context of error-tolerant applications, such as synthetic aperture radar (SAR) imaging and video decoding [see Palem and his colleagues (George 2006; Akgul 2006)]. Here, through the use of probabilistic arithmetic, device-level bit errors—which translate into image quality measured through signal-to-noise ratio (SNR) at the application-level—can be mitigated and traded for energy savings with minimal impact on application quality (George et al. 2006).

The rest of the chapter is organized as follows. Section 25.2 details the implementation example of a PCMOS switch, its behavior through a succinct analytical model, and the associated laws governing its behavior. These characterizations are validated through HSpice simulations for two technology generations. Section 25.3 introduces a PCMOS-based system-on-a-chip implementation to demonstrate the fact that well-characterized noise and, hence, well-quantified probability of error can be of value in realizing extremely efficient low-energy architectures in the context of probabilistic applications as well as in the context of error-tolerant applications. Section 25.4 summarizes the conclusions reached.

25.2 CHARACTERIZING THE BEHAVIOR OF A PCMOS SWITCH

This section comprehensively characterizes a switch rendered probabilistic due to thermal noise. First explained is a CMOS inverter realization of a probabilistic switch. We then develop an analytical model and the associated laws of PCMOS technology, crucial to understanding the probabilistic behavior of a PCMOS switch, and then validate the model via HSpice simulations. Through the study of the analytical model and its validation via simulations, the thermal noise source is initially assumed to be freely available. Finally, we present a practical realization of a PCMOS switch with limited available noise (rather than being freely available), wherein additional cost due to noise amplification is also considered.

25.2.1 INVERTER REALIZATION OF A PROBABILISTIC SWITCH

A CMOS inverter is a digital gate that executes the *inversion* function with one input and one output. The switching in this case is the invocation of the inversion function, which, in the context of the CMOS inverter, corresponds to the flow of the switching current through the output capacitance of the inverter. For a deterministic inverter, $Y(t_2) = \overline{X(t_1)}$, where X and Y denote the binary values of the input and the output of the inverter, respectively, t_2 denotes the point in time when the switching ends, and t_1 denotes the point in time when the switching starts. For a probabilistic inverter, on the other hand,

$$Y(t_2) = \begin{cases} \overline{X(t_1)} \text{ with probability } p \\ X(t_1) \text{ with probability } (1-p), \end{cases} \tag{25.1}$$

wherein p denotes the probability of correctness, such that $1/2 < p < 1$. In Equation (25.1), the probability p results from the noise coupled to the CMOS inverter. The particular noise sources coupled to the inverter include naturally occurring thermal noise sources as studied by Sano (2000), as well as power grid noise sources as studied by Heydari and Pedram (2000) and Pant et al. (2004), which are increasingly common in giga- and tera-scale integrated circuits (ICs) based on deep submicron technologies. In this chapter, thermal noise is considered as the available noise source coupled to the inverter.

As stated by Stein (1977), noise can be modeled as being coupled* to the output of the inverter as shown in Figure 25-1(a). To understand how noise induces probabilistic behavior, first consider the transfer characteristics of an ideal inverter shown in Figure 25-1(b). As shown in the figure, a single *switching* step of an ideal deterministic inverter is instantaneous and occurs at a value of

---

* For details on comparisons of output coupling, input coupling, as well as power supply noise coupling, please see Korkmaz et al. (2006).

$$\frac{V_{dd}}{2}.$$

The binary values of 0 and 1 correspond to a (measured) output voltage in the intervals

$$\left(-\infty, \frac{V_{dd}}{2}\right) \text{ and } \left(\frac{V_{dd}}{2} + \infty\right), \text{respectively.}$$

However, noise that is present at the output node interacts with the voltage values representing a deterministic 0 or 1 corresponding to the output signal. Noise that is characterized by a Gaussian distribution with a standard variation σ is superimposed on the output signal, and, therefore, it destabilizes the output of the inverter, causing incorrect switchings.

To characterize the erroneous behavior and establish a relationship between noise magnitude, signal magnitude, and the probability of correctness, p, refer to the digital 0 and 1 regions shown in Figure 25-1(c). This figure corresponds to the case in which the inverter is coupled with thermal noise at its output [see Figure 25-1(a)]. The thermal noise has a Gaussian distribution with a standard deviation of σ, also referred to as the root mean square (RMS) value of noise. Here, the curve on the left (right) has a mean of 0 V (1 V), and it corresponds to the case in which the input V_{in} is 1 (0). Thus, with an output value of 1, for example (measured to be V_{dd} in the ideal case), the instantaneous value will be determined by the noise distribution. Thus, while the output value ought to be V_{dd} because of additive noise, it can easily be in the interval

$$\left(-\infty, \frac{V_{dd}}{2}\right),$$

inherently yielding a value of 0. In Figure 25-1(c), the probability of error corresponding to an output value of 1 being erroneously treated as 0 is equal to the area **A**, and that of 0 being erroneously treated to be 1 corresponds to the area **B**. Note that, from symmetry, $\mathbf{A} = \mathbf{B} = (1 - p)$, where p is the probability of being correct [or, equivalently, $(1 - p)$ is the probability of error] in both cases.

To understand the behavior of a probabilistic inverter, compare the output waveforms of a deterministic inverter with those of a probabilistic inverter for the same input signal. The input signal waveform is shown in Figure 25-2(a). The corresponding output voltage waveform of a deterministic inverter is shown in Figure 25-2(b), and the corresponding output of a probabilistic inverter is shown in Figure 25-2(c). In Figure 25-2(c), the inverter is coupled with thermal noise at its input, and it is designed to switch correctly with a probability parameter $p = 0.87$. Because of the noise, the

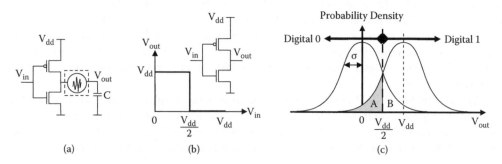

(a)   (b)   (c)

FIGURE 25-1   (a) An inverter coupled with thermal noise at its output; (b) an idealized inverter transfer curve; (c) probabilistic behavior determined by thermal noise coupled at the output of the inverter.

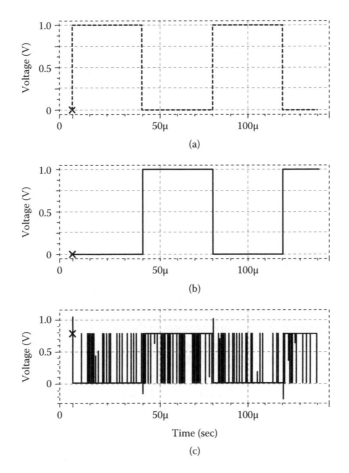

FIGURE 25-2 (a) Input voltage of a deterministic CMOS inverter; (b) output voltage of a deterministic CMOS inverter; (c) output voltage of a probabilistic CMOS inverter with probability parameter p = 0.87 for the same input.

output voltage of the inverter undergoes transitions to binary 0, while it should be at binary 1, and vice versa. Such a probabilistic inverter is of great value, and the utility of such a switch is presented in Section 25.3, wherein the benefits in terms of energy savings and performance improvements are studied via an example application with probabilistic workload.

Note that in Figure 25-2, the output voltage level for a deterministic inverter is higher (1 V) than it is for a probabilistic inverter (0.8 V). This is because the probabilistic behavior for the inverter is realized through varying two parameters: (1) the noise amount coupled on the inverter character-ized as its RMS value in volts and (2) supply voltage V_{dd} of the inverter. In Figure 25-2(c), the supply voltage value of 0.8 V corresponds to a probability value $p = 0.87$ with a noise RMS value of 0.4 V. The details of the effects of the two parameters, the amount of noise and the supply voltage, will constitute the two laws and are detailed in the following sections. Briefly, since the probability p results from noise destabilizing the inverter shown in Figure 25-1(a), the probability parameter p is decreased either by increasing the noise (rms) magnitude or by decreasing the operating supply voltage of the inverter, V_{dd}. As a result, incorrect switchings occur at the output of the inverter as shown in Figure 25-2(c).

With this as background, the next section presents the analytical model of the PCMOS inverter and then the associated laws deduced from this model.

25.2.2   ANALYTICAL MODEL AND THE THREE LAWS OF A PCMOS INVERTER

The *analytical model* of a PCMOS inverter is given in Equations (25.2), (25.3), and (25.4) below. This model is derived from the output voltage distribution [see Figure 25-1(c)] of an inverter coupled with thermal noise at its output [see Figure 25-1(a)] and summarizes the relationship between the probability p, the operating voltage V_{dd}, and the noise magnitude σ.

$$p = 0.5 + 0.5 \cdot er\,f\left(\frac{V_{dd}}{2\sqrt{2} \cdot \sigma}\right), \tag{25.2}$$

$$\sigma = \frac{V_{dd}}{2\sqrt{2} \cdot inver\,f\left(2 \cdot p - 1\right)}, \tag{25.3}$$

$$E = 4 \cdot C \cdot \sigma^2 \cdot \left[inver\,f\left(2 \cdot p - 1\right)\right]^2. \tag{25.4}$$

Note that in the above equations, *inver f* refers to the inverse of the well-known error function *er f* [see Strecok (1968)] and C corresponds to the output capacitance of the inverter. As demonstrated by Korkmaz and Palem (2006) and Akgul et al. (2006), the error function behavior can be approximated using asymptotic notions from algorithm analysis in computer science. Briefly, E, the energy to produce a probabilistic bit (as seen in Equation 25.4), grows with p and the order of this growth dominates an exponential. Such an approximation of the above analytical model allows us to deduce relationships between the probability parameter p, the voltage, or equivalently signal magnitude V_{dd}, the noise magnitude σ, and finally, the energy consumed per switching step denoted by E. The relationships can be extrapolated over successive technology generations by substituting the corresponding load capacitance value C from the ITRS road map (2005) in the model.

As seen from Equation (25.2), for a fixed value of σ, p increases as V_{dd} is increased. This fact can also be seen by referring back to Figure 25-1(c) as follows: as V_{dd} is increased, while keeping σ fixed, the area **A** (or **B**) decreases. This area corresponds to the probability of error $(1 - p)$. Thus, decreasing the area **A** (or **B**) increases p. We show the relationship between V_{dd} and p in Figure 25-3 for the 0.5 μm and 0.25 μm processes.

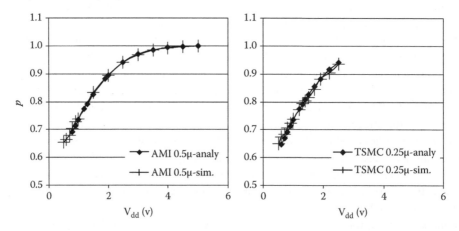

FIGURE 25-3   The relationship between V_{dd} and p with a noise RMS value $\sigma = 0.8$ volts across 0.5 μm and 0.25 μm processes.

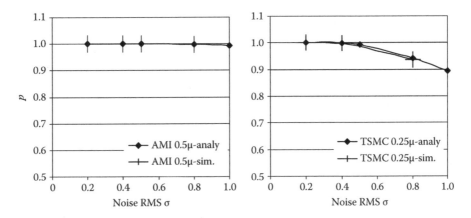

FIGURE 25-4   The relationship between σ and p across 0.5 μm and 0.25 μm processes.

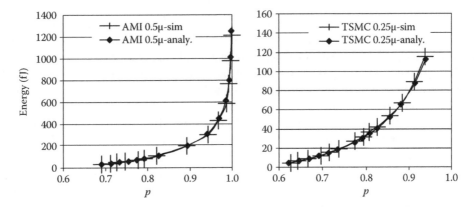

FIGURE 25-5   Law 1 validated over 0.5 μm and 0.25 μm processes.

On the other hand, for a fixed value of V_{dd}, p decreases as σ is increased. Similarly, this relationship can also be intuitively deduced from Figure 25-1(c): as σ is increased, while keeping V_{dd} fixed, the intersection area **A** (or **B**) increases and, hence, p decreases. Figure 25-4 shows this relationship between σ and p for the 0.5 μm and 0.25 μm processes.

Both Figures 25-3 and 25-4 give the analytical estimates of the relationships to the results from HSpice simulations. As seen from the figures, the analytical model matches very well with the simulation results, and the difference between the model and the simulations is less than 2.5% for the $V_{dd} - p$ relationship (Figure 25-3) and less than 1% for the σ – p relationship (Figure 25-4).

Based on the above relationships, the behavior of a PCMOS inverter is further characterized through the two laws that relate the energy consumed, the associated probability parameter p, and the noise RMS σ below.

Briefly, the *first law* relates the energy consumed per switching step to p, given a fixed amount of noise magnitude σ; whereas the *second law* relates the energy consumed per switching step to σ, given a fixed value of p. For a more complete description of the laws with asymptotic notions, please see the previous work of Palem and his colleagues (Akgul et al. 2006; Korkmaz and Palem 2006).

Law 1: For any fixed technology generation or feature size (which determines the capacitance C) and constant noise magnitude, the switching energy E consumed by a probabilistic switch grows with p. Furthermore, the order of growth of E in p is asymptotically bounded below by an exponential in p.

As shown in Figure 25-5, Law 1 captures the changes in E by varying probability p (resulting from varying V_{dd}) value determined by Equation (25.4) of the analytical model. As shown in this

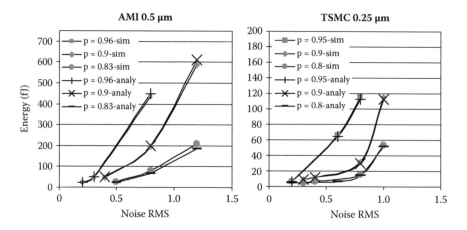

FIGURE 25-6   Law 2 validated over 0.5 μm and 0.25 μm processes.

figure, the analytically estimated values track simulated results well; and the difference between the model and the simulations is less than 4.5% in the case of the 0.5 μm process and less than 4% in the case of the 0.25 μm process. Note that the $E - p$ relationship in Figure 25-5 is obtained for a fixed amount of noise magnitude, wherein σ = 0.8 V. As an example to illustrate Law 1, for σ = 0.8 V, the energy consumed by a probabilistic inverter (designed in the 0.5 μm process) rises from 450 fJ to 1250 fJ, in going from a probability value of $p = 0.97$ to a (slightly) higher value of $p = 0.999$.

Law 2: For any fixed technology generation (feature size), the switching energy E consumed by a probabilistic switch increases quadratically with noise magnitude whenever p remains constant.

Law 2, illustrated in Figure 25-6, relates the quadratical changes in E to the variations in σ value determined by Equation (25.4) of the analytical model. This quadratical relationship between σ and E for a fixed value of p follows from the fact that the switching energy $E = 1/2 \cdot C \cdot V_{dd}^2$ is quadratically related to V_{dd}, which is linearly dependent on σ [as shown in Equation (25.3)]. Moreover, it is intuitive that as σ is increased, more energy will be consumed to realize the same value of p. Similar to the case of Law 1, and as seen from Figure 25-6, the analytical results are matched very closely with the simulation results; and again, the difference between the model and the simulations is less than 4.5% in the case of the 0.5 μm process and less than 4% in case of the 0.25 μm process. As an example to illustrate Law 2, for $p = 0.9$, the energy consumed by a probabilistic inverter (designed in the 0.5 μm process) rises from 50 fJ to 200 fJ by doubling the σ value from σ = 0.4 V to σ = 0.8 V (see Figure 25-6).

The two laws just presented relate p and σ to energy, and the relationships were obtained by varying the three independent parameters, namely, the operating voltage V_{dd} (referred to as the *signal*), the noise magnitude σ, and the technology generations (e.g., 0.25 μm and 0.5 μm considered here). These relationships were established through varying the value of V_{dd} and the value of σ.

A more succinct form of the analytical model that characterizes p is shown in Equation (25.5) below. Here, rather than specifying the values of V_{dd} and σ as two independent variables, they are presented as a single ratio, which we refer to as the noise-to-signal ratio (NSR). Given that $V_{dd} = 2\sqrt{2} \bullet \sigma \bullet inver f(2 \bullet p - 1)$, from Equation (25.3), we can express NSR as follows:

$$NSR = \frac{\sigma}{V_{dd}} = \left[2\sqrt{2} \cdot inver f \left(2 \cdot p - 1 \right) \right]^{-1}. \qquad (25.5)$$

Because NSR captures the two independent dimensions $(V_{dd}$ and σ) into a single dimension, the third law, stated below, is referred to as the "unifying" law of a probabilistic inverter.

Law 3: Independent of the technology generations (feature size), NSR uniquely determines the probability parameter p.

This form of the model of a probabilistic inverter is very interesting for the following reasons. First, it allows one to establish a succinct relationship between the independent parameters and the probability parameter p in a manner that does not depend on the CMOS technology generations; for example, there is no dependence on the capacitance C, which is typically determined by the feature size. Second, it is possible to simultaneously illustrate the invariance across technology generations and easily estimate p given an NSR value or vice versa. Figure 25-7 shows the NSR and p relationship. As seen from the figure, the analytical model

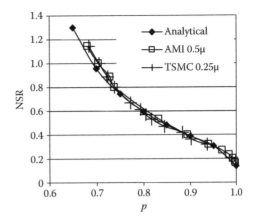

FIGURE 25-7 NSR-probability relationship validated across 0.25 μm and 0.5 μm processes.

closely follows the simulation results for both the 0.25 μm and the 0.5 μm processes; the difference between the simulated and the analytically estimated values is no more than 8%.

25.2.3 REALIZING A PROBABILISTIC INVERTER WITH LIMITED AVAILABLE NOISE

So far, the requisite amount of noise RMS, σ, has been considered to be freely available. Although noise is viewed as an ever-increasing impediment to reliable device operation as feature sizes approach the nanometer scale, currently available magnitudes of σ at coarser technologies (such as 0.25 μm here) are not adequate to yield particular and specific values of p. To validate the behaviors and benefits of PCMOS in technologies that are currently available, such as 0.5 μm and 0.25 μm studied in this chapter, we implement the probabilistic inverter with accompanying *amplification* of the available noise as shown in Figure 25-8. Hence, the requisite NSR value and the corresponding value of p are realized.

To produce a probabilistic device with $0.5 < p < 1$, a noise source with an RMS value of up to 0.8 V is needed. Because the available noise sources considered here are small, in the order of a few tens of millivolts (as in the case of thermal noise due to a resistor), an amplifier circuitry that amplifies this available noise is also needed. The amplifier is chosen to be a *subthreshold* amplifier to minimize the energy consumption of the circuit. The available thermal noise source, such as a resistor as shown in Figure 25-8, feeds the amplifier that amplifies the noise to the input of the inverter. The inverter, then, produces the probabilistic bits with a specific value of p with which it is desired to operate. Note that the two approaches to alter the probability parameter of a probabilistic inverter with the above amplifier circuitry involve varying the supply voltage of the inverter, or

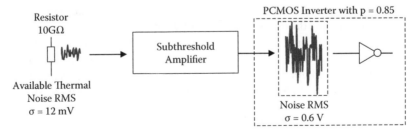

FIGURE 25-8 Realizing a probabilistic inverter with the help of a subthreshold amplifier, given limited available noise.

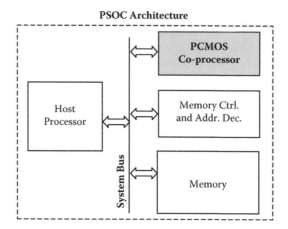

FIGURE 25-9 A "SOC-like" architecture for realizing low-energy computing architectures using PCMOS.

alternately the amplification factor, or gain of the amplifier, which in turn alters the effective noise RMS value σ.

Whereas the relationships discussed in the previous sections were based on noise being freely available, due to its limited availability in reality, the practical realization of a PCMOS inverter (switch) with the amplifier circuitry as presented in this section incurs additional energy and time costs. Therefore, in evaluating the architecture-level benefits of PCMOS switches for the currently available technologies, the cost of amplification is also considered. The particular low-energy subthreshold amplifier's structure and its attributes can be found in the work of Cheemalavagu et al. (2005). Section 25.3 presents these PCMOS switches used as building blocks and demonstrates the architecture-level gains of PCMOS-based system-on-a-chip (SOC) implementations in the context of an application with a probabilistic workload.

25.3 REALIZING PCMOS-BASED LOW-ENERGY ARCHITECTURES

In order to better understand the energy and performance benefits of PCMOS technology that can be derived at the architectural level, a SOC architecture as shown in Figure 25-9 is proposed. PCMOS is utilized in the design of an *application-specific* co-processor, and the probabilistic content of the computation is executed on this co-processor. Thus, early yet significant adoption of PCMOS is envisioned to be application-specific and evolving to a context that is domain specific; for details about the concept of a SOC and further details about custom-fit processors, please see Lyonnard et al. (2001), Tensilica (2007), and Wang et al. (2001). As shown in the figure, a low-energy host processor is used to compute the deterministic components of the application. A typical host processor will be a StrongARM (Intel), a MIPS (MIPS 2006), or an equivalent low-energy embedded processor, coupled to the co-processor through the system bus. Thus, the communication between the host and the co-processor is through memory-mapped input/output (I/O). The host could also be a custom-fit processor in its own right; thus, we also consider a host designed as a custom ASIC and analyze the impact of the efficiency of the host on the overall benefits of PCMOS (this is discussed in detail is subsequent sections). Section 25.3.1 introduces the metrics for evaluating PCMOS-based architectures, which will also serve as a basis for comparison with conventional CMOS-based architectures. In addition, Section 25.3.3 introduces additional metrics for application analysis to explain the gains seen in PCMOS-based architectures and presents insights into the design of efficient PCMOS-based SOCs.

25.3.1 METRICS FOR EVALUATING PCMOS-BASED ARCHITECTURES

The two basic characteristics of interest in realizing efficient application-specific SOC architectures are the performance (typically the running time of the application) and its energy consumption (or its derivative, power). Thus, the primary goal is to realize architectures that are significantly more efficient than is a conventional processor in terms of both of these characteristics. To compare the efficiency of conventional SOC architectures and PCMOS-based SOC architectures, metrics based on these criteria are introduced. The primary metric for consideration will be the *energy-performance* product of an architecture that implements a particular application. This is akin to the energy-delay product in circuit design and is defined below.

Energy-Performance Product (EPP): EPP is defined as the product of the application-level energy (measured in Joules) and performance (measured in number of cycles).

Given the EPP of two alternate implementations—for example, the case in which the entire algorithm is implemented as software executing on the host, referred to as the *baseline*, compared to the case in which the deterministic part of the algorithm is executed on the host with the probabilistic part executing on a PCMOS co-processor—they can be compared by computing the ratio of their individual EPP values. Since the goal is to compare the energy and performance gains realized through using PCMOS technology, this notion is refined and the metric, *EPP gain*, denoted as Γ, is defined as follows.

EPP Gain (Γ): EPP gain, denoted as Γ, is the ratio of the EPP of the baseline to the EPP of a particular implementation. The EPP gain of a particular implementation *I* is determined as

$$\Gamma_I = \frac{Energy_B \times Time_B}{Energy_I \times Time_I}. \tag{25.6}$$

In Equation (25.6), the *baseline* denoted as *B* refers to the case in which the entire application is realized using software on the host (e.g., a StrongARM SA-1100 processor) only, *without recourse to a co-processor*. Thus, the numerator of Γ_I is derived for the case in which the entire application executes deterministically on the host. The corresponding architecture realization is shown in Figure 25-10(a). While the baseline and, hence, the numerator of the EPP gain metric have been a purely deterministic realization of a deterministic algorithm corresponding to the case shown in Figure 25-10(a), in the context of applications that do *not* implement a deterministic algorithm, the software-based emulation [illustrated in Figure 25-10(b)] serves as the baseline. This approach is adopted whenever the deterministic realizations do not exist or are impractically inefficient. In this case, the probabilistic component of the application is "emulated" using pseudorandom bit generation in software [as shown in Figure 25-10(b)].

A further refinement of this approach is to consider a co-processor [Figure 25-10(c)] wherein the probabilistic parts of the application are emulated using a customized co-processor—typically using a pseudorandom number generator [PRNG; see Park and Miller (1988)]. Finally, as shown in Figure 25-10(d), the co-processor and, hence, the probabilistic computational component are realized using PCMOS. These cases (shown in Figure 25-10) capture all reasonable alternate implementation scenarios against which the benefits of PCMOS technology are compared.

25.3.2 EXPERIMENTAL METHODOLOGY

The utility of PCMOS technology at the application level for computing platforms will be demonstrated by using the metrics described in Section 25.3.1 and by considering a wide range of alternate implementations as illustrated in Figure 25-10. The experimental methodology described in this section is used to characterize these alternate implementations based on the metrics introduced. The primary metric of interest, the EPP metric, involves performance and energy estimation for each of the cases considered.

Performance Estimation: The performance of PCMOS- and CMOS-based SOC implementation and that of the baseline (where there is no co-processor) is estimated by using a modified version of the IMPACT simulator of the Trimaran infrastructure (Chakrapani et al. 2005; Trimaran). The modified simulator measures the performance of an application (the cycle count) executing on the StrongARM

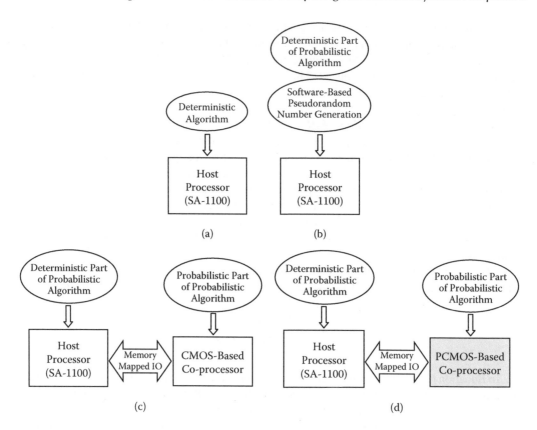

FIGURE 25-10   The four possible realizations of an application using an SOC platform wherein (a) a deterministic application is executed entirely on the host; (b) a probabilistic application is executed entirely on the host, using an emulation based on pseudorandom bits generated using software; (c) the emulation is realized using a custom CMOS co-processor; and (d) a functionally identical PCMOS co-processor is used to realize the probabilistic components of the application.

SA-1100 host. In addition, the simulator records a trace of the activity of the PCMOS and CMOS co-processors. This is combined with the performance models of co-processors, typically obtained through HSpice simulations, to yield performance in terms of execution time.

    Energy Estimation: Three components of energy consumption are estimated. The energy consumed by the host, the energy consumed by the PCMOS- (CMOS)-based co-processor(s), and the energy cost of communication between the host and the co-processor(s). Since the co-processors are memory mapped, communication is through load-store instructions executed on the host. To quantify the energy consumed by the SA-1100 host, the JouleTrack model introduced by Sinha and Chandrakasan (2001) is used. This model is reported to be accurate to within 3% of the energy measured on an actual SA-1100 core. The performance and energy modeling techniques applied to various components of the SOC architecture can be found in our earlier work (Chakrapani et al. 2006). The CMOS-based co-processor involves a 32-bit pseudorandom number generator (Park and Miller 1988) that is designed and synthesized into a TSMC 0.25 μm process and its energy cost is derived from HSpice simulations. In the context of extensions based on PCMOS, the energy cost of the co-processor is derived from HSpice simulations as well as chip measurements of functioning probabilistic switches realized in a TSMC 0.25 μm process.

25.3.3   METRICS FOR ANALYSIS OF PCMOS-BASED IMPLEMENTATIONS

Consider a probabilistic application and the three approaches to implementing it identified in Figures 25-10(b), (c), and (d)—wherein the baseline corresponds to the implementation shown in Figure 25-10(b). Clearly, the intention is to demonstrate progressive improvements quantified through the EPP gain metric, with PCMOS yielding the greatest benefits.

Preliminary observations indicate that for these benefits to be achieved, significant components of the application's overall computational effort ought to be probabilistic (and, hence, leverage the PCMOS-based co-processor). This central notion is extended to define the *flux* of an application that quantifies the "opportunity" in an application to leverage PCMOS technology. In this case, the total number of probabilistic operations in an application serves as an indicator of this opportunity. Normalized flux is here defined as the ratio of the number of probabilistic bits to the number of operations performed by (or the cycle count of) the application. In this sense, the normalized flux (hereafter referred to simply as flux) captures the distribution of the probabilistic operations over the deterministic content in an application. Intuitively, akin to the characterization of the opportunity for performance improvements popularly referred to as "Amdahl's Law" (Amdahl 1967), the opportunity for the overall EPP gain depends on the proportion of the probabilistic operations in an application and the gain per probabilistic operation. The former has been characterized as flux; the latter is captured through the *technology gain* metric.

Now consider the relative gains from two alternate realizations of the co-processor that, for convenience, are referred to as T_1 and T_2; T_1 can be identified as the case in which the host is coupled with a conventional co-processor realized using conventional technology [Figure 25-10(c)] and T_2 as a PCMOS-based co-processor [Figure 25-10(d)]. Now, considering the EPP gain metric, one can characterize the benefit of executing an operation on T_2 when compared to executing it on T_1. This metric, the normalized *technology gain*, is defined formally as follows: technology gain, denoted as $\delta(T_1, T_2)$, is defined as the ratio of the *energy* × *performance* of realizing an application-specific probabilistic core primitive using technology T_1 to that of realizing it using T_2.

The technology gain of PCMOS over CMOS in the context of the hyperencryption application is 2.03×10^3, indicating that PCMOS is at least three orders of magnitude more efficient than CMOS in terms of per-operation gains. These per-operation gains would, of course, be valuable at the level of an entire application only if the application embodies significant opportunity, characterized by its flux.

25.3.4 HYPERENCRYPTION APPLICATION AND PCMOS-BASED IMPLEMENTATION

As indicated by the flux metric, applications based on probabilistic algorithms benefit from PCMOS-based implementations. To demonstrate the energy and performance gains of PCMOS-based implementations over conventional CMOS-based implementations, an application based on a probabilistic algorithm that implements *hyperencryption* is considered (Ding and Rabin 2002).

Any SOC implementation of a probabilistic application involves *partitioning* the application between the host and the (application-specific) PCMOS-based co-processor. Even though the exact *host + co-processor* partition and the corresponding PCMOS-based co-processor architecture for these applications may vary, they follow a common theme. Common to almost all probabilistic algorithms is the notion of a *core probabilistic step* with its associated probability parameter p. For example, in the hyperencryption application that has been considered, this is the generation of the random string for the encryption pad generation. For each of the candidate applications, this core probabilistic step is manually identified and implemented in PCMOS. The deterministic parts of the application (for example, the actual encryption of the message) are implemented as software executing on the host processor. The following section describes the algorithm and the architecture for the hyperencryption algorithm.

Hyperencryption (HE) is a provably secure encryption technique in the bounded storage model (Ding and Rabin 2002). This scheme consists of generating an *encryption pad* based on a publicly available random string α and a shared (between the sender and the receiver) secret key. The secret key S is a sequence of whole numbers $S = s_1, s_2, s_3 \ldots s_k$ such that each number $0 \le s_i < |\alpha|$. If $\alpha [j]$ is

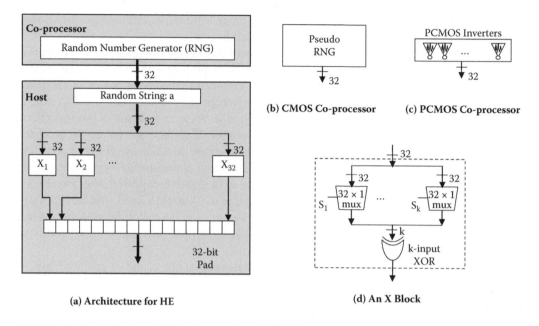

(a) Architecture for HE

(b) CMOS Co-processor

(c) PCMOS Co-processor

(d) An X Block

FIGURE 25-11 (a) Custom ASIC implementation of HE, including the host and the co-processor, (b) CMOS-based co-processor, (c) PCMOS co-processor, and (d) block diagram of an X hardware unit (residing in the host).

the jth bit of alpha, the encryption pad is generated by $\oplus \alpha [s_i]$, where $1 \leq i \leq k$. Message encryption is performed by a bitwise XOR operation of the encryption pad with the message.

Partitioning and Optimization: In the architectural implementation of hyperencryption, a co-processor (realized using CMOS or PCMOS) is used to generate the random string α, and the host is used to execute the rest of the code (the encryption pad generation and the encryption), which is deterministic. In the PCMOS-based architecture [see Figure 25-10(d)] the random string, α, is generated using PCMOS switches (with $p = 0.5$), and in the CMOS-based architecture [see Figure 25-10(c)] it is generated using a CMOS-based pseudorandom number generator (Park and Miller 1988).

For the host, two alternate design choices are used: (1) the StrongARM processor [see Figure 25-10(c) and (d)] and (2) custom ASIC. In the custom ASIC implementation shown in Figure 25-11(a), the entire algorithm is implemented so that the functionality of the deterministic host processor is realized using a full-custom hardware design. Figure 25-11(b) and Figure 25-11(c) show the CMOS and PCMOS co-processor(s) that serve as a random number generator for this particular example. The random string α is read by the host as shown in Figure 25-11(a), and using the dedicated hardware units denoted as X, the host generates the encryption pad. Each X unit selects any k bits from α and computes the XOR of these k bits. This is performed to compute every bit of the pad in parallel. Bits are selected based on the multiplexer select inputs, denoted as $S_1, S_2, \ldots S_k$ wherein each S_i is less than or equal to the bit width $|\alpha|$, of α (for $1 \leq i \leq k$). These select inputs are determined by the secret key S. Note that each X unit has its own select inputs, such that, for a message length of m, $S = \{S_1, S_2, \ldots S_k\} x_1, \{S_1, S_2, \ldots S_k\} x_2, \ldots \{S_1, S_2, \ldots S_k\} x_m$. The example custom implementation shown in Figure 25-11 used $m = 32$, $k = 64$, with a 32-bit random string, α.

25.3.5 RESULTS AND ANALYSIS

The EPP gain of PCMOS-based implementation Γ_{PCMOS} when the baseline has no co-processors but the StrongARM SA-1100 processor used as the host is 1.12. This is in spite of the fact that the hyperencryption application has a significant amount of flux and a high (over three orders of magnitude) technology gain, indicating that besides flux, the EPP gain also depends on the efficiency of the host serving as the baseline. If the energy consumed on the host to compute the deterministic part of an application is more dominant than the energy consumed to compute the probabilistic part on the co-processor, then the EPP gain would be very small. This fact is observed in the case of the HE application, in which the host SA-1100 energy is dominant and, hence, the resulting EPP gain is only 1.12. The effect of host efficiency in the study of PCMOS-based designs is discussed next.

Host efficiency is characterized through the *host energy ratio* metric and is defined as follows: the host energy ratio is the ratio of the energy consumed by the host through the execution of an application to the energy consumed due to one activation of a PCMOS-based co-processor in the system. The lower the host energy ratio, the higher the application-level EPP gains of PCMOS over the baseline and over CMOS-based implementations. In the inefficient regime where the dominant energy and performance bottleneck is the host—for example, the case in which the hyperencryption application is implemented using the StrongARM SA-1100 host—the EPP gain for a PCMOS-based SOC (over a baseline with no co-processor) is 1.12 and not different from the corresponding EPP gain value for a SOC with a CMOS-based co-processor (over a baseline with no co-processor), which is 1.12 as well.

Recall that the focus is on illustrating the (less obvious) impact of the efficiency of the host processor on the gain. Moving away from StrongARM to a host processor realized from custom ASIC logic as depicted in Figure 25-11, the gain of an implementation with a PCMOS-based co-processor over a baseline with a CMOS-based co-processor is 9.38. The architecture of this custom ASIC architecture is shown in Figure 25-11, and its performance and energy characteristics are estimated from HSpice simulations of an implementation in TSMC 0.25 µm technology. Thus, this increase in gains points to a favorable trend, wherein an increase in the efficiency of the host results in an increase in EPP gain and, hence, an increase in the energy and performance characteristics of PCMOS-based SOC, when compared to a conventional CMOS-based SOC.

25.3.6 PCMOS-Based Architectures for Error-Tolerant Applications

Previous work has shown that besides probabilistic designs targeted for inherently probabilistic applications (George et al. 2006), the PCMOS approach also offers significant benefits in the context of error-tolerant applications. Examples for these applications are those from the signal and image processing domain, in which hardware errors inherently reveal themselves as the SNR at the application level. Here, the application-level quality, measured through the SNR metric, can be traded against energy savings.

In the context of these error-tolerant applications, we have built PCMOS-based arithmetic primitives such as adders and multipliers, and signal processing elements such as finite impulse response (FIR) filters and fast Fourier transforms (FFTs) derived from them (George et al. 2006). The resulting architectures are probabilistic and yet they compute the end result with adequate application quality. Moreover, it was demonstrated that it is possible to trade the amount of error or equivalently the SNR degradation against energy through the novel approach referred to as *biased voltage scaling*, or BIVOS. In this approach, significant effort is expended in computing the "more" significant bits and, as a result, they are correct with higher probability—whereas the bits of lower significance are largely ignored. Conceptually, this "biased" probabilistic design methodology favors the most significant components of a computing primitive, such as a filter, that contribute to a critical path with longer delays and are more likely to affect the quality (accuracy) of its output through bit-significance. As an example, this can be accomplished by boosting the voltage (V_{dd}) in going from the bits of lower significance to those of higher significance. Thus, bits of lower significance are permitted to be erroneous with a higher probability. To compare the quality of the solution using this novel design methodology with a conventional voltage scaling approach, consider Figure 25-12. Here, the voltage is lowered uniformly across all of the bit positions of the arithmetic primitives. The associated energy savings and the associated image quality for both the H.264 application (using a PCMOS-based FIR element) and the SAR imaging (using a PCMOS-based FFT element) are

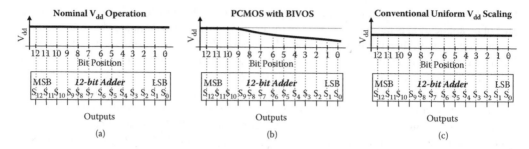

FIGURE 25-12   (a) Conventional digital design with very high (nominal) voltage levels, (b) the probabilistic BIVOS approach with significantly lower energy consumed and leading to minimal degradation of the quality of the image, and (c) conventional voltage scaling that achieves the same level of energy savings as (b) but with significantly lower image quality.

	H.264 Image Compression		SAR Imaging	
	PCMOS with BIVOS	Conventional Voltage Scaling	PCMOS with BIVOS	Conventional Voltage Scaling
Energy Savings	2X	2X	1.6X	1.6X
Image				

FIGURE 25-13   Comparison of the quality of the output images achieved through the novel BIVOS-based PCMOS to that achieved through conventional voltage scaling, and the corresponding energy savings of both approaches when compared to nominal operation at full-scale V_{dd}.

illustrated in Figure 25-13. It is easily seen from Figure 25-13 that the quality of the solution is significantly worse in the case of conventional voltage scaling when compared to our novel approach [Figure 25-12(b)], in which there is (almost) no discernible visual degradation and which is comparable to the original image computed using entirely conventional digital hardware with much higher energy consumption.

25.4   CONCLUSIONS

As device scaling continues into the nanometer regime, noise is emerging as an increasing threat to reliable computing. In addition, energy consumption and the accompanying heat dissipation also emerge as limiting factors. This chapter has introduced PCMOS as a novel technology to overcome both of these hurdles. A thorough characterization of PCMOS behavior and the utilization of noise in a controlled manner enable the design of architectural building blocks. Using such architectures that leverage PCMOS technology, we have demonstrated the impact of PCMOS at the application level. Several applications from the embedded domain, ranging over the domains of security and of speech and image processing, are based on probabilistic and error-tolerant algorithms and can readily leverage PCMOS technology. With the establishment of PCMOS utility in the context of hyperencryption applications, future investigation should expand into (1) demonstrating the impact of PCMOS technology for a wider suite of applications through application-specific SOC architectures; (2) studying domain-specific, rather than application-specific, architectures based on PCMOS technology; (3) studying the impact of voltage overscaling and, hence, associated propagation delays on probabilistic behavior; and (4) extending the PCMOS approach to encompass parameter variations envisioned to be treated as another source of noise and, hence, probabilistic behavior.

REFERENCES

Akgul, B.E.S., L.N. Chakrapani, P. Korkmaz, and K.V. Palem. 2006. Probabilistic CMOS technology: a survey and future directions. *Proceedings of IFIP International Conference on Very Large Scale Integration* 1–6.

Amdahl, G.M. 1967. Validity of the single-processor approach to achieving large-scale computing capabilities. *AFIPS Conference Proceedings* 30: 483–485.

Beinlich, I., G. Suermondt, R. Chavez, and G. Cooper. 1989. The ALARM monitoring system: a case study with two probabilistic inference techniques for belief networks. *Proceedings of the Second European Conference on AI and Medicine* 247–256.

Borkar, S., T. Karnik, S. Narendra, J. Tschanz, A. Keshavarzi, and V. De. 2003. Parameter variations and impact on circuits and microarchitecture. *Proceedings of Design Automation Conference* 338–342.

Bowman, K., X. Tang, J. Eble, and J. Meindl. 2000. Impact of extrinsic and intrinsic parameter fluctuations on CMOS circuit performance. *IEEE Journal of Solid-State Circuits* 35(8): 1186–1193.

Chakrapani, L.N., B.E.S. Akgul, S. Cheemalavagu, P. Korkmaz, K.V. Palem, and B. Seshasayee. 2006. Ultra efficient embedded SOC architectures based on probabilistic CMOS (PCMOS) technology. *Proceedings of Design Automation and Test in Europe* 1100–1105.

Chakrapani, L.N., J. Gyllenhaal, W.W. Hwu, S.A. Mahlke, K.V. Palem, and R.M. Rabbah. 2005. Trimaran: an infrastructure for research in instruction-level parallelism, in *Languages and Compilers for High Performance Computing,* Lecture Notes in Computer Science series 3602: 32–41. Berlin: Springer-Verlag.

Cheemalavagu, S., P. Korkmaz, and K.V. Palem. 2004. Ultra low-energy computing via probabilistic algorithms and devices: CMOS device primitives and the energy-probability relationship. *Proceedings of the International Conference on Solid State Devices and Materials* 402–403.

Cheemalavagu, S., P. Korkmaz, K.V. Palem, B.E.S. Akgul, and L.N. Chakrapani. 2005. A probabilistic CMOS switch and its realization by exploiting noise. *Proceedings of the IFIP International Conference on VLSI SoC* 452–458.

Ding, Y. Z. and M.O. Rabin. 2002. Hyperencryption and everlasting security. *Proceedings of the 19th Annual Symposium on Theoretical Aspects of Computer Science*, Lecture Notes in Computer Science series 2285: 1–26.

Fuks, H. 2002. Non-deterministic density classification with diffusive probabilistic cellular automata. *Physical Review E, Statistical, Nonlinear, and Soft Matter Physics* 66(066106).

Gelenbe, E. 1991. *Neural Networks: Advances and Applications.* New York: Elsevier Science Publishers.

Gelenbe, E. and F. Batty. 1992. Minimum graph covering with the random neural network model, in *Neural Networks: Advances and Applications*, vol. 2, E. Gelenbe, ed. New York: Elsevier Science Publishers.

George, J., B. Marr, B.E.S. Akgul, and K.V. Palem. 2006. Probabilistic arithmetic and energy efficient embedded signal processing. *Proceedings of the International Conference on Compilers, Architecture and Synthesis for Embedded Systems* 158–168.

Hegde, R. and N.R. Shanbhag. 2000. Toward achieving energy efficiency in presence of deep submicron noise. *IEEE Transactions on VLSI Systems* 8: 379–391.

Heydari, P. and M. Pedram. 2000. Analysis of jitter due to power-supply noise in phase-locked loops. *Proceedings of the IEEE Custom Integrated Circuits Conference* 443–445.

Intel. Moore's Law. Available online at http://www.intel.com/technology/silicon/mooreslaw/.

Intel. StrongARM-1100 Microprocessor Technical Reference Manual. Formerly available online at http://www.intel.com.

International Technology Roadmap for Semiconductors. 2005. Available online at http://www.itrs.net/common/2005itrs/execsum2005.pdf.

Kim, N., T. Austin, D. Blaauw, T. Mudge, K. Flautner, J. Hu, M. Irwin, M. Kandemir, and V. Narayanan. 2003. Leakage current: Moore's law meets static power. *IEEE Computer* 36: 65–77.

Kish, L.B. 2002. End of Moore's law: thermal (noise) death of integration in micro and nano electronics. *Physics Letters A* 305: 144–149.

Korkmaz, P., B.E.S. Akgul, and K.V. Palem. 2006. Ultra-low energy computing with noise: energy-performance-probability trade-offs. *IEEE Computer Society Annual Symposium on VLSI* 349–354.

Korkmaz, P., B.E.S. Akgul, K.V. Palem, and L.N. Chakrapani. 2006. Advocating noise as an agent for ultra-low energy computing: probabilistic CMOS devices and their characteristics. *Japanese Journal of Applied Physics, SSDM Special Issue Part 1* 3307–3316.

Korkmaz, P. and K.V. Palem. 2006. *The Inverse Error Function and Its Asymptotic "Order" of Growth Using O and ω*. Technical Report CREST-TR-06-02-01. Atlanta: Georgia Institute of Technology.

Lyonnard, D., S. Yoo, A. Baghdadi, and A.A. Jerraya. 2001. Automatic generation of application-specific architectures for heterogeneous multiprocessor system-on-chip. *Proceedings of the 38th Design Automation Conference* 518–523.

MIPS Technologies. 2002–2003 and 2006. MIPS32 M4K Processor Core Software Users Manual. Available online at http://www.mips.com.

Mudge, T. 2001. Power: a first-class architectural design constraint. *IEEE Computer* 34: 52–58.

— 2004. Keynote address: low power robust computing. *International Conference on High Performance Computing*, Bangalore, India.

Natori, K. and N. Sano. 1998. Scaling limit of digital circuits due to thermal noise. *Journal of Applied Physics* 83: 5019–5024.

Palem, K.V. 2003. Proof as experiment: probabilistic algorithms from a thermodynamic perspective. *Proceedings of the International Symposium on Verification (Theory and Practice)* 524–547.

— 2005. Energy aware computing through probabilistic switching: a study of limits. *IEEE Transactions on Computers* 54(9): 1123–1137.

Pant, S., D. Blaauw, V. Zolotov, S. Sundareswaran, and R. Panda. 2004. A stochastic approach to power grid analysis. *Proceedings of Design Automation Conference* 171–176.

Park, S.K. and K.W. Miller. 1988. Random number generators: good ones are hard to find. *Communications of the ACM* 31.

Pavlovic, V., A. Garg, J.M. Rehg, and T.S. Huang. 2000. Multimodal speaker detection using error feedback dynamic Bayesian networks. *Computer Vision and Pattern Recognition* 2: 34–41.

Rehg, J.M., K.P. Murphy, and P.W. Fieguth. 1999. Vision-based speaker detection using Bayesian networks. *Computer Vision and Pattern Recognition* 2: 110–116.

Sano, N. 2000. Increasing importance of electronic thermal noise in sub-0.1 mm Si-MOSFETs. *The IEICE Transactions on Electronics* E83-C: 1203–1211.

Sinha, A. and A. Chandrakasan. 2001. JouleTrack a web based tool for software energy profiling. *Proceedings of Design Automation Conference* 220–225.

Stein, K.-U. 1977. Noise-induced error rate as a limiting factor for energy per operation in digital ICs. *IEEE Journal of Solid-State Circuits* 12: 527–530.

Strecok, A.J. 1968. On the calculation of the inverse of the error function, mathematics of computation. *Mathematics of Computation* 22: 144–158.

Tensilica. 2007. Xtensa Microprocessor. Available online at http://www.tensilica.com.

Trimaran Consortium. Trimaran: an infrastructure for research in instruction-level parallelism. Available online at http://www.trimaran.org/ and http://web.mit.edu/rabbah/www/docs/chakrapani-lncs-2005.pdf.

Wang, J.-C., J.-F. Wang, A.-N. Suen, and Y.-S. Weng. 2001. A programmable application-specific VLSI architecture for speech recognition. *The Eighth IEEE International Conference on Electronics, Circuits and Systems* 1: 477–480.

26 Advanced Microprocessor Architectures

Janice McMahon and Stephen Crago, University of Southern California, Information Sciences Institute
Donald Yeung, University of Maryland

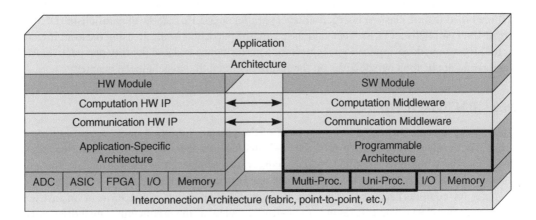

This chapter focuses on microprocessor technologies. It begins with a high-level summary of processor trends over the recent past and projects ahead on a wide range of architectures with a primary focus on on-chip parallelism.

26.1 INTRODUCTION

The past several decades have seen phenomenal performance improvements in commercially available microprocessors. Increases in performance have been accompanied by decreasing costs. This has not only made computers more affordable, but has also accelerated the development of additional software and hardware that have made even further performance improvement possible. Overall improvements have been enabled by a number of factors, including process technology, microarchitecture, architecture, and design and development tools (Ronen et al. 2001).

Technology improvements have been most commonly measured by conformance to Moore's Law, which states that the number of transistors per unit area on integrated circuits will double every 18 months since the invention of the integrated circuit. However, with these tremendous gains in complexity come challenges in a number of dimensions, including performance scaling, management of data and computation locality, power dissipation, fault and error handling, as well as software complexity and manufacturing cost. Furthermore, the high rates of performance growth in the recent past are rapidly becoming unsustainable in the face of technology limitations inherent in increased complexity. Examples of this are wire delay and its effect on clock rate, and bandwidth limitations and their effect on computational balance (Dally and Lacey 1999).

This chapter surveys current research architectures and discusses how they are meeting the challenges posed by today's technology environment. The primary focus is on-chip parallelism since that is the primary method for dealing with increases in complexity (Hammond, Nayfeh, and Olukotun 1997). However, manifestations of on-chip parallelism in microprocessor architecture can exhibit a wide variety of characteristics, examples of which are multiple cores, nonuniform memories, very long instruction words (VLIW), single-instruction multiple-data (SIMD) implementations, and tile-based architectures. The key aim of this chapter is to navigate the implementation space in order to attain a fundamental understanding of the nature of on-chip parallelism and an in-depth understanding of several of today's most promising new architectures.

The chapter begins with a high-level summary of processor trends over the recent past and a description of how architectural research has generally anticipated commercial offerings by 5–10 years. Key architectural innovations drawn from current designs, as opposed to technology improvements such as higher clock frequencies and transistor densities, are discussed. Key architectures are presented according to the taxonomy of on-chip parallelism techniques. This taxonomy is composed of three broad feature classes: instruction-level parallelism (ILP), data-level parallelism (DLP), and thread-level parallelism (TLP). Then addressed are the particular issues in using these architectures for real-time embedded applications, both from a performance and a scalability perspective. Architectural and application perspectives on the future of computing conclude the chapter.

26.2 BACKGROUND

Commercial general-purpose microprocessors of the past are all based on a sequential execution model. The highly successful microprocessor industry has succeeded by doubling performance every 18–24 months while only making gradual changes to a sequential execution model. The execution model says that a microprocessor executes one instruction at a time in the order that the programmer (or compiler) specifies. The execution model is specified in the form of an instruction set architecture (ISA). The ISA defines the software's view of the architecture, including machine state (registers and memory), instruction format, supported data types, and interrupt semantics. For example, the x86 instruction set architecture defines an execution model that has survived for decades. The x86 execution model has been extended, but the fundamental execution model has remained unchanged (Oklobdzija 2002).

The continuity of the instruction set architecture has been fundamental to the success of the microprocessor industry. It has allowed microprocessor companies to leverage technology improvements without disrupting software functionality. That is, users have been able to upgrade microprocessors without having to change existing software. While new instruction set architectures occasionally gain acceptance, there is a strong market influence to keep existing instruction set architectures.

26.2.1 ESTABLISHED INSTRUCTION-LEVEL PARALLELISM TECHNIQUES

While the execution model has stayed the same, the microprocessor architectures that implement that model have changed significantly. The high densities of transistors in recent processors have allowed a significant portion of chip resources to be devoted to structures that increase instruction-level parallelism. ILP allows multiple instructions to be issued or overlapped at one time as long as the resulting program is subject to the sequential constraints of the execution model. Since more instructions can be performed in parallel, the net effect is decreased latency, or execution time, of the program as a whole. The performance advances in microprocessors over the last few decades are due to ILP and clock rate increases.

There are many methods of achieving high ILP, any number of which may be employed by a given architecture at a given time. These include out-of-order execution, superscalar and pipelined architectures, cache architectures, as well as advances in instruction set architecture, such as reduced instruction set computers (RISC) and very long instruction word computers. These are briefly outlined in the paragraphs below.

Superscalar execution allows the microprocessor to issue instructions dynamically in order to achieve higher levels of parallelism. Out-of-order execution allows instructions to be reordered to increase instruction-level parallelism that can be exploited by superscalar instruction units. Much of the complexity of modern microprocessors lies in the mechanisms needed to support out-of-order execution, including instruction queues, register rename tables, memory disambiguation logic, and reorder buffers.

Pipelining breaks instruction execution into multiple stages, allowing clock speeds to be increased. Multiple instructions traverse the pipeline simultaneously, maintaining instruction throughput. In order to keep pipelines full, branch predictors must be implemented to prevent stopping the pipeline when branches and other control instructions are encountered.

As memory latency and bandwidth become more of a bottleneck, increasing numbers of on-chip transistors have been dedicated to caches. Caching of data allows for high ILP by ensuring that frequently used data are closer to the processor, i.e., accessible via lower latency than main memory. For a cache miss, i.e., an access to data that are not in the cache, memory latency can be hidden by overlapping instructions with memory accesses. Most high performance modern microprocessor chips have at least two levels of cache on chip.

While the basic execution model of microprocessors has remained unchanged, instruction set architectures have changed. RISC architectures have simplified instruction sets and allowed clock speeds and ILP exploitation to increase. Instruction extensions have been provided to improve the performance of floating-point instructions, multimedia applications, and even vector operations. These changes have been made within the constraint of maintaining compatibility with established instruction sets.

VLIW architectures achieve high ILP by executing long instructions that are composed of multiple operations. The architecture can interpret such instructions because it contains multiple functional units and high-bandwidth register files. However, generating code for these architectures is much more difficult than generating code for the previous architectures, and, hence, much of the burden of achieving high performance is placed on a compiler and scheduling technique.

The above discussion has shown that microprocessor architectures have grown much more complex as microprocessor developers have striven to maintain or gain performance advantage. However, most of the increase in microprocessor performance over the last 30 years has been due to very-large-scale integration (VLSI) process improvement. Improvements in process technology have allowed transistor feature sizes to shrink. Smaller transistors lead to faster clock speeds and more transistors (functionality) on a chip. Chip size has also grown to allow a further increase in on-chip functionality. Microprocessor performance has improved by about a factor of 1000 over the last 30 years. A factor of about 100 has been due to technology improvement, and about a factor of 10 has been due to architecture improvements.

26.2.2 Parallel Architectures

While a sequential execution model has dominated microprocessors, at the system level, parallel programming models have been developed and are successful in domains that require high performance. The success of high performance parallel computer architectures has fluctuated, but parallel programming models such as Message Passing Interface (MPI) have caught on amongst users that require high performance. Parallel computing architectures can be classified using the Flynn taxonomy, which defines two categories of parallel computers: SIMD (single-instruction multiple-data) and MIMD (multiple-instruction multiple-data).

SIMD computers have been developed, but have not been commercially successful. SIMD computers rely on a central instruction sequencer that broadcasts instructions to multiple processing elements. Each processing element executes the same instruction on different data elements. SIMD computers have not been successful for several reasons. First, the central instruction sequencer and broadcast bus can be a bottleneck. True broadcast buses are inherently slow because of the high fan-out (capacitance), which limits clock speed. Broadcast buses can be pipelined to increase clock speed, but pipelining introduces bubbles at startup and whenever data-dependent branches are executed. Second, because SIMD computers require their processors to execute in lock step (contrary to the way general-purpose multiprocessors operate), they cannot leverage the commercially successful general-purpose microprocessors. Finally, a relatively small set of applications can be easily mapped to SIMD architecture. While large parallel SIMD computers have not been commercially successful, the vector and media instruction set extensions adopted by modern microprocessors can be classified as SIMD processing.

Because of its execution constraints, SIMD has been most closely associated with data-level parallelism. Data-level parallelism is used to achieve high performance efficiently for applications that perform the same operations on many data elements. Much of the complexity of the ILP techniques discussed in the previous section is in finding instructions that can operate in parallel and in maintaining a sequential ordering of those instructions. Techniques that exploit data-level parallelism exploit the fact that many applications perform the same operation on many data elements. For example, a vector addition operation on vectors of length N performs an addition operation on N pairs of operations. An ILP-based architecture would require N addition instructions, each with two input operands and one output operand. A RISC-based load/store architecture would require four instructions for each addition: two load instructions, an addition instruction, and a store operation. The instruction bandwidth required for such an architecture is a source of inefficiency, placing unnecessary demands on the instruction memory or cache, fetch and decode logic, and instruction retiring logic. DLP techniques generally allow one instruction to operate on vector data instead of scalar data, reducing the number of instructions that need to be fetched, decoded, and executed.

The reduced instruction bandwidth can lead to increased density of functional units and increased efficiency. Chip area devoted to instruction caching or memory and instruction fetch and decode logic can be dedicated to functional units that directly perform useful computations. Instruction bandwidth can be reallocated to data bandwidth, leading to increased performance. Microprocessors have become increasingly limited by power dissipation and thermal density issues, so the efficiency of useful operations per unit of energy consumed has become a first-class design constraint. Reducing the number of instruction cache or memory accesses and the complexity of the fetch and decode logic increases the efficiency of a processor by amortizing the cost of each instruction over more operations.

Increasing the number of operations performed for each instruction increases the data bandwidth required to keep the functional units occupied, so data-parallel architectures often provide mechanisms for managing data bandwidth efficiently. These techniques are especially important because memory bandwidth is increasingly becoming a performance limiter for both microprocessor chips and parallel architectures. These issues will be explored in later sections.

In applications with DLP, the same operation is performed over a large data structure such as a vector or matrix. Such operations are usually performed in loops, and the loop iterations can be partitioned and assigned to processors that complete the computation in parallel. Another class of machines that has traditionally been designed to exploit high levels of DLP is the class of vector machines. Vector machines are characterized by parallel functional units with deep pipelines and large register files as well as memory systems that support vector-style processing.

Data-level parallelism can be exploited without making radical changes to the traditional sequential programming model. The only change necessary is that individual operations can be applied to multiple data elements, often thought of as vectors. The basic control flow is still sequential. Data-level parallelism can be identified by programmers or compilers. Data-level parallelism can be exploited in many signal processing applications, which are common in the embedded systems that are the focus of this book, and can also be exploited in many multimedia and game applications, which help drive the commercial market. However, data-parallelism is not common in many general-purpose applications, including desktop productivity tools or logic-based control applications.

The MIMD parallel computing model has been more successful than the SIMD model. The MIMD model is more general because each processing element can execute its own instruction sequence (program). High-end MIMD computers such as the Tera have been developed with their own microprocessors, but have struggled to maintain a performance advantage over those based on standard commercial microprocessors because of their smaller market. However, MIMD machines based on clusters of general-purpose microprocessors have been successful in domains that require high performance computing. Programming libraries, such as MPI, that support the MIMD programming model have become de facto standards. The coarse granularity of traditional MIMD computers restricts their usefulness to parallel applications with loosely coupled tasks.

Because of its flexibility, MIMD architecture can support multiple types of parallelism, including DLP and thread-level parallelism. Thread-level parallelism divides a program into threads, or flows of control, with an associated context or state. The context of a thread consists of a set of register contents (including program counter), a dynamic call/return stack, and heap storage for thread-private data. Programs are partitioned into threads that can be executed in parallel, either in a parallel architecture or a sequential processor that switches between multiple tasks or processes. In either case, the overhead of either context-switching in a sequential processor or communication and synchronization in a parallel architecture requires that threads be large enough to ensure that overall performance is improved. For this reason, thread-level parallelism usually results in a coarser granularity of processing than data- or instruction-level parallelism. The exact software and hardware trade-offs will be different for different architectures and implementations.

Partitioning of a program into parallel threads involves analyzing data and control dependencies in the program, balancing the computational load between threads, and minimizing the amount of communication between threads. Assignment of computation to threads can be performed explicitly by the programmer in the form of library calls, language extensions, or compiler directives. Alternatively, assignment of computation to threads can be performed by a compiler. Although compiler partitioning involves less programmer effort, its success is limited to applications in which data and control dependencies can be analyzed by the compiler. Typically, these applications are numeric, scientific applications. A final alternative is to include hardware in the processor design to perform partitioning at runtime. This final option allows for finer granularity in the threaded application, but at the expense of increased hardware design costs and processing overhead.

In a multithreaded model, parallel threads are instantiated with a "fork" operation and conclude with a "join" operation. Explicit synchronization between threads is performed using locks and/or barrier synchronization operations. Data are communicated between threads via shared memory or message passing. For shared-memory communication, coherence and consistency of shared data are a key aspect of thread implementation. For message-passing models, send and receive operations must be explicitly performed between threads. In either case, the effect of interthread communication on performance will depend heavily on the interprocessor communication network topology, bandwidth, and latency.

Multithreading in a multicore processor can take a variety of forms. In a conventional multithreaded machine, a processing core time-slices between threads, changing context at a coarse enough granularity to amortize overhead. In a superthreaded architecture, different threads execute at different stages within the processor's pipeline. In a hyperthreaded, or simultaneous multithreading architecture, different threads can execute simultaneously within the same stage in the processor pipeline by using different functional units within that processing stage. The cost of the extra flexibility as we increase the amount of threading is the extra hardware to manage the sharing of on-chip resources, including both execution units and on-chip cache (Tullsen, Eggers, and Levy 1995).

Because of the need to change context between them, threads by nature contain more overhead than do other mechanisms for exploiting parallelism. Since the threaded runtime and hardware architecture is responsible for scheduling and executing threads, the programming model is easier to use. Moreover, the dynamic execution of threads can result in more efficient scheduling of processing resources. However, not all loops in a program can be threaded, since assumptions must be made about data dependencies between loop iterations in order to perform them in a threaded manner. Additionally, partitioning of work between threads can result in load-balancing issues that can significantly affect performance.

26.3 MOTIVATION FOR NEW ARCHITECTURES

While microprocessors have been able to succeed for the last 30 years without radically changing the execution model, this paradigm is getting harder to maintain for two primary reasons. First, as transistor feature sizes continue to shrink, on-chip communications are beginning to dominate costs (in terms of both timing and chip area). This is already making it much harder to continue improving clock speed while providing the illusion that all instructions execute in order in a single clock cycle. Second, while shrinking transistors allow more functionality to be included on a chip, the pay-off of additional architecture improvements (while constrained by the old sequential execution model) has reached the point of diminishing returns. In order to provide significant performance improvements, microprocessor architectures will need to adopt new execution models.

26.3.1 LIMITATIONS OF CONVENTIONAL MICROPROCESSORS

Exploiting fine-grain instruction-level parallelism (ILP) is an important source of performance gain for microprocessors. In the past, the ILP technique responsible for the most performance gain is pipelining, which has contributed significantly to processor clock rate increases. Other techniques, such as branch prediction, multiple issue, and dynamic instruction scheduling, have also provided important performance gains by increasing the number of instructions that can execute in parallel (see discussion on conventional ILP techniques in Section 26.2.1). Along with device speed scaling, such conventional ILP techniques have accounted for the enormous performance improvement in microprocessors—roughly 60% per year—over the past few decades.

Several recent trends in both process technology and computer architecture will dramatically limit future performance gains provided by conventional microprocessors. Foremost, *wire delay* is becoming more significant relative to gate delay in deep submicron VLSI processes. Due to increased resistance and capacitive coupling effects, wire delay remains constant or even increases slightly as VLSI feature size scales downward (Ho, Mai, and Horowitz 2001). In contrast, gate delay reduces with scaling. Hence, while the access time of microarchitecture structures normally improves with technology scaling due to the gate delay reductions, such performance improvements will slow or reverse course as wire delay becomes dominant, especially for structures containing long wires.

Traditional microprocessors are particularly susceptible to wire delay effects because they are not designed with optimizing wires in mind. The problem is the implementation of several key ILP structures used in conventional microprocessors requires long wires. For example, Tomasulo's algorithm (Tomasulo 1967), an important technique for out-of-order execution, relies on broadcast networks to communicate results to instructions stalled in the instruction window. Such all-to-all communication within the instruction window leads to long wires heavily loaded by large numbers of comparators. Similarly, long wires are required to implement the forwarding paths that deliver completed results from functional units to the instruction window, or directly to other functional units. Forwarding paths become especially problematic as more functional units are supported on-chip. Not only are long wires needed in the communication networks that support conventional ILP techniques, they are also required in several key microarchitecture structures. One example is the register file. To exploit large amounts of ILP, high-bandwidth access (i.e., large number of ports) to a large register name space is necessary. These requirements increase the register file's physical size, lengthening the wires that traverse the register file.

While conventional microprocessors will suffer poor cycle time scaling due to wire delay effects, their performance will also be limited because certain techniques are approaching the *point of diminishing returns*. A prime example is pipelining. Increasing the pipeline depth of a central processing unit (CPU) results in less work performed per pipe stage, thus permitting a higher clock frequency. Unfortunately, pipeline depth cannot be increased indefinitely. The overhead of inserting additional pipe stages, as well as the increased hazards created by deeper pipelines, offsets the potential cycle time benefits of pipelining, reducing its effectiveness. Agarwal et al. argue that pipelining beyond 8–16 fan-out-of-four gate delays per pipe stage is problematic (Agarwal et al. 2000), a range that modern CPU pipelines are rapidly approaching. Similar to pipelining, other important ILP techniques are reaching or have reached the point of diminishing returns as well.

Finally, in addition to performance limitations, conventional microprocessors will also be limited by *design complexity*. As discussed previously, existing processors employ a host of ILP extraction techniques to squeeze as much ILP from applications as possible. Moore's Law has enabled this proliferation in ILP techniques by providing the necessary transistor budgets. However, the number and sophistication of conventional ILP techniques have made today's processors extremely complex. As complexity continues to increase, design time and time-to-market also increase. Worse yet, future designs may be so complex that complete verification becomes impossible. At some point, the costs associated with the design complexity of conventional ILP techniques will outweigh their performance benefits, which, as we have already argued, are themselves lessening due to wire delay and diminishing return trends.

The combined impact of wire delay, diminishing returns in performance, and design complexity makes continued use and scaling of conventional microprocessors unattractive. This motivates the need to develop new architectures.

The leading-edge processor architecture research of today has begun to explore new programming models for microprocessors. As we have seen earlier in this chapter, the capability of an architecture to support ILP is diminishing, not increasing, as densities get higher. Technology and architecture limitations require researchers to develop new ILP architectures. In addition, modern chips must find mechanisms to exploit new types of on-chip parallelism such as DLP and TLP within a variety of applications for even more performance. Additionally, modern chips must be flexible and reconfigurable enough to support a variety of workloads with different execution characteristics and in varying proportions. Research architectures are using on-chip parallelism to implement DLP and TLP in combination with ILP to achieve the levels of performance required by today's applications. This is made possible by the high number of transistors available from modern design processes and fabrication techniques. The resulting changes in execution model are already beginning to drive new programming models and languages, and the resulting gains in raw performance will ideally allow the gains of the last few decades to continue at the same fast pace that has benefited applications in all classes.

26.4 CURRENT RESEARCH MICROPROCESSORS

This section explores the different types of parallelism presented in the previous section and discusses how they are implemented in various classes of research architectures. For each class of parallelism in this taxonomy, examples are provided to illustrate how key architectures address that class of parallelism, and general performance trade-offs for different implementations are discussed.

26.4.1 INSTRUCTION-LEVEL PARALLELISM

In response to the limitations of existing ILP techniques in conventional microprocessors, the computer architecture research community has recently investigated several novel architectures for exploiting ILP. This section discusses Raw (Taylor et al. 2004), TRIPS (Nagarajan et al. 2001), and Wavescalar (Swanson et al. 2003), three examples of what we call *advanced ILP architectures*.

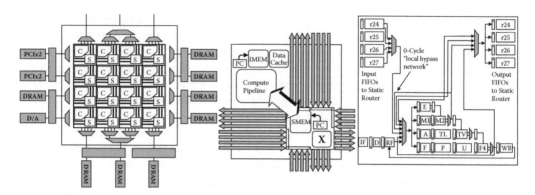

FIGURE 26-1 The Raw microprocessor employs coarse-grain tiles. Each Raw tile contains an in-order single-issue pipeline accompanied by instruction and data memories, as well as a switch for inter-tile communication. (From Taylor et al., Evaluation of the Raw microprocessor: an exposed-wire-delay architecture for ILP and streams, *Proceedings of the 31st International Symposium on Computer Architecture* 2–13, 2004. With permission. © 2004 IEEE.)

The discussion presents three novel features exhibited by advanced ILP architectures: tile-based organization, explicit parallelism model, and scalable on-chip networks. When appropriate, examples from individual architectures are used to illustrate the ideas.

26.4.1.1 Tile-Based Organization

Advanced ILP architectures employ a *tile-based organization* in which a computation resource, known as a tile, is replicated across the microprocessor die to form a two-dimensional computation fabric. Tiles can vary in size and functionality, but are significantly less complex than a modern CPU. Although individual tiles provide only modest computing resources, large amounts of ILP can be exploited by mapping instructions onto multiple tiles and leveraging their collective resources simultaneously. (The mapping process is a critical component of advanced ILP architectures and will be discussed later.)

Tile-based architectures are based on exposing two fundamental premises in the instruction set: (1) memory accesses are expensive and single large monolithic caches cannot operate at high frequencies and (2) local communication is faster than global communication and can be exploited if exposed in the instruction set architecture. The distinguishing feature of tile-based architectures is that they divide a microprocessor chip into a grid of processing units. Each processing unit executes its own instruction stream, and very-low-latency, high-bandwidth connections exist between adjacent tiles. The low-latency, high-bandwidth connections allow fine-grain parallelism to be exploited, thereby allowing scalable performance for a wider range of applications. These architectures have been shown to achieve order of magnitude speedups over traditional microprokessors, and prototype systems for both chips are in development to physically demonstrate the performance gains.

An important distinction between different advanced ILP architectures is tile size or granularity. Two tile sizes have been studied to date. The Raw microprocessor (Taylor et al. 2004), illustrated in Figure 26-1, uses coarse-grain tiles consisting of simple processing pipelines. Each Raw tile contains an in-order single-issue pipeline accompanied by instruction and data memories, as well as a switch for inter-tile communication. Each Raw tile implements an in-order single-issue MIPS-style execution core, with an eight-stage integer pipeline and a four-stage floating-point unit (FPU). Accompanying the execution core is a 32 KB instruction and 32 KB data memory, as well as a network switch that facilitates inter-tile communication via network channels that connect near-neighbor tiles. In Raw, each tile contains enough compute resources to support a single thread of execution; however, the intent is for ILP from a single application to be decomposed into multiple flows of control that can then be mapped onto separate Raw tiles.

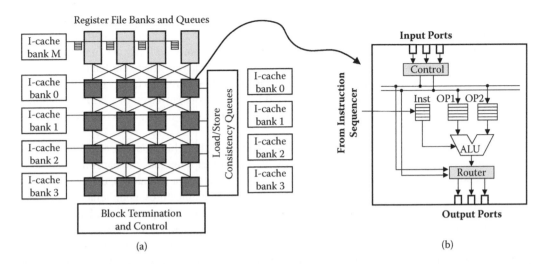

FIGURE 26-2 The TRIPS architecture employs fine-grain tiles. Each TRIPS tile contains a functional unit controlled by an instruction sequencer and fed by operand queues, as well as a router for inter-tile communication. Tiles form a processing grid that shares instruction and data caches, and a register file. (From Nagarajan et al., A design space evaluation of grid processor architectures, *Proceedings of the 34th Annual International Symposium on Microarchitecture* 40–51, 2001. With permission. © 2001 IEEE.)

In contrast to Raw, the TRIPS (Tera-ops Reliable Intelligently-adaptive Processing System) architecture (Nagarajan et al. 2001), illustrated in Figure 26-2, uses fine-grain tiles consisting of individual functional units. Accompanying each functional unit is an instruction sequencer that specifies what operation is performed in each tile on any given cycle, and a pair of operand queues that feed the functional unit with data values. As in Raw, each TRIPS tile also contains a network switch or router that facilitates inter-tile communication via network channels that connect neighboring tiles. Multiple TRIPS tiles form a processing grid onto which ILP can be mapped (in Figure 26-2, a 4 × 4 grid is shown). The entire grid shares an instruction cache (to the grid's left), a register file (above the grid), and a data cache (to the grid's right). The caches and register file are banked, with each bank connected to a row or column of the grid. Similar to TRIPS, the Wavescalar architecture (Swanson et al. 2003) also uses fine-grain functional unit tiles. Due to its similarity, a diagram of the Wavescalar organization was omitted.

Tile-based organizations address the limitations of conventional ILP processors discussed in Section 26.3. Because they are modular, tile-based organizations exhibit lower design complexity, particularly when tiles are kept simple. Once a tile is designed and verified, it can be replicated to form large processors. Furthermore, processors can be scaled up by stamping out additional tiles, saving a costly redesign and reverification cycle. In addition to addressing design complexity, tile-based organizations also address wire delay. Since the on-chip network provides near-neighbor connections only, the longest wire in a tiled architecture cannot exceed the dimension of a single tile.

26.4.1.2 Explicit Parallelism Model

Conventional microprocessors rely on the hardware to analyze and identify independent instructions, and to manage their parallel execution. This process is completely hidden from the programmer or compiler, which instead sees a purely sequential execution model. Advanced ILP architectures, in contrast, off-load the discovery of ILP to the programmer or compiler. By making parallelism *explicit* and performing its extraction in software, advanced ILP architectures eliminate many complex hardware structures found in conventional microprocessors.

Because software, rather than hardware, is responsible for finding ILP, the performance of advanced ILP architectures relies critically on the algorithms for extracting ILP from applications.

Hence, research on advanced ILP architectures involves compiler innovation in addition to hardware innovation. For example, the Raw project has developed Rawcc (Taylor et al. 2004), a C compiler for the Raw microprocessor. Rawcc distributes code and data onto the Raw tiles, and uses several VLIW-like techniques to schedule instructions for execution across tiles simultaneously (Lee et al. 1998). Similar to Rawcc, the compiler for the TRIPS architecture also uses VLIW-inspired ILP discovery. The TRIPS approach forms predicated hyperblocks (Mahlke et al. 1992) from application code to create large regions of dataflow execution. These data-driven code blocks can then be mapped onto the TRIPS tiles.

In addition to compiler-based ILP discovery, advanced ILP architectures also rely on the compiler to manage ILP exploitation at runtime. Again, this further simplifies the hardware by moving more functionality into the compiler. Of all the runtime management techniques, the most critical is managing inter-instruction communication. In advanced ILP architectures, communication between dependent instructions is performed explicitly by the compiler. For example, in TRIPS and Wavescalar, computation "flows" across a grid of functional unit tiles in a data-driven manner. At each tile, a computed value must be forwarded to its successor instructions on remote tiles. Unlike conventional microprocessors in which such data forwarding is performed via a broadcast to **all** waiting instructions, the compiler for TRIPS and Wavescalar explicitly names the consumer instructions, and forwards the data directly to the named instructions only. This process eliminates the need for a broadcast network and its associated long wires. Such dataflow execution also saves having to write-back compiler-forwarded results to the register file, thus removing the register file as a centralized bottleneck.

Similar to TRIPS and Wavescalar, Raw also relies on the compiler to perform data forwarding. In Raw, the inter-tile network ports appear to each execution core as registers within the core's pipeline, as illustrated in Figure 26-1. The compiler performs communication by specifying these network port registers as source and destination registers in the appropriate fields of communicating instructions. Explicit synchronization between communicating instructions is not necessary due to timing and ordering guarantees provided by the network (see Section 26.4.1.3). In essence, inter-tile communication in Raw is supported by extending conventional register forwarding mechanisms to include forwarding between pipelines on separate tiles.

As mentioned earlier, compiler-driven ILP discovery and management simplify the hardware. Along with tile-based organization, this helps to further address the design complexity problems associated with conventional microprocessors. Using an explicit compiler-centric approach to parallelism also has the potential to address diminishing return effects as well. Because a compiler can analyze much larger code regions than can hardware, which is limited by the size of its instruction window, advanced ILP architectures have the potential to exploit greater amounts of ILP than can conventional microprocessors.

26.4.1.3 Scalable On-Chip Networks

The last novel feature of advanced ILP architectures discussed is the *on-chip interconnection network*. Sometimes referred to as the *scalar operand network* (SON) (Taylor et al. 2003), this on-chip interconnect carries the communication load between tiles and facilitates the explicit compiler management of inter-instruction communication discussed in Section 26.4.1.2. SONs have two important design criteria. First, they must be scalable to permit processing chips with large numbers of tiles. And second, they must provide low-latency communication for effective ILP exploitation.

In conventional microprocessors, the equivalent to a SON is the bypass network that connects functional units and the register file within the CPU pipeline. Typically, this interconnect is centralized, supports broadcasts, and, as such, is unscalable. In contrast, SONs for advanced ILP architectures are point-to-point networks with distributed topologies. For example, Raw's SON employs a two-dimensional mesh topology. In TRIPS, the SON is also a two-dimensional mesh, but with diagonal channels between "catty-corner" tiles. And in Wavescalar, the SON is hierarchical, with shared buses connecting groups of tiles, called clusters, and a two-dimensional mesh connecting separate clusters. Because communication bandwidth in point-to-point networks increases with node count, using point-to-point networks to implement SONs allows advanced ILP architectures to scale to large configurations.

Achieving low latency is another important design goal for SONs. Given the fine-grain nature of inter-instruction communication, the amount of ILP an architecture can exploit is highly sensitive to communication latency across the on-chip interconnect. Tile-based organizations with point-to-point SONs provide low-latency communication between near-neighbor tiles; however, global communication may incur high latency since multiple hops are necessary across distant tiles. One way to minimize latency is to exploit *communication locality*. Researchers have observed most instructions communicate with a small fixed set of dependent instructions throughout a program's lifetime (Swanson et al. 2003). Given such communication locality, it is possible to identify frequently dependent instructions and place them onto physically close tiles. In addition to managing inter-instruction communication as discussed in Section 26.4.1.2, compilers for advanced ILP architectures also explicitly place instructions on tiles to exploit communication locality, leading to shorter communication routes and reduced communication latency.

Another way to reduce communication latency is by performing static routing, an approach taken by the Raw microprocessor. In Raw, network routers within the SON are programmable. This provides a mechanism for the compiler to specify communication routes that are known statically. (Raw also provides dynamically routed networks for communication that is known only at runtime.) Static routing reduces communication latency because routing decisions made by the compiler no longer need to be computed in hardware, thus streamlining the switch time at routers. Furthermore, compared to hardware, the compiler has global knowledge about communication patterns and can apply more sophisticated routing algorithms. Such intelligent routing can mitigate network congestion, further reducing communication latency.

26.4.2 DATA-LEVEL PARALLELISM

In addition to new ILP techniques, the computer architecture research community has recently investigated several novel architectures for exploiting DLP using on-chip parallelism. This section looks at Clearspeed, DIVA, VIRAM, and Imagine, all examples of architectures that are particularly suited to exploiting DLP. The discussion presents three key features exhibited by these advanced DLP architectures: SIMD architectures, vector processing, and streaming architectures. When appropriate, examples from individual architectures are used to illustrate the ideas.

26.4.2.1 SIMD Architectures

SIMD architectures were developed in the 1980s to exploit large-scale parallelism. The first two Connection Machine architectures, CM-1 and CM-2, were massively parallel SIMD computers. A Connection Machine of up to 64K 1-bit processors was developed, and the processors were very simple, bit-sliced elements with special floating-point units attached to groups of 32-bit slice processors. The processors were connected via a hypercube, and later a fat tree, network. Special languages were developed for the Connection Machine to allow parallelism to be identified by the programmer and exploited. The MASPAR architecture was also a large-scale SIMD architecture and had 32-bit processors connected in a two-dimensional mesh. These large-scale SIMD architectures eventually failed commercially because the market was not large enough for the limited number of applications that could exploit them, they were too hard to program, and the performance improvements of sequential microprocessors were too much to keep up with.

More recently, the search for ways to improve performance while improving power efficiency has led to the development of new SIMD architectures. The ClearSpeed CSX600 architecture, shown in Figure 26-3, has been developed as a co-processor designed to accelerate signal processing and scientific applications (Reddaway et al. 2005). Each CSX600 processing chip has 96 data

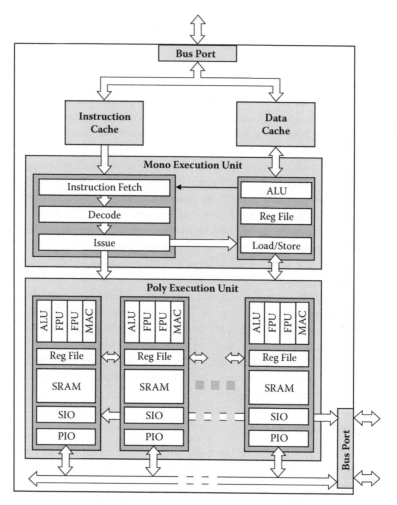

FIGURE 26-3 Clearspeed CSX600 chip architecture. (From Reddaway et al., Ultrahigh performance, low-power, data parallel radar implementations, *Proceedings of the IEEE International Radar Conference* 822–826, 2005. With permission. © 2005 IEEE.)

processing units that each can execute a 16-bit integer multiply-addition operation or two 64-bit floating-point operations. The data processing units are controlled by a single controller unit and are connected by a packet-switched interconnect. The parallelism exploited by the SIMD architecture allows the chip to run at a slower clock speed while still achieving high performance. The low clock speed leads to much better power efficiency.

General-purpose commercial microprocessors added SIMD vector extensions to exploit parallelism in multimedia applications as a way to exploit parallelism with little overhead and minimal changes to the instruction set architecture. The Intel x86 architecture added the MMX instruction extensions, and the PowerPC architecture added Altivec, both of which allowed four single-precision floating-point operations to execute simultaneously using SIMD instructions.

The IBM Cell architecture exploits data-level parallelism in a high performance design developed for video-game processing (Gschwind et al. 2006). The Cell architecture includes a general-purpose processor, called the power processing element (PPE) and eight vector processors, called synergistic processing elements (SPEs). While the Cell architecture also supports instruction-level parallelism within both the PPEs and the SPEs and thread-level parallelism since each processing element executes its own thread, the Cell architecture derives most of its performance from

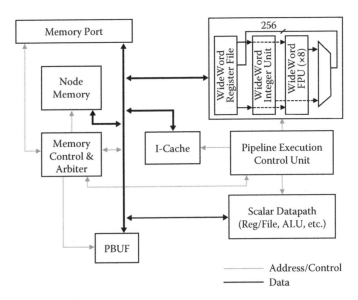

FIGURE 26-4 DIVA PIM node organization. (From Barrett et al., A double-data rate (DDR) processing-in-memory (PIM) device with wideword floating-point capability, *Proceedings of the IEEE International Symposium on Circuits and Systems* 1933–1936, 2006. With permission. © 2006 IEEE.)

data-level parallelism. Each of the eight SPEs can execute four single-precision, floating-point multiply-accumulates per cycle, making a total of 64 floating-point instructions per cycle (counting a multiply-accumulate as two operations). The Cell also has register files and a memory system to support these high-bandwidth vector operations.

The DIVA (Data IntensiVe Architecture) chip is a processor-in-memory (PIM) developed to overcome the performance bottleneck caused by the bandwidth limitation between processor and dynamic random access memory (DRAM) memory chips (Barrett et al. 2006). The DIVA architecture shown in Figure 26-4 integrates processing units and DRAM onto the same chip, increasing the memory bandwidth to the processor and decreasing memory latency. In order to leverage that data bandwidth, the DIVA architecture implements wide-word operations, which are essentially SIMD arithmetic instructions. These SIMD instructions are well suited to a PIM architecture because the DRAM modules on the chip typically access many words simultaneously when fetching a row from the DRAM. A typical DRAM chip must sequentially return these words because of the limited width of the pins available to the DRAM/processor interface (because inter-chip communication is expensive). These limitations are not present within the chip of a PIM architecture, so SIMD instructions can exploit the wide words returned by the DRAM. The DIVA architecture allows wide words to be broken into data elements of various sizes. For example, a 256-bit word can be broken into 8 words of 32 bits, 16 words of 16 bits, or 32 words of 8 bits to allow various degrees of precision and parallelism, depending on application requirements. The DIVA architecture is relatively power-efficient for two reasons. First, sending data from on-chip DRAM to processing elements and back consumes less power than driving external pins between memory and processor chips. Second, the SIMD wide-word instructions can efficiently exploit parallelism in those applications in which data-parallelism is available.

26.4.2.2 Vector Architectures

Vector architectures exploit data-parallel operations with a sequential instruction stream that is supplemented by vector instructions that operate on multiple sets of operands. These operands stream through highly pipelined memory systems and functional units. While the line between

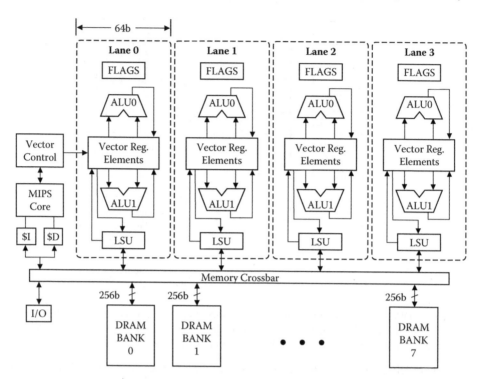

FIGURE 26-5 The microarchitecture of the VIRAM vector processor chip. (From Kozyrakis and Patterson, Vector vs. superscalar and VLIW architectures for embedded multimedia benchmarks, *Proceedings of the 35th International Symposium on Microarchitecture* 283–293, 2002. With permission. © 2002 IEEE.)

SIMD and vector architectures can be blurry, vector architectures generally have more support for vector (sequential, strided, and indexed) memory accesses and long pipelines in place of the spatial parallelism of SIMD architectures. Vector architectures were pioneered for supercomputers of the 1970s and 1980s by Cray Research, NEC, Fujitsu, and Hitachi. These architectures had very fast clocks and high memory bandwidths for their time and were implemented using emitter-coupled logic (ECL), which provided a performance advantage at the time. These expensive supercomputers eventually lost out to commercial microprocessor-based systems. However, vector architectures are being re-examined for the same reason as are SIMD architectures: vector architectures can provide high performance with improved efficiency. Vector architectures can also improve utilization of memory bandwidth, which often limits performance in modern architectures, especially for applications in which the memory behavior does not match caches well.

Like the DIVA architecture, the Vector Intelligent Random Access Memory (VIRAM) architecture is also a PIM architecture that exploits the increased on-chip memory bandwidth available in a power-efficient way (Kozyrakis and Patterson 2002). The VIRAM, as shown in Figure 26-5, exploits the bandwidth available from the on-chip DRAM using vector instructions. The VIRAM supports traditional vector load/store instructions that support sequential, strided, and indexed memory accesses and that move vectors from the DRAM to vector registers. Vector elements are interleaved across four lanes that support parallel operations, and each lane performs the same operation during any given clock cycle. Each lane has its own memory interface, slice of the vector registers, and functional units. Vector elements are interleaved between lanes, so that vector operations are parallelized spatially across lanes and temporal within the lanes. VIRAM has instructions to perform permutations and reduction operations across vector elements. VIRAM hides memory latency using a fixed-length pipeline between the DRAM and the functional units.

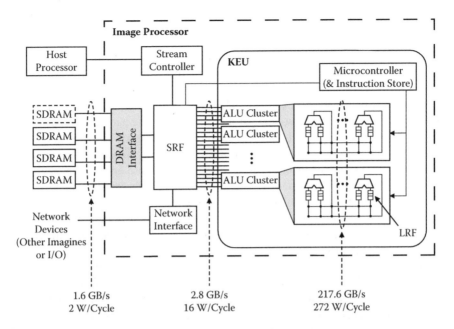

FIGURE 26-6 Diagram of the Imagine processor. (From Ahn et al., Evaluating the Imagine stream architecture, *Proceedings of the 31st Annual International Symposium on Computer Architecture* 14–26, 2004. With permission. © 2004 IEEE.)

26.4.2.3 Streaming Architectures

Streaming is the newest of the techniques for exploiting data-level parallelism discussed in this section. Stream processing can be thought of as a generalization of vector processing with two primary differences. First, stream architectures provide local registers associated with functional units to increase the number of operands available to the functional units without using expensive global storage elements (such as shared register files). Second, streaming architectures allow multiple instructions/operands to be applied to each vector element (or more generally each stream record) before sending the result back to the register file, thus reducing the number of intermediate operands that must be stored in the global register file. These modifications can lead to increased performance and improved efficiency.

The Imagine architecture shown in Figure 26-6 is the first implementation of a stream processor (Ahn et al. 2004). The key ideas behind Imagine are (1) that arithmetic functional units are inexpensive and plentiful and (2) that an architecture should leverage bandwidth where it is available [e.g., between local register files and arithmetic logic units (ALUs)] and minimize bandwidth where it is scarce (e.g., between the processor chip and main memory). The Imagine architecture uses a stream register file (SRF) to hide memory latency and to make operands available to eight parallel ALU clusters. The SRF decouples the memory accesses from the arithmetic operations and allows memory accesses to be scheduled to maximize use of the off-chip memory bandwidth. The ALU clusters operate on data elements in parallel using a SIMD instruction stream that comes from the microcontroller, and within an ALU cluster, multiple functional units perform parallel operations using a VLIW instruction stream. The combination of VLIW and SIMD parallelism is designed to provide maximize parallelism. The compiler or programmer must schedule operations to leverage that parallelism.

26.4.3 THREAD-LEVEL PARALLELISM

This section discusses advanced architectural techniques for exploiting thread-level parallelism (TLP) via on-chip support for multithreading. The discussion proceeds with three main thrusts: (1)

the performance challenges of exploiting TLP and reasons why on-chip threading is of interest to the research architecture community; (2) an overview of several new on-chip threading techniques, highlighting performance trade-offs in terms of both on-chip area and processing overhead; and (3) representative research processors employing new structures and programming models that relate to on-chip TLP.

The issues described earlier in exploiting TLP are equally important for multichip and single-chip parallel processors; however, on-chip parallelism changes the nature of some of the trade-offs present in multithreading in general. On-chip parallel processors are likely to have smaller cache memory. Additionally, they will have to streamline the hardware provided for dynamic thread management and control. On-chip networks will also need to be dense and low latency to support the required data communication and memory consistency for multithreading. Finally, context switching must be performed rapidly and independently for each on-chip processor. For these reasons, TLP in a chip multiprocessor is likely to occur at a finer granularity than with multichip parallel processors. This finer granularity will enable both higher levels of parallelism while challenging the programmer and compiler to find ways to exploit TLP in applications.

In response to the need to exploit TLP on-chip while simultaneously exploiting the ILP and DLP gains as outlined in the previous sections, the computer architecture research community has recently investigated several novel architectures for exploiting TLP. This section looks at five examples of architectures that contain support for threading on-chip—Raw (Taylor et al. 2004), TRIPS (Sankaralingam et al. 2003), Smart Memories (Mai et al. 2000), Scale (Krashinsky et al. 2004), and TCC (Hammond et al. 2004). The discussion presents novel features exhibited by these on-chip parallel architectures: variable granularity, flexible memory, and speculative threading support. The section then examines how these novel features are implemented in each processor and how TLP will be exploited in each case.

26.4.3.1   Multithreading and Granularity

A key feature of new multicore chip multiprocessors is the ability to support different granularities of threads within the same architecture. If an application exhibits primarily coarse-grain parallelism, then a chip multiprocessor can be organized into larger units on chip with larger threads assigned to these larger units. If an application exhibits a large amount of fine-grain parallelism, then the on-chip cores can be configured to support ILP or DLP. Hybrid approaches in which larger threads contain ILP or DLP within a thread are also possible. Architectures that support varying levels of parallelism using different processing paradigms are called polymorphous. Tiled architectures allow these hybrid approaches within the same architecture, either by synthesizing larger thread processing units from multiple small tiles aggregated together or by partitioning larger on-chip processing units into smaller units with finer-grain parallelism.

The TRIPS architecture described in Section 26.4.1 supports TLP using a partitioning approach to polymorphous processing. The TRIPS architecture contains large, coarse-grained processing cores to achieve high performance on single-threaded applications with high ILP, but allows the cores to be subdivided to achieve TLP on applications with lower ILP. The prototype TRIPS chip contains four such cores, each of which consists of an array of homogeneous execution nodes. Each execution node contains an ALU, an FPU, reservation stations, and router connections. To support multiple threads, the TRIPS processing core can be partitioned into either row processors or frame processors. Row processors space-share the ALU array, allocating one or more rows per thread. Each thread has cache bandwidth and capacity proportional to the number of rows, but threads allocated to the bottom rows have higher latency to the register file. Frame processors time-share the ALU array by allocating threads to unique sets of physical frames in the set of reservation stations and maintaining multiple program counters and global history registers. By supporting these two modes of operation, the TRIPS core can achieve high performance on workloads with varying levels of ILP, DLP, and TLP.

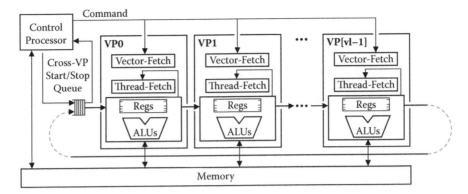

FIGURE 26-7   Vector-thread architecture. (From Krashinsky et al., The vector-thread architecture, *Proceedings of the 31st International Symposium on Computer Architectures* 52–63, 2004. With permission. © 2004 IEEE.)

In contrast to the partitioning approach, the Raw processor, also covered in Section 26.4.1, supports TLP by providing coarse-grain tiles that each implement an in-order single-issue MIPS execution core. Since each tile contains enough compute resources to support a single thread, TLP is supported by allocating threads to tiles. The dynamic communication network can be used for inter-thread communication, and DRAM banks external to the chip can be used for shared memory.

The Scale processor is an example of a vector-thread architecture that combines vector and threaded control mechanisms to achieve performance and efficiency by exploiting both application parallelism and locality. An abstract model of a vector-threaded architecture is shown in Figure 26-7. In this architecture, there is a control processor and a number of virtual processors organized into an ordered sequence (vector). To exploit DLP, each virtual processor can execute a loop iteration while the control processor is responsible for managing overall loop execution. Virtual processors execute blocks of code that are fetched either by the control processor in a vector-fetch operation or by the virtual processor itself in a thread-fetch operation issued when the virtual processor finished its current block. Since virtual processors can execute thread-fetches independently of each other and of the control processor, multiple threads can be active at any time.

The examples in this section show how today's research architectures support multithreading on chip via tiled, on-chip multiprocessors that can be organized in multiple ways to exploit varying levels of parallelism. The capability to support different levels of parallelism in the same architecture allows for high performance and efficiency over a variety of workloads and processing granularities.

26.4.3.2   Multilevel Memory

Traditionally, multithreaded processors have been shared multiprocessors, consisting of a number of processors all accessing a shared memory over a high-bandwidth bus or interconnect. These shared-memory processors (SMPs) provide uniform memory access for the sharing processors; however, the bus or interconnect can create a performance bottleneck. Hence, the scalability of such processors has been limited. High performance architectures that incorporate SMPs typically exhibit a hierarchical architecture in which clusters of processors organized as SMPs are connected by a message-passing network that typically provides lower bandwidth but less contention due to the higher granularity.

For these SMP-based architectures, the memory system is a key determinant of processing performance. Implementation of a multilevel cache memory hierarchy requires that consistency and coherence be maintained at all levels in the hierarchy, and if any particular cache level is distributed and shared, then consistency and coherence must be maintained over an interconnection network. This is another factor that has limited the scalability of SMPs; however, on-chip tiled architectures

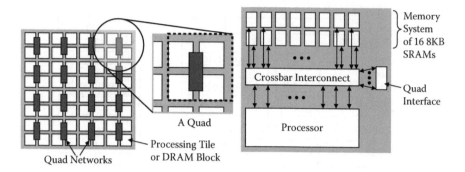

FIGURE 26-8   Smart Memories. (From Mai et al., Smart memories: a modular reconfigurable architecture, *Proceedings of the 27th Annual International Symposium on Computer Architectures* 161–171, 2000. With permission. © 2000 IEEE.)

have changed the performance trade-offs for these memory architectures as well as for the processing units that use them.

In the TRIPS processor, memory is organized in an array of memory tiles interspersed among the processor cores and connected by a routed network, with distributed memory controllers for channels to external memory. These memory tiles can be configured to behave in different ways, including level 2 cache banks in the style of nonuniform cache architecture. In nonuniform cache architecture, elements of a dataset are spread across multiple banks. Data can migrate between banks on the two-dimensional switched network that connects all memory banks on chip. To support multithreading, cache tags have per-thread identifiers, and data structures are provided in hardware for thread-private data protection.

In the Raw processor, the nonuniform cache architecture could be implemented by allocating levels of cache to regions in the Raw topology. For example, successive levels in the cache hierarchy can be allocated to columns of tiles on the Raw chip. As the levels of cache move to the right in the Raw mesh, they are closer to DRAM. Code execution is primarily mapped to the left side of the Raw mesh; then the cache levels to the left are closer to the processor. This spatial locality creates a nonuniform architecture in which data can migrate in a manner similar to that of the TRIPS case.

A different kind of tiled, reconfigurable memory architecture is found in the Smart Memories processor (Mai et al. 2000). An overview of the Smart Memories processor is shown in Figure 26-8. This chip consists of an array of processor tiles and on-die DRAM memories connected by a packet-based, dynamically routed network. Each quad in the smart memories chip consists of four tiles connected by a quad network. Each tile in a quad consists of a processor, a crossbar interconnect, and an array of 16 memory mats. Each memory mat is small (8 KB) and can perform independent accesses. The data paths for the mats contain configurable logic that can be used to implement a number of different memory access modes. The mats can thereby be configured to implement a wide variety of caches from simple, direct, mapped structures to more complicated set-associative designs. Mats can also be configured to implement local scratch-pad memories and vector or stream register files.

To map a cache memory hierarchy, the set of mats on each tile can be divided and portions allocated to level 1 and level 2 caches. The level 1 cache mats are private to the tile, and the level 2 cache is distributed and shared by the tiles in the quad. Because access to memory mats is over a high-bandwidth, low-latency interconnect network, the level 2 cache is accessible by all tiles. Because of its flexibility and morphability, the Smart Memories architecture has been shown to emulate both the Imagine streaming processor and the Hydra threaded processor, other research architectures that have been tailored for specific execution models. For multithreading, the Smart Memories chip can allocate threads to processors in the quad; these threads can utilize their own level 1 cache or the shared level 2 cache distributed over the quad.

For cache management, a strategy that promises success for tiled on-chip multiprocessors is called victim replication (Zhang and Asanovic 2005). In a tiled multiprocessor, the local memory of a tile can implement either a shared or private level 2 cache. Sharing the level 2 cache results in a higher capacity, but access latency and, hence, performance may be poor in some cases. However, a private level 2 cache results in redundant copies distributed over the chip, reducing the effective capacity of the level 2 cache. Victim replication has the benefit of both schemes by keeping a copy of level 1 cache victims within the local level 2 cache slice. If the level 2 cache is shared, hits to these copies reduce the effective latency of the level 2 cache. For nonuniform cache architectures in which the level 2 cache is distributed over multiple memory banks, victim replication has been shown to have performance advantages over purely private and distributed schemes.

As shown in this section, tiled, multicore, chip multiprocessor research architectures implement the cache hierarchy in a variety of ways, each of which supports on-chip multithreading as well as other modes of operation. The common theme in these processors is the nonuniformity of cache access and the inclusion of hardware structures to support a variety of set-associative cache structures. Because of the high bandwidth and low latency of on-chip communication, memory access latency and communication can be overlapped with processing in this distributed context.

26.4.3.3 Speculative Execution

To combat the difficulty of parallel programming while exploiting as much parallelism as possible in a program, a key thrust in modern processors is speculative multithreading. In speculative multithreading, multiple functional units execute a program by dividing it into threads that are not guaranteed to be free of control or data dependencies. These threads are speculative in that if their execution cannot complete due to an unforeseen dependency, the thread is terminated. However, the goal of these architectures is to have the speculation exhibit a high percentage of correctness, so that, most of the time, threads execute in parallel and high performance is achieved overall. With on-chip multiprocessing, the capability for speculative multithreading is enhanced. This is due to the availability of both extra processing units to perform speculative threads and on-chip memory and communication bandwidth to support speculation.

The TRIPS processor supports multiblock speculation by providing an instruction window size much larger than average block size. The hardware fills empty frames with speculatively mapped blocks. The speculative blocks are marked in the frame buffer. As the execution hardware cycles through the blocks, speculative blocks are marked *nonspeculative* as they are executed. When execution completes, the frame is released and filled with a new speculative block. On a misprediction, all blocks beyond the offending prediction are squashed and restarted.

Another paradigm emerging from the research community and showing great promise for speculative execution is *transactional coherence and consistency* (TCC). In TCC systems, transactions serve as the fundamental unit of parallel work, communication, and coherence. A transaction is a sequence of instructions that is guaranteed to execute and complete only as an atomic unit. Each transaction produces a block of writes, called the *write state*, that are committed to shared memory only as an atomic unit after the transaction completes execution. As each transaction completes, it atomically writes all of its newly produced state to shared memory (called a commit), while restarting other processors that have speculatively read stale data. Therefore, data synchronization is handled correctly, without programmer intervention. A sample TCC system is shown in Figure 26-9.

Consistency is maintained in a TCC system by imposing sequential ordering rules on transaction commits rather than on individual memory operations. This reduces the number of arbitrations and synchronization events in program execution and, hence, incurs less latency for these overhead operations. Processors that have read data that are subsequently updated by another processor's commit are forced to violate and roll back in order to enforce the ordering rules. Coherence is maintained in a TCC system by broadcasting state to all processors after every commit operation. If required, processors will invalidate or update their cache lines as required. If they have read and

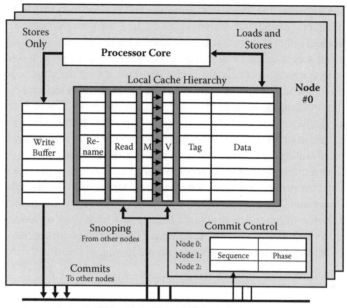

FIGURE 26-9   Transactional coherence and consistency. (From Hammond et al., Transactional memory coherence and consistency, *Proceedings of the 31st Annual International Symposium on Computer Architectures* 102–113, 2004. With permission. © 2004 IEEE.)

used data too early, they must restart and reload data. The use of transactions both increases the amount of parallelism in the application and simplifies parallel programming since the programmer does not need to explicitly parallelize the application. Instead, the architecture assumes everything is parallel and only incurs overhead if it is wrong, i.e., if it has speculated incorrectly. In those cases, threads must restart and reload data; however, correct program operation is ensured by the system and its data protocols. These protocols are implemented in hardware and made possible by the architectural innovations present in tiled on-chip multiprocessor chips.

26.5   REAL-TIME EMBEDDED APPLICATIONS

This chapter examines new processors focused on using on-chip multiprocessing to exploit new levels of parallelism in applications. The use of these processors in real-time applications offers the high performance embedded processing community some new challenges, which are touched on briefly in this section.

26.5.1   SCALABILITY

The architectures presented in this chapter represent key offerings of the research architecture community and may be regarded as in their infancy. Performance results published to date for these architectures have primarily been gathered via simulation or analysis and at smaller scales than would be exhibited by production chips using advanced fabrication technologies. Because of this, performance analysis has not been conducted at a scale commensurate with the requirements of future high performance real-time applications. As prototype chips and more advanced emulation facilities become available, more performance analysis and benchmarking must be performed to ascertain both the scalability of these architectures and the effects on performance of increases in both problem and machine sizes (particularly as systems are built from multiple chips). This research should be conducted in the context of processing and algorithm requirements of future applications. In this endeavor, the application and architectures must work together to produce a thorough analysis and assessment so that the exact effect on future processing platforms can be predicted.

26.5.2 INPUT/OUTPUT BANDWIDTH

Although much attention has been paid to on-chip memory and communication bandwidth in the architectures described in this chapter, much has yet to be done with input and output (I/O) channels. Specifically, embedded application data rates have been steadily increasing over the last few decades. This trend is anticipated to continue well into the future, especially with the advent of advanced applications such as multisensor fusion and knowledge-aided signal processing. New architectures must ensure that they maintain a balance of processing and I/O so that the raw performance improvements of new processors will not be overshadowed by the rising data rates of future sensors.

26.5.3 PROGRAMMING MODELS AND ALGORITHM MAPPING

The research architectures presented in this chapter have implications for expressing parallelism in an application and the mapping of that application to on-chip parallel units. In the case of transactional coherence and consistency, programming is envisioned to be much easier, with much of the burden of parallelization and synchronization removed from the programmer. In the case of TRIPS and Raw, the compiler is projected to assume much of the burden of finding ILP, DLP, and TLP within an application. The development of compilers to automate parallel mapping is critical to lowering the cost of programming these architectures. The feasibility of building these compilers is an open research question, particularly with respect to increasing problem and machine size. The application and architecture community must work together on programming models to ensure that applications are expressed completely and compactly, performance is achieved without significant programmer "gymnastics," and programming models scale into future architectures and real-time problems. Programmer productivity is becoming increasingly important to the high performance embedded processing community and may take its place beside Moore's Law as a key metric of architectural longevity for systems of the future.

26.6 SUMMARY

Architectural techniques that exploit data-level parallelism have been around for over 30 years. While these techniques were originally used in expensive supercomputers implemented in exotic technologies, they are now being used in architectures that need efficient ways to increase performance for signal processing and multimedia applications. General-purpose commercial microprocessors have integrated small-scale SIMD extensions into their legacy instruction-set architectures, but new architectures that are not restricted by legacy instruction sets are being developed that achieve much higher levels of performance at much greater efficiencies. Architectures that use SIMD, vector, and streaming architectural techniques have been presented, and the authors expect these techniques will appear in future HPEC architectures.

Multithreaded architectures that exploit coarse-grain parallelism have been around for some time although their practical use has been limited by scalability and bandwidth issues that arise from interthread synchronization and communication. On-chip multiprocessing enables the performance advantages of multithreading at new levels of granularity and parallelism. Combining TLP with techniques for exploiting ILP and DLP on chip shows the greatest promise for dealing with the performance challenges of maintaining Moore's Law performance improvements at ever higher processing densities while providing uniform performance improvements over a larger variety of workloads and applications. This chapter has shown a number of research processors that showcase new techniques for exploiting these levels of parallelism.

Architectural research has shifted from pure density/clock-speed-based innovation to more structural innovation, with flexibility and programming models becoming as important as raw processor speed. Future goals will most likely include better performance on a more complicated set of algorithms, enabling more advanced applications such as those requiring cognitive processing and artificial intelligence techniques in a real-time processing stream.

Cognitive processing is a natural fit for tile-based architectures for several reasons. First, tile-based architectures are parallel collections of relatively simple processors. Since cognitive processing workloads have traditionally been modeled as collections of simple processes, they can be easily mapped onto tiled architectures. Second, the high-bandwidth, low-latency connections available between tiles allow fine-grain parallelism to be exploited, making it easy to spawn parallelism to gain performance. Third, the memory in tile-based processors is distributed and each local memory can be accessed simultaneously, allowing the fast retrieval of a lot of data. Finally, the number of processing elements available allows for the programming of robust algorithms, an important feature of cognitive processing.

REFERENCES

Agarwal, V., M.S. Hrishikesh, S.W. Keckler, and D. Burger. 2000. Clock rate versus IPC: the end of the road for conventional microarchitectures. *Proceedings of the 27th Annual International Symposium on Computer Architecture* 248–259.

Ahn, J.H., W.J. Dally, B. Khailany, U.J. Kapasi, and A. Das. 2004. Evaluating the Imagine stream architecture. *Proceedings of the 31st Annual International Symposium on Computer Architecture* 14–26.

Barrett, T., S. Mediratta, T. Kwon, R. Singh, S. Chandra, J. Sondeen, and J. Draper. 2006. A double-data rate (DDR) processing-in-memory (PIM) device with wideword floating-point capability. *Proceedings of the IEEE International Symposium on Circuits and Systems* 1933–1936.

Dally, W.J. and S. Lacy. 1999. VLSI architecture: past, present, and future. *Proceedings of the 20th Anniversary Conference on Advanced Research in VLSI.*

Gschwind, M., H.P. Hofstee, B. Flachs, M. Hopkins, Y. Watanabe, and T. Yamazaki. 2006. Synergistic processing in Cell's multicore architecture. *IEEE Micro* 26(2): 10–24.

Hammond, L., B.A. Nayfeh, and K. Olukotun. 1997. A single chip multiprocessor. *IEEE Computer* 30(9): 79–85.

Hammond, L., V. Wong, M. Chen, B.D. Carlstrom, J.D. Davis, B. Hertzberg, M.K. Prabhu, H. Wijaya, C. Kozyrakis, and K. Olukotun. 2004. Transactional memory coherence and consistency. *Proceedings of the 31st Annual International Symposium on Computer Architectures* IEEE: 102–113.

Ho, R., K.W. Mai, and M.A. Horowitz. 2001. The future of wires. *Proceedings of the IEEE* 89(4): 490–504.

Kozyrakis, C. and D. Patterson. 2002. Vector vs. Superscalar and VLIW architectures for embedded multimedia benchmarks. *Proceedings of the 35th International Symposium on Microarchitecture.* IEEE: 283–293.

Krashinsky, R., C. Batten, M. Hampton, S. Gerding, B. Pharris, J. Casper, and K. Asanovic. 2004. The vector-thread architecture. *Proceedings of the 31st Annual International Symposium on Computer Architectures.* IEEE: 52–63.

Lee, W., R. Barua, D. Srikrishna, J. Babb, V. Sarkar, S. Amarasinghe, and A. Agarwal. 1998. Space-time scheduling of instruction-level parallelism on a Raw machine. *Proceedings of the Eighth International Conference on Architectural Support for Programming Languages and Operating Systems* ACM: 46–57.

Mahlke, S.A., D.C. Lin, W.Y. Chen, R.E. Hank, and R.A Bringmann. 1992. Effective compiler support for predicated execution using the hyperblock. *Proceedings of the 25th Annual International Symposium on Microarchitecture.* IEEE: 45–54.

Mai, K., T. Paaske, N. Jayasena, R. Ho, W.J. Dally, and M. Horowitz. 2000. Smart memories: a modular reconfigurable architecture. *Proceedings of the 27th Annual International Symposium on Computer Architectures.* IEEE: 161–171.

Nagarajan, R., K. Sankaralingam, D. Burger, and S.W. Keckler. 2001. A design space evaluation of grid processor architectures. *Proceedings of the 34th Annual International Symposium on Microarchitecture.* IEEE: 40–51.

Oklobdzija, V.G., ed. 2002. *The Computer Engineering Handbook.* Boca Raton, Fla.: CRC Press.

Reddaway, S., P. Bruno, R. Pancoast, and P. Rogina. 2005. Ultrahigh performance, low-power, data parallel radar implementations. *Proceedings of the IEEE International Radar Conference*. IEEE: 822–826.

Ronen, R., A. Mendelson, K. Lai, S.-L. Lu, F. Pollack, and J.P. Shen. 2001. Coming challenges in microarchitecture and architecture. *Proceedings of the IEEE* 89(3).

Sankaralingam, K., R. Nagarajan, H. Liu, C. Kim, J. Huh, D. Burger, S.W. Keckler, and C.R. Moore. 2003. Exploiting ILP, TLP, and DLP with the polymorphous TRIPS architecture. *IEEE Micro* 23(6): 46–51.

Swanson, S., K. Michelson, A. Schwerin, and M. Oskin. 2003. WaveScalar. *Proceedings of the 36th International Symposium on Microarchitecture*. IEEE: 291–302.

Taylor, M.B., W. Lee, S. Amarasinghe, and A. Agarwal. 2003. Scalar operand networks: on-chip interconnect for ILP in partitioned architectures. *Proceedings of the Ninth International Symposium on High Performance Computer Architecture*. IEEE: 341–353.

Taylor, M.B., W. Lee, J. Miller, D. Wentzlaff, I. Bratt, B. Greenwald, H. Hoffmann, P. Johnson, J. Kim, J. Psota, A. Saraf, N. Shnidman, V. Strumpen, M. Frank, S. Amarasinghe, and A. Agarwal. 2004. Evaluation of the Raw microprocessor: an exposed-wire-delay architecture for ILP and streams. *Proceedings of the 31st International Symposium on Computer Architecture*. IEEE: 2–13.

Tomasulo, R.M. 1967. An efficient algorithm for exploiting multiple arithmetic units. *IBM Journal* 11: 25–33.

Tullsen, D.M., S.J. Eggers, and H.M. Levy. 1995. Simultaneous multithreading: maximizing on-chip parallelism. *Proceedings of the 22nd Annual International Symposium on Computer Architectures*. IEEE: 392–403.

Zhang, M. and K. Asanovic. 2005. Victim replication: maximizing capacity while hiding wire delay in tiled chip multiprocessors. *Proceedings of the 32nd Annual IEEE International Symposium on Computer Architecture*. IEEE: 336–345.

Glossary of Acronyms and Abbreviations

ABF	adaptive beamforming
ABS	adaptive beamforming subsystem
AC	alternating current
ACE	adaptive coherence estimator
ADC	analog-to-digital converter
AESA	active electronically scanned antenna
ALU	arithmetic logic unit
AMF	adaptive matched filter
AMS	analog mixed signal
AMTI	airborne moving-target indicator
API	application programming (programmer) interface
APP	*a posteriori* probability
APU	auxiliary processing unit
ASIC	application-specific integrated circuit
ASIP	application-specific instruction processor
ASSP	application-specific standard products
ATE	automatic test equipment
ATPG	automatic test pattern generation
BCJR	algorithm named after its inventors: Bahl, Cocke, Jelinek, and Raviv
BF	beamformer
BGA	ball grid array
BiCMOS	bipolar CMOS
BIOS	basic input/output system
BIST	built-in self-test
BIVOS	biased voltage scaling
BJT	bipolar junction transistor
BLAS	basic linear algebra subprograms or subroutines
BW	bandwidth
CAD	computer-aided design
CAF	co-array Fortran
CCD	charge-coupled device
CDL	Common Data Link
CDMA	code-division multiple-access
CFAR	constant false alarm
CLB	configurable logic block
CMOS	complementary metal oxide semiconductor: complementary MOSFET
CMP	chemical-mechanical polishing
COI	community of interest
CONOPS	concept of operations
CONUS	continental United States
CORBA	Common Object Request Broker Architecture
COTS	commercial off-the-shelf
CPI	coherent processing interval

CPLD	complex programmable logic devices
CPM	continuous phase modulation
CPU	central processing unit
CRC	cyclic redundancy check
DA	distributed arithmetic
DAC	digital-to-analog converter
DARPA	Defense Advanced Research Projects Agency
dB	decibel
dBc	decibels relative to carrier level
dBFS	decibels relative to full scale
DC	direct current
DDR&E	Director of Defense Research and Engineering
DDS	Data Distribution Service
DEM	digital elevation model
DET	detection and estimation subsystem
DFM	design for manufacturability
DFS	digital filtering subsystem
DFT	discrete Fourier transform; design for testability
DIQ	digital in-phase and quadrature
DIS	data input subsystem
DLP	data-level parallelism
DMA	direct memory access
DMUX	demultiplexer
DNL	differential nonlinearity
DoD	Department of Defense
DOF	degrees of freedom
DPE	detection processing element
DRAM	dynamic random access memory
DRC	design rule check
DSP	digital signal processor or processing
DTED	digital terrain elevation data
ECCM	electronic countercountermeasures
ECL	emitter-coupled logic
ECM	electronic countermeasures
EDA	electronic design automation
EDIF	Electronic Design Interchange Format
EIB	element interconnect bus
ELINT	electronic intelligence
ENOB	effective number of bits
EO	electro-optical
EPP	energy-performance product
EQ	effective quantization level
ERC	electric rule check
ESD	electrostatic discharge
FDP	Fast Digital Processor
FEC	forward error correction
FEP	front-end processor
FFT	fast Fourier transform
FFTW	"Fastest Fourier Transform in the West"
FIFO	first-in-first-out
FIR	finite impulse response

FLOPs	floating-point operations
FLOPS	floating-point operations per second
FPGA	field programmable gate array
GaAs	gallium arsenide
Gbyte	gigabyte
GFLOPs	giga (one billion) floating-point operations
GFLOPS	gigaFLOPS (one billion floating-point operations per second)
GHz	gigahertz
GIG	Global Information Grid
GIG-BE	Global Information Grid–Bandwidth Expansion
GMTI	ground moving-target indication (indicator)
GOPS	billions of operations per second
GPIO	general-purpose input/output
GPL	General Public License
GPP	general programmable processor
GPS	global positioning system
GPU	graphics processing unit
GSPS	billions of samples per second
HAIPE	High Assurance Internet Protocol Encryptor
HBA	host bus adapters
HDL	hardware description language
HE	hyperencryption
HPC	high performance computing
HPEC	high performance embedded computing
HPF	high performance Fortran
Hz	hertz
IAC	intelligence analysis cell
IC	integrated circuit; integrated chip
IDE	integrated development environment
IEEE	Institute of Electrical and Electronics Engineers
IF	intermediate frequency
IFFT	inverse fast Fourier transform
IIR	infinite impulse response
ILP	instruction-level parallelism
IMU	inertial measurement unit
InGaAs	indium gallium arsenide
INL	integral nonlinearity
I/O	input/output
IOB	input/output block
IP	intellectual property; Internet protocol
IQ	in-phase quadrature
IR	infrared
IRAM	intelligent RAM
ISA	instruction set architecture
ISDS	integrated sensing and decision support
ISI	intersymbol interference
ITA	intermediate time average
ITRS	International Technology Roadmap for Semiconductors
JMS	Java Messaging Service
JNS	jammer-nulling system
JTRS	Joint Tactical Radio System

JVM	Java virtual machine
K	thousand
KASSPER	Knowledge-Aided Sensor Signal Processing and Expert Reasoning
KB	kilobyte
Kbyte	kilobyte
kHz	kilohertz
KOPS	thousands of operations per second
L	liter
LADAR	laser detection and ranging
LAPACK	Linear Algebra Package
LDPC	low-density parity check
LFSR	linear feedback shift register
LOSC	lines of source code
LSB	least significant bit
LST	layered space-time
LTA	long time average
LUT	lookup table
LVDS	low-voltage differential signaling
LVS	layout versus schematic
MAC	multiply and accumulate
MB	megabyte
Mbyte	megabyte
MCM	multichip module
MEMS	microelectromechanical system
MFLOPs	mega (one million) floating-point operations
MFLOPS	megaFLOPS (one million floating-point operations per second)
MHz	megahertz
MIMD	multiple-instruction multiple-data
MMI	man-machine interface
MMU	memory management unit
MOPS	millions of operations per second
MOS	metal oxide semiconductor
MOSFET	metal oxide-semiconductor field-effect transistor
MOSIS	Metal Oxide Semiconductor Implementation Service
MPI	Message Passing Interface
MPMD	multiple-program multiple-data
MPU	microprocessor unit
MPW	multiproject wafer
MSPS	millions of samples per second
MTAP	multithreaded array processor
MTI	moving-target indicator (indication)
MUD	multiuser detection
MUX	multiplexer
mW	milliwatt
NCES	Net-Centric Enterprise Services
NP-hard	nondeterministic polynomial-time hard
NRE	nonrecurring expenses
NSR	noise-to-signal ratio
OASIS	Open Artwork System Interchange Standard
OCM	on-chip memory
OE	output enable

OFDM	orthogonal frequency division multiplexing
OMG	Object Management Group
OPB	on-chip peripheral bus
OpenMP	Open Multiprocessing
ORB	Object Request Broker
OS	operating system
OSA	open system architecture
OSI	Open System Interconnection
PC	personal computer
PCA	Polymorphous Computing Architecture
PCB	printed circuit board
PCC	parallel concatenated code
PCMOS	probabilistic CMOS
PDA	personal digital assistant
PE	processing element; poly-execution
PETE	portable expression template engine
PFA	probability of false alarm
PGA	pin grid array
PGAS	Partitioned Global Address Space
PIC	parallel interference cancellation
PIM	processor in memory
PIP	programmable interconnect point
PLB	processor local bus
POOMA	Parallel Object-Oriented Methods and Applications
POSIX	Portable Operating System Interface
PPE	power processor (processing) element
PPF	polyphase filter
PRF	pulse repetition frequency
PRI	pulse repetition interval
PRNG	pseudorandom number generator
PSK	phase shift keying
PSM	programmable switch matrix
PSP	programmable signal processor
PVL	Parallel Vector Library
PVM	Parallel Virtual Machine
QAM	quadrature amplitude modulation
QoS	quality of service
RADAR (radar)	radio detection and ranging
RAM	random-access memory
RAPTOR	Reconfigurable Adaptive Processing Test bed for Onboard Radars
RET	resolution enhancement technique
RF	radio frequency
RFP	Request for Proposal
RISC	reduced instruction set computer
RMI	remote method invocation
RMS	root mean square
ROI	region of interest
ROM	read-only memory
ROSA	Radar Open System Architecture
RPC	remote procedure call
RPM	relationally placed macro

RSC	recursive systematic convolutional
RTL	register transfer level
RTOS	real-time operating system
RTP	real-time processor
s	second
S^{3}P	Self-Optimizing Software for Signal Processing
SAH	sample and hold
SAR	synthetic aperture radar
SATA	serial ATA (advanced technology attachment)
SBC	single-board computer
SCC	serial concatenated code
SCSI	small computer system interface
SDG	scalable data generator
SFDR	spurious free dynamic range
SHA	sample-and-hold amplifier
SiC	silicon carbide
SiGe	silicon germanium
SIL	system integration laboratory
SIMD	single-instruction multiple-data
SINAD	signal to noise and distortion
SIP	signal and image processing
SLOC	source lines of code
SMI	sample matrix inversion
SMP	shared-memory processor
SNDR	signal-to-noise-plus-distortion ratio
SNR	signal-to-noise ratio
SOA	service-oriented architecture
SOAP	Simple Object Access Protocol
SoC	system on chip
SOI	silicon-on-insulator
SON	scalar operand network
sonar	sound navigation and ranging
SoP	system on package
SPE	synergistic processor element
SPICE	Simulation Program with Integrated Circuit Emphasis
SPN	systolic processing node
SQL	structured (standard) query language
SRAM	static random access memory
SRF	stream register file
STA	short time average
STAP	space-time adaptive processing
STAPL	STAP library
STC	space-time coding
SWAP	size, weight, and power
TCC	transactional coherence and consistency
TDMA	time-division multiple-access
TIN	triangular irregular network
TLP	thread-level parallelism
TOPS	trillions of operations per second
T/R	transmit/receive
TRIPS	Tera-ops Reliable Intelligently-adaptive Processing System

TSMC	Taiwan Semiconductor Manufacturing Company
TTCM	turbo trellis-coded modulation
TTL	transistor-transistor logic
UAV	unmanned aerial vehicle
UDCA	ubiquitous and distributed computing architecture
UDDI	Universal Description, Discovery and Integration
UGS	unmanned ground system
UHF	ultrahigh frequency
UML	Universal Modeling Language
UPC	Unified Parallel C
UUV	unmanned underwater vehicle
V	volt
VHDL	very-high-speed integrated circuit hardware description language
VLIW	very long instruction word
VLSI	very-large-scale integration
VMAP	vector smart map (also vector map)
VMEBus	Versamodule Eurocard bus
VSIPL	Vector, Signal, and Image Processing Library
VXS	VME switched serial
W	watt
WGN	white Gaussian noise
WSDL	Web services description language
XML	Extensible Markup Language
XOR	exclusive *or*

Index

A